AF412886

SEDIMENT ROUTING SYSTEMS

This cutting-edge summary combines ideas from several sub-disciplines, including geology, geomorphology, oceanography and geochemistry, to provide an integrated view of Earth surface dynamics in terms of sediment generation, transport and deposition. Introducing a global view of fundamental concepts underpinning source-to-sink studies, it provides an analysis of the component segments which make up sediment routing systems. The functioning of sediment routing systems is illustrated through calculations of denudation and sedimentation as well as the response to external drivers; with the final sections focusing on the stratigraphic record of sediment routing systems. Containing quantitative solutions to a wide range of problems in Earth surface dynamics, this book is suitable for graduate students as well as academic and professional researchers.

PHILIP A. ALLEN is Emeritus Professor of Sedimentary Geology at Imperial College London and a process-oriented Earth scientist with a particular interest in the interactions and feedbacks between the solid Earth and its 'exosphere' through the critical interface of the Earth's surface. He has received the Royal Society Wolfson Research Merit Award for 2006 to 2011 and the Lyell Medal from the Geological Society of London in 2007. He served on the Council of the Geological Society from 2008 to 2012, and was Secretary of the Science Committee from 2009 to 2012.

In this new book, Philip Allen has distilled a lifetime of insightful study of the Earth's surface into a wide ranging and rigorous synthesis of planetary sediment processes. *Sediment Routing Systems* is the first to use the idea of global sediment routing – 'following the sediment' – to provide a framework for synthesis across environments and scales, to integrate the source and sink sides of the routing system, and to link geochemical and particulate fluxes. It manages to do this in a quantitative framework that is carefully formulated, accessible, and perfectly pitched in clarity and detail. *Sediment Routing Systems* is a landmark and masterpiece; for many Earth scientists, it will be all they need in terms of global sediment dynamics.

Chris Paola
University of Minnesota

If reading the sedimentary record is the destination, then this book is a brilliant companion for the road, ranging widely from bedload to organic carbon and providing thorough detail on processes and methods at every turn. Philip Allen turns the sediment routing system from Pandora's box into a dynamic source-to-sink cascade.

Niels Hovius
GFZ German Research Centre for Geosciences

Sediment Routing Systems is the first complete, quantitative process-based account of sediment generation, transport and deposition in book format. In true style, ahead of anyone else, Philip Allen gives an extremely thorough, comprehensive view of all process aspects of source-to-sink systems. This book combines theoretical with practical aspects and will be an obvious choice as an advanced text book in universities and as a key reference book for professional geologists dealing with energy and Earth systems.

Ole J. Martinsen
Chief Geologist and Vice President, Statoil

SEDIMENT ROUTING SYSTEMS

The Fate of Sediment from Source to Sink

PHILIP A. ALLEN

Imperial College London

CAMBRIDGE
UNIVERSITY PRESS

University Printing House, Cambridge CB2 8BS, United Kingdom

One Liberty Plaza, 20th Floor, New York, NY 10006, USA

477 Williamstown Road, Port Melbourne, VIC 3207, Australia

4843/24, 2nd Floor, Ansari Road, Daryaganj, Delhi – 110002, India

79 Anson Road, #06–04/06, Singapore 079906

Cambridge University Press is part of the University of Cambridge.

It furthers the University's mission by disseminating knowledge in the pursuit of
education, learning, and research at the highest international levels of excellence.

www.cambridge.org
Information on this title: www.cambridge.org/9781107091993
DOI: 10.1017/9781316135754

First published 2017

Printed in the United Kingdom by TJ International Ltd. Padstow Cornwall

A catalogue record for this publication is available from the British Library.

Library of Congress Cataloging-in-Publication Data
Names: Allen, Philip A., author.
Title: Sediment routing systems : the fate of sediment from source to sink /
Philip A. Allen, Imperial College, London.
Description: Cambridge, United Kingdom ; New York, NY : Cambridge University
Press, 2017. | Includes bibliographical references and index.
Identifiers: LCCN 2017014847 | ISBN 9781107091993 (Hardback : alk. paper)
Subjects: LCSH: Sediments (Geology)
Classification: LCC QE471.2 .A43 2017 | DDC 551.3/54–dc23 LC record
available at https://lccn.loc.gov/2017014847

ISBN 978-1-107-09199-3 Hardback

Contents

Preface

By far the most powerful way to find out about Earth's history is to study the sediments and sedimentary rocks comprising the stratigraphic record. With this in mind, Doyle and Bennett (1998) stated confidently that

Stratigraphy is the key to understand the Earth, its materials, structure and past life. It encompasses everything that has happened in the history of the planet.

Chris Paola (www.esci.umn.edu/people/chris-paola) observed lyrically and cautiously

Rachel Carson wrote that 'The sediments are a sort of epic poem of the Earth'. Unfortunately this poem is written in a language we don't understand

Part of this lack of understanding of the language of the epic poem of the Earth has its origins in the imperfect recording of events in the stratigraphic record, and particularly in the difficulty of deciphering what the English poet Tennyson (1809–1892) called the 'long result of Time', since the familiar processes responsible for the liberation, transport and deposition of sediment are best known at the very short timescales of human observation. As James Hutton wrote (1785),

it is not in human record, but in natural history, that we are to look for the means of ascertaining what has already been.

The other part of our lack of understanding derives from the difficulty of seeing the big picture with all of its feedbacks and linkages, which allows us to effectively 'join the dots' (Allen, 2014) (ch.1). This big, integrated picture is that of the sediment routing system.

Sediment routing systems are the dynamical systems that link the fate of sediment from source to sink and integrate the processes taking place at or near the surface of the Earth, and the resultant depositional products, at the timescales relevant to the understanding of stratigraphy. They involve a 'cascade' of sediment from primarily erosional source areas to depositional sinks, and therefore are responsible for upland landscapes and their sediment and solute effluxes, transport across the continental surface and delivery to the ocean, and dispersal into long-term depositional sites in the deep sea. The processes acting in sediment routing systems therefore shape erosional and depositional landscapes and extract sediment to build stratigraphy, the primary archive of the 'epic poem of the Earth'.

Sediment routing systems are important from the point of view of basin analysis, landscape evolution and the building of the stratigraphic record. They are also the context for understanding impacts of climate change and human activities on the environment. These impacts might include landscape sensitivity, ecological changes to rivers, restoration, soil loss and landslides in upland terrains, siltation of reservoirs, hazards and risk, starvation and compaction of river delta regions, as well as management of future water resources. Study of sediment routing systems is an integral part of a 'unified science of the Earth's surface' (Paola et al., 2006) (p.W03S10), as expressed in the research agenda of the National Center for Earth-Surface Dynamics, an NSF Science and Technology Center headquartered at St Anthony Falls Laboratory, University of Minnesota. In addition, I am confident that sediment routing system-type thinking will form the basis for a next generation of stratigraphic analysis, with profound implications for subsurface exploration and production. Source-to-Sink (S2S) methodologies are increasingly used in the hydrocarbon sector.

The scientific community has recognised the importance of joined-up thinking on the dynamics of the thin surface skin of the planet. The CSDMS (Community Surface Dynamics Modeling System) is building a community model for the quantitative prediction of material fluxes from source to sink (see http://csdms.colorado.edu/w/images/CSDMS_lecture7.pdf). A white paper on 'Building a Community Surface Dynamics Modeling System' is found at www.nsf-margins.org/S2S/S2SWhitePaper.pdf. CSDMS states that it 'will be a community-built and freely available suite of integrated, ever-improving software modules predicting the transport and accumulation of sediment and solutes in landscapes and sedimentary basins over a broad range of time and space scales'.

The same inclusive view of Earth surface dynamics was the motivation for a topical session at the American Geophysical Union (AGU) Fall Meeting in San Francisco in 2001 entitled 'Source to Sink: Production, Transport and Accumulation of Sediment with a Special Focus on Climate Signals and Impacts'. Papers originating at this meeting were later published as a set in *Sedimentary Geology* in 2003 edited by Steven Goodbred and Steve Kuehl. Several papers stressed the role of climate change in forcing trends in sediment discharge and continental margin evolution in contrasting regions from alluvial plains to the deep sea.

Integrated source-to-sink research programmes have been supported by a number of national and transnational funding agencies. The National Science Foundation (NSF) MARGINS initiative recognised that a predictive capability for sediment dispersal system behaviour has critical implications for understanding geochemical cycling, ecosystem change and resource management (www.nsf-margins.org/S2S/S2S.html). The MARGINS initiative sponsored the AGU Chapman Conference in 2011 (24–27 January), entitled 'Source-to-Sink Systems around the World and through Time: Recent Advances in Understanding Production, Transfer and Burial of Terrestrial and Marine Materials on the Earth Surface', at Oxnard, California, convened by Charles Nittrouer and Steven Kuehl. The programme is at http://csdms.colorado.edu/wiki/Chapman_Source_to_Sink.

The European margin strata formation (EuroSTRATAFORM) programme was funded under the 5th Framework Programme of the European Union for 2002–2005.

EuroSTRATAFORM set out to study marine sediment dispersal from source to sink on contrasted European margins, with the aim of understanding how geological strata are generated. Investigations were carried out at a variety of timescales in order to bridge the gap between currently active processes and the geological record. EuroSTRATAFORM participants contributed to special issues of the journals *Marine Geology* and *Oceanography*. The guest editors of the *Oceanography* special issue, published in 2004, James P. M. Syvitski, Philip P. E. Weaver, Serge Berné, Charles A. Nittrouer, Fabio Trincardi and Miguel Canals, stated that the five goals of EuroSTRATAFORM were

(1) To evaluate the influences on continental-margin sediment flux, including the characteristics of sediment sources and their temporal variability due to climatic evolution and human impacts. (2) To understand the oceanic processes that erode, transport and deposit sediment in the margin system, including short-term (i.e., hours to weeks) processes that produce event beds and the longer-term variability (e.g., seasonal, interannual) of those processes. (3) To quantify the physical and biological processes responsible for post-depositional modification of strata. (4) To understand the creation of the preserved stratigraphic architecture and sedimentary facies on continental margins as the product of processes acting with spatial and temporal heterogeneities. (5) To explore the nested expression of sedimentary successions, including the geologic identities of seismic properties, sequence boundaries, and intervening sequences.

The European Geosciences Union (EGU) sponsored conference sessions on 'Tectonics, Sedimentation and Surface Processes' in 2012, 2013 and 2014. A special issue of the journal *Earth Surface Processes and Landforms* collects a number of papers presented in these conference sessions, fronted by a commentary that highlighted three active challenges (Castelltort, Whittaker, and Verges, 2015). These challenges are to make progress in answering the questions of how tectonic and climatic drivers are recognised in landscapes; how signals, such as pulses in sediment discharge, are propagated through sediment routing 'systems' with different internal dynamics; and how the coupled processes of tectonics and surface processes are best modelled. Another thematic set, entitled 'Source-to-Sink Systems: Sediment and Solute Transfer on the Earth Surface', was published in *Earth Science Reviews* in 2015, edited by John Walsh, Patricia Wiberg and Rolf Aalto.

Considerable attention has been paid to the delivery of sediment to deltas. Continually updated information can be found at the Global River and Delta Systems SourcetoSink Information Center, which uses a Google map on an open platform, coded and edited by Paul Liu of North Carolina State University (www.meas.ncsu.edu/sealevel/s2s/).

In Part I of this book, sediment routing systems are introduced from a global perspective, starting with a broad view of how sediment routing systems function from source to sink. Whereas the terrestrial segments of sediment routing systems are relatively easy to map, since they correspond primarily to river drainage basins, the submarine segments are more elusive to constrain. Part I concludes with a global perspective on the role of the sediment routing system in biogeochemical cycling of the Earth. The controls on the chemistry of river water and the fate of particulate organic carbon carried by rivers to the ocean emphasise the important role of bedrock geology and sediment dynamics in mediating ocean chemistry and global climate.

Part II delves more deeply into the main segments that make up sediment routing systems, comprising the catchment-fluvial segment, separated from the continental shelf segment by the highly dynamic coastal zone, and the deep marine segment. In each segment, the controlling factors behind erosion, sediment transport and deposition are considered, principally using well-documented examples from the present day, or with the temporal perspective of the time since the last glaciation (10^4 yr).

Part III turns attention to how sediment routing systems function dynamically, with an emphasis on their sediment fluxes and budgets. Fluxes and budgets are best explored in a mass balance framework. An indicator of system dynamics is the migration of 'moving boundaries' such as the gravel front or the coast. Sediment moves intermittently during its cascade through the sediment routing system, so that the sediment transit time is largely governed by the time spent in temporary storage. The 'conveyor' or 'capacitor' properties of sediment routing systems determine the way in which such systems respond to a perturbation in the processes driving sediment flux. Signals may be transmitted quickly and without significant modification, or may be propagated as modified signals, or may be lost or 'shredded' by strong interaction with the system's internal dynamics. Sediment routing systems are mediated or driven by tectonics, which control rock uplift in source regions and accommodation generation in depositional basins. Tectonic processes operate at a range of scales from individual faults to linked extensional and contractional arrays, to flexural downwarps and the changing dynamic topography caused by mantle circulation.

Part IV of the book concerns the sedimentary archive of sediment routing systems. The formation of sediment by weathering is the start of a chain of dispersal processes leading to long-term deposition. Source and sink can be connected in a number of possible ways using provenance tools. Petrographic, compositional and isotopic information, heavy mineral analysis, palaeocurrents and U-Pb geochronology allow, in combination, an identification of sediment source areas. Sediment is transformed and fractionated from the site of weathering to eventual deposition in terms of composition and grain size. Tectonic processes and climate change have a fundamental impact on the down-system fining of grain size. Part IV concludes with a review of the significance of the sediment routing system approach for the interpretation of sequence stratigraphic architectures. Source-to-sink studies of the Quaternary are particularly informative of how depositional architectures and fluxes respond to changes in climate, sea level and sediment dynamics. Taking the integrated approach of sediment routing systems emphasises that stratigraphic architectures are generally non-unique, or non-diagnostic, in terms of forcing mechanisms, with the precise roles of autocyclicity, sea level change, climate, sediment supply and accommodation generation difficult to disentangle. As a result, the compilation of a global sea level chart from a stratigraphic database is problematical.

In summary, in this book I highlight the emergence of sediment routing systems as a compelling new way to understand Earth surface dynamics and as a tool to understand Earth history through the archive of stratigraphy. The book is not intended as a synthesis of thinking in a mature subject area, but instead as an early assessment of a rapidly emerging field that sets out some general principles and philosophies that may guide future research activity.

Acknowledgements

I am pleased to acknowledge the generous funding of Statoil for a generic research programme on sediment routing systems from 2006 to 2010. I am grateful to William Helland-Hansen, Peter Clift, James Syvitski, Brian Romans, Jake Covault, Ron Steel, Chris Paola, Andrew Miall, Fritz Schlunegger, Gert-Jan Weltje, John Holbrook and Tor Sømme for their help with literature, illustrations and spreadsheet data. I am also grateful for the ongoing collaboration and friendship of Hugh Sinclair, Peter Burgess, Alex Whittaker, Alex Densmore, Niels Hovius and Sanjeev Gupta, and for a long list of former graduate students, postdoctoral workers and former colleagues over many years. In the field of sediment routing system research, I have particularly benefitted from interaction with Guy Simpson, Sébastien Castelltort, Miriam Dühnforth, Sonia Scarselli, Alex Whittaker, Rob Duller, Amy Whitchurch, John Armitage and Nikolaos Michael. Working together has been an inspiration and a joy. To all of you, a fond farewell.

Part I
A Global View of Sediment Routing Systems

1

Sediment Routing Systems: First Concepts

1.1 How Sediment Routing Systems Function

Sediment routing systems link the fate of particulate sediment from source to sink, and essentially frame the problems of denudation, sediment transport and deposition as a box model characterised by sources, reservoirs and sinks with connecting fluxes (Figure 1.1). Sediment routing systems, or denudation-accumulation systems (Einsele, Ratschbacher, and Wetzel, 1996; Hinderer and Einsele, 2001), are integrated, dynamical systems connecting regions of erosion, sediment transfer, temporary storage and long-term deposition (Meade, 1972, 1982; Schumm, 1977; Castelltort and Van Den Driessche, 2003; Allen and Allen, 2013; Romans and Graham, 2013; Sadler and Jerolmack, 2015). They involve a sediment cascade (Burt and Allison, 2010) from single or multiple source regions to long-term depositional sinks via a series of geomorphic environments characterised by intermittent storage (Section 1.2). The sediment routing system philosophy therefore firmly places geomorphology, sedimentology and stratigraphy within an Earth system context, but also forms the framework for allied investigations of, for example, global biogeochemical cycles.

Sediment routing systems represent a vigorous way in which the Earth recycles mass, and their dynamics are fundamental to global responses to, for example, supercontinental assembly and dispersal, mountain building and climate change (Whipple, 2009). They also provide thoroughfares for the transmission of chemical signals from mountains to the ocean (Meybeck, 1987; Hay, 1998; Galy et al., 2007). Sediment routing systems therefore participate strongly in many geochemical cycles, such as that of carbon (Leithold, Blair, and Wegmann, 2015), including in the drawdown of atmospheric CO_2 mediated by rates of chemical weathering (Raymo and Ruddiman, 1992), in the delivery of particulate organic carbon to the ocean and its removal from the short-term carbon cycle by rapid burial in deltas and sediment fans (France-Lanord and Derry, 1997; Galy et al., 2007), in the charging of coastal waters with nutrients (Orive, Elliott, and de Jong, 2002) and in the catalysing of ocean anoxia in sheltered and semi-enclosed seas by changes in freshwater discharge of rivers (Beckmann et al., 2005).

The concept of sediment routing systems was not until recently part of mainstream geological thinking, despite the fact that sediment provenance, based on sediment mineralogy (Boswell, 1933; Pettijohn, Potter, and Siever, 1987), bulk composition (Dickinson and

(a) Box model

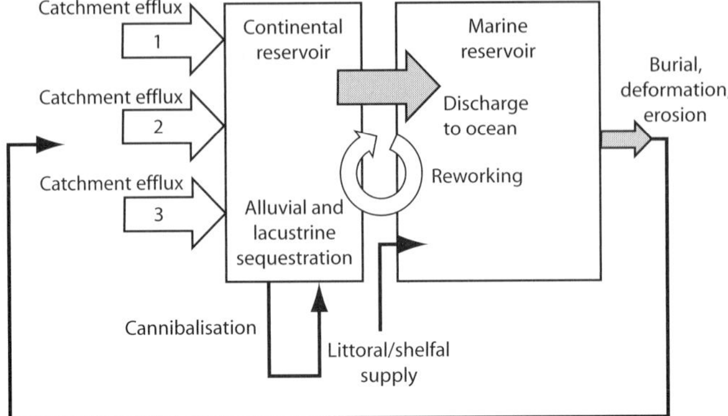

(b) Network

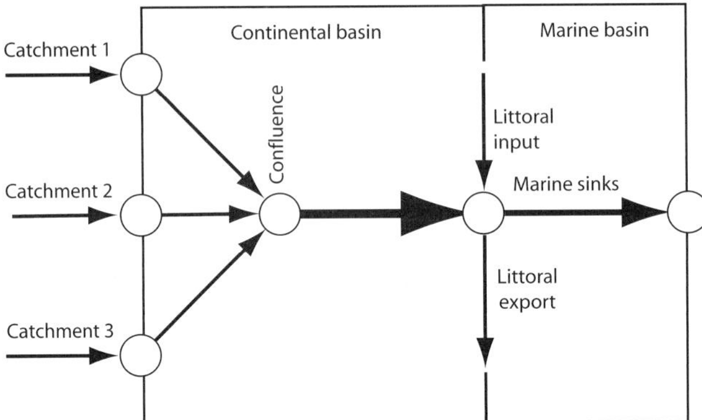

Figure 1.1 (a) Representation of a sediment routing system as a box model involving 3 sources and continental and marine sediment reservoirs. As sediment passes through the continental reservoir en route to the ocean, some of it is sequestered as alluvial and lacustrine sediment. Sequestered sediment may be cannibalised and enter the transport system again, and continental and marine reservoirs may be reworked as a result of base-level change. (b) The same sediment routing system as a network, showing schematic transport pathways. Discharge from the continent to the ocean is added to and subtracted from by marine sediment fluxes such as littoral transport.

Suczek, 1979) and palaeocurrents (Potter and Pettijohn, 1977), has been studied systematically for many years. The linkage between source and sink can be made by forward modelling the cascade of sediment (Section 1.2) from erosional landscapes primarily in terms of volumes and fluxes, but also in terms of sediment composition and grain size (Johnsson, 1993; Weltje and von Eynatten, 2004), and inversely by reconstructing source area characteristics from detrital mineralogy, thermochronology and geochronology.

Sediment dynamics in sediment routing systems is rarely smooth, continuous and uninterrupted like a relentless conveyor belt. Instead, the release of particulate sediment and dissolved solids from erosional source regions is neither uniform nor steady. These effluxes tend to be transient in relation to the forcing mechanisms. Some systems may respond quickly and linearly to a perturbation, in which case they can be termed 'reactive' (Allen, 2008a). Other systems may be sluggish to respond, and the signal may be transformed markedly during down-system propagation. Large alluvial systems have the ability to buffer incoming signals, which smooths out their amplitude and wavelength (Castelltort and Van Den Driessche, 2003; Allen, 2008a; Romans et al., 2015). The passage of signals through sediment routing systems therefore depends on the internal dynamics of sediment routing system segments and the 'transfer zones', 'transition zones' or 'energy fences' between them (Carvajal and Steel, 2009), some signals being effectively lost, or 'shredded', during transmission (Jerolmack and Paola, 2010). Sediment dynamics and hydrodynamics commonly change markedly at these transition zones, as takes place for example between river systems and the shoreface.

It is the passage of signals through the sediment routing system that causes different parts to be 'teleconnected'. Most commonly, signals are communicated down-system, so that, for instance, a climate change in an erosional source region passes down to the ocean in the form of a change in particulate sediment discharge. Alternatively, a base-level change at a river mouth, for example caused by global sea level fall, passes up-system as an erosional wave that generates enhanced erosion rates in the source region. In this way, the entire sediment routing system is dynamically connected.

Buffering of signals is caused by the interruption of sediment transport by periods of intermittent storage (Malmon, Dunne, and Reneau, 2003; Allen, 2008b), which causes an increase in transit time for a grain-size population. As Philip Allen 2008a (p.274) put it:

One can imagine tracking the trajectory of a single grain of sand from its source in mountain headwaters to its sink in a river flood plain, delta or the deep sea. Each grain would have a different trajectory and a different time in transit. The integration of a multiplicity of such trajectories defines a sediment-routing system, and an integration of the different transit times, were it possible, would provide information on the ability of the routing system to buffer incoming sediment flux signals.

The release of sediment from source regions carries a fingerprint that is potentially recognisable in basin stratigraphy in terms of mass fluxes, grain-size characteristics and sediment composition (Ibbeken and Schleyer, 1991, 2003). In a simple catchment-fan system, for example, where there is a single bedrock lithology such as granite, there will be little temporal change in sediment composition but a lateral compositional gradient caused by the hydraulic sorting of the different mineral components of the source region lithology. In such a case, particular diagnostic minerals, or index minerals, used in thermochronometric and cosmogenic dating, can be unambiguously connected to their hinterland source. Most basin-fills, however, are derived from multiple sources (Vezzoli, Garzanti, and Monguzzi, 2004; Weltje and Brommer, 2011) consisting of a range of bedrock lithologies, which induces a lateral (spatial) and vertical (temporal) variation in sediment composition in basin stratigraphy. The heterogeneous distribution of lithologies in the hinterland combined with the complexities of unroofing history produce a mixing structure in basin stratigraphy that is challenging to decipher. Specifically, the clast compositions of proximal conglomerates have been used to monitor the progressive erosion of source regions in tectonically active hinterlands (Graham et al., 1986; DeCelles, 1988). Complex mixing structures in basin stratigraphy make it more difficult to relate diagnostic minerals to their source region.

If stratigraphic units in the basin can be linked to source area lithologies, then the relative contribution of different source terrains can be assessed and sediment budgets estimated for the time span of interest. However, basin sediments may be recycled from older stratigraphy, chemical sediments must be accounted for, basin stratigraphy must be corrected to porosity-free solid volumes, and catchments serving as source areas for sediments must be correctly identified and delineated. Present-day sediment routing systems commonly benefit from the presence of well-defined and quantitatively characterised catchment areas but may suffer from a poor database on stratigraphy buried beneath the basin surface. Ancient sediment routing systems may have a well-known stratigraphy but source area parameters may be unknown or speculative. In some cases, the history of drainage evolution and sediment supply to adjacent basins can be evaluated on a continental scale (Simoes, Braun, and Bonnet, 2010). For example, the changing total sediment input to the Gulf of Mexico Basin over the Cenozoic (since 65 Ma) has been calculated from the supply from 8 fluvial axes draining a high proportion of the North American continental surface (Galloway, Whiteaker, and Ganey-Curry, 2011) (Figure 1.2). Peaks in sediment delivery, evaluated from depositional volumes, of 150×10^3 km^3 Myr^{-1} occurred in the Palaeocene and Pleistocene. The Mississippi River delivers the bulk of the sediment entering the Gulf of Mexico at the present day. The modern TDS (total dissolved solids) of 400 Mt yr^{-1}, equivalent to 150 km^3 Myr^{-1}, is in agreement with the long-term value derived from the volumes of depositional episodes in the Gulf of Mexico Basin. In well-constrained examples, sediment volumes may be inverted to obtain mean catchment erosion rates over geological timescales that can be compared with estimates derived from thermochronometric methods (Rouby et al., 2009; Carvajal and Steel, 2012; Michael et al., 2014b).

A major goal within the broader remit of sediment routing systems research is to gain predictive capability of subsurface compositional and granulometric trends. Harnessing such capability would be highly beneficial to the understanding of the flow of fluids through the porous media represented by basin sediments, with applications to hydrocarbon production, contaminant fluxes in groundwater and drawdown of aquifers as well as to basin analysis. Yet it is problematical to sample subsurface stratigraphy at the resolution required to build a reliable database. To make progress demands more than an inversion of field observations. A holistic analytical approach was proposed by Heins and Kairo (2007) in which all of the factors controlling sand composition and texture were analysed together, chief of which are tectonics, geomorphology, lithological make-up of the source region, weathering and transport processes. These factors are constrained by quantitative input on source region lithologies, the regional topographic gradient, climate (weathering and transport potential), transport distance in the source catchment and in the depositional basin, basin subsidence rate and depositional facies (Heins and Kairo, 2007).

The total volume of sediment deposited in the sediment routing system and the spatial extent of its footprint, or 'fairway' (Michael, Whittaker, and Allen, 2013; Michael et al.,

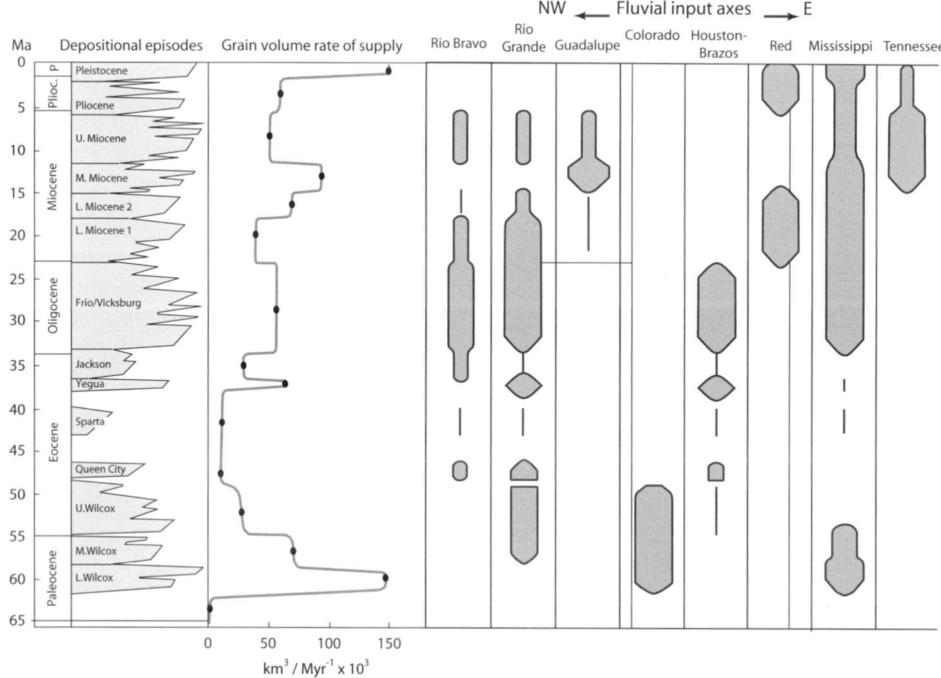

Figure 1.2 Chronology of Cenozoic depositional episodes in the Gulf of Mexico Basin, total (porosity free) sediment supply (in km^3 Myr^{-1} × 10^3) as a function of time, and the individual contributions from 8 fluvial axes calculated from volumes of depositional episodes. Width of bar indicates volumetric importance of supply to depocentre. Modified from Galloway, Whiteaker, and Ganey-Curry (2011) (fig.1, p.940; fig.3, p.943), with permission of Geological Society of America.

2014a), provides the basis for a sediment budget. Sediment budgets allow a connection to be made between hinterland erosion and basin filling, and also potentially explain the downstream evolution of sedimentary facies, stratigraphic architectures and the partitioning of grain size (Cross and Lessinger, 1998; Granjeon and Joseph, 1999) (Figure 1.3). Sediment is fractionated by particle size during transport (Cross and Lessinger, 1998), principally by selective extraction from surface fluxes to build stratigraphy (Strong et al., 2005; Allen, 2008a; Paola and Martin, 2012; Michael et al., 2013), producing a down-system trend in the relative proportions of grain-size classes and in the mean grain size of gravel and sand (Fedele and Paola, 2007; Duller et al., 2010; Whittaker et al., 2011; Parsons et al., 2012; Michael et al., 2013).

Laboratory experiments (Toro-Escobar, Parker, and Paola, 1996; Sheets, Hickson, and Paola, 2002; Strong et al., 2005; Paola et al., 2009; Paola and Martin, 2012; Rohais, Bonnet, and Eschard, 2012), numerical models (Allen and Densmore, 2000; Densmore, Allen, and Simpson, 2007a; Duller et al., 2010; Jerolmack and Brzinski, 2010; Armitage et al., 2011) and field studies (Duller et al., 2010; Whittaker et al., 2011; Carvajal and Steel, 2012; Parsons et al., 2012) have concluded that the main controls on stratigraphic architectural

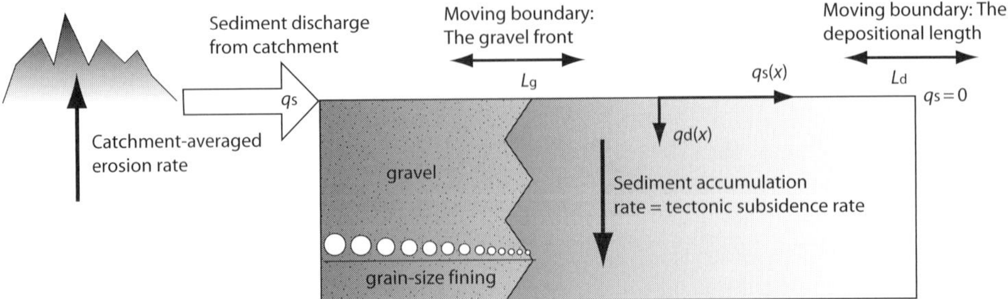

Figure 1.3 Sediment dispersal and deposition in a sediment routing system. Sediment is discharged from a source region q_s into a basin undergoing subsidence. Sediment is selectively extracted from the surface flux $q_s(x)$ to build stratigraphy $q_d(x)$, causing a down-system fining of mean grain size. The dynamics of sediment dispersal and deposition are reflected in the migration of moving boundaries such as the gravel front (Section 8.1.1). From Allen and Allen (2013) (fig.7.49, p.274) with permission of John Wiley & Sons, Inc.

trends in continental basins are: (1) the sediment supply to the basin, which Allen et al. (2013) referred to as the 'Qs problem', (2) the characteristic grain-size mix of this sediment (Allen et al., 2015b), (3) the spatial distribution of subsidence and base-level changes controlling accommodation and (4) hydraulic parameters and sedimentary processes in the basin. One of the most important aspects underpinning the sediment routing system concept and aiding downstream stratigraphic prediction is therefore the volumetric or mass sediment budget.

Sediment budgets, despite their generic value in illustrating how the sediment routing system functions, are unique to the case being studied (Hinderer, 2012). This problem can be overcome by setting the surface sediment discharges and depositional fluxes of sediment routing systems within a mass balance framework (Strong et al., 2005; Paola and Martin, 2012; Michael et al., 2013). In the mass balance framework, the cumulative deposited volume of sediment in the down-system direction is normalised by the total sediment volume and the total depositional length (Figure 1.4), and thereby illustrates in a dimensionless fashion the interplay between the depositional flux and the bypassed surface sediment discharge. This not only allows systems of different shapes and lengths, but also of different total sediment volumes, to be compared, and potentially permits the upscaling of small laboratory tests to real stratigraphic problems (Strong et al., 2005; Paola and Martin, 2012).

A mass or volume balance of the entire system allows a sediment budget to be evaluated that balances the particulate sediment derived by physical weathering against sediment volumes in depositional sinks (Clift et al., 2001a; Slaymaker, 2003; Tinker, de Wit, and Brown, 2008; Carvajal and Steel, 2012; Hinderer, 2012; Michael et al., 2014b). Quantification of the sediment budget, its component fluxes and the down-system evolution of grain size therefore provides essential information to calibrate and test predictive models of basin filling and sedimentary architecture (Paola, Heller, and Angevine, 1992; Marr et al., 2000; Armitage et al., 2011; Kim et al., 2011), including those derived from sequence stratigraphy.

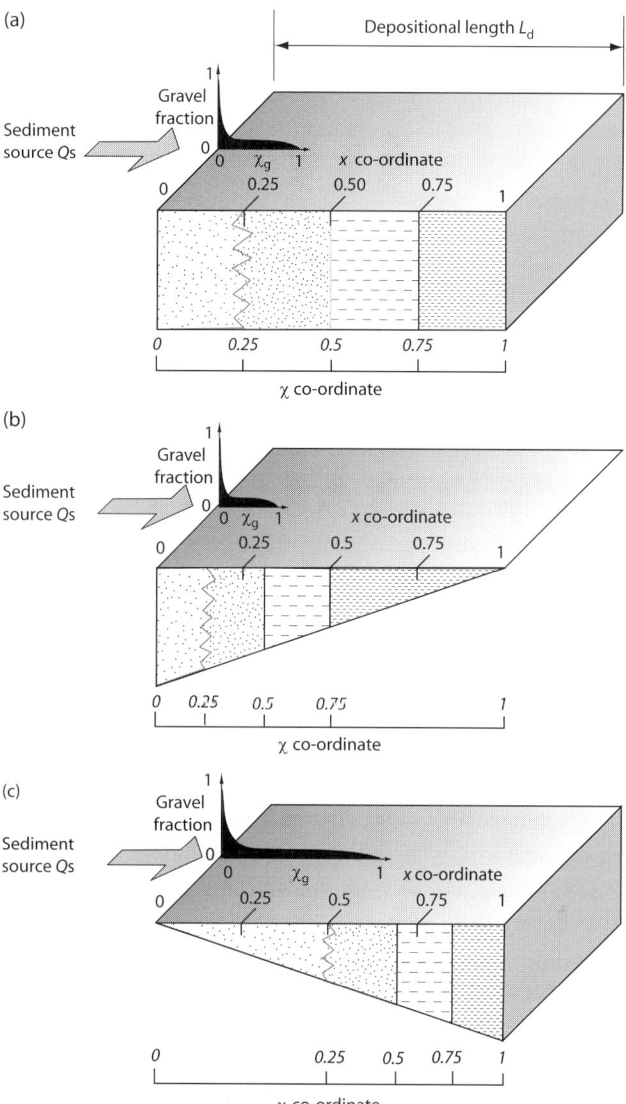

Figure 1.4 The transformation from Cartesian co-ordinates into a mass balance framework for different basin cross-sectional shapes. The down-system co-ordinate x is transformed by normalising the depositional volume (or mass) by the total volume (or mass) deposited in the entire sediment routing system. The mass balance co-ordinate χ therefore varies between 0 (at the origin, or apex of the depositional system) to 1 at the depositional length L_d where sediment is exhausted. This allows sediment routing systems with different sediment budgets, accommodation and size to be directly compared. Modified from Michael et al. (2013) (fig.4) with permission of University of Chicago Press.

Strong et al. (2005) and Paola and Martin (2012) showed from both laboratory and subsurface studies that fluvial and deep marine sedimentary architectures and grain-size patterns are governed primarily by mass extraction by selective deposition over geological timescales. In other words, for the same ratio of sediment supply to total accommodation space in any basin, the same facies assemblages and the same bulk grain-size characteristics are expected to be present, regardless of system length and spatial pattern of tectonic subsidence. Consequently, extraction of certain grain sizes by selective deposition provides a fundamental control on the availability of finer grain sizes downstream, thereby controlling depositional environments, facies and stratigraphic architectures.

Sediment routing systems might be categorised and named from their source, their main artery of sediment transport or their main depocentre. There are difficulties in whichever method of categorisation in chosen. Most natural sediment routing systems have multiple sediment sources, more than a single sediment transport system and occasionally different depocentres (multi-source and multi-sink systems). For example, the western Adriatic mud wedge is fed by the Adige, Po and several Apennines rivers (Weltje and Brommer, 2011). Coastal currents supply the macrotidal Gulf of Kachchh in western India with sediment derived from both the perennial Indus River as well as from a number of small ephemeral 'dryland' streams (Prizomwala, Bhatt, and Basavaiah, 2014). The Golo sediment routing system of the western Mediterranean is fed from a single catchment in Corsica, but its efflux is joined by the contribution of a littoral cell; sediment is stored on the continental shelf and slope and intermittently discharged through shelf-edge canyons to lobes on the deep sea floor of the Corsican Trough (Sømme et al., 2011). At a larger scale, the Orange River drainage basin supplies sediment to a number of passive margin basins stretching from the Falkland-Agulhas fracture zone in the south to the Walvis Ridge in the north (Rouby et al., 2009). A number of independent fluvial systems draining the North American continent, such as the Brazos, Colorado and Mississippi, have supplied large volumes of sediment to the Gulf of Mexico during the Cenozoic (Galloway, Whiteaker, and Ganey-Curry, 2011). Each river system has its own coastal delta at times of sea level highstand, but river systems merge on the continental shelf at times of sea level lowstand (Blum and Aslan, 2006; Blum and Garvin, 2010) (Figure 10.9). The deep marine fan of the Bay of Bengal is supplied with sediment from multiple sources, including the rivers of the eastern Ghats of peninsular India and the Ganges, Brahmaputra, Salween and Irrawaddy river systems draining the Himalaya-Tibetan region. The networks of sediment routing systems, as shown generically in Figure 1.1, may therefore be complex (Figure 1.5). In many cases, the sediment routing system can be named according to its main river artery. These main arteries correspond closely to those rivers with outlets into oceans, inland seas and lakes, so the categorisation of sediment routing systems is essentially the same as the challenge facing the authors of global databases of sediment delivery to the ocean (Milliman and Farnsworth, 2011). This option of naming sediment routing systems from their main river arteries was adopted by Helland-Hansen et al. (2016) in their review of 'Earth's natural hourglasses'.

Sediment routing systems are termed 'Earth's natural hourglasses' (Helland-Hansen et al., 2016) since they commonly involve a fairway that is broad in the hinterland

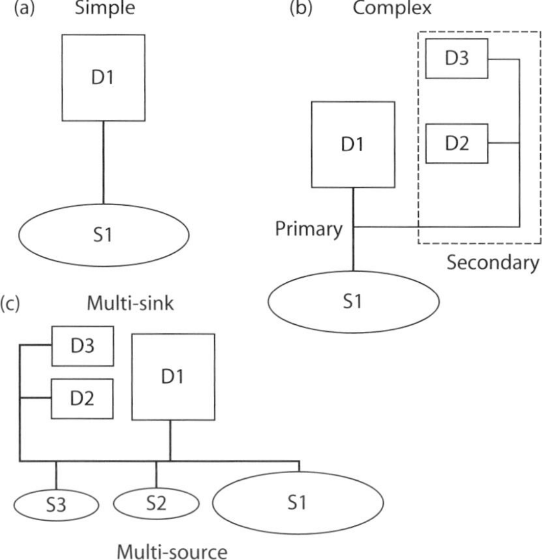

Figure 1.5 Sediment routing system networks, which may be (a) simple or (b) complex, with primary or secondary pathways, and comprise (c) multiple sources and sinks. S, Source; D, Depositional sink.

catchment and broad in the depositional zone, but relatively narrow in the transitional zone between. Source-to-sink systems are viewed in a framework with three end-member types: (1) 'steep, short and deep', which occur along active plate boundaries, and are 'reactive' (Allen, 2008a), such as the Var and Golo of the Mediterranean Sea and Monterey and Astoria of the California Borderland; (2) 'wide and deep', which are characteristic of passive continental margins, and are 'buffered', such as the Rhône, Danube, Niger, Zambezi, Congo, Mississippi, Amazon and Nile; and (3) 'wide and shallow', which are commonly located on normal thickness continental crust such as in foreland basins and cratonic basins, as in the Po and Ganges. Clearly, there is a strong tectonic control on the occurrence of this variability in sediment routing systems (Section 8.6).

The networks of sediment routing systems may thus be simple, involving a single source area linked unambiguously to a single depositional sink, or complex, with additional sources, sinks and sediment transport pathways (Figure 1.5). Complex sediment routing systems may involve a primary route for sediment dispersal and a secondary route carrying lower sediment discharges. Many sediment routing systems contain multiple sources and, less commonly, multiple sinks (Table 1.1).

1.2 The Sediment Cascade

Particulate sediment and dissolved solids cascade through the sediment routing system with characteristic fluxes, which allows it to be treated as an essentially closed system in terms of volume or mass (Figure 1.6). Closure of the sediment budget is a powerful tool, but

Table 1.1. *General attributes of sediment routing system networks.*

Attribute	Examples
Simple	Catchment-fan systems (Death Valley, Owens Valley)
	Eel River system, Oregon Shelf
	Var system and deep-water fan, Nice
Complex	Bengal Fan
Primary	Nile delta
	Rhône Fan
	Indus Fan
Secondary	Levant margin littoral cell
	Roussillon-Languedoc littoral current
	Eastern Ghats sources for Bengal Fan
Multi-sink	Amazon outflow
	Orange, southwest Africa
	Golo, Corsican Trough
Multi-source	Bengal Fan
	Tarim Basin
	Western Adriatic mud wedge
	Newport Fan, Santa Ana system
	Gulf of Mexico
	Gulf of Kachchh, Arabian Sea, India

there are a number of reasons why this is problematical. The main reason is that solute fluxes enter the ocean to join a globally mixed pool. In addition, marine depositional sinks may be difficult to identify, constrain and sample, and very fine-grained sediments may be transported far into the open ocean. Third, part of the sediment routing system budget may involve sediment transported by wind. Nevertheless, notwithstanding these problems, to describe a sediment routing system ideally requires the identification of regions of sediment generation, transport and deposition, together with the estimation of the fluxes of particulate sediment and solutes that connect them.

In a hypothetical unglaciated sediment routing system, sediment is released from hill-slope regolith by soil creep, debris flows and landslides and is transferred to catchment channels. A new set of processes is involved in the evacuation of this stored sediment from the catchment into the fluvial system. Further processes are responsible for down-system advection, principally by floods, and intermittent storage in channel bars, levees, splays and floodplains. Subsequent floods may or may not entrain these deposits. The result is that sediment takes a large number of stochastic steps towards the coast. At the coast, new processes take over to transport sediment by waves, tides, jets, geostrophic flows and gravity flows to storage sites as shorelines, estuaries, coastal marshlands, deltas, shallow marine sand sheets, shelfal mud belts, and coastal and shelf-edge clinoforms and sediment prisms. In some cases, the ultimate sink is these coastal and shallow marine deposits, and

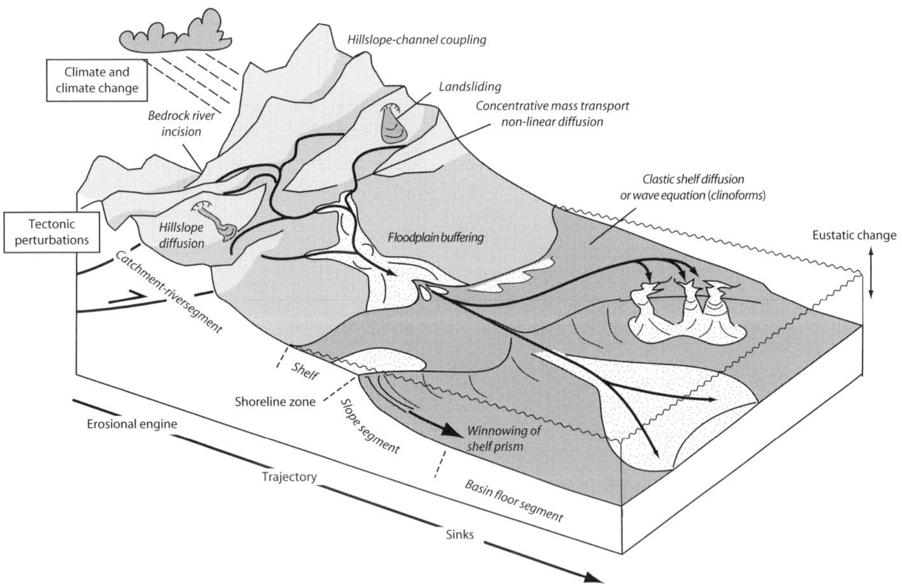

Figure 1.6 Concept of the sediment routing system from source to sink. The sediment routing system is driven and perturbed by tectonics and climate. It comprises a number of morphodynamic, mutually coupled subsystems or segments. The erosional engine feeds sediment to a predominantly transportational system (the trajectory) and to long-term sinks. From Allen and Heller (2012) (fig. 6.2, p.113), with permission of John Wiley & Sons, Inc.

no sediment escapes onto the continental shelf and into the deep sea, as in Chesapeake Bay, on the Atlantic coast of the United States (Meade, 1982). However, a final step may remobilise some of the riverine or marine sediment and transport it to depositional sinks on the continental slope and deep basin floor (Figure 1.6). When viewed at the medium to long timescale (more than 10^3 yr), sediment is simply released from upland sources and delivered to marine sinks. At the shorter timescale, however, more important sources and sinks are the storage sites along the sediment cascade, such as valley bottoms, terraces and floodplains. The delivery of sediment to the marine basin floor (Covault and Graham, 2010) is therefore the end of a series of stochastic steps from sites of storage, known as 'transient states' or 'staging areas,' to the 'absorbing state' (Malmon et al., 2005) of long-term deposition at the terminus of the sediment routing system (Figure 1.7). Sediment routing systems, or individual segments within them such as the continental shelf, may act as 'conveyors' or 'capacitors' in terms of the transmission of sediment (Covault and Fildani, 2014).

Consequently, although in some instances sediment pulses may pass rapidly through the routing system and in one cycle arrive on the deep marine basin floor, bypassing intervening transient states, it is more common for signals originating in the source region, such as the grain-size distribution of the sediment efflux, to be subject to transformation during propagation through the sediment routing system (Weltje, 2012; Romans et al., 2015).

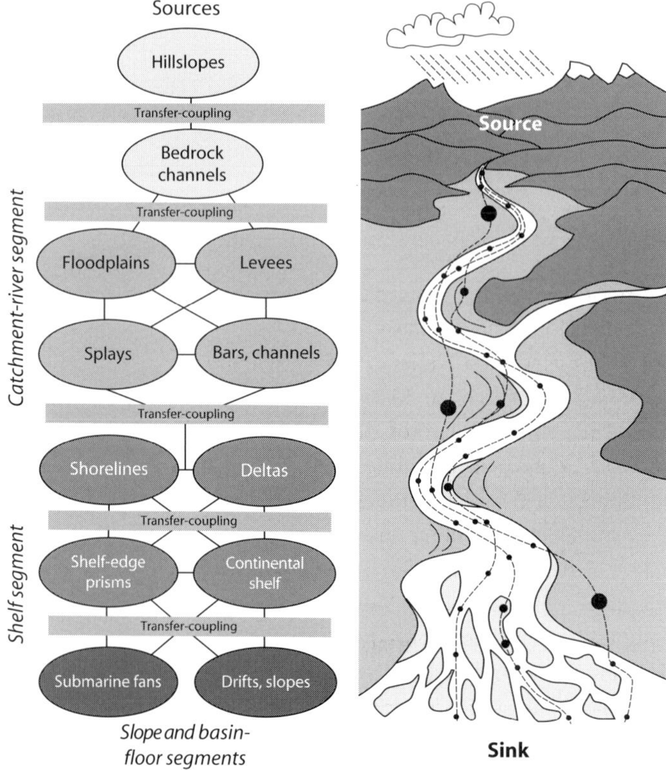

Figure 1.7 Right: sediment is transferred from a source region to a sink along trajectories (dashed lines). Some trajectories involve short transit times with brief periods of storage in the sediment routing system (small circles), whereas others involve long transit times with prolonged periods of storage in transient states (large circles). Long periods of storage of sediment in transient states implies buffering of sediment supply signals. Modified from Malmon, Dunne, and Reneau (2003) (fig.1, p.526). Left: sources, sites of temporary storage (transient states) and sites of permanent storage (absorbing states). The transfer of sediment between macrogeomorphic segments may be complex and involve dynamic feedbacks. From Allen and Heller (2012) (fig.6.1, p.112) with permission of John Wiley & Sons, Inc.

Within a sediment routing system undergoing tectonic subsidence, part of the surface flux in the sediment cascade is selectively sequestered at any point along the routing system, until the surface flux becomes exhausted at the downstream depositional limit (Paola and Voller, 2005; Fedele and Paola, 2007; Allen, 2008a; Paola et al., 2009; Duller et al., 2010; Allen and Heller, 2012; Paola and Martin, 2012; Michael et al., 2013, 2014a,b). Over geological time, depending on the available accommodation and position of base level (Muto and Steel, 2000), a portion of this depositional flux is locked into stratigraphy (Allen and Heller (2012), fig.6.3), a phenomenon sometimes broadly referred to as 'Landscapes into Rock' following the William Smith meeting with this name at the Geological Society of London in 2010 (see http://romania-rocks.blogspot.co.uk/2010/09/landscapes-into-rock-part-1.html). The sequestration of merely a portion of the surface flux of sediment to build stratigraphy, the intermittency and spatial limitations of depositional events and the obliterating effects of erosion together imply that the stratigraphic record is a far from complete recorder of time. Allen (2008a) (p.275) wrote:

Time transforms sediment routing systems into geology, and like history, selectively samples from the events that actually happened to create a narrative of what is recorded.

The simplest sediment routing system that can be imagined comprises a single upland catchment that feeds sediment to a basin-margin fan, acting in a closed system, with no sediment escape (Humphrey and Heller, 1995; Allen and Hovius, 1998; Allen and Densmore, 2000; Carrétier and Lucazeau, 2005; Densmore, Allen, and Simpson, 2007a; Pépin, Carrétier, and Herail, 2010). In a prototype such as this, sediment can be unambiguously linked to lithological sources in the hinterland catchment, and thermochronological and cosmogenic signatures in the depositional sink of the fan can be reliably connected to erosion in the source (Stock, Ehlers, and Farley, 2006). Commonly, however, sediment routing systems are more complex and involve far-field sediment transport, well-developed alluvial sequestration (Castelltort and Van Den Driessche, 2003; Allen, 2008b; Covault et al., 2013), complex sediment transport dynamics in the coastal and continental shelf areas (Covault and Fildani, 2014) and multiple sources and/or multiple sinks (Vezzoli, Garzanti, and Monguzzi, 2004; Weltje and Brommer, 2011) (Table 1.1) (Figure 1.5).

Some sediment routing systems are continental in scale, involving sources in strongly contrasting climatic and tectonic settings, such as the Ganges and Brahmaputra systems draining into the Bay of Bengal (Einsele, Ratschbacher, and Wetzel, 1996; Clift et al., 2001a; Curray, Emmel, and Moore, 2002). In large, complex examples, sediment can be traced back to source regions using a range of provenance methods (analysis of pebble lithologies, heavy minerals, isotopic fingerprinting and geochronology of detrital grains) (Section 9.5), but mixing of fingerprints from different sources during sediment transport and down-system attenuation of source area signals introduce considerable uncertainties in linking source(s) to sink(s). Identification of source areas for detrital minerals in the basin is important in understanding where sediment ultimately comes from, but is only a part of understanding the sediment routing system. As Allen (2008a) (p.275) writes:

No matter how we make this match between erosional source area and depositional sink, provenance studies cannot help us to fully understand the dynamics of the sediment routing system that conveyed [the sediment] from source to sink. It is rather like being present at the birth of a baby and the funeral of a man, but missing out on the life story.

Most studies of the sediment budget have addressed relatively modern systems, where catchments serving as source regions and depositional sinks, such as dams, natural lakes and floodplains, as well as the continental shelf, are well defined (Einsele and Hinderer, 1997; Hinderer, 2012). Most present-day sediment routing systems are strongly influenced by human activities. Longer-term budgets have also been estimated, particularly in the context of climate change and tectonic activity in mountain belts acting as source regions, commonly using sediment isopachs derived from seismic reflection data, borehole information and stratigraphic outcrops (Barnes and Heins, 2009; Brommer, Weltje, and Trincardi, 2009; Rouby et al., 2009; Lopez-Blanco et al., 2010; Weltje and Brommer, 2011; Carvajal and Steel, 2012), by palaeohydrological analysis of upstream fluxes from cross-sections of alluvial deposits (Holbrook and Wanas, 2014), from inventories of lake basin, valley and delta volumes (Kuhlemann et al., 2002; Walford, White, and Sydow, 2005; Tinker, de Wit, and Brown, 2008), and from geochemical indices (Galy and France-Lanord, 2001). A range of studies have addressed the long-term sediment delivery to the ocean from rivers at the basin scale using seismic reflection and borehole data in the depositional sink (Métivier and Gaudemer, 1999; Clift and Gaedicke, 2002; Walford, White, and Sydow, 2005; Tinker, de Wit, and Brown, 2008; Rouby et al., 2009; Galloway, Whiteaker, and Ganey-Curry, 2011; Guillocheau et al., 2012).

By connecting source to sink, and by keeping an inventory of the grain-size fractionation, depositional volumes and sediment calibre can be integrated throughout the sediment routing system fairway, which solves concurrently two challenging problems: (1) the total volume of particulate sediment derived from mountain catchments serving as 'erosional engines' averaged over geological timescales can be calculated (Carvajal and Steel, 2012) and compared with erosion estimates derived from thermochronology (Tinker, de Wit, and Brown, 2008; Rouby et al., 2009; Michael et al., 2014b), and (2) the 'global' size distribution of sediment released from mountain catchments at long timescales can be obtained (Michael et al., 2014a) and used as a boundary condition in models of sediment dispersal. For example, the 'global' grain-size mix fluxed out of mountain catchments to supply the mid-upper Eocene Escanilla palaeo-sediment routing system of the wedge-top zone of the southern Pyrenees was 9% gravel, 25% sand and 64% finer than sand, based on about 4,000 km^3 of sediment released over a 7.7 Myr period (Michael et al., 2013, 2014a).

Erosion and deposition rates, sediment fluxes and grain sizes can be calculated in modern sediment routing systems using a variety of techniques targeted at short-term observational periods, but these calculated values cannot be easily upscaled to the longer timescales of geological sediment routing systems. Ancient sediment routing systems, however, are difficult to recognise and investigate: they are likely to be only partially preserved as stratigraphy, making the definition of their fairway problematical; depositional sinks are commonly disconnected from erosional source regions; there may be multiple sources, some of which may be difficult to identify using conventional provenance methods; and the chronological resolution may be inadequate to allow a confident correlation scheme to be developed. Constraining the sediment budget of a geological sediment routing system is therefore a deceptively challenging task, but one that is vital if we are to couple measurements of rock exhumation and erosion rates in mountain belts with realistic estimates of the timing, locus and magnitude of sediment supply to basins (Sinclair et al., 2005; Beamud et al., 2010; Whitchurch et al., 2011; Carvajal and Steel, 2012; Filleaudeau, Mouthereau, and Pik, 2012; Parsons et al., 2012).

Beyond the river mouth, the continental shelf can either act as a capacitor (storage or staging area) or as a conveyor for sediment from the land to the deep sea (Covault and Fildani, 2014). Shelves are conventionally regarded as capacitors during periods of rising and high relative sea level (Vail, Mitchum, and Thompson, 1977; Jervey, 1988), when the shelf offers accommodation for sediment preservation, and low river gradients during times of elevated sea level result in lower fluxes of sediment delivered to the shelf. On the other hand, shelves may act as conveyors, particularly where canyons are incised into continental shelves and intersect nearshore sediment transport pathways, or where high sediment discharges from the land cause submarine deltas to migrate to the shelf edge.

Off-shelf transport to the deep sea is therefore affected by a number of factors:

1 Relative sea level: transport rates may be reduced at times of sea level highstand due to a reduction in river gradients (Posamentier and Vail, 1988; Burgess and Hovius, 1998; Covault and Graham, 2010).
2 Shelf width: off-shelf transport may reduce as shelf width increases, which increases shelf accommodation (Posamentier, Erskine, and Mitchum, 1991; Walsh and Nittrouer, 2003).
3 Sediment supply: transport to the deep sea may increase by the driving of deltas to the shelf edge by high sediment discharges (Burgess and Hovius, 1998; Covault et al., 2007; Carvajal, Steel, and Petter, 2009).
4 Connection of canyon heads: effective conveyance of the shelf is critically dependent on the connection of canyon heads to high river and littoral sediment discharges (Covault et al., 2011).

Although Eel Canyon, offshore California, is currently disconnected from large fluvial or littoral sediment sources, as it has been over the last thousands of years of rising and high sea level (Burger, Fulthorpe, and Austin, 2001), over the hundred year timescale, sediment budget studies indicate that a large percentage (approximately 80%) of sediment has been exported from the shelf into the deep sea (Sommerfield and Nittrouer, 1999). Covault and Fildani (2014) therefore point out that the Eel shelf acts as a conveyor over the short timescale of the last century while acting as a capacitor at the timescale of the last million years.

The Washington–Oregon shelf is wider (25–60 km) than the Californian shelf (average of 5 km) and is fed principally by the large sediment discharge of the Columbia River (5–10 Mt yr^{-1}), half of which is sequestered on the shelf (Sommerfield and Nittrouer, 1999). Maxima in the delivery of sediment from the Columbia River are out of phase with periods of high wave and current reworking, which promotes shelf storage. However, routing of sediment to the Astoria canyon and fan system was most efficient during periods of post-glacial marine transgression, when climate systems caused intense terrestrial flooding. The Washington–Oregon shelf therefore acts as a capacitor at the short timescale of the present day, but as a conveyor during periods of post-glacial climate instability at the timescale of glacial-interglacial cycles (Covault and Fildani, 2014).

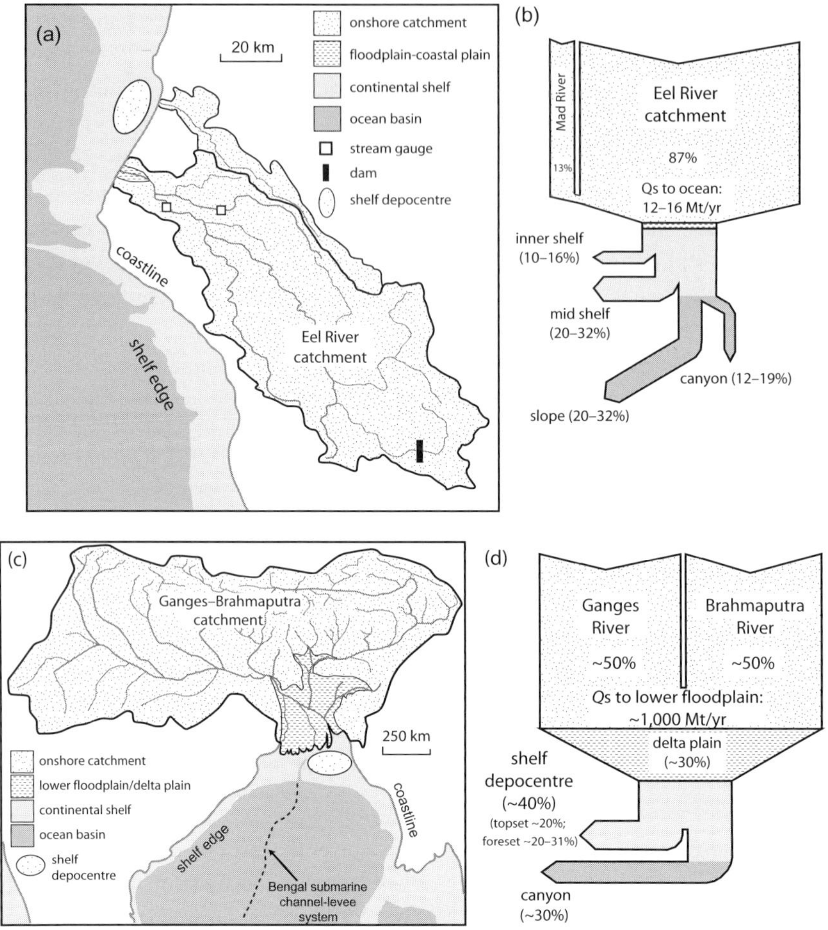

Figure 1.8 (a), (b) Combined Eel-Mad sediment routing system, northern California, showing upstream catchment area and storage areas in the coastal floodplain, shelf, continental slope and canyon leading to the deep sea. Sediment budget estimates are from Sommerfield and Nittrouer (1999) and Warrick (2014). (c), (d) Combined Ganges–Meghna–Brahmaputra sediment routing system showing upstream catchments and storage areas of the delta plain, shelf and Bengal submarine fan channel-levee system. Budget estimates are from Kuehl et al. (2005). Modified from Romans et al. (2015) (figs.3, 4) with permission of Elsevier.

The varied styles of routing of sediment in the cascade from source to sink can be illustrated from two examples (Figure 1.8). The previously mentioned Eel sediment routing system has a source region (area of 9,400 km^2) in the mountainous topography of northern California (Nittrouer et al., 2007) with a sediment discharge to the Pacific Ocean of 12–16 Mt per year. Sediment discharges, most importantly in major floods, construct a prism on the continental shelf, but most sediment is exported beyond the shelf edge to the deep sea (Wheatcroft and Borgeld, 2000). Export to deep water is enhanced by storm wave action superimposed on sediment gravity flows.

In contrast, the large Ganges–Brahmaputra system (catchment area 1,656,000 km^2) exports sediment to the vast Bengal Fan. At the historical timescale, about 30% of the sediment released from source areas is trapped onshore, principally in tectonically subsiding alluvial areas (Allison, 1998; Goodbred and Kuehl, 1999). Approximately 20% of the sediment delivery to the ocean builds topsets of subaqueous clinoforms, and 20–31% is stored in clinoform foresets, but along-shelf currents transport the remaining sediment to the head of a major canyon (Swatch of No Ground) and thereby to the Bengal submarine fan. Cyclones are particularly effective in advecting sediment to the canyon, and trigger mass transport events that travel down to the Bengal Fan.

A realistic assessment of the nature of the sediment cascade from source to sink in most real-world sediment routing systems clearly demands a sound understanding of the surface dynamics of sediment transport in a broad range of environmental settings. This is challenging enough, but to interpret the stratigraphic record, and to invert it in order to understand Earth history, requires us to also appreciate the 'long result of Time'. Until the geomorphic and oceanographic perspectives of the modern Earth and its land-ocean linkages (Romans and Graham, 2013) are more closely married with the geological perspective of deep time, we will find that we are like the fortune-teller in Shakespeare's *Antony and Cleopatra*, who says:

In Nature's infinite book of secrecy, A little I can read.

The sediment routing system philosophy offers an avenue of making progress through such a marriage.

2

The Global Character of River Basins

This chapter provides an overview of the global distribution of river catchments at the present day, since they are the powerhouses for the particulate and dissolved fluxes of sediment routing systems. Although some sediment routing systems are confined to the land surface, most export sediment to the ocean, where it is commonly mixed with sediment from other sources. Sediment routing systems are therefore a network connecting sources with sinks.

2.1 Mapping of the Terrestrial Segment of Sediment Routing Systems

Sediment routing systems connect erosional source regions with depositional sinks. At the present day, erosional source areas are relatively easy to recognise, since they comprise the upstream parts of river drainage basins. To map the erosional source areas of sediment routing systems is therefore essentially a task of delineating the net erosional parts of river catchments at a global scale. This process starts with an investigation of global catchment maps available from online databases (www.cambridge.org/milliman, which contains an environmental characterisation of 1,534 rivers; the Global Runoff Data Centre, www.grdc.bafg.de, which contains information from more than 9,000 gauging stations; major watersheds of the world from the WRI (World Resources Institute), http://worldmap.harvard.edu/maps/4960). Such maps are based on elevation data obtained from orbiting satellites or from NASA's Shuttle Radar Topography Mission (e.g. http://hydrosheds.cr.usgs.gov).

The number of rivers with drainage basins larger than a certain area follows a power law for the range of basin area from 0.001 to 1×10^6 km^2, suggesting that the distribution of river basin size may be fractal (Turcotte, 1997). Using the Milliman and Farnsworth (2011) database,

$$N_r(A_d > A) = aA_d{}^m \tag{2.1}$$

where N_r is the total number of river basins with an area A_d (in millions of km^2) greater than a certain size A, m is an exponent equal to -0.7903 and a is a coefficient equal to 16.92 (Figure 2.1). The Milliman and Farnsworth (2011) database includes most large drainage basins of area greater than 3,000 km^2, but like other global databases, is less comprehensive

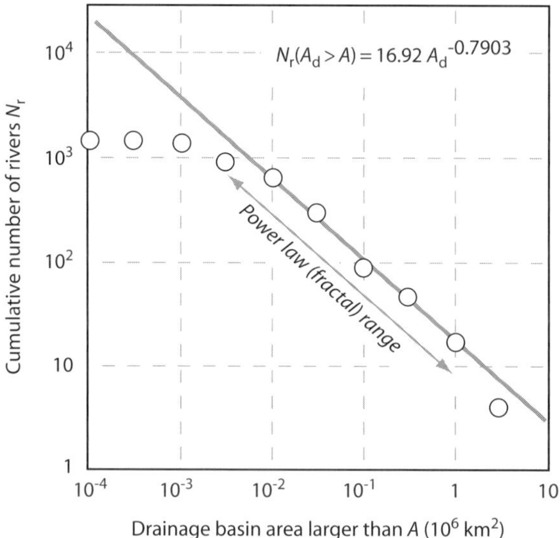

Figure 2.1 Number of rivers in the Milliman and Farnsworth (2011) global database with a drainage area greater than a certain size in millions of square kilometres based on 1,361 examples. Regression over the range 10^{-3} to 1×10^6 km^2 is a power law indicating a fractal distribution of drainage basin area.

for smaller drainage basins, although small, mountainous catchments deliver a disproportionately high amount of sediment to the ocean (Milliman, 1995).

It is impractical to estimate the intercept a from field data, since very small drainage basin areas are not measured, whereas the maximum drainage basin area recorded is easily available. If we define the maximum drainage basin area as A_{max}, then equation (2.1) becomes

$$\log(N_r(A_d > A)) = -m\log\left(\frac{A_{max}}{A}\right) \tag{2.2}$$

As an example, if the slope m is -0.5 and the maximum drainage basin area A_{max} is 10^7 km^2, then we should expect to find 10 drainage basins with an area greater than 10^5 km^2 and 100 drainage basins with an area greater than 10^3 km^2.

The largest drainage basin area possible is set by the size of the continental land mass, whereas the minimum area is constrained by the resolution of the particular technology being used to map. The coefficient in the power law of equation (2.1), or the maximum drainage basin area A_{max} in equation (2.2), is a parameter whose value describes the overall *scale* of drainage basin areas, whereas the exponent describes the *shape* of the distribution of drainage basin area. Continents may differ from other continents in the value of a (or A_{max}) and m, and a single continental landmass may also have different values depending on climate and plate tectonic setting (Table 2.1).

Table 2.1 *Maximum drainage basin area* A_{max} *and slope* m *of the power law segment of bi-logarithmic plots of number of drainage basins versus drainage basin area, using the Milliman and Farnsworth (2011) database. Data are binned at intervals of an order of magnitude of basin area.*

Continent/plate	Receiving ocean	Slope m	Maximum area A_{max} ($\times 10^6$ km^2)
Africa	Atlantic	−0.63	3.8 (Congo)
	Indian	−0.74	1.3 (Zambesi)
North America	Atlantic/Arctic	−0.73	3.3 (Mississippi)
	Pacific	−0.76	0.85 (Yukon)
South America	Atlantic	−0.61	6.3 (Amazon)
	Pacific	−0.85	0.032 (Guayas)
	Caribbean	−0.57	0.26 (Magdalena)
Eurasia	Atlantic/Arctic/Indian	−0.88	3.0 (Ob)
	Pacific/Mediterranean	−0.82	0.98 (Ganges)
Global	Passive margins	−0.97	6.3 (Amazon)
	Active margins	−0.87	0.98 (Ganges)

Taking Africa as an example, the distribution of catchment size is different according to whether rivers are draining towards the Atlantic Ocean, Indian Ocean, or Red Sea and Mediterranean (Figure 2.2a). The slopes m of the straight line segments of the bi-logarithmic plots of river drainage basins flowing into the Atlantic Ocean and the Indian Ocean are -0.63 and -0.74, whereas the maximum river drainage basin areas are approximately 3.8×10^6 and 1.3×10^6 km^2 respectively. The Nile is a very large river draining to the Mediterranean (2.9×10^6 km^2), but fitting a straight line to Mediterranean/Red Sea data is problematical because of the relatively small number of drainage basins sampled. The same procedure can be followed for North America (Figure 2.2b). River drainage basins flowing to the Atlantic and Arctic Oceans have a slope of -0.73, and those draining to the Pacific Ocean have an almost identical slope of -0.76. The corresponding maximum drainage basin areas are 3.3×10^6 and 0.85×10^6 km^2 respectively. Once again, there are too few data for drainage basins entering the Caribbean to draw meaningful conclusions. The same picture emerges for river drainage basins in South America and in Europe and Asia (Table 2.1, Figures 2.3 and 2.4).

Clearly, there are factors causing the different distributions of drainage basin area. A primary factor is the topography associated with active, convergent margins compared with the topography of passive, trailing-edge, continental margins. Passive margins included in Figures 2.3 and 2.4 and Table 2.1 are the Arctic margin of North America, the U.S. Eastern Seaboard, Gulf of Mexico, the Atlantic margin of South America, the Atlantic and Indian Ocean margins of Africa, the Indian Peninsula, the Yellow Sea and South China Sea northeast of the Red River Fault Zone, the Arctic Ocean margin of Eurasia and the Atlantic

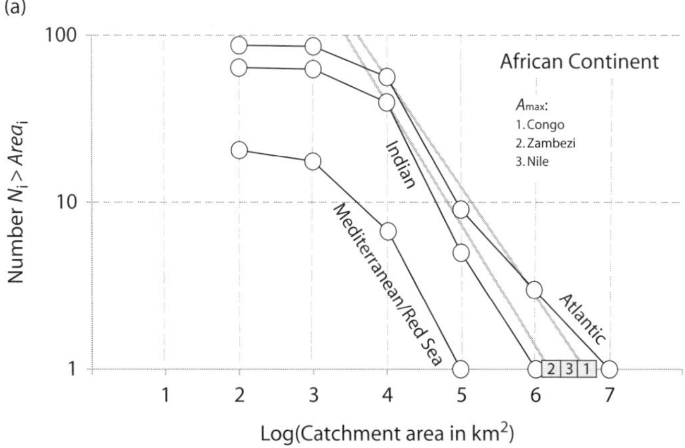

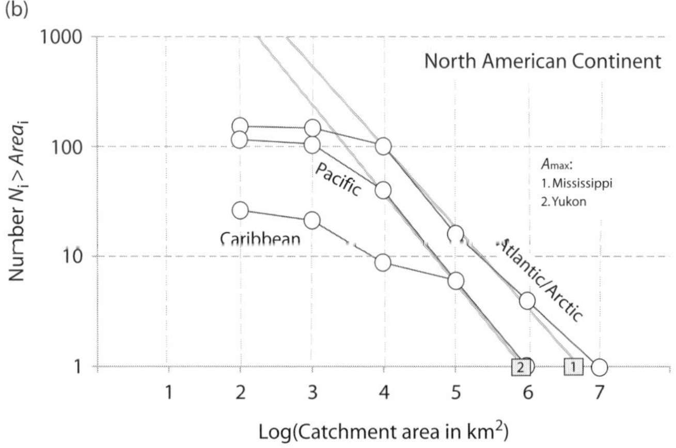

Figure 2.2 Number of rivers in the Milliman and Farnsworth (2011) database with a drainage area greater than a certain size in millions of square kilometres for Africa (a) and North America (b). In (a), different curves are shown for rivers draining into the Indian Ocean, Atlantic Ocean, or Mediterranean and Red Sea. In (b), different curves are shown for rivers draining to the Atlantic/Arctic Ocean, Pacific Ocean and Caribbean. Thick grey lines are linear fits to power law segments. Maximum river drainage basin areas A_{max} are shown.

margin of Western Europe. Active margins include the ocean-continent convergent margins of western Canada, the United States, Central America and South America, the ocean-ocean convergent boundaries of the western Pacific, the subduction and arc-related terrains of the Caribbean and Mediterranean, and the Alpine-Himalayan continental collision zone. The global compilation of active and passive margins have similar slopes of approximately unity (Figure 2.5), but different values of maximum drainage basin area. Rewriting equation (2.2) gives

$$\log(N_r(A_d > A)) = -0.97 \log\left[\frac{6.3 \times 10^6}{A}\right] \tag{2.3}$$

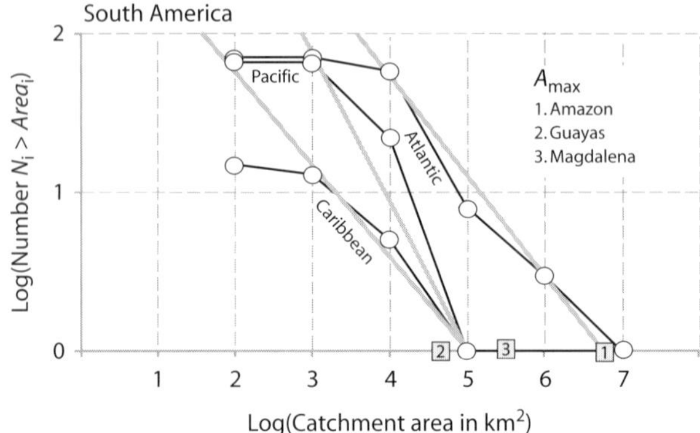

Figure 2.3 Bi-logarithmic plot of number of drainage basins exceeding a certain area in South America. Plots are shown for river basins draining to the Atlantic and Pacific Oceans and the Caribbean Sea. Thick grey lines are linear fits to drainage basin area data.

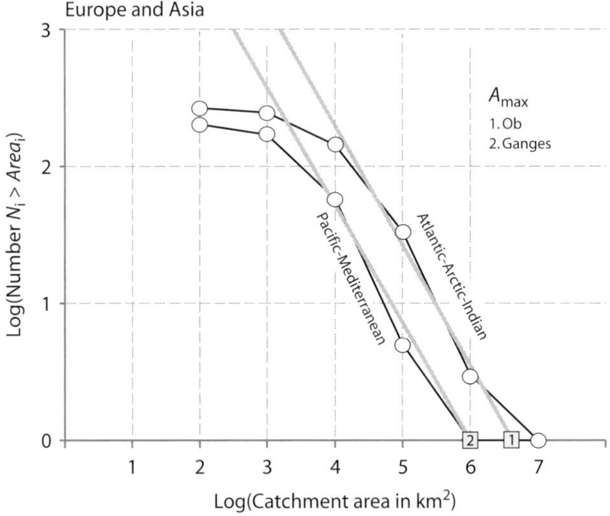

Figure 2.4 Bi-logarithmic plot of number of drainage basins exceeding a certain area in Europe and Asia. Data for Atlantic, Arctic and Indian Oceans include rivers draining mainland China and the Indian Peninsula. Data for Pacific-Mediterranean include rivers draining the Alpine-Himalayan mountain chains and the SE Asian forearc and backarc. Thick grey lines are linear fits to drainage basin area data.

for passive margins and

$$\log(N_r(A_d > A)) = -0.87 \log\left[\frac{0.98 \times 10^6}{A}\right] \tag{2.4}$$

for active margins, where the drainage basin area A is in km$^2 \times 10^6$. Equations (2.3) and (2.4) might potentially be used in modelling the distribution of drainage basins in ancient

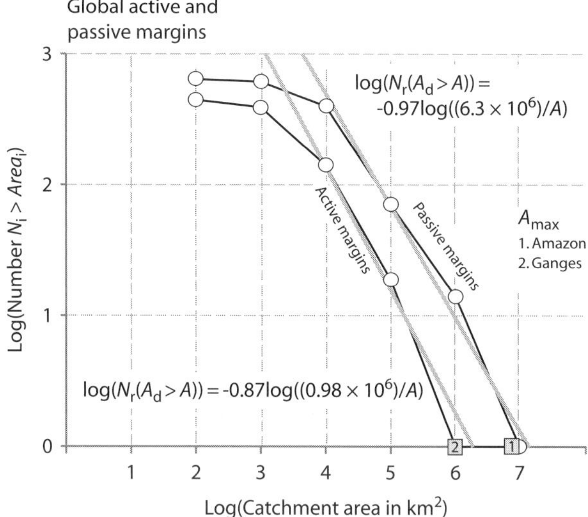

Figure 2.5 Bi-logarithmic plot of number of drainage basins exceeding a certain area for a global compilation of active and passive margins. The global Milliman and Farnsworth (2011) database is sampled for rivers draining passive and active continental margins. Thick grey lines are linear fits to drainage basin area data, with equations shown.

palaeogeographies where the rough dimensions of a continental land mass are known. In this case, the appropriate values of A_{max} and m could be used to gain insights into the likely distribution of drainage basin areas.

Drainage basin area has a well-known relationship to channel length, giving an aspect ratio for drainage basin shape. In the case of simple transverse drainage from the flank of linear mountain belts, there is a regularity in the aspect ratio of catchment spacing and straight-line catchment length, with a ratio of approximately 2.2 (Hovius, 1996). Taking the global database compiled by Milliman and Farnsworth (2011), there is a power law relationship between drainage basin area A_d (km^2) and channel length L in km (Figure 2.6):

$$L = 2.28A_d^{\,0.508} \tag{2.5}$$

The power law in equation (2.5) has implications for average thalweg sinuosities in drainage basins of different shape. For an equant, square-shaped catchment, the average sinuosity must be 2.46 to satisfy equation (2.5). For a circular catchment, the average thalweg sinuosity must be 2.20. Taking the Hovius relationship indicating rectangular catchments with an aspect ratio of 2.2, the average thalweg sinuosity is 1.67. A global plot of basin length versus river thalweg length reveals the average sinuosity for all drainage basins (Figure 2.7). The majority of drainage basins of rivers discharging to the ocean have sinuosities between 1 and 3. The linear trendline for these data gives an average global sinuosity of 1.72 ($R^2 = 0.895$).

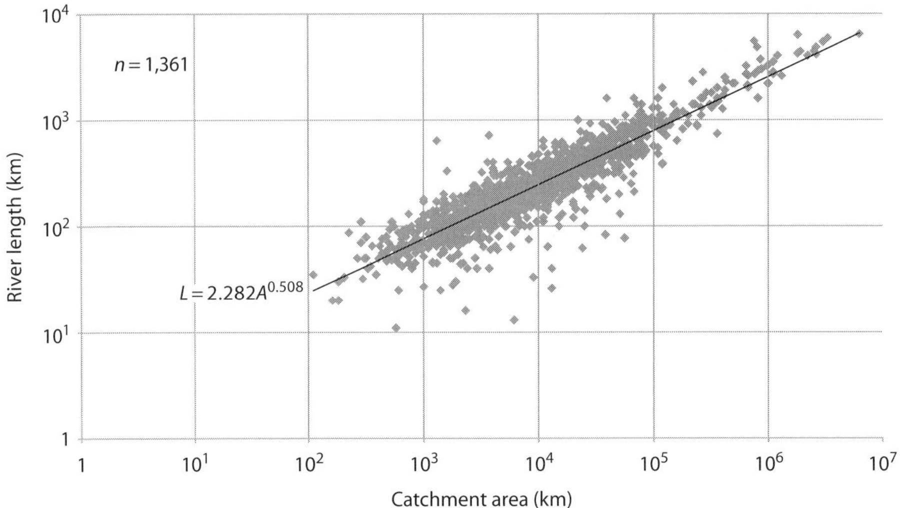

Figure 2.6 Regression of river length measured along the thalweg against drainage basin area based on 1,361 rivers from the Milliman and Farnsworth (2011) global database. The regression shows that river length scales on the square root of basin area. The coefficient in the power law reflects channel sinuosity and drainage basin planform shape.

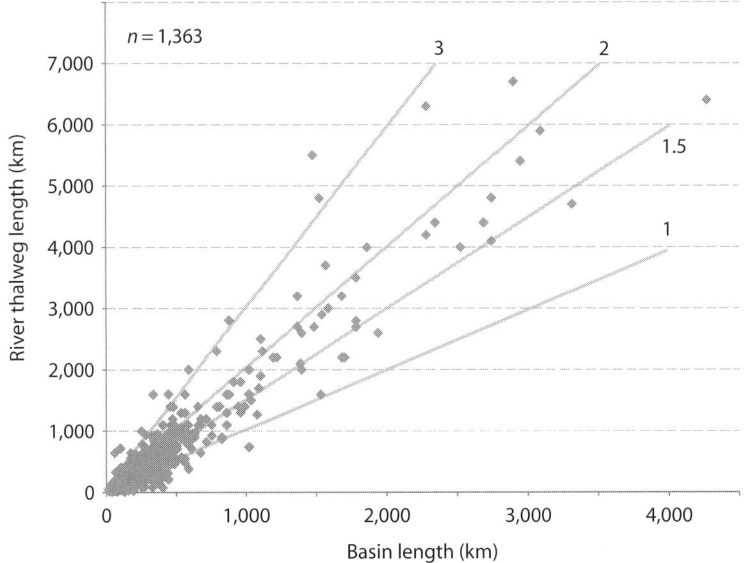

Figure 2.7 Global plot of basin length versus river thalweg length, which approximates the average sinuosity for each river drainage basin, based on 1,363 examples in the Milliman and Farnsworth (2011) database. The few examples where basin length exceeds river thalweg length ($S < 1$) are anomalous and result from ambiguities such as rivers flowing out of lakes, rivers beginning at the confluence of tributaries and river termini being located at the most seaward gauging station but located considerable distances inland from the coast. The global average sinuosity is 1.72.

The present-day global distributions of catchment size, channel length and aspect ratio are therefore well known and provide very useful guidance in the estimation of the drainage of continental land masses in the geological past.

2.2 Topology of River Basins

Rivers are flow pathways that empty into either an ocean (exorheic), or inland lake or other topographic depression (endorheic). Using a simulated topological network (STN) of the non-glacierised continental surface at 30-min resolution, potential flow pathways of rivers spanning orders 1 to 6 (Strahler, 1957) have been constructed (Vörösmarty et al., 2000). Only two river basins are at order 6, the Amazon and Lena. The resolution of the STN determines a lower limit of river basin size of about 25×10^3 km^2, which reduces the number of stream orders compared to those revealed by detailed topographic maps. There are 6,152 individual watersheds represented in the STN for the 133.1×10^6 km^2 of non-glacierised land area of the Earth. A plot of topological order i versus the logarithm of the number of river basins in each continent of order greater than i is shown in Figure 2.8. Although there are some clear differences between continents, the slopes for each continent are similar, indicating a common pattern of river branching over the Earth's surface.

There is more variation in the distribution of river order when viewed in terms of the world's major ocean basins into which the river enters (Figure 2.9). Since the world's continents have very different underlying tectonic and thermal processes controlling their margins (Allen and Allen, 2013), the distribution of river order for different receiving

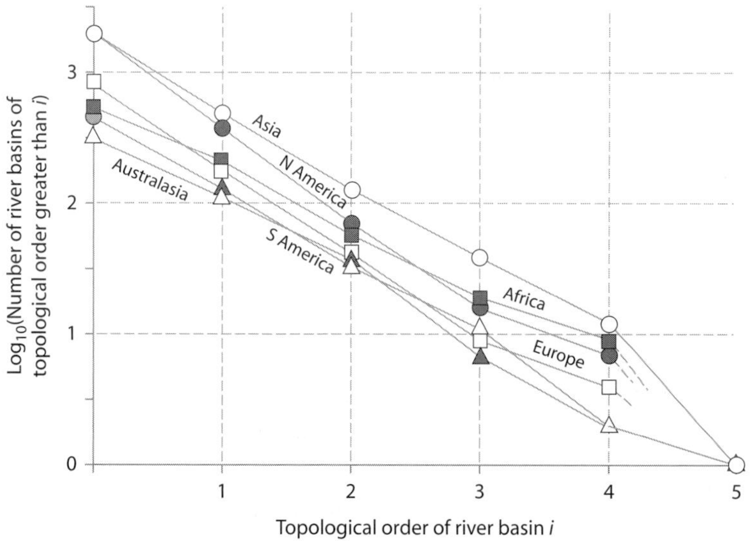

Figure 2.8 Logarithmic plot of the number of basins of topological order greater than i (integers 1 to 6) for different continental land masses, from the STN of Vörösmarty et al. (2000).

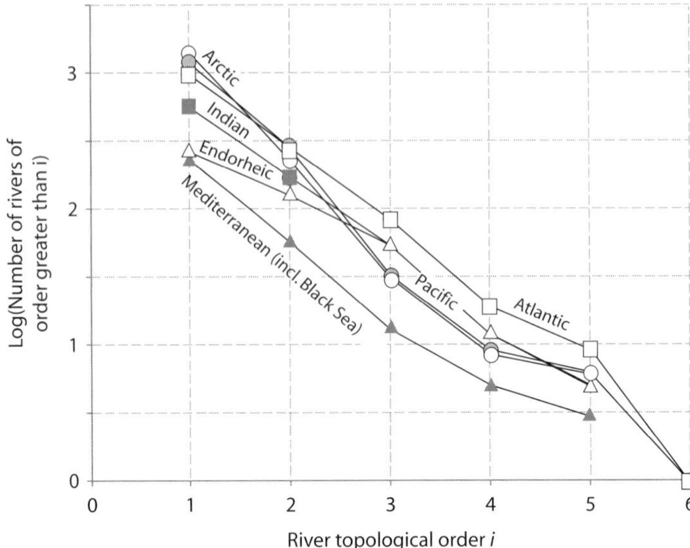

Figure 2.9 Logarithmic plot of the number of basins of topological order greater than i (integers 1 to 6) for different receiving ocean basins, from the STN of Vörösmarty et al. (2000). Regions of internal drainage (endorheic) are included.

basins reflects these tectonic styles and offers predictive capability in assessing drainage basin development and particulate delivery of palaeo-continents in the geological past. The Atlantic Ocean, which is fringed by passive margins, stands out as having a high proportion of contributing river basins of high order. That is, passive margins are fed by large drainage basins, potentially carrying high particulate and solute loads. Rivers terminating in endorheic basins also have a relatively high proportion of high order examples. In contrast, the Pacific Ocean is fringed by active margins involving continent-ocean and ocean-ocean subduction. There is a high proportion of low order river basins entering the Pacific Ocean. Although many of the rivers entering subduction-type margins are small in catchment area, they are commonly high-gradient, and may therefore contribute high unit-width discharges of sediment into the ocean (Milliman and Syvitski, 1992), with outlet points closely spaced along the tectonically active continental margin.

The mean drainage basin area spans several orders of magnitude, reaching greater than 10^6 km^2 for orders 5 and 6. River order is not directly related to mean basin area, but globally there is a progressive increase in mean basin area with ascending topological order. River basins of order 5 and 6 drain 45% of the entire land area. About half of the terrestrial globe is drained by the 50 largest river basins in the STN at 30-min spatial resolution. Highly skewed distributions of mean river basin area are found in each continental landmass. The mean length of the mainstem increases across river basin order from less than 100 to greater than 4,300 km. The 50 largest rivers (ranked by river basin area) all have mainstem lengths in excess of 1,000 km. The distribution of mainstem length is potentially a valuable parameter in modelling catchment development and sediment discharge in geological sediment routing systems where only broad palaeogeographies are known.

River topological order, catchment area and sediment discharge are all related to the mean catchment length. The mean distance to the ocean for individual continents derived from the STN shows that the catchments draining across active and passive continental margins are strikingly different. In South America, the mean distance to the Atlantic Ocean is 1,429 km, whereas the mean distance to the Pacific Ocean is just 134 km. Likewise, in North America, the mean distance of rivers draining to the Atlantic Ocean is 1,030 km compared with mean distances of 793 km and 716 km for the Pacific and Arctic Oceans respectively. The continental drainage divide is therefore displaced towards the active margin of continents due to the mountain building associated with plate convergence. As a result, 89% of the continent of South America and 56% of the area of North America drain into the Atlantic Ocean. In contrast, only 7% of South America and 23% of North America drain into the Pacific Ocean. Significant portions of Asia (20%) and Australasia (28%) are occupied by river systems that drain internally, and which therefore do not contribute to the global flux of sediment to the ocean.

The global mean travel distance to the river mouth across all STN flow pathways is 1,050 km. Run-off therefore travels a mean distance of 1,050 km to the sea, with important variation between individual continents and receiving ocean basins. If the mean speed of channelised water is 0.5 m s^{-1}, and neglecting lakes, reservoirs and wetlands, the mean travel time of run-off for the world's 50 largest river basins is about 60 days. Although this may approximate the travel time of solutes in pristine rivers, particulate sediment load makes slower progress through the terrestrial sediment routing system due to intermittent deposition (Malmon et al., 2003). The effect of alluvial sequestration on sediment transit times is discussed in Section 4.5, within a probabilistic framework in Section 8.3, and against the background of the transient response of landscapes in Section 8.5.

The Land2Sea database compiled by Bernhard Peucker-Ehrenbrink contains data on the sizes of 1,519 river basins draining to the ocean, covering approximately 79% of the exorheic land area (Peucker-Ehrenbrink, 2009) (Table 2.2). The river drainage basins are grouped into regions following the categorisation of Graham, Famiglietti, and Maidment (1999) (Figure 2.10). Although these regional continental watersheds provide a useful basis for initialisation of hydrologic, terrestrial ecosystem and climate change models, they are not directly related to the plate tectonic underpinning of continental topography.

The western flanks of North and South America, which are dominated by active plate tectonic settings characterised by subduction, fall close to the origin in Figure 2.11, indicating small average river basin areas. In comparison, extensional, trailing-edge passive margins, such as those of eastern South America, the Russian Arctic, the North American Arctic and West Africa, have moderate to high average basin areas.

Table 2.2 *Land2Sea database for world's large-scale drainage regions showing total river drainage area, average catchment size and annual discharge of water, after Peucker-Ehrenbrink (2009) (tab.1, p.3) with permission of American Geophysical Union.*

Drainage region	Total drainage area (km^2)	Number of basins N	Average catchment area (km^2)	Water discharge (km^3 yr^{-1})
Russian Arctic	11,444,776	40	286,119	2,472
N American Arctic	2,830,837	18	157,269	663
E North America	7,667,969	107	71,663	2,246
Western Europe	1,585,093	246	6,443	633
E South America	14,083,468	100	14,0835	9,599
West Africa	9,712,254	69	140,757	2,077
East Africa	4,823,881	58	83,170	359
Arabia, India, SE Asia	5,991,365	62	96,634	2,668
East Asia	9,585,259	204	46,987	4,629
W North America	3,806,397	94	40,494	1,197
W South America	661,283	86	7,689	597
Australia-NZ	3,744,781	207	18,091	369
Mediterranean	4,108,421	123	33,402	386
Caspian Sea	4,164,523	8	520,565	508
Black Sea	2,175,984	25	87,039	365
Baltic Sea	1,553,992	56	27,750	426
Hudson Bay	2,792,775	24	116,366	619

Drainage area is total area of all river basins within the drainage region for which the Land2Sea database contains data; N is the number of river basins within the drainage region for which the Land2Sea database contains estimates of basin size, annual suspended sediment flux and annual discharge of water. Such data are unavailable for the Red Sea and Antarctica drainage regions.

2.3 Global Particulate Sediment and Solute Delivery to the Ocean

The bulk of particulate and dissolved solids is transported across continents and into the oceans by running water. Rivers discharge approximately 36,000 km^3 of fresh water into the ocean each year (Figure 2.12), carrying more than 20×10^9 tons of dissolved and particulate material (Milliman and Farnsworth, 2011). The primary source of information on the sediment fluxes across the surface of the Earth therefore comes from gauging stations on rivers. There have been a number of databases containing global information on rivers and their sediment loads. The relatively recent compilation by John Milliman and Elizabeth Farnsworth (Milliman and Farnsworth, 2011) covers 1,534 rivers draining 86.6×10^6 km^2 of watershed, which can be compared with a total land area of 148×10^6 km^2 for the Earth, and 105×10^6 km^2 for the total area draining to the ocean, the remainder being internally drained, or *endorheic*.

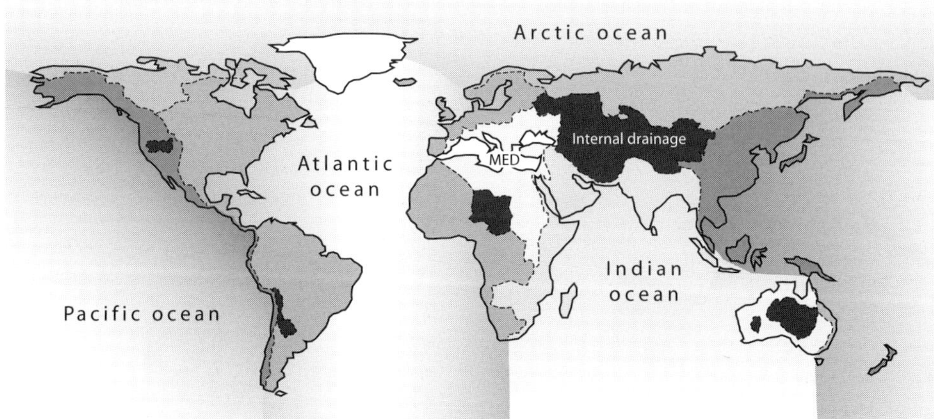

Figure 2.10 Division of the STN land mass into areas draining to the world's ocean basins, including the Mediterranean Sea and regions of internal drainage. Continental drainage divides are shown in dashed lines. From Vörösmarty et al. (2000) (fig.4) with permission of American Geophysical Union.

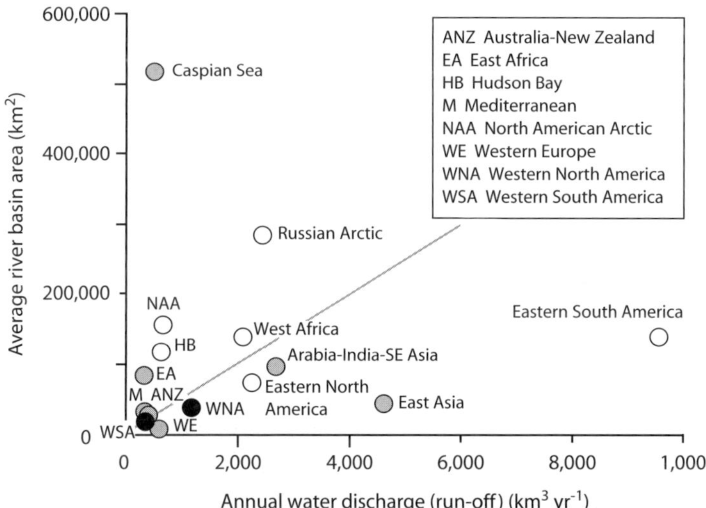

Figure 2.11 Average river basin area in global drainage regions using categorisation of Graham, Famiglietti, and Maidment (1999) versus annual water discharge of global drainage region, using data in Peucker-Ehrenbrink (2009). Open circles: extensional trailing edge passive continental margins; filled circles: contractional subduction-related active continental margins; grey circles: mixed plate tectonic settings. Diagonal line separates global drainage regions with high run-off for a given average river catchment area (for example, eastern South America) from low run-off for a given river catchment area (for example, Caspian Sea).

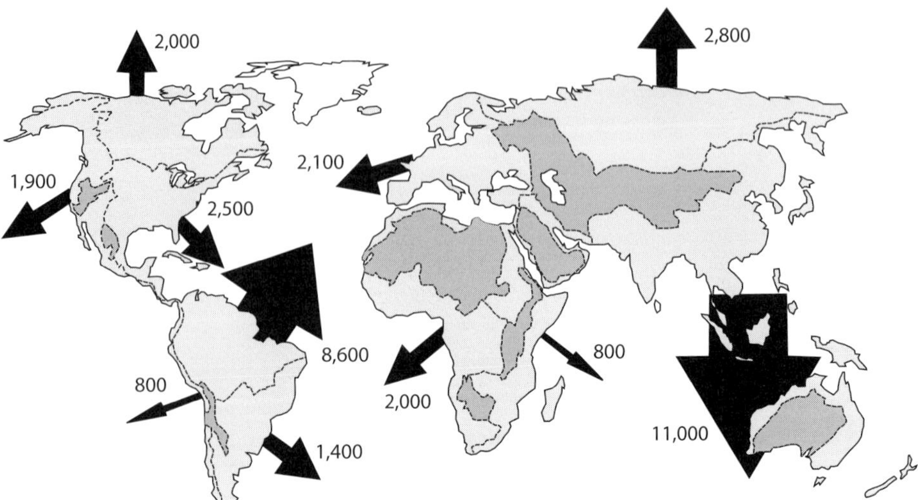

Figure 2.12 Discharge of fresh water into the coastal ocean by rivers. Numbers are mean annual discharge of water in km^3 yr^{-1}. Arrows are proportional to these numbers. The total global discharge of fresh water to the coastal ocean is 36,000 km^3 yr^{-1}. Note the very high cumulative discharge from southeast Asia and Oceania and from equatorial South America. Dark grey areas are endorheic. Modified from Milliman and Farnsworth (2011) with permission from Cambridge University Press.

2.3.1 Run-off

The transfers of solid particulate and dissolved loads are driven by the run-off of precipitation. In total, the river discharge of fresh water into the ocean is equivalent to a mean annual run-off of about 350 mm yr^{-1} (Milliman and Farnsworth, 2011). River hydrology is a reflection of the topographic relief, vegetation and climate of the drainage basin, but the primary factor correlating with water discharge is drainage basin size. This is illustrated by a log-log plot of drainage basin area versus mean annual discharge (Figure 2.13). In the hydrological budget, groundwater is relatively unimportant in the total freshwater discharge to the ocean, but is more important in the seaward transfer of dissolved ions, including nutrients (Moore, 1996). One billion tons of dissolved solids are thought to be discharged to the global ocean each year through groundwater (Dzhamalov and Safronova, 2001), which is one third of the rate discharged by rivers.

Precipitation varies very widely globally from effectively nil in hyperarid regions to 10 m yr^{-1} in some equatorial mountain belts (Section 9.2). It can also vary spatially and temporally within a drainage basin, particularly where the catchment has rugged topography. Global maps of precipitation are available from internet databases (e.g. Climate Research Unit, East Anglia University, UK) to inform hydrological models of sediment routing systems. There are water losses due to evapotranspiration, which generally reduces

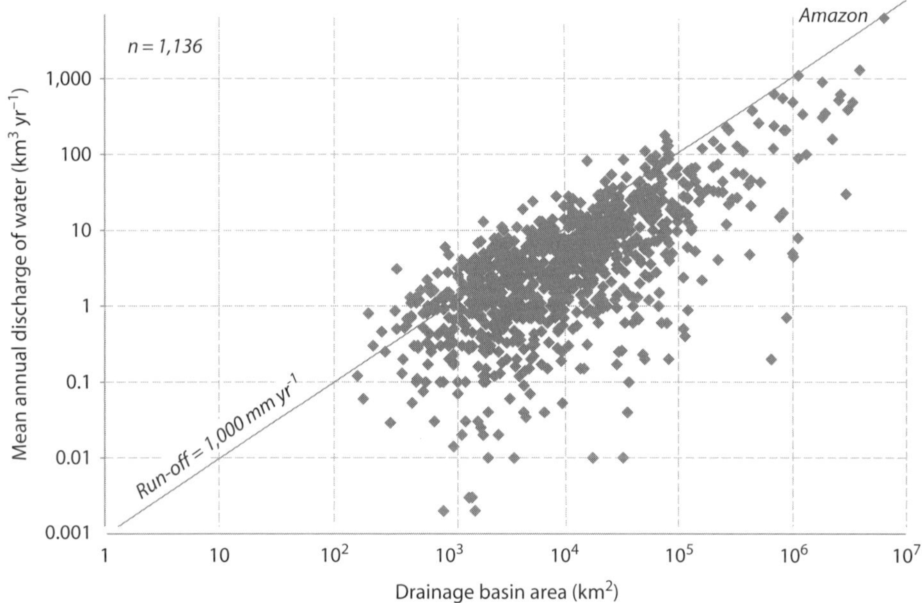

Figure 2.13 Bi-logarithmic plot of river discharge to the ocean versus basin area, based on data on 1,136 rivers in the Milliman and Farnsworth (2011) global database.

with increasing latitude, storage (as in recharging aquifers and lakes) and consumption for human uses, leaving a reduced discharge as run-off. Run-off is the principal agent for geomorphic change.

The 12 rivers that individually discharge more than 400 km^3 of water annually into the ocean occupy about 25% of the total land area of the Earth and in total discharge 13,800 km^3 of fresh water annually (Table 2.3). Since the total freshwater discharge is about 36,000 km^3 annually, these 12 rivers are responsible for about 35% of the total global discharge. The global chemistry of the ocean is therefore strongly influenced by a relatively few major rivers (Section 3.1.1). The global distribution of river discharge to the ocean depends on catchment size and precipitation regime (neglecting ice-melt run-off from Antarctica and Greenland). At the present day, rivers draining northern South America and southern Asia/Oceania, with annual precipitation rates of more than 2,000 mm yr^{-1}, dominate discharge to the ocean.

Of the 12 rivers listed in Table 2.3, most drain across passive margins into the ocean, and several have hinterlands situated in wet mountainous topography caused by tectonic convergence. Although rivers such as the Amazon, Orinoco, Congo and Changjiang are in regions of high run-off, the Asian rivers draining into the Arctic Ocean are semi-arid, and it is the very large drainage basin areas of these rivers that contribute to the high freshwater discharges at their mouths.

Clearly, a thorough understanding of the fluxes of present-day sediment routing systems requires a detailed knowledge of the patterns and discharges of run-off, but the investigation

Table 2.3 *Rivers that discharge annually more than 400 km³ of fresh water into the global ocean, after Milliman and Farnsworth (2011) (tab.2.3, p.21).*

River	Country	Basin area ($\times 10^3$ km³)	Discharge (km³ yr^{-1})
Amazon	Brazil	6,300	6,300
Congo	DR Congo	3,800	1,300
Orinoco	Venezuela	1,100	1,100
Changjiang	China	1,800	900
Brahmaputra	Bangladesh	670	630
Yenisei	Russia	2,600	620
Mekong	Vietnam	800	550
Lena	Russia	2,500	520
Ganges	Bangladesh	980	490
Mississippi	USA	3,300	490
Parana	Argentina	2,600	460
Irrawaddy	Myanmar	430	430
Totals		27,000	13,800
Global		108,000 (25%)	36,000 (35%)

of palaeo-sediment routing systems demands a different line of attack. Not only are palaeo-catchments difficult to identify and characterise, but the prevailing climate at the time is also unknown at the resolution characteristic of the present day. There need to be rules of thumb to aid hindcasting of drainage basin hydrology. First, palaeoclimatic information can be obtained from soil types, chemical sediments in lakes, isotopic composition of cave deposits, and palaeofloral and palaeofaunal assemblages. Second, palaeogeographic position and palaeotopographic character can be used to make general statements about the likely hydrology of river drainage basins. There is a broad latitudinal variation in run-off, with high values in equatorial-tropical latitudes, low values in the subtropics, and high values again in the temperate zone, especially where ocean circulation and topography combine in effect.

Today, the greatest run-off (mm yr^{-1}) is in short, steep catchments where mountainous topography causes orographic rain, whereas low run-off rivers not surprisingly tend to drain regions in arid and semi-arid climates. The pattern of run-off at the present day can be used to make inferences of run-off in the past, particularly if the hydrology can be associated with plate tectonic background. For instance, the western margin of North and South America is an ocean-continental convergent margin associated with oceanic subduction and magmatism. Where the high topography along this margin from Alaska to Tierra del Fuego coincides with high convective or frontal precipitation, as in Alaska, British Columbia, Washington, Central America and southern Chile, high run-off rivers are situated. However, the discharge of fresh water to the ocean is considerably greater where precipitation is collected in very large drainage basins that transport water eastwards to the trailing-edge passive margin, such as the Amazon-Orinoco in South America, and the Mississippi and St Lawrence rivers of North America. The interrogation of global databases therefore requires geological insights to be used that are transferable to the palaeo-sediment routing systems of the geological record.

2.3.2 Particulate Loads

The study of the particulate loads of rivers is a powerful way of understanding the way in which sediment routing systems function at the present day. Particulate loads are a direct indication of surface sediment fluxes and a means of estimating denudation rates in the river catchments feeding sedimentary basins.

Particulate sediment can be transported by rivers in suspension or in contact with the bed, largely dependent on the grain size of the sediment in relation to the velocity of the flow. Consequently, sediment grains may change in their mode of transport within a single flood event and along the course of a river. A simple but useful concept is that of the 'transport stage' (Francis, 1973). The dimensionless ratio of the shear velocity driving downstream sediment transport u_* and the settling velocity of the sediment grain w gives an indication of whether grains will move as bedload ($u_*/w < 1$) or as suspended load ($u_*/w > 1$). Discharge data from river gauging stations generally refer to suspended loads rather than total loads. That is, in general any bedload transport is ignored.

Yield and Denudation Rate

The sediment loads of pristine rivers (unaffected by man) allow the average denudation rate, or sediment yield, of the contributing catchment area to be calculated. If A is the area of the catchment, Q_{total} is the discharge of sediment and solutes at the exit of the drainage basin, there are no additional sources or sinks of solutes and particulate sediment, and there is no change in storage of sediment within the drainage basin over time, long-term average denudation rate is simply

$$\frac{\partial h}{\partial t} = \frac{(1 - \phi)}{\rho} \frac{Q_{total}}{A} \tag{2.6}$$

where h is the elevation of the Earth's surface, t is time and ϕ is the porosity of rocks of density ρ undergoing erosion in the catchment. Discharge is measured in mass per unit time (kg yr^{-1} or t yr^{-1}). Measurement of both the particulate and solute loads of rivers is needed to calculate the total denudation rate. 'Sediment yield' refers to the component of the denudation that generates solid particulate sediment:

$$Y = \frac{Q_p}{A} \tag{2.7}$$

where the sediment yield is Y, with units of mass flux (kg m^{-2} yr^{-1}) and Q_p is the particulate sediment discharge. As an example, the annual solid particulate load of the

Amazon River at its mouth is 1,150 Mt yr^{-1} (1 megaton is 10^9 kg), and the annual solute load is 223 Mt yr^{-1}, making a total load of 1,373 Mt yr^{-1}. The drainage area of the River Amazon is 6,150,000 km^2. The sediment yield is therefore 187 t km^{-2} yr^{-1}, whereas the total average denudation rate is 223 t km^{-2} yr^{-1}. If rock density ρ is 2,700 kg m^{-3} and the average rock porosity ϕ is 5%, the denudation rate averaged across the entire Amazon basin is 79 mm kyr^{-1}. It is important to appreciate that sediment yield refers to the total particulate discharge averaged over the upstream contributing drainage area. Consequently, if sediment is stored in downstream reaches, as is common in large drainage basins, sediment delivery to the ocean measured at gauging stations close to the river outlet gives a value of sediment yield that bears no relation to the average denudation rate of sediment source regions. The sequestered sediment in floodplains not only drops out of the oceanward flux, but also may be subjected to intense chemical weathering, thereby releasing enhanced fluxes of solutes (Johnsson and Meade, 1990; Willenbring, Codilean, and McElroy, 2013). The lower basin of the Amazon traps a net 1–2 Gt of sediment per year, which does not reach the Atlantic Ocean (Figure 2.14). The bulk of the sediment carried by the Amazon River and its tributaries comes from steep, rapidly incising rivers in the Andes (Aalto, Dunne, and Guyot, 2006) (2,300–3,100 Mt yr^{-1}). The Madeira river system alone provides more than 700 Mt yr^{-1} of suspended sediment to the Amazon trunk stream. In the Andes, average physical erosion rates are far in excess of the averaged rate of 0.08 mm yr^{-1} previously quoted. The lower parts of major river basins therefore may sequester large volumes of particulate sediment. In the conterminous United States, it is estimated that only about 10% of the eroded sediment reaches the oceans (Holeman, 1980). On the Atlantic coast of United States, rivers draining the Piedmont, such as the Susquehanna River, deliver water and sediment (3 Mt yr^{-1} enters Conowingo Dam) into large estuaries such as Chesapeake Bay. However, coastal zone sinks are fed with sediment not only from rivers but also from shore erosion and from the continental shelf. In cases such as Chesapeake Bay, no sediment bypasses the coastal zone en route for the continental shelf and deep sea (Schubel and Carter, 1976).

The extent of storage is likely to increase in larger drainage basins, which explains the well-established inverse relationship between drainage basin area and sediment yield (Milliman and Meade, 1983; Milliman and Syvitski, 1992) (Figure 2.15). The regression of all river data in the Milliman and Syvitski (1992) database gives

$$Y = \exp(9.18 - 0.46\ln A) \tag{2.8}$$

where Y is the yield in kg m^{-2} yr^{-1} and A is drainage basin area in m^2.

The sediment and solute discharges of rivers entering the ocean are therefore an indicator of catchment-averaged physical and chemical denudation rates using equation (2.6), but the denudation rate actually experienced by a source region for sediment depends on the amount of storage in the alluvial valley Q_a and the percentage area of the river basin that is depositional rather than eroding, F_a. Equation (2.6) can therefore be modified to give the physical denudation rate averaged over the eroding fraction of the river basin

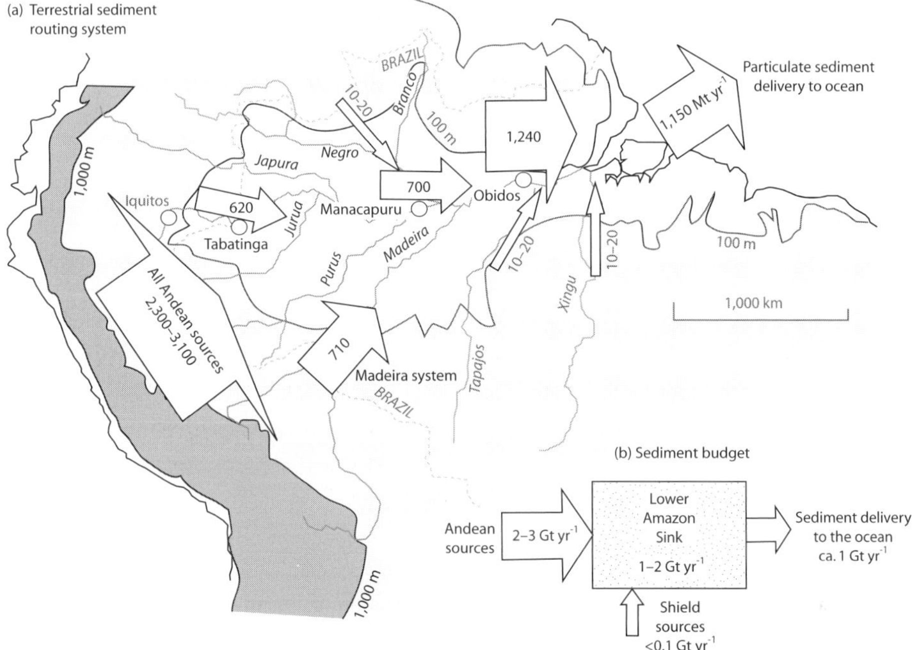

Figure 2.14 (a) The terrestrial compartment of the Amazon sediment routing system, showing the sediment discharge of Andean and Guyanan-Brazilian Shield sources, suspended load data from gauging stations and sediment delivery to the ocean. Sediment discharges in Mt yr^{-1} at Tabatinga (Sao Paulo de Olivenca), Manacapuru, Madeira tributary and Obidos from Dunne et al. (1998) based on the period 1974–1989. Sediment delivery to the ocean from Milliman and Farnsworth (2011). Sediment discharge from Andean rivers is from Aalto, Dunne, and Guyot (2006). (b) Outline sediment budget illustrating importance of the Lower Amazon drainage basin as a sediment trap. Sediment discharges do not include solutes (223 Mt yr^{-1} at Obidos).

$$\frac{\partial h}{\partial t} = \frac{(1-\phi)}{\rho}\frac{(Q_{total}+Q_a)}{A(1-F_a)} \tag{2.9}$$

or the combined physical and chemical denudation rate over the eroding fraction of the river basin

$$\frac{\partial h}{\partial t} = \frac{(1-\phi)}{\rho}\frac{(Q_{total}+Q_a-Q_{sol})}{A(1-F_a)} \tag{2.10}$$

where Q_{sol} is the solute gain through alluvial weathering. Returning to the Amazon River basin, the discharge at the river mouth plus the Lower Amazon alluvial sink is 2–3 Gt yr^{-1}, the area of the entire river basin is 6,150,000 km^2, and we assume that $\rho = 2,700$ kg m^{-3} and $(1-\phi)$ is 0.95. If the fraction of the river basin that is alluvial is 0.8, the average physical denudation rate of the eroding source region is 0.6–0.9 mm yr^{-1}, compared with the rate obtained for the entire river basin of 0.08 mm y^{-1} given previously. If the alluvial fraction is 0.9, the calculated average physical denudation rate over the eroding part of the

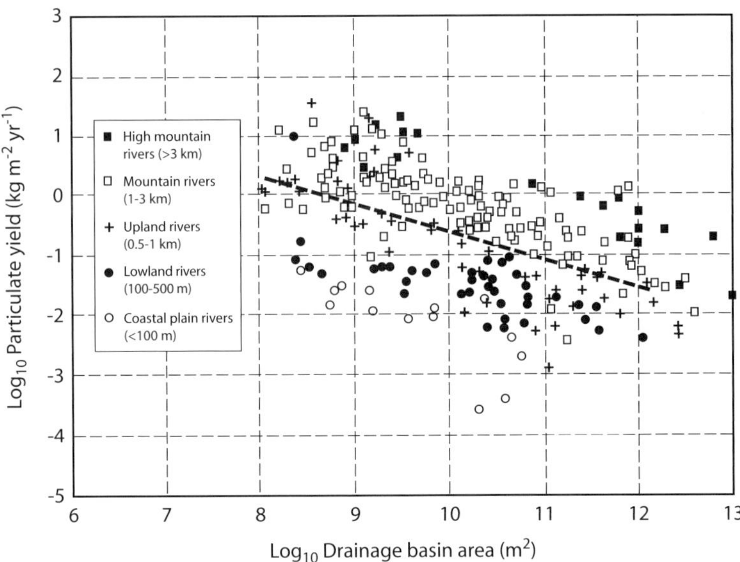

Figure 2.15 Plot of drainage basin area A versus particulate sediment yield Y for the elevation classes of Milliman and Syvitski (1992). The regression of all river data has the form $Y = \exp(9.18 - 0.46\ln A)$. Redrafted after Hay (1998), published with permission of Elsevier.

river basin becomes 1.2–1.8 mm yr^{-1}. The discharge into the alluvial sink and the fractional area of the alluvial tract are expected to co-vary. Inspection of equation (2.9) shows that an increase in alluvial storage increases the calculated denudation rate in the source region necessary to explain the rate of delivery at the river mouth.

A number of processes affect the relationship of the sediment and solute effluxes leaving the channel-hillslope systems of the catchment area Q_{hc} and the sediment and solute fluxes entering the ocean Q_o (Figure 2.16). Within the alluvial system there are sequestration losses of the riverine load to sediment deposition Q_s, counteracted to some extent by gains due to reworking of alluvial sediment Q_r. The riverine solute load is reduced by losses due to cementation of alluvial sediment and evaporation Q_f, counteracted to some extent by gains due to the generation of solutes through chemical weathering Q_{cw}. The alluvial valley of sediment routing systems is therefore a dynamic zone involving gains/losses and transformations of particluate sediment and solutes rather than a simple conveyor belt of detritus. The sediment and solute delivery to the ocean is therefore not a direct recorder of catchment effluxes, particularly in sediment routing systems with large alluvial tracts.

The sediment release of catchment areas acting as source regions can be compared with sediment volumes in depositional basins. Depositional sediment volumes are generally estimated from isopachs and cross-sections derived from borehole and seismic reflection data (Métivier et al., 1999; Clift, 2006). For example, the average solid phase accumulation rate for Asian sedimentary basins since the beginning of the Tertiary has been calculated

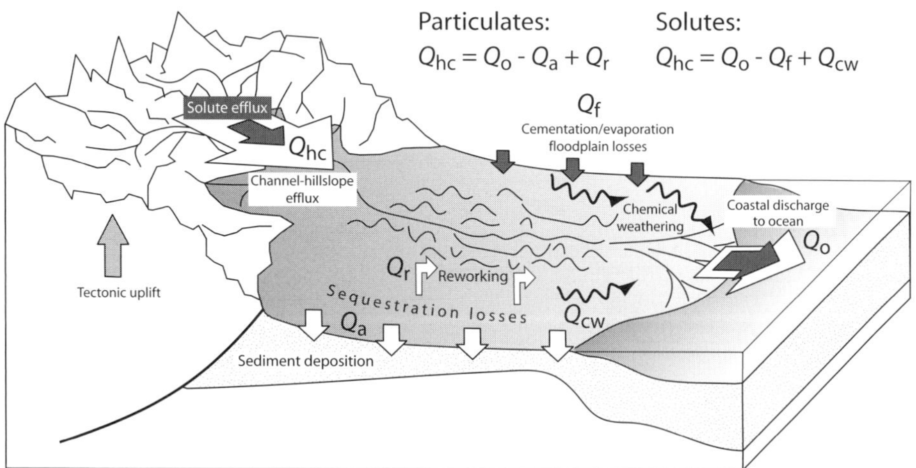

Figure 2.16 Schematic illustration of the fluxes of sediment and solutes from an upland catchment to a depositional basin. For particulates (white arrows), Q_{hc}, channel-hillslope efflux of particulate load from hinterland or catchment; Q_a, sequestration of sediment by deposition on alluvial floodplain; Q_r, reworking of fluvial sediment; Q_o, sediment discharge to the ocean. For solutes (black arrows), Q_{hc}, channel-hillslope efflux of solutes in run-off; Q_f, loss of solutes by cementation and evaporation on floodplains; Q_{cw}, generation of solutes by chemical weathering reactions; Q_o, discharge of solutes to the ocean.

after correction of stratigraphic thicknesses for the effects of compaction (Métivier et al., 1999). The Bengal Basin contains a vast amount of Upper Tertiary sediments (locally up to 21 km in thickness) deposited primarily from southwestward prograding deltas feeding a large deep-sea cone. The Bengal Basin has accumulated 12×10^6 km^3 solid volume during the Tertiary, with a rate of 0.45×10^6 km^3 Myr^{-1} over the last 2 Myr. The Bengal Basin has received sediment from the Ganges, Brahmaputra and the rivers of the eastern platform of India (Mahanadi, Krishna and Godavari), which have a total drainage area of approximately 2.266×10^6 km^2. The average mechanical denudation rate for these Indian subcontinent catchments during the Quaternary is about 200 mm kyr^{-1}, with an average sediment yield of 546 t km^{-2} yr^{-1} (using $\rho = 2{,}750$ kg m^{-3}). The Ganges, Brahmaputra, Mahanadi, Krishna and Godavari rivers have a combined river mouth discharge of 0.49×10^6 km^3 Myr^{-1}. This is in excellent agreement with the estimate derived from the volume of preserved stratigraphy, and indicates that present-day discharges in large, buffered systems may be closely representative of longer-term (geological) rates (Einsele et al., 1996; Métivier et al., 1999; Métivier and Gaudemer, 1999).

Suspended Sediment

Calculations of the total annual sediment discharge to the world ocean range from 14×10^9 tons (Syvitski et al., 2005) to 16×10^9 tons (Milliman and Meade, 1983) and 19×10^9 tons (Milliman and Farnsworth, 2011). The highest discharges are found in rivers in

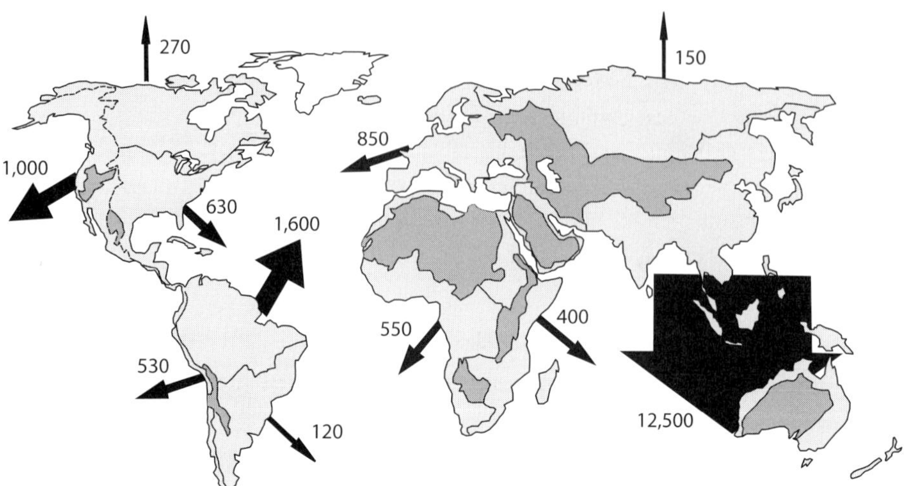

Figure 2.17 Annual discharge of riverine suspended sediment into the coastal ocean. Numbers are the sediment discharges in t yr^{-1} × 10^6. Black arrows are proportional to these numbers. The global total discharge is 19,000 ×10^6 t yr^{-1}. Dark grey areas are endorheic. Modified from Milliman and Farnsworth (2011) with permission of Cambridge University Press.

southeast Asia and Oceania, which together account for approximately two-thirds of the global delivery (Figure 2.17).

Compilations of suspended sediment concentrations (Milliman and Farnsworth, 2011) (tab.2.5, p.27), which show that the global average for suspended sediment concentration is 500 mg l^{-1}, are strongly influenced by climate. Rivers with the highest concentrations are generally found in subtropical arid to semi-arid climates. In contrast, the lowest concentrations (1–2 mg l^{-1}) are found in low-gradient shield terrains, as in northern Europe. Compilations of suspended sediment loads (Milliman and Farnsworth, 2011) (tab.2.6, p.28) indicate the importance of the elevation class of the drainage basin. Compilations of sediment yield based on suspended loads (Milliman and Farnsworth, 2011) (tab.2.7, p.29), which show a global annual average for suspended sediment yield of 190 t km^{-2}, also suggest a dependence on elevation class. Most high-yield rivers drain mountains with headwaters at elevations greater than 3,000 m. All high-yield rivers are small, with areas less than 3,000 km^2.

As seen in relation to run-off, sediment discharges at river mouths vary strongly according to the tectonic style of plate margin. Active margins characterised by subduction, such as the western margin of North and South America, have river basins with high discharges relative to their size that deliver sediment quickly to the deep sea via canyons incised into narrow continental shelves. Passive margins, however, are typified by large, low-gradient rivers that despite their very high discharges have low to moderate yields. They commonly border wide continental shelves, and much of their sediment load is trapped in estuaries, deltas and the continental shelf. During sea level lowstands, sediment may bypass the coastline and shelf and be fed directly to the deep sea, as in the case of the Amazon system (Section 3.2.5).

As in the case of water discharge, suspended sediment loads also increase with drainage basin area, though there is considerable variation. The reasons for some of this variation can be appreciated by categorising river basins according to the maximum elevation of their headwaters (Milliman and Syvitski, 1992; Hay, 1998) (Figure 2.15). Mountainous rivers have higher sediment loads than lowland rivers by three orders of magnitude, for the same drainage basin area. A global database (Summerfield and Hulton, 1994) revealed an exponential relationship between the average denudation rate E (mm yr^{-1}) and the maximum elevation h_{max} (m):

$$E = 0.0072\exp(0.0015h_{max}) \tag{2.11}$$

for $h_{max} > 100$ m, which can be modified to give a relationship for annual sediment load Q_s (t yr^{-1}), neglecting for the moment porosity in the rocks and regolith undergoing erosion, the contribution of bedload to the annual sediment load and the denudation attributable to chemical weathering releasing solutes:

$$Q_s = 0.0072(10^6)\exp(0.0015h_{max})\rho \tag{2.12}$$

where ρ is the density of rock, regolith and sediment. If $\rho = 2,700$ kg m^{-3}, $E = 0.5$ mm yr^{-1} and the drainage basin area $A = 10$ km^2, the sediment load becomes 13,500 t yr^{-1}.

The influence of maximum elevation on sediment load can be understood from the dependence of erosion rate on catchment area relief (Ahnert, 1970; Pinet and Souriau, 1988), since high relief implies high river gradients that cause more fluid stress to be exerted on the river bed. High catchment area relief is also commonly generated where the landscape is built by earthquakes associated with active tectonics. Ground shaking is highly effective at releasing mass flows on mountain hillslopes (Hovius, Stark, and Allen, 2007). Consequently, sediment loads of rivers are directly linked to the tectonic environment of the catchment area (Hovius, 1998) (Figure 2.18). High elevations are also likely to increase rates of precipitation through orographic effects (Reiners et al., 2003).

Glaciated terrains appear to have a different and more complex relation between sediment discharge and catchment size than unglaciated river basins (Hallet, Hunter, and Bogen, 1996) (Section 7.2.1). Ice and snow cover in mountains may increase physical and chemical weathering, particularly where the glacier or snowpack is warm-based. Consequently, Vezzoli, Garzanti, and Monguzzi (2004) found a positive relationship between sediment yield and glaciated area in the Western Alps. Sediment delivery from the upland part of glaciated catchments may be particularly high when pluvial conditions promote the rapid evacuation of loose morainic material from a previously glaciated region. There may be peaks in sediment yield, therefore, following glacial maxima at times of glacial-interglacial cyclicity.

Some currently glaciated basins, such as the steep and wet catchments of Alaska and the Southern Alps of New Zealand, have very high sediment loads and yields, in excess of those of most nonglaciated settings. A global compilation of about 18,000 thermochronological

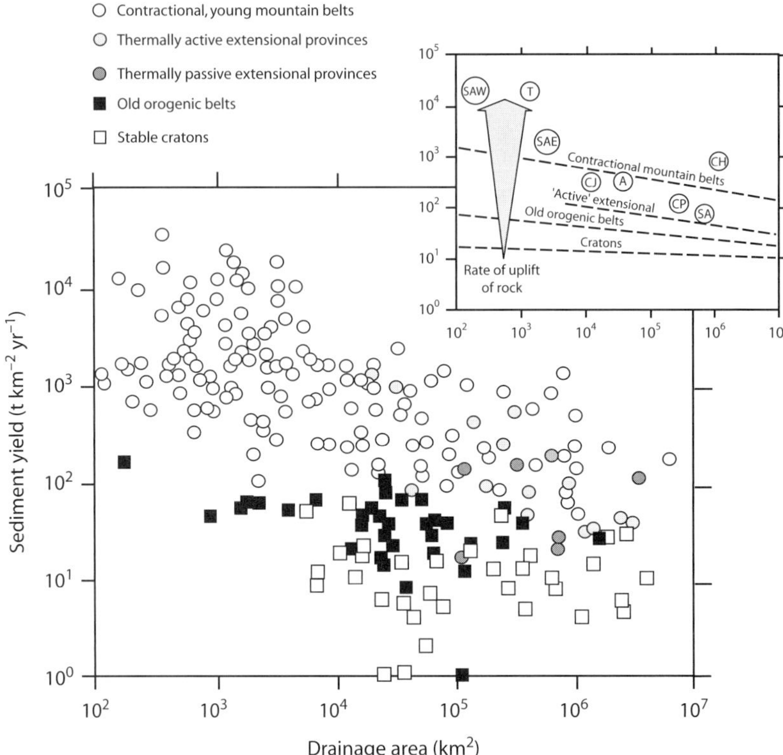

Figure 2.18 Plot of sediment yield versus drainage basin area for five categories of tectonic setting, derived by Hovius (1998) using the sediment yield data of Milliman and Syvitski (1992). Inset shows the approximate trends for various tectonic settings in the same existence field of drainage area (in km^2) versus sediment yield (in t km^{-2} yr^{-1}). SAW, Southern Alps New Zealand western flank; SAE, Southern Alps New Zealand eastern flank; T, Taiwan; CJ, Central Japan; A, Alps; CP, Colorado Plateau; SA, Southern Africa; CH, Central Himalaya. Published with permission of Wiley & Sons Inc.

measurements suggest that erosion rates have increased significantly over the last 2–4 Myr coincident with the occurrence of northern hemisphere glaciation (Herman et al., 2013). Little is known about the sediment budget of high-latitude, glacierised landmasses such as Greenland and Antarctica. The annual sediment discharges from wet-based glaciers range widely, from 100–10,000 t km^{-2} yr^{-1} (Gurnell, Hannah, and Lawler, 1996).

Bedload Fluxes

Bedload fluxes are generally not included in the sediment loads measured at river mouths. In the downstream reaches of large, low-gradient rivers, such as the Mississippi River and Amazon River, there is little transport of bedload, but in steep and some ephemeral catchments the coarse bedload is potentially very important, most of which is transported during intermittent major floods (Turowski, Rickenmann, and Dadson, 2010). Bedload transport rates are commonly taken to be a fixed percentage of the total particulate load, such as 10–20% in general, or 20–40% for mountain streams. The empirical basis of these estimates is, however, uncertain. The fraction of the total load transported as bedload varies widely, to smaller or greater extent dependent of catchment geology, degree of forest cover or glaciation, grain size of channel bed (gravel versus sand) and channel slope. The partitioning of bedload and suspended load has also been expressed by a regression against drainage area A in km^2, principally using data from rivers in the European Alps (Schlunegger and Hinderer, 2003):

$$F_{bed} = 0.525 - 0.0506\ln(A) \tag{2.13}$$

and

$$F_{sus} = 0.475 + 0.0506\ln(A) \tag{2.14}$$

where F_{bed} and F_{sus} are the bedload and suspended load fractions respectively.

Inspection of equations (2.13) and (2.14) shows that the bedload fraction ranges from 0 to 0.525 and the suspended load fraction ranges from 0.475 to 1 for drainage basin areas up to 835,304 km^2, but the scatter of field data shows that these equations only provide a good fit for small catchments of up to 5,000 km^2 (Turowski, Rickenmann, and Dadson, 2010).

Suspended load transport rates can also be expressed as a power law function of bedload transport rates (Métivier et al., 2004; Meunier et al., 2006). The median relationship is

$$Q_{bed} = aQ_{sus}^b \quad \text{for} \quad Q_{sus} \le (a/c)^{1/(d-b)} \tag{2.15}$$

and

$$Q_{bed} = cQ_{sus}^d \quad \text{otherwise} \tag{2.16}$$

where the best-fit values of the constants a, b, c and d for the median fit are 0.833 ± 0.052 [kg sec^{-1}]$^{1-b}$, 1.3240 ± 0.079, 0.437 ± 0.210 [kg s^{-1}]$^{1-d}$ and 0.647 ± 0.076 respectively, and sediment transport rates Q are in kg s^{-1}. The maximum value of the bedload transport rate as a percentage of the total transport rate occurs at $Q_{sus} = (a/c)^{1/(d-b)}$, which for the values of the constants given earlier gives 38% bedload at a suspended sediment transport rate of 0.386 kg s^{-1}.

A combination of the relationships in equations (2.13) and (2.14) and in equations (2.15) and (2.16) is potentially valuable in carrying out simulations of sediment transport in catchments of known size.

Equations (2.13) and (2.14) indicate that the fraction of bedload in the surface discharge should become negligible in rivers draining large catchments, which further suggests that these rivers should be discharging primarily silt and mud into the ocean. In the Solimoes-Amazon system, mean annual fluxes of sand and of silt-clay are known for the period 1974–1989 (Dunne et al., 1998). The sand flux as a percentage of the total flux stays constant at about 20% over a downstream distance of 2,750 km, and 248 Mt yr^{-1} of sand passes through the gauging station at Obidos. An increase in the sand flux of the Solimoes-Amazon system is caused by the joining of the River Madeira at 2,300 km downstream of Iquitos. Despite the enormous size of the Amazon system, therefore, a relatively high proportion of the total sediment discharge is made up of sand. To reconcile the predicted bedload fraction with the observed discharges of sand and silt-clay requires that the bulk of the sand travels in suspension and that bedload transport rates are negligible where the Amazon enters the Atlantic Ocean.

The bedload as a percentage of the total sediment load delivered to 51 globally distributed deltas (Table 2.4) was calculated using a Bagnold (1966) bedload transport equation (Syvitski and Saito, 2007):

$$Q_b = \frac{\rho_s}{\rho_s - \rho_f} \frac{\rho_f g Q_w^\beta D_g e_b}{\tan\phi} \tag{2.17}$$

when $u \geq u_{cr}$, where ρ_s and ρ_f are the densities of sediment and water respectively; g is the acceleration due to gravity; D_g is the gradient of the delta plain measured along the main channel; e_b is the dimensionless bedload efficiency; β is a bedload rating term, assumed to equal 1 by Syvitski and Saito (2007); ϕ is the limiting angle of repose of sediment grains lying on the river bed; u is the stream velocity; and u_{cr} is the critical velocity for bedload sediment transport. The bedload calculations show an average value of 6.6%, with a range of 0.7% (Mississippi) to 30% (Kolyma). Elevated bedload fractions may result from a number of factors: values may be high because of proximity of the delta to steep highlands, or because of high, flashy discharges of streams draining glaciers, or where the suspended sediment load is particularly low.

The regression of a log-log plot of suspended sediment discharge (kg s^{-1}) versus bedload sediment discharge (kg s^{-1}) for 51 deltas gives

$$Q_b = 0.8 Q_s^{0.5926} \tag{2.18}$$

with $R^2 = 0.58$ (Figure 2.19). However, the bedload fraction does not correlate well with either catchment area or length of the main stem using equation (2.17), in contrast to equation (2.13). Comparing equation (2.18) with equation (2.16), the exponent d of 0.5926 based on the delta database compares favourably with the value of 0.647 ± 0.076 based on the database in Métivier et al. (2004). The coefficient c of 0.8 in equation (2.18) compares with the value of 0.437 ± 0.210 based on the database in Métivier et al. (2004). The difference in the value of the power law coefficient, which corresponds to the intercept in cross plots of Q_b versus Q_s, is illustrated in Figure 2.19. The Bagnold (1966) equation slightly overestimates the bedload discharge relative to equation (2.16).

2.3.3 Dissolved Solids

The database on dissolved solids (solutes) is less extensive than for suspended particulate sediment, but databases such as GEMS-GLORI (Meybeck and Ragu, 1996, 1997) provide data from nearly 1,000 rivers. The total discharge of dissolved solids (TDS) is known

Table 2.4 *Bedload sediment discharge to 51 of the world's deltas calculated using equation (2.17) of Syvitski and Saito (2007).*

River	Water discharge	Suspended sediment discharge	Bedload discharge	Percentage bedload
	Q_w $m^3\ s^{-1}$	Q_s $kg\ s^{-1}$	Q_b $kg\ s^{-1}$	%
Amazon (Brazil)	198,676	37,843	618	1.61
Arno (Italy)	57	70	2	2.78
Brazos (Texas USA)	189	290	18	5.84
Ceyan (Turkey)	222	173	7	3.89
Chao Phraya (Thailand)	963	349	6	1.69
Colorado (California USA)[1]	694	3,803	69	1.78
Colorado (Texas USA)	76	44	5	10.20
Copper (Alaska USA)	1,240	2,218	184	7.66
Danube (Romania)[2]	6,420	2,123	98	4.41
Ebro (Spain)[3]	1,400	576	71	10.98
Eel (California USA)	235	558	58	9.42
Fly (PNG)	2,510	2,219	120	5.13
Fraser (Canada)	3,560	634	57	8.25
Ganges–Brahmaputra (BNG)	31,000	34,870	689	1.94
Godavari (India)	2,650	5,387	174	3.13
Homathko (Canada)	253	136	14	9.33
Huanghe (China)	1,480	34,857	4	0.01
Indigirka (Russia)	1,734	346	64	15.61
Indus (Pakistan)	3,171	7,884	94	1.18
Irrawaddy (Myanmar)	13,558	8,119	258	3.08
Klamath (California USA)	473	315	15	4.55
Klinaklini (Canada)	330	158	27	14.59
Kolmya (Russia)[4]	3,784	186	78	29.55
Krishna (India)[5]	1,890	2,028	218	9.71
Lena (Russia)	16,240	631	28	4.25
Limpopo (Mozambique)	835	1,046	111	9.59
MacKenzie (Canada)	9,750	3,153	120	3.67
Magdalena (Colombia)	7,530	6,970	186	2.60
Mahanadi (India)	2,112	1,901	52	2.66
Mekong (Vietnam)	17,345	3,090	86	2.71
Mississippi (Louisiana USA)	15,452	12,614	88	0.69
Niger (Nigeria)	6,130	1,268	106	7.71
Nile (Egypt)[6]	3,484	3,800	76	1.96
Orange (South Africa)	442	2,806	109	3.74
Orinoco (Venezuela)	34,500	5,493	170	3.00

    The Global Character of River Basins

Table 2.4. *(cont.)*

	Water discharge	Suspended sediment discharge	Bedload discharge	Percentage bedload
	Q_w $\mathrm{m^3\ s^{-1}}$	Q_s $\mathrm{kg\ s^{-1}}$	Q_b $\mathrm{kg\ s^{-1}}$	%
River				
Parana (Argentina)	14,506	2,838	217	7.10
Pechora (Russia)	4,099	193	14	6.76
Pescara (Italy)	29	60	6	9.09
Po (Italy)[7]	1,525	545	16	2.85
Rhone (France)	1,700	1,861	45	2.36
Song Hong (Red) (Vietnam)	3,784	3,469	187	5.11
Squamish (Canada)	250	57	8	12.31
Tigirs-Euphrates (Iraq)	1,500	1,680	82	4.65
Var (France)	41	48	8	14.29
Vistula (Poland)	1,050	79	6	7.06
Volga (Russia/Ukraine)	8,200	602	38	5.94
Waiapaoa (New Zealand)	41	295	47	13.74
Yana (Russia)	1,009	95	6	5.94
Yangtze (China)	28,278	15,210	391	2.51
Yukon (Alaska USA)	6,620	1,702	180	9.56
Zhujiang (Pearl) (China)	8,199	2,459	124	4.80

[1] 1904–1923; [2] 1931-1955, [3] 1913–1962 [4] 1942–1965; [5] 1901-; [6] 1871–1898 [7] 1933–1939

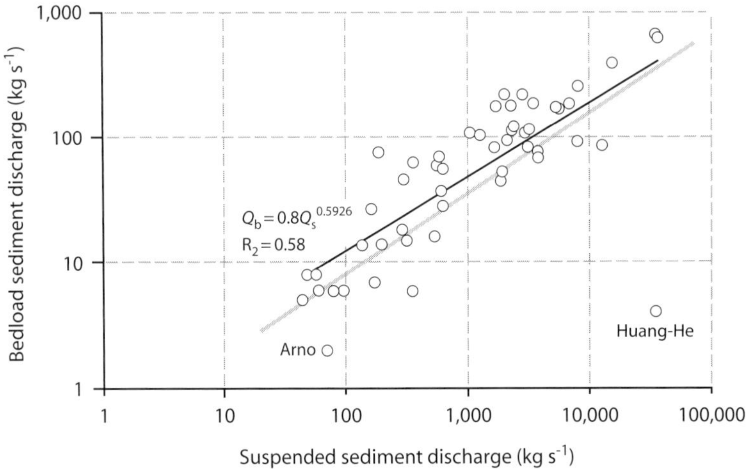

Figure 2.19 Plot of bedload versus suspended sediment discharges for 51 of the world's major deltas. Q_b is calculated using a Bagnold (1966) bedload sediment transport formula. Data from table 1 in Syvitski and Saito (2007). The trendline for equation (2.16) is shown in grey.

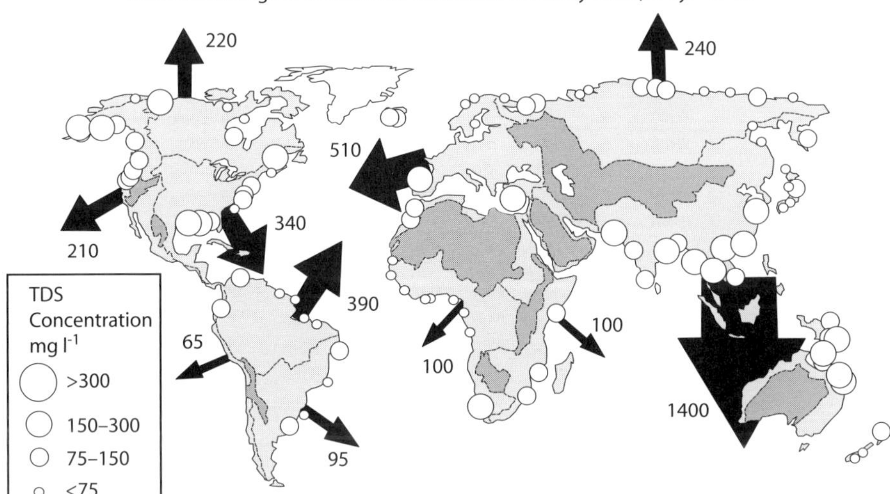

Figure 2.20 Annual delivery of dissolved solids into the coastal ocean by rivers using data from Meybeck and Ragu (1996). Circles are average solute concentrations for selected rivers. Numbers are the solute discharges in t yr^{-1} × 10^6. Black arrows are proportional to these numbers. The global total discharge of solutes is 3,800 × 10^6 t yr^{-1}. Data from western South America are particularly speculative. Grey areas are endorheic. Modified from Milliman and Farnsworth (2011). Published with permission of Cambridge University Press.

for many river basins, but information on individual chemical species, such as carbon, phosphorus, nitrogen and silica, all important as nutrients, is more patchy. Rivers discharge annually about 3.8 billion tons (3.8 × 10^9 tons) of dissolved solids to the global ocean, the highest fluxes being in Asian rivers and lowest in those in Africa and the Eurasian Arctic (Figure 2.20). There are contrasting views on the controls on the chemistry of rivers, with those in favour of climate (temperature, run-off) and those in favour of bedrock lithology as the dominant control (Section 3.1.1).

Ions are released in solution during chemical weathering. For chemical weathering to be effective, precipitation must penetrate to deep levels in the regolith in order to promote the chemical breakdown and flushing away of solutes into the groundwater system (Sections 9.1 and 9.2). For a given bedrock, infiltration and flushing depend on climate and topography. For instance, in semi-arid regions there is a scarcity of water available for flushing, and solutes are drawn to the surface in dry periods and reprecipitated, reducing chemical weathering of bedrock. Topography affects solute yield through the effect of slopes on the infiltration and flushing of water. On steep slopes, flushing rates are high, which reduces the time available for chemical weathering of regolith and bedrock. On low slopes, flushing rates are low, which limits the removal of solutes. Optimum conditions for solute removal are found in regions with moderate precipitation and gentle upland slopes.

Table 2.5 *Rock types at the surface of the Earth and their relative rate of chemical weathering. Data from Meybeck (1987). Published with permission of* American Journal of Science.

Class	Rock type	% by area	Chemical weathering	Global contribution
Plutonic	Granite	10.4	1	10.4
	Gabbro/ultrabasics	0.6	1.3	0.78
Metamorphic	Marble	0.4	1.3	0.52
	Amphibolite	1.9	1.3	2.47
	Mica schist	1.5	5	7.5
	Gneiss	10.4	1	10.4
	Quartzite	0.8	1.3	1.04
Volcanic	Basalt	4.15	1.5	6.23
	Andesite	3.0		
	Rhyolite	0.75	1.3	0.98
Sedimentary	Quartz arenite	12.6	1.3	16.38
	Arkose	0.8	2	1.6
	Greywacke	2.4	2	4.8
	Shales	33.1	2.5	82.75
	Carbonates	15.9	12	190.8
	Evaporites	1.3	40–80	52–104

Chemical weathering rate is relative to that of granite. The global contribution of each rock type is the product of the percentage surface area and the relative weathering rate. Amphibolite has been given the same relative weathering rate as ultrabasics; arkose and greywacke are attributed relative weathering rates of 2; rhyolite and quartzite are given the same relative weathering rate as quartz-arenite since all are dominated by SiO_2; evaporites is given as a range from gypsum to halite. The percentage of outcrop of different rock types is based on early compilation exercises (Ronov and Yaroshevskiy, 1972, 1976; Blatt and Jones, 1975; Sibley and Wilband, 1977).

The chemical denudation of regolith caused by removal of solutes does not, however, equate strongly to the total dissolved solids in rivers. This is partly because of non-denudational inputs (precipitation, wind-blown dust, salt aerosols, mineralisation of organic matter, plant metabolism and, today, human pollution). The total dissolved solids in rivers may also reflect chemical weathering of fluvial sediment during storage in foodplains, splays and bars (Johnsson and Meade, 1990). Solutes may in addition be lost through evaporation and precipitation in soils and groundwater. Nevertheless, there are clear trends in the total dissolved solids (TDS) of river water draining certain catchment lithologies (Meybeck, 1987). The relative global contribution of solutes from pristine rivers draining certain bedrock lithologies is the product of the percentage of the Earth's terrestrial surface covered by those lithologies and an index of the chemical weathering rate relative to granite (Table 2.5). By far the greatest contribution of solutes comes from shales, principally because of their large outcrop area; evaporites (gypsum, halite) due to their high rate of dissolution; and carbonates (limestones and dolomites) resulting from a combination of high weathering rate and high outcrop area. Consequently, notwithstanding the effects of climate and topography previously described, rivers draining from an exhumed, carbonate-dominated passive margin or fold-thrust belt are likely to carry high solute loads. Rates of denudation in these hinterlands are strongly impacted by solute mass transfer. In contrast, rivers draining cratonic shields composed of granites and gneisses are likely to have low solute loads, and chemical denudation rates are a small percentage of the total denudation rate. The ratio of chemical denudation measured as the solute yield derived from TDS in river water, and physical denudation measured from the total suspended sediment of river water, is shown for 28 of the world's largest rivers in Table 2.6 (Summerfield and Hulton, 1994) and in Figure 2.21.

The chemical denudation caused by solute mass transfer as a percentage of the total denudation varies across a very wide range, from less than 3% (Brahmaputra) to nearly 95% (St Lawrence). It is notable that the large, low-gradient rivers draining the cratonic continental interior in cold, semi-arid climates, such as the Ob, Lena and Yenisei in northern Eurasia, and the St Lawrence of Canada, have very high solute loads compared to particulate loads. In contrast, rivers draining steep topography in regions of active mountain building with high amounts of orographic rainfall, such as the Ganges and Brahmaputra, have particulate loads that are far greater than solute loads.

Long-term fluxes of solutes in rivers vary widely with geographic region and therefore climate (Table 2.7). The tropical zone dominates the world's run-off and is a major source of, for example, dissolved silica (Section 3.3) and organic carbon (Section 3.2). The high level of dissolved silica reflects the intensity of chemical weathering processes in tropical regions, while the large amount of organic carbon is due to the high productivity of tropical vegetation.

In essence, waters draining catchments made of carbonates (principally limestones and dolomites) contain high amounts of total dissolved solids (TDS, measured in mg litre^{-1} or alternatively ppm), particularly Ca^{2+} and HCO_3^-, but small particulate sediment loads. Waters draining varied igneous, metamorphic and sedimentary terrains have a correspondingly varied chemical composition and higher particulate loads. We explore the chemistry of river water in more detail in the chapter on global biogeochemical cycles (Chapter 3) that follows.

2.4 Chemical Weathering Fluxes Associated with Glaciation

It is expected that chemical weathering rates will be high where physical erosion rates are high (Gaillardet, Millot, and Dupré, 2003), since physical breakdown increases mineral surface areas susceptible to chemical attack (Stumm and Morgan, 1996) (Section 9.1). Temperate-zone, glaciated basins have some of the highest erosion rates on Earth, and the surface area production rate is high due to the fragmentation of glacial materials into fine-grained products. However, the fluxes of solutes, particularly silica fluxes, produced

Table 2.6 *Particulate and solute yields for 28 of the world's largest river drainage basins and equivalent denudation rates. Data from Summerfield and Hulton (1994); published with permission of American Geophysical Union.*

River basin	Area (10^6 km^2)	Sediment yield t km^2 yr^{-1} (mm kyr^{-1})	Solute yield t km^2 yr^{-1} (mm kyr^{-1})	Chemical denudation as % of total
Amazon	5.88	221 (82)	29 (11)	11.6
Amur	2.04	28 (10)	6 (2)	17.6
Brahmaputra	0.64	1,808 (670)	49 (18)	2.6
Changjiang	1.73	281 (104)	72 (27)	20.4
Colorado	0.70	239 (89)	19 (7)	7.4
Columbia	0.67	48 (18)	32 (12)	40.0
Danube	0.79	94 (35)	45 (17)	32.4
Dniepr	0.54	2 (1)	12 (4)	85.7
Ganges	0.98	694 (257)	42 (16)	5.7
Huang He	0.79	127 (47)	18 (7)	12.4
Indus	0.93	323 (120)	42 (16)	11.5
Kolyma	0.65	9 (3)	4 (1)	30.8
La Plata (Parana)	2.86	30 (11)	9 (3)	23.1
Lena	2.45	7 (3)	22 (8)	75.9
Mackenzie	1.77	62 (23)	23 (9)	27.1
Mekong	0.76	232 (86)	36 (13)	13.4
Mississippi	3.20	189 (70)	20 (7)	9.6
Murray	1.14	30 (11)	6 (2)	9.7
Niger	2.16	19 (7)	4 (91)	17.4
Nile	3.63	28 (10)	3 (1)	9.7
Ob	2.98	6 (2)	11 (4)	64.7
Orange	0.89	65 (24)	11 (4)	14.5
Orinoco	0.92	179 (66)	23 (9)	11.4
Rio Grande	0.63	48 (18)	4 (1)	7.7
Shatt al-Arab	0.89	56 (21)	14 (5)	20.0
St Lawrence	1.05	2 (1)	34 (13)	94.4
Yenisei	2.55	5 (2)	18 (7)	78.3
Yukon	0.84	94 (35)	23 (9)	19.7
Zaire (Congo)	3.63	14 (5)	6 (2)	30.0
Zambesi	1.41	34 (13)	6 (2)	15.0

by chemical weathering in basins with active glaciers are low when compared with non-glaciated basins with similar water discharges (Anderson, Drever, and Humphrey, 1997; Anderson, 2005) (Figure 2.22).

The surface area production rate P_s is given by

$$P_s = N_g A_g \tag{2.19}$$

Table 2.7 *Dissolved loads of different climatic regions, expressed as percentages of the total input to the oceans. TOC, total organic carbon. After Allen and Allen (2013) (tab.7.3, p.243), with permission from John Wiley & Sons, Inc.*

	Geographic region			
	Cold regions	Temperate	Tropical	Arid
Area (%)	23.4	22.4	37.0	17.2
Run-off (%)	14.7	27.5	57.2	0.65
Dissolved SiO$_2$ (%)	5.4	19.9	73.6	1.0
Ions (%)	15.5	39.9	41.8	2.8
TOC (%)	17.5	28.5	52.0	1.3
Suspended matter (%)	2.7	56.5	34.2	6.6

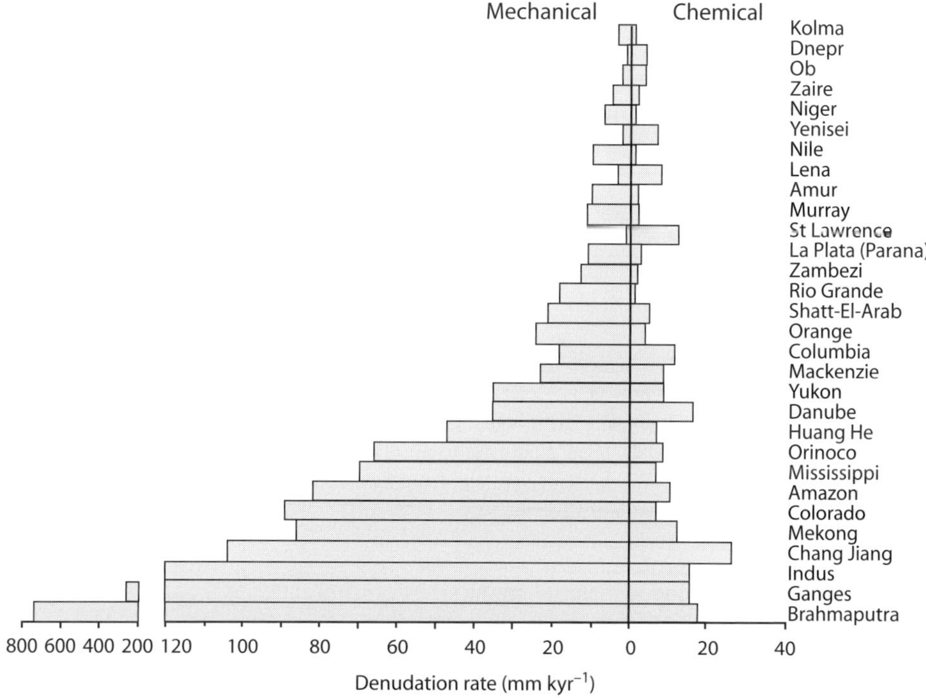

Figure 2.21 Histogram comparing mechanical and chemical denudation in the world's major externally drained river basins. After Summerfield and Hulton (1994) with permission from American Geophysical Union.

where N_g is the number of grains produced per unit time and A_g is the surface area per grain. The number of grains produced is given by

$$N_g = V/v_g \tag{2.20}$$

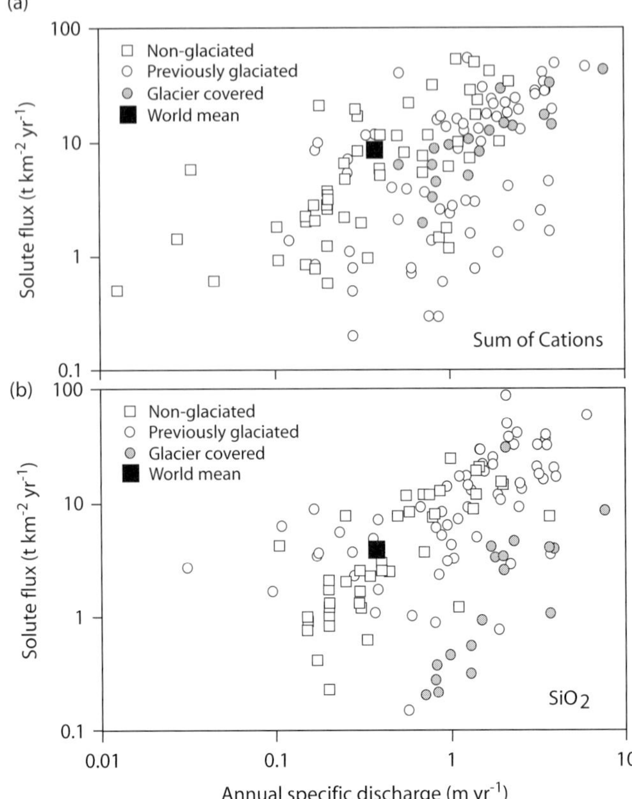

Figure 2.22 Chemical denudation rates viewed as total solute flux (a) and dissolved silica flux (b) in t km^{-2} yr^{-1} from drainage basins with active glaciers compared with a global data set of non-glacier covered watersheds. Data from Anderson et al. (1997) supplemented by data from Arctic cold-based glaciers, non-glacial Siberian rivers and from Kennicott Glacier, compiled by Anderson (2005). Published with permission of Elsevier.

where V is the total volume of material produced by erosion and v_g is the volume per grain. For the case of uniform grain size, such as spheres or cubes, the surface area production rate becomes

$$P_s = 6(E/D)A \qquad (2.21)$$

where E is the erosion rate, D is the particle size and A is the basin area. For a constant basin size and erosion rate therefore, the surface area production rate varies inversely with mean grain size. Grain-size reduction of sediment might result from comminution (fragmentation, Section 9.3) caused by glacial erosion of bedrock. Glacial erosion is likely to be accelerated under conditions of high mean annual water discharge, as found in temperate-zone glaciers.

Solute fluxes in both glacierised and non-glacierised catchments show a positive relationship with specific annual discharges of water. Whereas total cations data overlap between glacierised and non-glacierised catchments (Figure 2.22a), data for dissolved silica show a distinct separation (Figure 2.22b). Low silica fluxes are caused by the Arrhenius-type effect of temperature on weathering rates. The lack of a differentiation between glacierised and non-glacierised catchments in terms of total cations is because much of the cation flux is derived from the weathering of carbonates exposed by glacial abrasion.

Solute fluxes are also dependent on the mineral weathering reaction rate (mol per mineral surface area per time): basins with outcropping sedimentary rocks have low silica concentrations, whereas basins with igneous and metamorphic rocks have high silica concentrations.

3

Global Biogeochemical Cycles

3.1 Biogeochemistry of World Rivers

The geochemist Abraham Lerman stated (Lerman, 1988) (p.1) that

A global geochemical cycle is a conceptual model of the pathways and flows of individual chemical elements and their compounds in the surface environment of the Earth.

A major goal of global biogeochemistry is to understand the transport processes responsible for these pathways and flows at the present day, to apply this understanding to the past and to predict the future. The most important pathway is riverine transport from source areas on the continental surface to the ocean. Rivers are the primary agents responsible for the fluxes of dissolved solids across the Earth's surface and therefore participate strongly in the functioning of biogeochemical cycles (Garrels and Mackenzie, 1971). The solutes delivered to the world ocean by rivers control its steady-state composition and the nutrient levels necessary for biological productivity.

Although solutes are derived from the weathering of soil and bedrock, they also originate from non-denudational sources such as the atmosphere (precipitation, dust and other aerosols), the mineralisation of organic matter, plant metabolism and man-made pollution. Solutes from non-denudational sources may make up to half of the total dissolved loads of rivers.

The most important denudational source of solutes is from weathering of soils, regolith and bedrock. The different lithologies present in catchments strongly influence their solute loads (Table 3.1). Combining typical water analyses of major rock types with their relative abundance (Table 2.5) allows the solute delivery associated with different rock types to be evaluated. This can be done globally (Meybeck, 1976, 1987; Gaillardet et al., 1999) or for individual catchments (Miller, 2002).

In Meybeck's classic study of the impact of bedrock lithology on river water chemistry (Meybeck, 1986, 1987), water analyses were derived from small, unpolluted catchments in France, constituting the Temperate Stream Model. Water chemistry analyses were based on representative watersheds under different climates with negligible contribution from oceanic salts, where only one bedrock lithology is present and where the stream is perennial. The French watersheds are generally forested and situated in lowland regions (median

Table 3.1 *Weathering products of different rock types as percentage of total amount released, and proportion of solute load of rivers with outlets at the coast, compared to abundance of rock type at the continental surface. From Meybeck (1987) (tab.5, p.418) with permission of* American Journal of Science.

Rock type	SiO_2	Ca^{2+}	Mg^{2+}	Na^+	K^+	Cl^-	SO_4^{2-}	HCO_3^-	Σ^+ (meq l^{-1})	Outcrop (%)	Load (%)
Granite	10.9	0.60	1.2	5.9	4.9		2.0	1.6	1.6	10.4	2.6
Gabbro	0.75	0	1.1	0	0		0.3	0.3	0.3	0.6	0.2
Gneiss & mica-schist	11.7	1.1	2.8	6.6	8.2		4.6	2.0	2.4	12.7	3.3
Misc. metamorphic	1.5	4.7	3.2	0.3	1.6		1.7	4.7	3.7	2.3	2.6
Volcanic rocks	11.1	1.8	4.9	5.4	6.6		0.5	4.2	3.1	7.9	2.8
Sandstone	16.6	2.1	3.8	5.2	19.6		9.6	2.3	3.2	15.8	5.0
Shale	35.1	19.9	30.7	22.6	41.0	7.4	30.3	22.5	23.1	33.1	20
Carbonates	11.3	60.4	39.3	3.5	13.1		8.6	59.5	46.7	15.9	45.1
Gypsum	0.5	7.2	7.2	4.9	1.6	10.0	32.7	1.7	6.8	0.75	8.1
Rock salt (halite)	0.4	2.2	5.8	45.6	3.3	82.6	9.6	1.2	9.1	0.5	10.2
Total (10^{12} g y^{-1})	320	504	118	132	24	120	280	1950			

Σ^+ is the sum of cations in meq l^{-1} or milliequivalents of solute per litre of solvent; load % is the percentage of the dissolved load derived from weathering, excluding atmospheric CO_2; HCO_3^- values include bicarbonate originating from atmospheric and soil CO_2, which is approximately half of the total HCO_3^-.

altitude of 650 m) and are relatively small (median drainage area 7.8 km^2). The results are therefore not directly applicable to global patterns of solute generation, but nevertheless give a useful background on the varied influence of bedrock lithology on the chemistry of river waters.

The basis for making the link between bedrock type and water chemistry is the chemical weathering pathways associated with the attack of silicate and carbonate minerals by carbonic and sulphuric acids generated by the reaction of atmospheric and soil CO_2 with water and by the oxidation of pyrite respectively.

The sources of major elements found in river waters draining to the ocean can be attributed as follows (Meybeck, 1987):

- Silica and potassium are derived mostly from silicate weathering, which in crystalline rocks is due to the breakdown of K-micas and orthoclase feldspars
- Sodium, in contrast, has an important source in the weathering of halite, the remaining Na coming from silicate weathering
- Magnesium originates from dolomites and silicates, with the remainder from Mg-rich evaporites
- Calcium mainly comes from carbonate weathering, with a subordinate amount from silicate weathering of sedimentary rocks
- Chloride concentrations are very sensitive to pollution and contamination with ocean salts; in pristine rivers it derives from halite weathering and less importantly from crystalline rocks
- Sulphate is derived from the weathering of evaporite sulphate (e.g. gypsum), pyrite and organic sulphur compounds
- Bicarbonate originates mainly from the atmosphere and secondarily from the chemical weathering of carbonate rocks.

3.1.1 Global Water Chemistry

On a global basis, the major elements dissolved in river water are in the following order of abundance (Chester, 1990):

$$Cl^- > SO_4^{2-} > Ca^{2+} = Na^{2+} > Mg^{2+} > HCO_3^- > SiO_2 > K^+$$

The concentrations of major elements are expressed in cumulative form in Figure 3.1. The two principal water types are those dominated by the cation Ca^{2+} for *fresh water* and the cation Na^+ for *highly saline waters*, and the anion HCO_3^- for fresh water and the anion Cl^- for highly saline waters. The observed differences in the chemical composition of river water have been investigated in terms of both the chemistry of rainwater and the dissolved weathering products as a function of the total dissolved solids (TDS).

It has been proposed that water compositions in rivers fall along two diverging arms when total salinity (expressed as TDS) is plotted against the cationic weight ratio

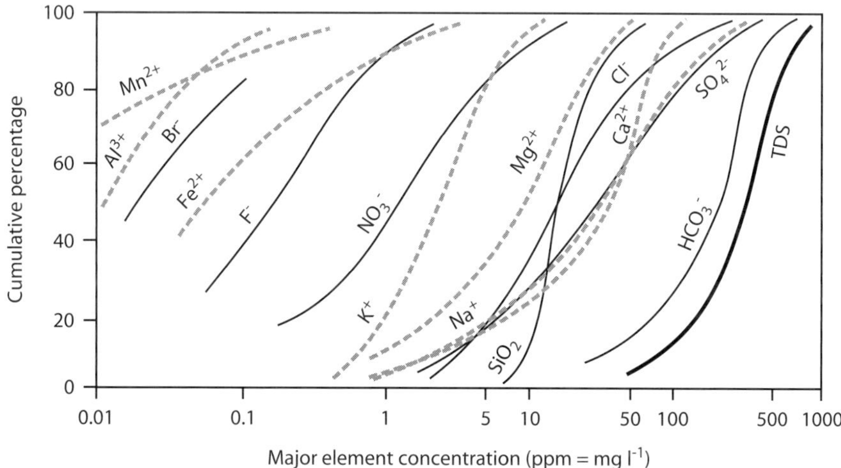

Figure 3.1 Cumulative probability curves for the concentrations of major elements in natural terrestrial river waters unaffected by anthropogenic influences. Dashed lines, cations; solid lines, anions. TDS, Total Dissolved Solids. After Stumm and Morgan (1996) (fig. 15.1), with permission from Wiley-Interscience.

$Na^+/(Na^++Ca^{2+})$ (Figure 3.2) or the anionic weight ratio $Cl^-/(Cl^-+HCO_3^-)$ (Gibbs, 1970).

In the scheme proposed by Gibbs (1970), water compositions are controlled by three main mechanisms (Figure 3.2):

1. Control by precipitation, where the chemistry is dominated by rainwater rich in Na^+, producing a cationic ratio of nearly 1, and low TDS (20–30 mg l^{-1}) since evaporated seawater falling as rain is dilute. Rivers draining areas of low relief with well-weathered bedrock, plentiful rainfall and low evaporation are typically controlled by precipitation. The tropical lowland rivers of Africa and South America, such as the Negro tributary of the Amazon, fall into this class. The low TDS values may also be affected by the large biomass of the lowland tropical forests effectively immobilising chemicals from release into rivers.

2. Control by weathering, where the chemistry of river water is controlled by weathering reactions with soil, regolith and bedrock within their catchments. The cationic ratio depends on the type of rock being weathered. Limestone bedrocks produce the lowest cationic ratio. TDS levels are higher than in rivers dominated by precipitation (30–1,000 mg l^{-1}), reflecting the increased role of weathering. Solutes are dominated by calcium bicarbonate. In the Amazon River system, 85% of the solute load is derived from a relatively small area of intense weathering in the Andes, whereas the lowland part of the catchment, particularly that draining the low-relief Brazilian Shield, is dominated by precipitation. Waters dominated by rock weathering are found in regions of high weathering intensity and low rates of evaporation.

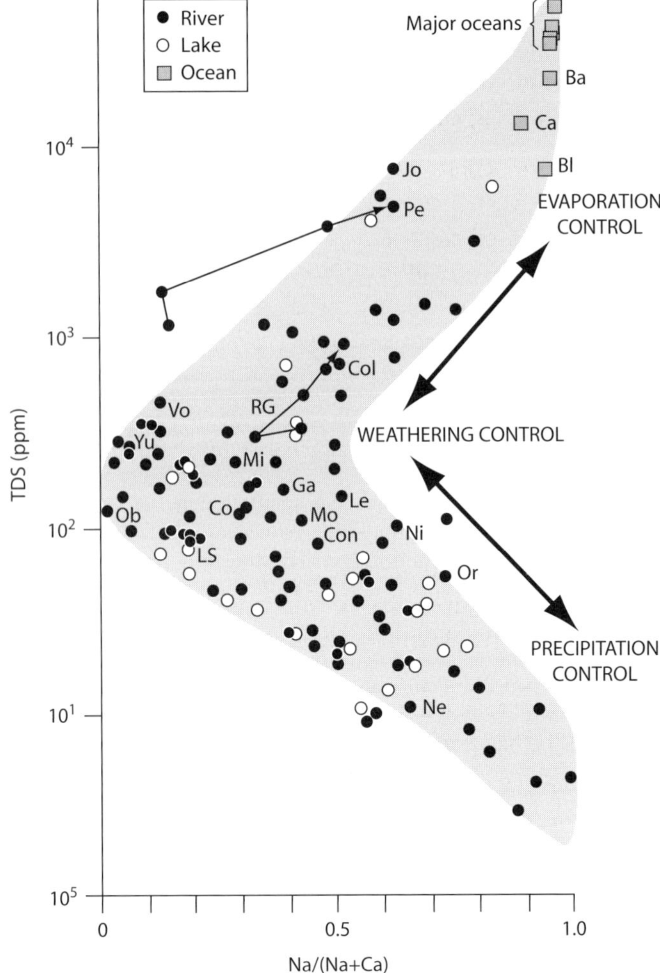

Figure 3.2 Scheme for global water chemistry based on plot of cationic weight ratio versus total dissolved solids (TDS) for rivers, lakes and oceans. Arrows connecting data points show the geochemical downstream evolution of river waters of the Pecos and Rio Grande. Ba, Baltic; Bl, Black Sea; Ca, Caspian Sea; Col, Colorado; Con, Congo; Ga, Ganges; Jor, Jordan; LS, Lake Superior; Le, Lena; Mi, Mississippi; Mo, Mobile; Ne, Negro; Ni, Niger; Ob, Ob; Or, Orinoco; Pe, Pecos; RG, Rio Grande; Vo, Volga; Yu, Yukon. From Gibbs (1970)(fig. 1). Reprinted with permission from American Association for the Advancement of Science (AAAS).

3. Control by evaporation and crystallisation, common in hot and arid climates where early precipitation of calcite ($CaCO_3$) increases the cationic ratio due to the relative concentration of Na^+ and Cl^-, and evaporative losses of river water increases the TDS (1,000–2,000 mg l^{-1}). This control produces a series of waters extending from the Ca-rich rock-weathering end to the Na-rich high-salinity end, typified by sea water.

A pathway of low to high salinity may occur as a river drains towards the ocean, as in the case of the Rio Grande, southwest United States and its major tributary, the Pecos River. Using this approach, river waters dominated by evaporation-cystallisation are favoured by high weathering intensity combined with high evaporation rates.

The scheme proposed by Gibbs (1970), however, does not account for the impact of the inflow of groundwater brines, as may occur in the Pecos River (New Mexico, United States), and the dissolution of carbonates and evaporites in the watershed, as may be important in the case of the Amazon River. The dominance of the control of precipitation in some Amazonian tributaries has also been questioned (Stallard and Edmond, 1983). Instead of two diverging arms of water composition suggested by Gibbs (1970) therefore, the major elements of river waters may depend essentially on different rates of weathering of variable parent rock types (Meybeck, 1987), so that almost all rivers are weathering-dominated. Using this approach, river waters may be categorised according to their total ionic contents based on the climate, relief and bedrock geology:

1. Rivers with relatively low total ionic contents, typical of those draining highly leached areas of low relief where rainfall is small, as occurs in some tropical regions of South America and Africa, and in rivers that drain crystalline cratons such as the Canadian, Brazilian and Congo shields. Waters as dilute as 19 mg l^{-1} are found in rivers draining the Canadian Shield.
2. Rivers with intermediate total ionic contents due to increasingly important rock weathering. The water of the Mackenzie River of northwest Canada, which drains a mixture of sedimentary and crystalline rock types, has a total ionic content of approximately 200 mg l^{-1} and a ratio of Ca^{2+}/Na^+ of 4.7.
3. Rivers with high total ionic contents and high Na/(Na+Ca) ratios. The Colorado River of southwest United States, for example, has a total ionic content of 200 mg l^{-1} and Na concentration close to that of calcium (a Na/(Na+Ca) ratio close to 0.5). In the case of the Colorado River, the major element chemistry may be strongly influenced by the influx of saline groundwater.

The chemical weathering of bedrock, regolith and soil is therefore of paramount importance in explaining the major element chemistry of river waters. Sedimentary rocks generate much higher quantities of ions in solution than do the plutonic and high-grade metamorphic rocks of crystalline shields. Clear relationships between catchment geology and water chemistry have been demonstrated from a number of large river basins, such as the Amazon (Stallard and Edmond, 1981, 1983, 1987). Stallard and Edmond analysed rivers in the Amazon Basin in terms of their total cation content or TZ^+ (units of μeq l^{-1}). Rivers with TZ^+ values less than 200 drain intensely weathered materials, and are enriched in silica relative to other major species, with cation ratios similar to the underlying bedrock. Rivers with TZ^+ values between 200 and 450 μeq l^{-1} drain siliceous bedrocks (quartzites, sandstones, felsic volcanics) and are rich in silica relative to other solutes. Rivers with TZ^+ values of between 450 and 3,000 μeq l^{-1} drain sedimentary rocks comprising carbonates, shales and minor evaporites, with relatively high concentrations of Ca^{2+} and Mg^+, high alkalinity, and high SO_4^- where rivers drain terrains with reduced shales and sulphate evaporites such as gypsum. Even higher values of more than 3,000 μeq l^{-1} are found in rivers draining giant evaporite successions. Their waters are rich in Na^+ and Cl^- derived from highly evolved salt minerals.

The chemistry of river input into the ocean therefore varies from continent to continent, depending on topography, tectonics, climate and rock types.

3.1.2 Chemistry of the Particulate Load

Solids transported by rivers as suspended particulate load also carry a chemical signature. Suspended particles reflect the mineralogy of the host bedrock, regolith or soil. Clay minerals of various types are derived directly by fragmentation of the host rock, but the action of acid hydrolysis of aluminosilicate minerals during chemical weathering causes an abundance of clay minerals in the sediment loads of many rivers. During progressive chemical weathering clay minerals are stripped of their metal cations, and eventually of their silicon and iron, leading to a stable aluminium-rich residue. In areas of less intense leaching, the cations released by the breakdown of bedrock promote the formation of cation-bearing clay minerals such as illites and smectites. A large throughput of water is necessary for advanced stages of leaching, so clay minerals reflect in part the climatic conditions experienced by regolith (Strakhov, 1967):

- In regions of high precipitation rates such as the equatorial and humid tropics, advanced leaching leads to the formation of gibbsite, kaolinite and aluminium oxides and hydroxides (laterisation). Intense chemical weathering is typical of equatorial-tropical South America (including the Amazon and Orinoco basins), Africa (including the Congo basin) and southeast Asia and northeastern Australia.
- In regions of moderate precipitation rates, such as the temperate, peri-humid midlatitudes, chemical weathering results in the formation of illite and smectite.
- In arid and semi-arid climates, such as the subtropics and the interiors of continental land masses such as Asia and Australia, low rates of chemical weathering result in little accumulation of weathering products, which are dominated by carbonate and salt duricrusts and concretions, and unweathered bedrock.
- The subpolar regions of northern Eurasia and northern North America experience low temperatures and consequently low rates of chemical weathering. Deep weathering is possible in regions of minimal local relief.
- Mountainous regions with highly variable rates of chemical weathering, such as the Alpine-Himalayan chain, typically have relatively little weathering products due to rapid removal by erosion.

The clay minerals in regolith also depend on depth within the weathering profile (Strakhov, 1967), since the flux of water decreases with depth towards the bedrock interface

Table 3.2 *Dissolved Transport Index (DTI) in global river water. Data from Martin and Meybeck (1979) (p.193). Published with permission of Elsevier.*

DTI	Elements
90–50%	Br, I, Cl, Ca, Na, Sr
50–10%	Li, N, Sb, As, Mg, B, Mo, F, Cu, Zn, Ba, K
10–1%	P, Ni, Si, Rb, U, Co, Mn, Cr, Th, Pb, V, Cs
1–0.1%	Ga, Tm, Lu, Gd, Ti, Er, Nd, Ho, La, Sm, Tb, Yb, Fe, Eu, Ge, Pr, Al

due to decreasing permeability. Residual minerals such as kaolinite and gibbsite are found near the top of weathering profiles, whereas smectite and illite are concentrated near the base.

The mineralogy of the particulate load of rivers is sensitive to particle size. Quartz and feldspar grains are mostly greater than 2μm in diameter, whereas mica and clay minerals are concentrated in the less than 2μm size fraction. Taking the full grain-size range, a survey of 20 of the world's major rivers indicates that there are a number of trends in the chemistry of the particulate loads of rivers (Martin and Meybeck, 1979). Individual river systems have concentrations of major elements that are fairly uniform, but rivers differ strongly from one to the other, depending on their history of weathering.

The amount of an element transported in solution as a percentage of the total discharge (particulate and dissolved) can be expressed by the *dissolved transport index* DTI (Martin and Meybeck, 1979). Some elements, on a global scale, are transported almost entirely in solution (DTI of 50–90%), which includes Na, Ca, Cl and S, whereas others are transported almost entirely in particulate form (DTI of less than 1%), including Al and Fe (Table 3.2). Those elements with low DTI are prone to deposition along the sediment routing system, particularly in deltas, and are therefore commonly buried and taken out of the biogeochemical cycle, whereas elements with high DTI are flushed through the transport system into the open ocean where they are mixed rapidly.

3.1.3 The Estuarine Filter

River-transported chemical signatures are modified, or filtered, at the land-sea interface, particularly in estuaries where seawater is mixed with and diluted by freshwater. Estuaries act as sediment traps. The estuarine filter is selective in terms of its impact on different elements, some dissolved species being carried quickly out to sea, whereas others undergo complex reactions that involve exchanges between dissolved, colloidal and particulate states. Reactions between the dissolved, colloidal and particulate loads are especially important near the *turbidity maximum* that occurs at the head of the saltwater intrusion. Components may be removed from the dissolved phase and transferred to the particulate phase by flocculation, adsorption, precipitation and biological uptake. Some components may be added to the dissolved phase by desorption from particle surfaces and breakdown of organics. Components may be stabilised in the dissolved stage by complexation and chelation reactions with inorganic and organic ligands. Dissolved chemical species tend to be flushed out of the estuary into the ocean, whereas particulate species may be retained within the estuarine sediment trap.

Much of the particulate sediment that enters estuaries and delta-tops from rivers is trapped and does not enter the ocean on short timescales (Chester, 1990). As examples, it is estimated that only approximately 8% of the sediment that enters the upper zone of the Scheldt estuary (northern France, western Belgium and southwestern part of the Netherlands) is transported to the North Sea. Similarly, more than 90% of the particulate matter carried by the Mississippi River is deposited in the delta region. About 95% of the suspended solids of the Amazon River is deposited in river mouth channels and shoals. Of the particulate sediment transported to its estuary and gulf by the St Lawrence River of eastern Canada and northeastern United States, approximately 90% is trapped.

3.2 The Fate of Organic Carbon

Sediment routing systems are the sites of oxidation of organic carbon contained in sedimentary rocks undergoing weathering and of the burial of organic carbon in newly deposited sediments. These are key processes in the global carbon cycle and thereby represent important controls on long-term trends in atmospheric carbon dioxide and oxygen levels, which mediate global climate (Berner, 1999; Galy et al., 2011). Based on the work of Blair, Leithold, and Aller (2004), Leithold et al. (2015) (p.33), stated that the sediment routing system

> ... can be viewed as a network of biogeochemical reactors that facilitate the cycling of organic carbon and are interconnected by processes of sediment generation, transport and deposition.

Highland reactors comprise bedrock and upland soil, which supply organic carbon to rivers and lowland stores (such as floodplains, soils and wetlands). River transport delivers organic carbon to continental shelf and deep sea reactors. Each reactor is characterised by a residence time (age) of organic carbon. The bedrock reactor is dominated by long-residence time fossil C (kerogen), and residence times decrease in the stages toward the deep sea reactor as younger, more reactive C with terrestrial or marine sources is included in the cycling of organic carbon. Particulate organic carbon is therefore transmitted through the sediment routing system, each reservoir receiving inputs from upstream reactors and from in situ production. A large proportion of the particulate organic carbon in the sediment routing system is derived from mountain and upland regions.

Continental shelf areas and margins adjacent to major river systems are currently receiving very high amounts of riverine organic carbon (Ludwig, Probst, and Kempe, 1996; Leithold, Blair, and Wegmann, 2015). Much of this terrigenous organic matter is buried close to the river mouth (Keil et al., 1997). The total flux of terrestrial organic carbon to the world ocean is of the order of 0.1 to 1×10^{12} kg C yr^{-1}, the main repositories being the deltas of major rivers and adjacent continental shelves, which together store

0.104×10^{12} kg C yr^{-1} (Berner, 1989). The largest supplier of terrestrial organic carbon to the oceans is the Amazon River (31 to 55×10^9 kg TOC yr^{-1}).

The fluxes of solids from the continents to the oceans is critical to the operation of many biogeochemical cycles, including that of carbon. Yet only a minute proportion of the net primary production of carbon is exported by rivers, since ca. 99% is oxidised in soils and returned to the atmosphere as CO_2, and at the global scale a further ca. 70% of organic carbon delivered by rivers is oxidised in the ocean before burial in sediments. Nevertheless, the burial flux of organic carbon in the world's sediment routing systems is high enough to determine, in part, the global budget of CO_2 sequestration from the atmosphere to long-term burial. The most important of such sediment routing systems are those involving a conveyor belt of high rates of erosion and efficient burial in the ocean. The Ganges–Brahmaputra system linking Himalayan erosion to deposition in the Bengal Fan accounts alone for about 15% of the total global burial flux of organic carbon (Galy et al., 2007).

3.2.1 Particulate Organic Carbon

Much carbon is transported in rivers as particulate organic carbon (POC), the bulk of which is sorbed to the surfaces of suspended mineral material. Consequently, within a range of organic loading, the amount of organic matter varies in direct proportion to the mineral surface area (Mayer, 1994). In river water, POC can originate from autochthonous organic matter production of phytoplankton and from mineralisation of dissolved organic carbon, and more importantly from allochthonous sources such as biomass, soils and regolith, or from the atmosphere. The organic matter ranges from small free molecules to large aggregates and organisms, and is dominated by the size range greater than 0.45 μm.

POC is made up of both living (plankton, bacteria) and non-living (detritus) components. Refractory compounds have low solubility and resist decay, whereas labile compounds have high solubility and are used rapidly in aquatic ecosystems. Some 5–35% of the POC carried in rivers is labile and therefore prone to oxidation (Degens and Ittekot, 1985; Ittekot, 1988), so the remaining more refractory POC dominates the flux into the oceans.

Having entered the sediment routing system, the fate of particulate organic carbon depends on the physico-chemical and biological processes influencing mineralisation, disaggregation and sedimentation. These processes change strongly as the river enters coastal estuaries and the open ocean. As a result, the average organic loadings of the Amazon River (0.67 ± 0.14 mg C m^{-2}) are approximately twice that of sediments in the delta region (ca. 0.35 mg C m^{-2}) (Keil et al., 1997) (Figure 3.3). Allowing for the fact that two-thirds of organic material in delta sediments is terrestrial in origin, only less than 30% of the riverine POC supplied by the river is buried in the delta. If this rate is typical of other river-delta systems, the global loss of POC in the delta regions at river mouths amounts to approximately 0.1×10^{15} g C yr^{-1}.

The present-day flux of POC is thought to be 0.17–0.30 Gt C yr^{-1} (Sarmiento and Sundquist, 1992), or 0.07–0.2 Gt yr^{-1} (Meybeck, 1982). Using data collected during the

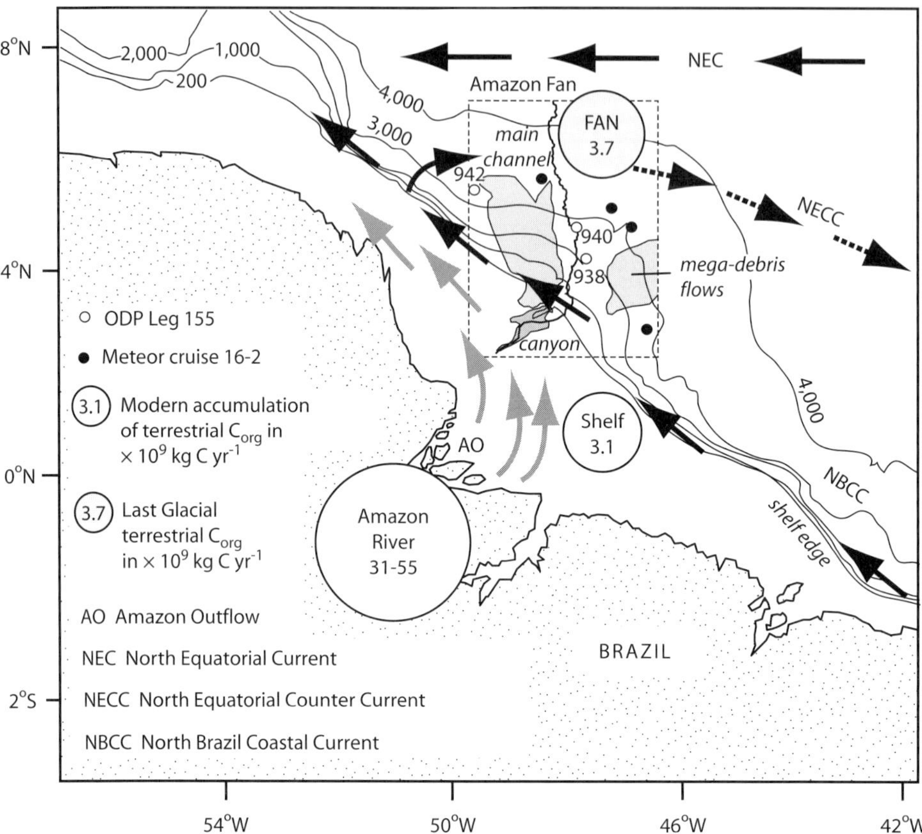

Figure 3.3 Location map of Amazon River mouth, continental shelf and deep sea fan, with shelf currents and location of core positions, main channel and debris flows on the Amazon Fan, adapted from Flood et al. (1995). Fluxes of terrestrial organic carbon of the Amazon source, modern accumulation on the continental shelf and burial flux on the fan during the Last Glacial (15−35 ka) are in units of $\times 10^9$ kg C yr^{-1}, reported in Schlünz et al. (1999).

SCOPE (Scientific Committee on Problems of the Envronment) study, the annual POC transport to the ocean by rivers with total suspended matter concentrations of $1-1,500$ mg l^{-1} is 0.19 Gt yr^{-1} (Ittekot and Laane, 1991). If rivers with total suspended matter concentrations of greater than $1,500$ mg l^{-1} are included, the annual figure becomes 0.23 Gt of particulate carbon. On a global scale, only a small amount of the total flux of carbon to the ocean is buried in deep sea sediment, which is 0.006 Gt yr^{-1} (Berner, 1982).

As a percentage of the total suspended matter, POC is generally in the range $1.3-8.4\%$, corresponding to concentrations in milligrams of carbon per litre of $0.6-14.2$ mg l^{-1}. As the flux of total suspended matter increases, the concentration of POC increases in concert as expected, but the POC as a percentage of the total suspended matter decreases (Meybeck, 1982). Wang et al. (2016) found a power law relationship between particulate organic carbon and suspended sediment concentration for river gauging stations downstream of the main area of landsliding following the Wenchuan (China) earthquake of 2008. For the Zagunao and Sangping gauging stations respectively,

$$POC = 8.946C_s^{0.952} \qquad (3.1)$$

and

$$POC = 8.813C_s^{1.008} \qquad (3.2)$$

where POC is the particulate organic concentration in mg l^{-1} and C_s is the suspended sediment concentration in g l^{-1}.

The rate of burial of terrestrial carbon supplied by rivers is considerably larger than the burial rate of organic carbon from marine biogenic sources (Berner, 1982). Particulate carbon delivered to the continental slope and deep ocean is extracted from the short-term carbon cycle and preserved for long timescales within marine sediment. Approximately two-thirds of particulate solids delivered to the ocean by rivers is transported beyond the shelf-edge of continental margin clinoforms (Petter et al., 2013) (Chapter 5). As a first approximation, it is expected that the deposition rate of POC is directly related to the depositional flux of sediment. Consequently, the deposition rate of POC in the ocean beyond the shelf edge is likely to be $0.11-0.20$ Gt C yr^{-1}. Deep water storage of carbon is likely to be most important where deltas of major rivers are close to the shelf edge, in other words, where the shelf width is small and indented deeply by canyons (Section 6.2).

3.2.2 Dissolved Organic Carbon

Carbon is also transported to the ocean as inorganic carbon dissolved in river water. The dissolved organic carbon (DOC) in river waters at the present day comes from three main sources: soils, in which the organic matter is derived from plant and animal material via microbial activity, fluvial production and anthropogenic inputs. DOC therefore originates from both aquatic and terrestrial sources. Between 40 and 80% of fluvial DOC consists of combined humic substances regarded as refractory and consequently capable of avoiding degradation en route to the ocean. The autochthonous production of carbon includes labile (easily metabolised) biochemical material that is degraded in the river or estuarine system and does not escape into the ocean. The weighted global average river water concentration of DOC is 5.75 mg l^{-1} (Ertel et al. 1986), but there are strong variations between rivers draining different environments.

The relationship between DOC and POC appears to depend on the flux of total suspended matter. Values of the ratio DOC/POC are higher in rivers with low suspended matter concentrations and lower in rivers with high total suspended matter concentrations (Table 3.3). Most rivers carry much more DOC than POC, up to a ratio of about 10. The major rivers that drain the Himalayan mountain range (Indus, Ganges, Brahmaputra, Irrawaddy) together account for about one-third of the global delivery of suspended sediment to the ocean (Table 3.4). The equivalent transport of particulate organic carbon by these rivers

Table 3.3 *Particulate (POC) and dissolved organic matter (DOC) in major world rivers, for ranges of total suspended matter (TSM). Compiled from data in Wetzel (1975); Meybeck (1982); Ittekot and Laane (1991).*

TSM (mg l^{-1})	POC (%)	POC (mg l^{-1})	DOC/POC
5–15	8.4	0.6	10.8
15–50	3.6	1.1	5.8
50–150	2.2	1.7	3.4
150–500	1.3	3.7	2.3
500–1500	1.6	14.2	0.9

Table 3.4 *Hydrological data for rivers draining the Himalayan mountains. Data from Milliman and Meade (1983) and Subramanian and Ittekkot (1991).*

River	Area ($\times 10^6$ km^2)	Run-off (km^3 yr^{-1})	Sediment discharge ($\times 10^6$ t yr^{-1})	POC (%)
Indus	1.17	238	100	2.2
Ganges	0.97	459	573	0.6–2.3
Brahmaputra	0.70	511	597	1.0–6.97
Ganges–Brahmaputra (Bangladesh)	1.48	971	1670	-
Irrawaddy	0.43	428	265	-
Totals	4.67	3578	3205	-

is 0.057 Gt yr^{-1}. Adding the flux of dissolved organic carbon (DOC) of 0.012 Gt yr^{-1}, the total carbon flux annually is 0.069 Gt, which constitutes a large percentage of the global total.

3.2.3 Burial of Organic Carbon and Global Climate

The burial of organic carbon in marine sediments is the second largest sink of atmospheric CO_2, after silicate weathering coupled with carbonate precipitation in the ocean. Yet the delivery of organic carbon to the ocean as a result of erosion of the continental land surface, and its burial in marine sediment, is complex for a number of reasons:

1. The organic carbon carried by rivers is composed of both recent organic matter and fossil refractory carbon derived by erosion of carbon-rich sedimentary rocks. Burial of fossil refractory carbon has no effect on the long-term carbon cycle.

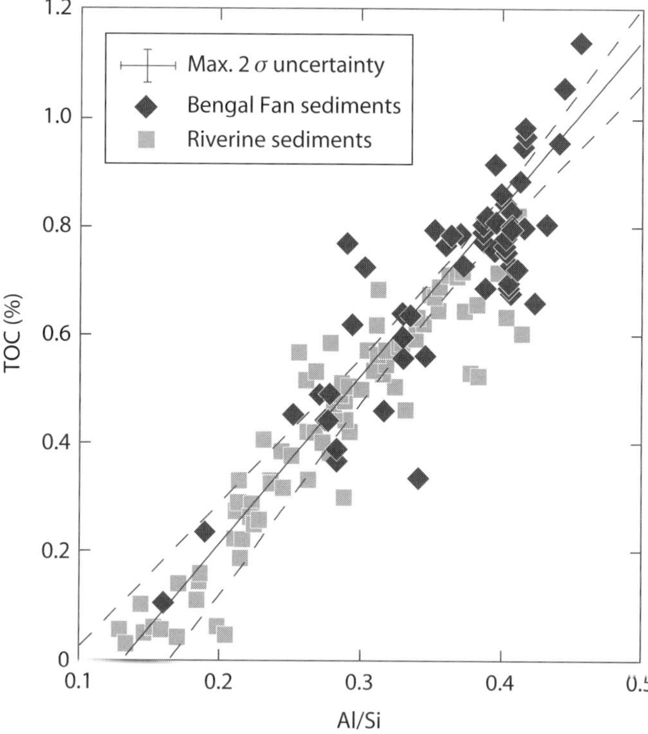

Figure 3.4 TOC of riverine and deep sea fan sediments in the Himalayan-Bengal sediment routing system. Solid line is linear best-fit for Bengal Fan sediments, dashed lines are 95% confidence interval. From Galy et al. (2007) (fig.4, p.409) with permission of Nature Publishing Group.

2. A large proportion (about 70%) of the carbon exported into the ocean from the continents is oxidised on the continental shelf and returned to the atmosphere.
3. Measurement of the organic carbon fluxes of rivers is made difficult by the segregation of particles during transport.

It is evident from Table 3.4 that rivers draining the Himalayan mountains export billions of tons of sediment, some of which is buried in the deep-water Bengal Fan (Curray, Emmel, and Moore, 2002). Over the past 15 Myr approximately 15% of the global burial flux of carbon has been sequestered in the sediments of the Bengal Fan (France-Lanord and Derry, 1997). Understanding the Himalayan-Bengal sediment routing system, and others like it, is therefore critical to assessment of the role of the carbon cycle in global climate.

All sediments in the Ganges–Brahmaputra river system have a positive correlation between TOC and the Al/Si ratio (Figure 3.4), which expresses a covariation of TOC with grain size and mineralogy, since Al is concentrated in fine-grained minerals such as clays. An identical relationship is found in sediments of the Bengal Fan, suggesting that the loads of organic carbon in river and submarine fan sediments are identical. Both fluvial and

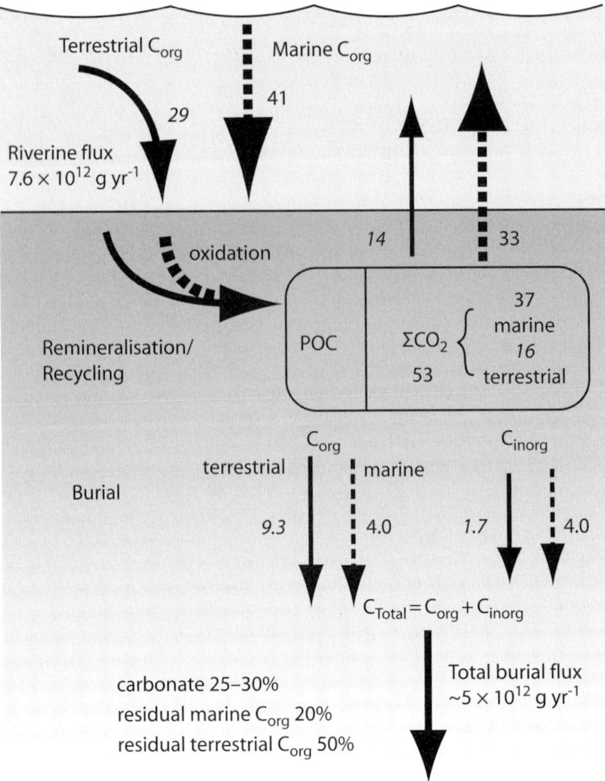

Figure 3.5 Sedimentary carbon cycle for deltaic regions of the Amazon continental shelf. Units of flux are in mmol m^{-2} day^{-1}, corresponding to 7.6 $\times 10^{12}$ g C yr^{-1} for the riverine flux of organic carbon, and a total burial flux of 5 $\times$ 10^{12} g C yr^{-1}. Of the total carbon buried, 70–75% is C$_{org}$ and 20–25% is C$_{inorg}$. The organic carbon is 70% terrestrial in origin and 30% is marine, whereas the inorganic carbon has a reciprocal relationship; it is dominated by marine (70%) compared to terrestrial (30%). Numbers in italics and solid arrows refer to fluxes with entirely terrestrial sources; dashed arrows and roman numbers refer to marine sources. The rectangle shows the sedimentary remineralisation part of the cycle. After Aller et al. (1996) with permission of Pergamon Press.

fan sediments are also dominated by terrestrial carbon. The Bengal Fan therefore seems to efficiently preserve and bury terrestrial carbon provided by the Ganges–Brahmaputra system, as opposed to the much lower burial efficiency of other examples, such as the Amazon (Figure 3.5). Oxidation of organic carbon is limited by rapid transport and burial and low availability of O$_2$ in the stratified waters of the Bay of Bengal, whereas decomposition of organic carbon on the sea bed in water depths of 10–40 m is important on the Amazon shelf. The Amazon shelf and deep sea fan therefore contrast strongly with the Bengal Fan as a carbon sink.

Most tropical shelves, such as that of the Amazon, act as organic carbon *incinerators*, burning off nearly all the organic matter supplied by rivers, estuaries and oceanic waters, with minimal preservation by burial in sediments. Inorganic carbon in carbonate is drawn down in coral reef building, but almost all biogenic $CaCO_3$ in ocean waters is dissolved and recycled below the calcite and aragonite compensation depths. Although operating as incinerators at the present day, tropical shelves have been sites of prolific burial of carbon at times when oxygen levels were low.

While it may be true that deltas and continental margins are the main storage reservoirs for organic matter and most terrestrial weathering products (Berner, 1982), continental margins are also sites of much enhanced biological respiration, microbial decomposition and redox cycling of the elements, especially those of the tropics, where temperatures are high and plentiful nutrient and light levels favour microbial processes. If so, many tropical margins may be carbon neutral or net sources of carbon dioxide to the atmosphere. Nevertheless, sites where major rivers such as the Amazon and Ganges–Brahmaputra debouch into the ocean are carbon sinks. Sustained high carbon fixation on continental shelves may require N and P sources, for instance from oceanic upwelling combined with high sediment burial rates (Section 3.3).

3.2.4 Organic Carbon in the Amazon Sediment Routing System

The best documented sediment routing system in terms of the fate of organic carbon is that of the Amazon. Each year the Amazon River transports about 1.2×10^9 tons of suspended sedimentary material to its lower reaches (Meade et al., 1985). Of this amount, roughly half accumulates in the coastal waters of the subaqueous delta where it builds seaward-prograding clinoforms. The impact of this sediment delivery on biogeochemical cycling is determined in large part by the diagenetic reactions and sediment-water exchange processes taking place at shallow depths in the sediment pile. The early diagenetic processes taking place on the Amazon shelf are typically dominated by Fe and Mn reduction/cycling, in contrast to other tropical deltas where the cycling of S is dominant.

Turbidity limits biological productivity nearshore, while depletion of nutrients limits it offshore, so productivity in restricted to an intermediate zone marked by its green colour, compared with the brown nearshore waters and the blue offshore waters. Sustained primary production is highest in water depths of 40–80 m northwest of the Amazon River mouth. Average grain size decreases seaward and along-shelf. Sediments of the shelf are intensely reworked and homogenised by strong tidal currents and waves. Away from and seaward of this high energy corridor, the extent of bioturbation is the dominant influence on diagenetic processes in the sediment beneath the sea bed.

Despite the very high input of sediment from the Amazon River and a total organic carbon percentage of only 0.6–0.8%, surficial sediments on the inner shelf have ΣCO_2 remineralisation rates that are comparable to other nearshore regions with much higher organic carbon content. Sea bed sediments are diagenetically highly reactive, leading to high ΣCO_2 production and the formation of authigenic minerals such as carbonates. Relative to the primary productivity of organic carbon in the water column above the subaqueous delta, the proportion remineralised in bottom sediments is 20–30%, which is typical to the proportions found in many other estuarine and shelf environments. Terrestrial organic carbon inputs also contribute to the remineralisation of sea bed sediments on the Amazon shelf.

The generation of organic carbon in the water column by phytoplankton production increases seaward away from the turbid Amazon plume to maximum values in water depths of 50–70 m. However, this trend in phytoplankton productivity is not mirrored by remineralisation rates in surficial sediments, which show little variation over water depth ranges of 10–40 m, despite the fact that most of the remineralised organic matter in bottom sediments is of marine origin. This is most likely a result of cross-shelf transport and homogenisation of sea bed particles. There is ample evidence of cross-shelf tidal currents and estuarine circulation (Kuehl, DeMaster, and Nittrouer, 1986) to achieve this homogenisation.

Part of the particulate and remineralised organic carbon is recycled into the water column, or lost from the shelf, but the remainder is buried within the sediments of the Amazon delta. There is a net diagenetic conversion of remineralised organic carbon to precipitated inorganic carbon within and below the surface mixed layer of sediment, leading to a relatively constant rate of burial of total organic carbon of 0.70% (Figure 3.5). Much of the inorganic carbon in Amazon shelf sediments is $FeCO_3$ (siderite). The anoxic diagenetic reaction affecting remineralised carbon storage by the formation of siderite is:

$$4Fe(OH)_3 + C_{org} + 3H_2CO_3 \rightarrow 4FeCO_3 + 9H_2O$$

Assuming an annual sediment accumulation on the Amazon shelf of approximately $0.63 \pm 0.2 \times 10^9$ tons (Kuehl et al., 1986), and a percentage total carbon content of sediment derived from piston cores of 0.79%, then 5×10^9 kg yr^{-1} of total carbon must be buried, of which 25–30% is derived from remineralised inorganic carbon having a predominantly marine origin (Aller et al., 1996) (Figure 3.5). On the basis of the carbon isotopic values of surface sediment over most of the subaqueous delta, 70–80% of the organic carbon has a terrestrial origin. On average, less than 10% of the ΣCO_2 remineralised from marine organic carbon is buried, so more than 90% must be eventually recycled into the water column.

The Amazon River brings 14×10^9 kg yr^{-1} of particulate organic carbon (POC) and 22×10^9 kg yr^{-1} of dissolved organic carbon (DOC) to its lower reaches. The bulk of this POC and DOC is refractory under riverine conditions (Hedges et al., 1986), that is, it is essentially inert. Assuming that most DOC transits the shelf or is remineralised and that POC is not strongly fractionated from other sedimentary particles, then approximately half of the particulate flux of riverine POC should be deposited in Brazilian shelf waters, with a proportion of C_{org} of 1.2% POC inherited from the terrestrial carbon in river water. However, the proportion of terrestrial POC in shelf muds is on average 0.4%, indicating a substantial loss of 60–70%. A large percentage of the refractory terrestrial organic carbon is most likely remineralised or lost on the shelf due to oxidation (degradation).

3.2.5 Organic Carbon through Glacial-Interglacial Cycles

The fluxes between reservoirs in the carbon cycle vary significantly in response to global environmental change, such as, for example, the glacial-interglacial cycles of the Quaternary. Relative to the Last Glacial Maximum (LGM) approximately 18 kyr ago, the present interglacial phase in pre-industrial times is characterised by higher fluxes of carbon dioxide from the surface ocean layer (euphotic zone) to the atmosphere, its uptake by land biomass and humus, CO_2 consumption during weathering of rocks, and transport of inorganic and organic carbon in rivers draining the continental land mass.

It has previously been noted that only a small fraction of the organic carbon contributed to the ocean by rivers makes its way to the deep sea fans at the termini of sediment routing systems. This fraction may vary over time, particularly as a result of climate and sea level forcing associated with glacial-interglacial cycles (Schlünz et al., 1999) (Figure 3.6). Sediment supply to the deep sea Amazon Fan, together with its fraction of terrestrial organic carbon, changed dramatically at times of sea level change (Figure 3.7). At times of interglacial highstand, such as the present day, the organic carbon buried on the fan surface is dominated by marine sources. During glacial sea level lowstands, when the continental shelf was incised and Amazon River sediment was transported directly via a canyon to the deep sea fan, the organic carbon accumulating on the fan was almost entirely terrigenous, representing 7–12% of the modern terrestrial organic carbon input from the Amazon River.

The Amazon Fan extends 700 km seawards to water depths of 4,800 m in the Demerara Abyssal Plain. The last glacial-interglacial cycle can be recognised in fan sediment by a change from light brown pteropod-foraminifera marl in the top 50–120 cm, underlain by a black, organic-rich mud deposited during the last glacial phase. Carbon isotope ratios of organic matter show a change from a terrigenous supply during the last glacial to marine phytoplankton production during the Holocene (Showers and Bevis, 1988). Carbon isotope ratios δC_{org} from piston cores recovering glacial phase sediments are very similar to Amazon river and river mouth values of POC whereas Holocene values are close to ratios found in continental shelf waters. Schlünz et al. (1999) estimate that during glacial conditions, 60–90% of the organic carbon in Amazon Fan sediments is terrestrial in origin, whereas during the Holocene, up to 100% of the organic carbon is marine. This sedimentological and carbon isotopic pattern is also found in older glacial-interglacial cycles sampled in cores.

The alternations seen in cores indicates rapid switchovers between glacial and interglacial phases. The rapid and massive input of terrigeneous organic carbon heralding the glacial phase may be due to erosion of sediment deposited in continental shelf depths before the Amazon River directly contributed riverine C_{org}. Any autochthonous marine production would have been swamped by the terrigenous input during glacials. During sea level rise in the following interglacial, organic matter deposited on the fan was quickly replaced by autochthonous marine material (Figure 3.6). During interglacial sea level highstands such as the present-day, terrigenous organic matter supplied by the Amazon River is transported northwestwards as the river plume is diverted by the North Brazil Coastal Current.

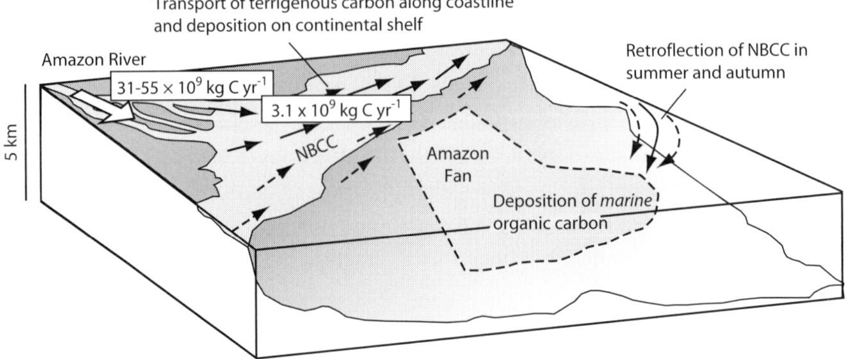

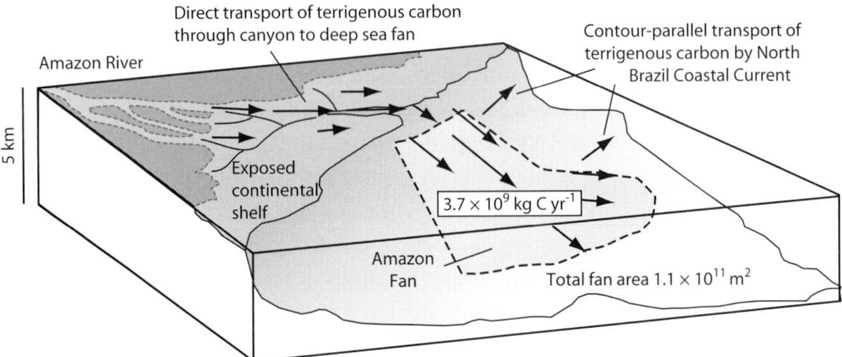

Figure 3.6 Model of the history of the Amazon Fan through glacial-interglacial climate cycles following Milliman, Summerhayes, and Barretto (1975). (a) Interglacial (sea level highstand) accumulation of organic carbon offshore the Amazon River. The terrestrial organic carbon discharged by the Amazon River is deflected to the northwest by the North Brazil Coastal Current (NBCC). The Amazon Fan only receives terrestrial organic carbon by the reflection back of the NBCC during the summer and autumn. Deposition of organic carbon on the fan is from marine sources. The present-day supply of terrestrial organic carbon by the Amazon River and the estimated accumulation of terrestrial organic carbon on the shelf are from Richey et al. (1980) and Showers and Angle (1986). (b) During glacial lowstand periods, terrestrial organic carbon is discharged directly across the exposed shelf and onto the Amazon Fan. The sea level drop during glacial phases was 30–50 m. Rate of accumulation of terrestrial organic carbon on the fan is from Schlünz et al. (1999). The exposed shelf and canyon are drawn speculatively. Modified from Schlünz et al. (1999), published with permission of Elsevier.

Marine organic carbon accumulation rates were highest during the interglacial isotope stages 1 and 5 (0.1 and 0.17 g C m^{-2} yr^{-1}) respectively (Figure 3.7), with minimal contribution by terrestrial sources to the TOC. During glacial phases, the accumulation rate of terrigenous organic carbon varies according to position on the fan, being highest (up to 150 g C m^{-2} yr^{-1}) close to the main submarine channels and levee systems.

(a) Suspended particulate sediment

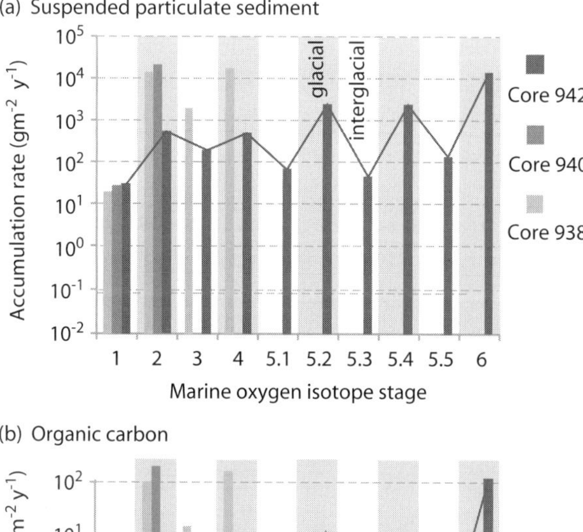

(b) Organic carbon

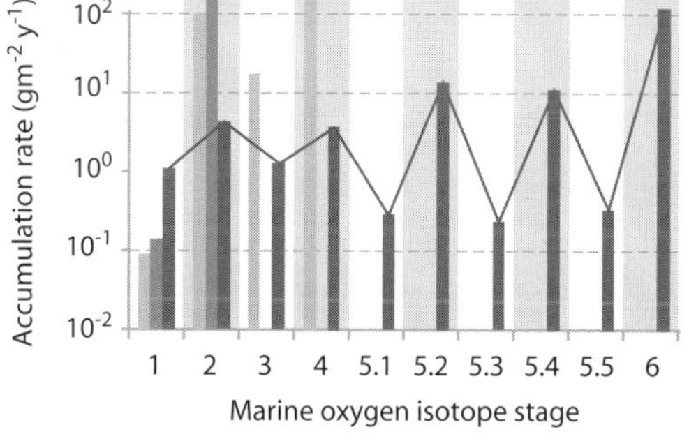

(c) Terrigenous organic carbon

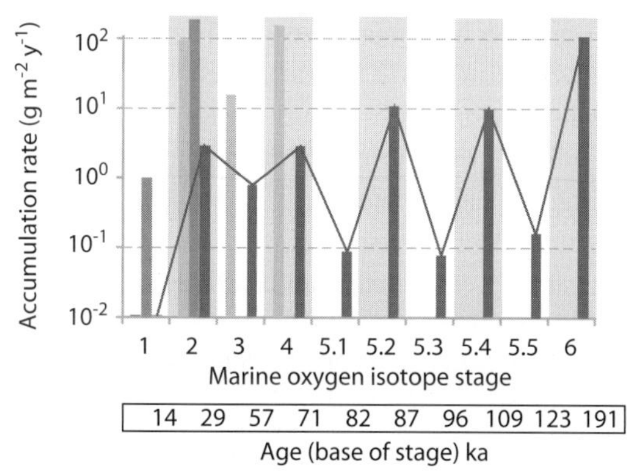

Figure 3.7 Accumulation rates of (a) suspended particulate load, (b) organic carbon and (c) terrestrial organic carbon at ODP sites 938, 940 and 942. Data from Schlünz et al. (1999). Zig-zag (up-down) trend through glacial-interglacial alternations is shown for Core 942.

The discharge of organic carbon by the River Amazon ranges between 31 and 54.8 $\times$ 10^{12} g C_{org} yr^{-1}, which if constant over time would amount to 0.62 to 1.1×10^{18} g of terrestrial organic carbon over 20 kyr (15–35 ka). This implies that only 7–12% of the terrestrial organic carbon delivered to the river mouth accumulated on the Amazon Fan. However, the estimated accumulation rate of terrestrial organic carbon on the fan over 20 kyr (equivalent to 3.7×10^{12} g yr^{-1} of terrestrial organic carbon) is of the same order as the magnitude of the accumulation rate of terrestrial organic carbon on the Amazon continental shelf today (3.1×10^{12} g C yr^{-1}), which is 6–10% of the riverine discharge. Although this represents a rather inefficient burial rate of terrestrial carbon, the time-integrated burial over 20 kyr from this single, albeit large, river system is a significant fraction (ca. 10%) of the entire atmospheric store of carbon. Viewed at a global scale, the loss of organic matter derived from the continents by burial in deltas, continental shelves and deep sea fans worldwide in just 6,000 years represents the entire store of atmospheric carbon.

3.2.6 Particulate Organic Carbon at Active and Passive Margins

The organic carbon composition varies according to the fluxes involved in the various reservoirs or reactors in the sediment routing system. In mountainous regions, sediment is delivered by hillslopes to channel networks, and the same processes transport particulate organic carbon from sources as above-ground biomass as well as soils and sedimentary rocks. Near-surface litter and above-ground biomass tends to be more labile (reactive) than the older organic matter in deeper soil horizons and in sedimentary rocks. The relative importance of these sources and processes therefore controls the amount and character of organic matter transported through headwater streams. The fraction of particulate organic carbon in total riverine suspended solids (TSS) decreases with increasing TSS concentrations (Figure 3.8), signifying that major erosional events excavate organic carbon-poor, deeper soil horizons and sedimentary bedrock. Consequently, there is a clear relationship between the radiocarbon age of particulate organic matter and the catchment-averaged sediment yield for a range of the world's rivers (Figure 3.9).

Highly erosive events in steep, wet mountainous catchments are responsible for a disproportionately large fraction of the total sediment and particulate organic carbon fluxes, as in the Li-Wu catchment of Taiwan (Hilton et al., 2008). Where bedrocks are competent, slopes are moderate and vegetation is thick, high POC fluxes are principally derived from recent biomass, but in steep catchments with less competent bedrocks, such as in the Southern Alps of New Zealand, debris flows and landslides deliver organic carbon from deeper soil horizons and from bedrock. In such cases, fluxes of biospheric organic carbon are much smaller than those of petrogenic (fossil) organic carbon. Some petrogenic carbon is oxidised at source during weathering, releasing CO_2, the remainder being transported to lower reaches of the sediment routing system.

Sediment and POC leaving mountainous headwater regions is delivered to low-gradient river systems with variable sizes of floodplains that act as traps. Small, steep rivers at

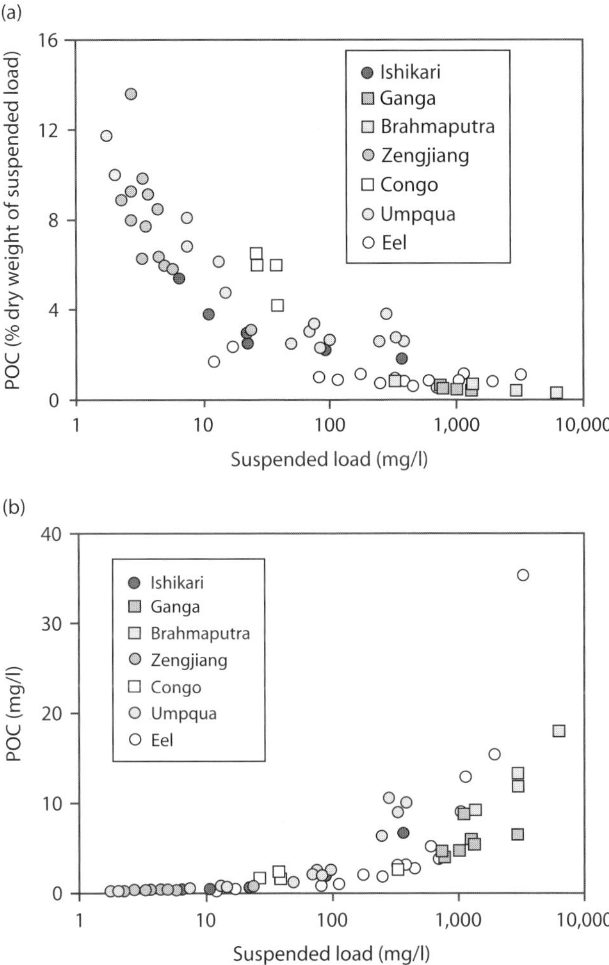

Figure 3.8 Suspended sediment concentration versus particulate organic carbon in percentage dry weight of suspended load (a) and in mg/l (b). (a) shows the decreasing weight percent POC with increasing suspended sediment loads, signifying the dilution of organic matter derived from near-surface and above-surface positions compared to mineral matter eroded from deeper soil horizons and sedimentary rocks. (b) shows the increase in POC export with increasingly high suspended sediment concentrations. From Leithold, Blair, and Wegmann (2015) (fig.4, p.35) with permission of Elsevier.

active plate margins have small floodplains providing minimal possibilities for the storage of sediment and POC. The transit time of particulates through the river system is rapid, so organic carbon is negligibly degraded. Large, low-gradient rivers flowing from mountainous source regions across extensive cratonic and foreland basins, however, have widespread, flat, floodplains, where sediments and POC undergo cycles of erosion and deposition over long periods of storage. Sediment and POC is deposited on floodplains and

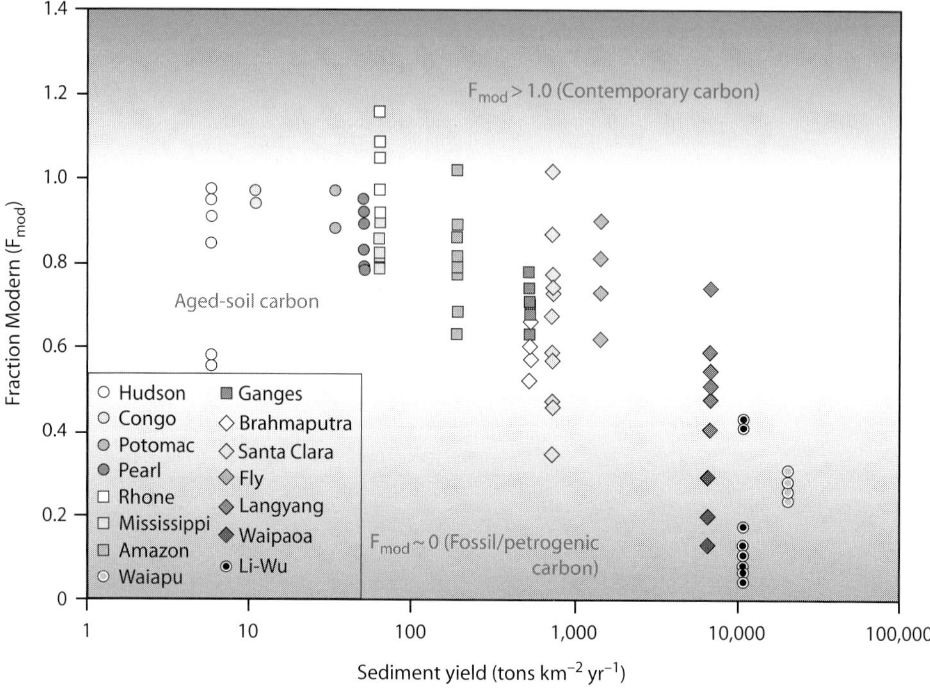

Figure 3.9 The radiocarbon age of suspended POC expressed as the Fraction Modern (F_{mod}) versus the catchment-averaged sediment yield for a range of rivers in tropical and temperate environments. Contemporary (biospheric) carbon has $F_{mod} > 1.0$, whereas ancient, fossil or petrogenic carbon has F_{mod} of close to 0, and aged-soil has intermediate values. Consequently, the trend indicates that with progressively high rates of erosion, organic carbon is mobilised from deeper, older soil horizons and bedrock. From Leithold, Blair, and Wegmann (2015) (fig.5, p.35) with permission of Elsevier.

in floodplain lakes and abandoned channels during floods and subsequently decomposed and aged. Although new biospheric organic carbon is added to the floodplain reservoir during subsequent floods, repeated cycles of flooding, deposition and erosion cause the material entering the ocean to be significantly aged, mean radiocarbon ages of $10^3 - 10^4$ yr being common at the mouths of large catchments such as the Amazon, Ganges–Brahmaputra, Yenesei and Lena.

The organic carbon of the continental shelf is partly degraded and oxidised. Shelves with low oxygen levels have high efficiencies of carbon burial, whereas in normal shelf waters, there is a positive relationship between burial efficiency and sediment accumulation rate (Figure 3.10). Sediment accumulation on the continental shelf depends on the supply rate, mostly from nearby river mouths, shelf accommodation driven by thermal subsidence, and sediment redistribution by waves, tides, plume outflows and gravity currents (Walsh and Nittrouer, 2007). Sediment redistribution is particularly important on the narrow shelves off small mountainous rivers, where sediment is advected to depocentres on the mid and

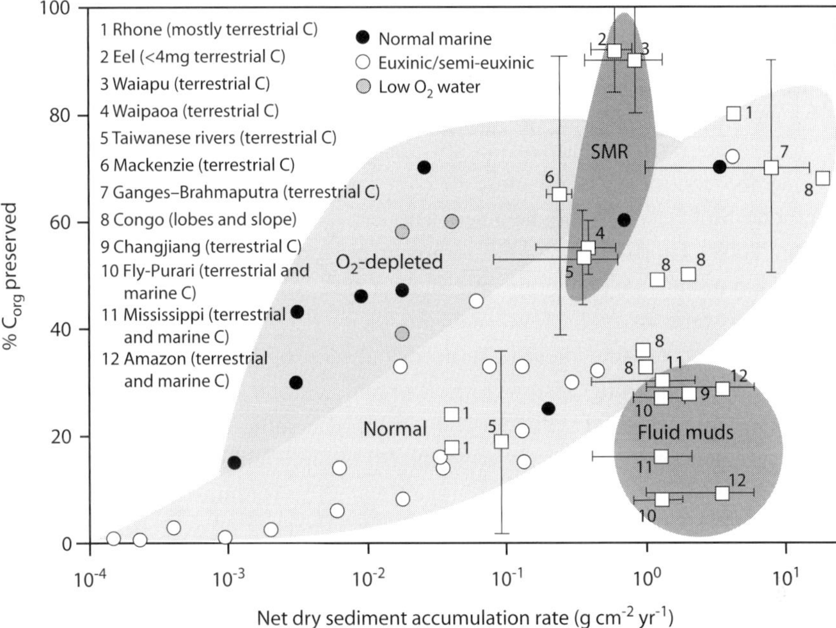

Figure 3.10 The percentage of organic carbon deposited and buried in sediments on the continental shelf as a function of sediment accumulation rate. The continental shelves adjacent to small mountainous rivers are denoted by SMR. Data sources and compilation in Leithold, Blair, and Wegmann (2015) (fig.8, p.37), published with permission of Elsevier.

outer shelf, continental slope and deep sea. Particulate organic carbon is therefore rapidly and efficiently buried on active margins. The Waiapu shelf, off North Island, New Zealand, for example, experiences burial efficiencies of about 90% of riverine carbon in a mid-shelf depocentre (Blair and Aller, 2012).

In contrast, continental shelves are relatively wide along passive margins. Commonly, large rivers deliver enormous quantities of sediment to the ocean, constructing thick sediment prisms built internally of clinoforms. The burial efficiency of organic carbon is variable in such systems. In the clinoform topset region, repeated resuspension of fine grained sediments produces dense suspensions known as fluid muds, in which much of the terrestrial organic carbon and locally produced marine organic carbon is lost by remineralisation. In other locations where sediment accumulation rates are high, degradational losses of terrestrial organic carbon may be minimal.

Sediment and organic carbon is also fluxed from continental shelves into the deeper sea by nepheloid layers, cascades of dense water, and gravity-driven slides, slumps, debris flows and turbidity currents. Burial efficiencies of organic carbon are low where sediment influxes are highly infrequent, sedimentation rates are low and exposure to oxygenated water and pore fluid prolonged, as in the Madeira Abyssal Plain of the deep (5,400 m) eastern Atlantic Ocean. Where turbidite activity is more frequent and accumulation rates are higher, as in proximal fan settings fed by river flood discharges, organic carbon burial is more efficient.

Active and passive margins differ in their dynamics of delivery of sediment and organic carbon to the deep sea. On active margins, submarine canyon heads commonly incise deeply into the continental shelf close to river entry points (Section 6.2), so terrestrial OC is rapidly transferred to the deep sea. The Eel River and Canyon is a good example (Puig et al., 2004). On passive margins, sediment and terrestrial organic carbon is typically stored for long periods on the continental shelf at times of sea level highstand such as the present day. In some cases, however, canyons incise deeply into the continental shelves of passive margins, eroding previously deposited, consolidated sediments of the upper continental slope containing aged terrestrial organic carbon (Tesi et al., 2010). In other cases, canyons funnel fluvial sediment and terrestrial organic carbon to deep sea fans by bypassing the shelf segment entirely, as in the Mississippi and Ganges–Brahmaputra systems. Consequently, the organic carbon in the Bengal Fan is almost identical to that of the lower reaches of the river system (Galy et al., 2007). The resulting exceptionally high burial efficiency of organic carbon on the Bengal Fan, which is close to 100%, is caused by rapid accumulation rates combined with low oxygen concentrations in the waters of the Bay of Bengal.

3.3 Nutrient Fluxes

By far the greatest species diversity and most vigorous biogeochemical cycling takes place in the relatively shallow waters of the world's ocean margins, particularly those of the tropics, in part at least due to the input of high amounts of dissolved and particulate loads from rivers. The river loads include nutrients that limit biological productivity. In addition, oceanic water is advected onto the continental shelf at levels of tens to hundreds times the river input, where it is scavenged of elements by riverine particles and resuspended shelf sediments. Major bio-limiting nutrients carried in river waters are carbon, nitrogen and phosphorus. The C-N-P system is easily perturbed by human activity. Anthropogenic influences have caused an increase in the delivery of C, N and P to coastal waters, particularly since the mid-1900s.

Dissolved silica can be taken as an illustration of the role of the sediment routing system in the operation of global biogeochemical cycles. A brief introduction is provided here, and more comprehensive and detailed analysis can be found in Beusen et al. (2009), Dürr et al. (2009) and Bernard et al. (2010). Although the study of the chemistry of river waters draining different bedrock lithologies shows wide variations of the major anions and cations (Section 3.1), levels of dissolved silica appear to be independent of bedrock lithology, with the exception of rivers draining volcanic rocks and soils, as in Iceland (Meybeck and Ragu, 1996). Dissolved silica has therefore been used to estimate chemical denudation rates since its variation is independent of catchment area geology. The fact that dissolved silica

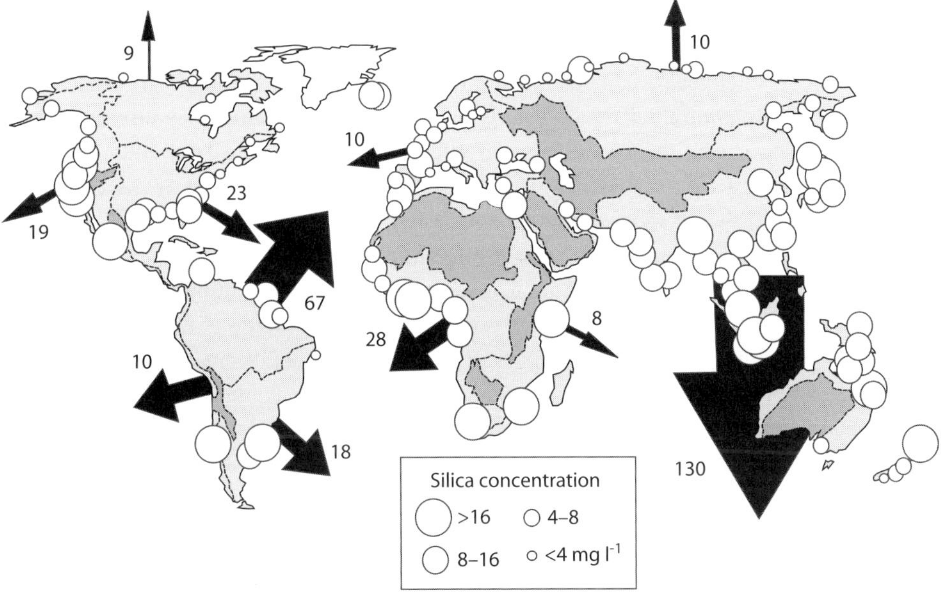

Figure 3.11 Discharge by rivers of dissolved silica to the global coastal ocean, using data in Meybeck and Ragu (1996). The total annual discharge of dissolved silica is 330×10^6 t yr^{-1}. River entry points are hotspots in dissolved silica concentration in the waters of the world's continental shelves. From Milliman and Farnsworth (2011) (fig.2.49, p.63) with permission of Cambridge University Press.

discharges and yields are very high in rivers in tropical, wet, tectonically active settings supports this view. The high-rainfall tropical rivers of southern Asia and Oceania account for 60% of the chemical weathering in watersheds draining into the global ocean. As a result there are strong regional variations in the dissolved silica fluxes to the coastal ocean (Figure 3.11).

Since silica is a nutrient, these regional variations contribute to the patterns of biological primary production in continental shelf seas. Dissolved silica is particularly important for the growth of diatoms, which form a large part of the phytoplankton biomass. The delivery of dissolved silica to the coastal ocean is therefore important in the carbon cycle and forms a critical part of the marine food web. Delivery of riverine dissolved silica and uptake of dissolved silica by diatoms is compensated for by the burial of biogenic silica (opal) in seabed sediments. Shallow continental shelves shorten the sink time of biogenic opal, preventing dissolution before reaching the sea bed, and promoting its storage on the continental shelf for long time periods. In deep continental shelf waters, and in the open ocean, the sink time is long and dissolution is more significant.

Hotspots in nutrient concentration are related to major river inputs, such as the Amazon plume, the Arctic, where rivers have high ratios of dissolved to particulate solids, and areas where human activity (crop fertilisation, urban sewage) has caused large increases in the release of nutrients, such as in southeast Asia (Dürr et al., 2009). Continental shelves are the sites of much-enhanced primary production, making up 25% of global ocean production, yet occupying just 8% of the ocean surface (Ver, Mackenzie, and Lerman, 1999).

Global biogeochemical ocean general circulation models are available that couple riverine nutrient supply to productivity, ocean dynamics and sedimentation (e.g. Bernard et al. (2011)). Models can be used to assess the effects of riverine dissolved silica input on the concentration of silicic acid in surface waters (μmol l^{-1}). A map of the computed difference in [dSi] due to riverine inputs reveals four major hotspots – the Arctic Ocean, eastern Gulf of Guinea (west Africa), Amazon plume and southeast Asia.

Input of riverine dissolved silica increases concentrations in continental shelf waters, whereas use by phytoplankton decreases concentrations, and silica is exported from the euphotic zone as biogenic opal. Hotspots of riverine silica input are generally in areas of high organic productivity, for two reasons: (1) productivity is high because of nutrient availability directly from the river input, as in the case of the Amazon plume, and (2) productivity is high due to the combining of riverine input and local upwelling of nutrients, as in the southwest coast of Africa or the southwest coast of United States–Mexico.

The silica cycle is likely to be significantly perturbed by climate change. Precipitation and temperature control the rate of chemical weathering of the continents and thereby the dissolved silica flux of rivers. Climate also controls wind circulation patterns and thereby upwelling of nutrient-rich waters along continental margins. The combination of river input and upwelling therefore is sensitive to global climate change.

Part II

The Segments of Sediment Routing Systems

4

The Catchment-Fluvial Segment

Sediment routing systems consist of one or more morphodynamic zones within which a coherent set of sediment transport processes operate. These morphodynamic zones, or segments, are genetically related and dynamically connected, since sediment is transferred across their boundaries in a cascade from source to sink. Sømme et al. (2009) proposed a framework where source-to-sink systems may comprise up to three segments:

1. *Catchment-fluvial segment*, which includes the source region of sediment and the river system connecting the catchment to the ocean, lake or endorheic basin. The morphology and dynamics of this segment are characterised by parameters describing the catchment area, the maximum elevation of the headwaters, the length of the main catchment trunk stream, the water discharge and the sediment load, all of which are impacted by tectonic history of uplift and subsidence, climate, vegetation and bedrock lithologies. Some areas acting as sediment sources may be glaciated. The sediment released from ice sheets and valley glaciers varies according to thermal regime. Sediment is dispersed at the base, within and above the ice mass, as well as by icebergs, and is commonly reworked on the continental shelf and transported to the deep sea by gravity flow processes.

2. *Continental shelf segment*, which is separated from the fluvial sediment conveyor by the complex dynamics of the coast and is highly susceptible to changes in sea level. Processes on the shelf determine sediment storage potential and the export of particulate sediment to the deep sea. Shelves vary in their width, slope and water depth at the shelf-slope break, the extent to which canyons incise towards the coast, and the complex dynamics of river plumes, waves, tides and geostrophic flows in driving sediment transport.

3. *Deep marine segment*, comprising the continental slope and deep basin floor. The continental slope is the fully subaqueous, overall progradational slope constructed from the spilling over the shelf of primarily suspended sediment. The continental slope varies in bathymetric gradient, slope length and amplitude, dependent on the distribution of tectonic subsidence and sediment supply. The basin floor is the absorbing state in deep water, receiving sediment primarily from gravity-driven debris flows and turbidity currents travelling down canyons incised into the continental slope and from deep ocean currents in the form of contourite drifts. The amount of sediment bypassing the shelf en route to the deep sea varies strongly from close to nil to more than 90%.

The catchment-fluvial segment contains morphological elements of hillslopes and bedrock channels, basin-margin fans at catchment outlets, and alluvial river systems with their floodplains and lakes. The catchment-fluvial segment acts both like a fire hose delivering sediment to the ocean and also like a sponge storing sediment in alluvial floodplains and lakes.

The majority of the land surface of the Earth is composed of areas that are experiencing net erosion over the long term. As a result, these areas have no potential for the preservation of stratigraphy. Areas experiencing net subsidence over long timescales, that is, *sedimentary basins*, comprise only 16% of the Earth's terrestrial land surface (Nyberg and Howell, 2015). Furthermore, 60% of the area of modern sedimentary basins has an arid climate, compared to 27% of the Earth's land surface, implying that sediments deposited in dry climates are over-represented in the stratigraphic record relative to their occurrence on the present-day Earth's surface. Whereas river systems in areas experiencing net erosion commonly show a tributive channel pattern, those in terrestrial sedimentary basins without marine influence are commonly distributive. Distributive fluvial systems commonly show a downstream decrease in channel size, are unconfined in their alluvial tract and over time build a radial depositional pattern from a distinct apex (Weissmann et al., 2010) (Figure 4.1). The apex of distributive systems is generally at the exit point of a river from a confined, highland valley.

Foreland basins, which contain sediments derived from a bordering mountain belt deposited in a flexural depression (DeCelles and Giles, 1996; Allen and Allen, 2013), contain very large distributive systems. For instance, the Pilcomayo fluvial system of the Andean retro-foreland basin in Bolivia and Argentina is more than 700 km long. Distributive fluvial systems greater than 10^3 km^2 in area are termed *megafans* (Geddes, 1960; DeCelles and Cavazza, 1999). Smaller alluvial fans commonly fill in the area between megafans along the basin edge (Figure 4.1). Distributive fluvial systems also dominate wedge-top (piggyback) basins, but they are smaller than in foredeeps. In wedge-top settings, alluvial fans, supplied with sediment by bordering thrust-related anticlines, dominate the sedimentary basins, and commonly pass downstream into axial river systems and playas.

Extensional fault block basins (Gawthorpe and Leeder, 2000), such as those of the Basin and Range province in the United States, are occupied by distributive fluvial systems in the form of debris-flow and streamflood-dominated basin-margin fans, with playas, perennial lakes and braided axial rivers in the deepest part of the hangingwall basin. The position of the axial river or playa depends on the sediment influxes and tectonic accommodation on the two opposing sides of the extensional basin (Kim et al., 2011; Connell et al., 2012; Allen and Allen, 2013). A similar arrangement of fluvial elements is found in strike-slip basins, their horizontal scale being controlled by the amount of offset of the main fault strands.

One of the most important arrangements of fluvial elements therefore is the deposition of transverse, distributive fans along basin edges together with an axial trunk stream (Figure 4.1). The flow paths of river systems are controlled by the regional topographic

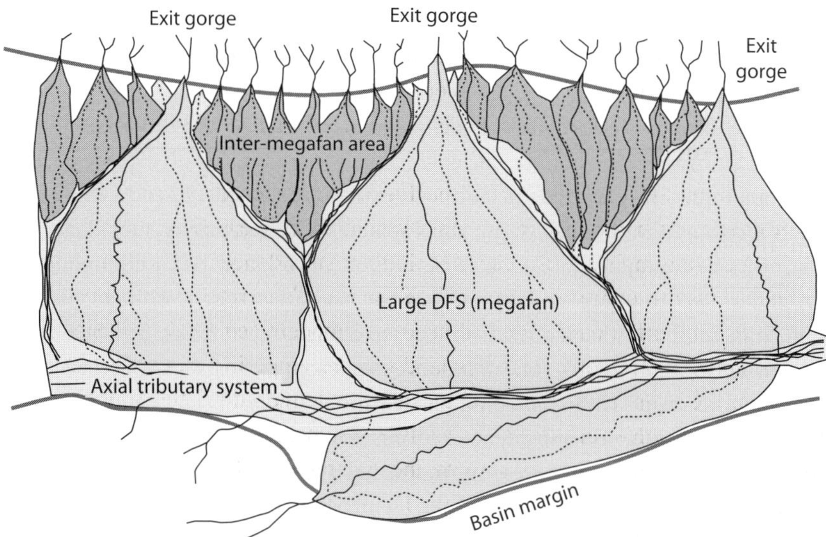

Figure 4.1 Schematic diagram of the arrangement of fluvial elements in a sedimentary basin between two uplifting terrains, from Weissmann et al. (2015), modified from DeCelles and Cavazza (1999), with permission of Elsevier.

slopes set up by tectonic uplift and subsidence and by the sediment supply from the basin margins. The arrangement of fluvial elements is commonly asymmetrical. In extensional half-graben, sediment is commonly supplied transversely from rapidly uplifting footwalls and slowly rotating hangingwalls. In foreland basins, the transverse supply is predominantly from the orogenic wedge (Sinclair et al., 1991; Sinclair, 2012) and to a lesser extent from the flexural forebulge (Crampton and Allen, 1995). In the wedge-top region, sediment may be shed backwards from growing contractional folds and thrust culminations.

Uplifting footwall blocks are typically fringed by coalesced alluvial fans (Whipple and Trayler, 1996; Allen and Hovius, 1998; Gawthorpe and Leeder, 2000), as is classically seen in the arid basins of the southwest Basin and Range province. Progradation or retrogradation of the coalesced fans is determined by the mass balance of sediment supply and accommodation. It is also influenced in some cases by erosion at the toes of the fans by incising axial rivers, a process known as toe-cutting, which boosts the sediment discharge of the axial system (Leeder and Mack, 2001). Toe-cutting represents a different distal boundary condition in numerical models of fan development (Pépin et al., 2010) to those involving a closed fan budget (Allen and Densmore, 2000; Densmore et al., 2007a).

Sediment dispersal patterns away from contractional orogens are strongly affected by the balance between the sediment supply q_s and the distribution of accommodation $\sigma(x)$ (Allen et al., 2013). Where this ratio is low, steep depositional systems such as fan deltas build into underfilled lacustrine or marine basins. At intermediate values of $q_s/\sigma(x)$, larger depositional systems fill structurally confined basins, but sediment dispersal is guided by tectonic structures into an axial flow. At higher values of the ratio, large transverse megafans bury defunct structures and follow the regional palaeoslope from the mountain belt to the flexural foredeep.

Sedimentary basins may evolve in their tectonic subsidence and sediment discharge in a predictable fashion. For example, an extensional basin may evolve from an initiation stage with a number of independent faults with small offset, to a stage of interaction and linkage, to a through-going fault stage, accompanied with increasing subsidence rate and increasing surface topography (Cowie et al., 2006). A foreland basin commonly evolves from an underfilled foredeep and wedge-top basins, to a wider overfilled basin associated with high sediment discharges from the orogen (Covey, 1986; Sinclair and Allen, 1992; Allen and Allen, 2013). The temporal evolution of basin-margin fans and axial systems, therefore, responds to changing patterns of sediment discharge and accommodation. An increase in tectonic subsidence rate in an extensional basin causes a shrinkage of footwall-generated fans, a widening of the axial fluvial belt and underfilling. Later overfilling, as sediment discharge 'catches up' with accommodation generation, causes fan progradation and export of large sediment discharges down-system (Allen and Allen, 2013) (appendix 49, pp.543–544).

4.1 Hillslopes and Bedrock Channels

Looking down on the upland landscapes of the Earth's surface from an airplane, particularly in dry regions where vegetation is nearly absent, one is struck by its repeated patterns and its crenulations. These patterns, produced by the erosion of water, comprise channel networks that are typically dendritic. Drainage networks comprise the 'erosional engines' (Whittaker, Attal, and Allen, 2010) or 'clastic factories' (Leeder, 1999) of the landscape, acting as source regions where sediment is formed and conveying it to sedimentary basins as part of a sediment cascade. Dendritic channel networks involve a hierarchical order of stream segments (Horton, 1945; Strahler, 1952). The drainage density D_d of a channel network is the total length of channels L per unit area of drainage basin A and can be defined as

$$D_d = \frac{\sum_{i=1}^{n} L_i}{A} \tag{4.1}$$

where i is the number of channels. High drainage densities imply short hillslope lengths, while the reverse is true of low drainage densities. Since hillslopes transmit water slowly compared to channellised flows, short hillslopes cause flood events to be quickly registered downstream, as in the badlands landscapes of dry regions experiencing flashy rainfall events.

4.1.1 Outlet Spacing of Transverse Drainage Basins

The upstream limit of a drainage basin is marked by a channel head region. The location of channel heads, which may be initiated by landsliding (Montgomery and Dietrich, 1988, 1992) or seepage erosion ('sapping') (Abrams et al., 2009) sets the drainage density of a landscape. The spacing of channel heads influences the lateral length scale of the drainage basin and thereby, in some instances, the lateral spacing of sediment entry points into sedimentary basins. River entry points are generally the apices of depositional systems.

The spacing of outlets (S) of transverse catchments along a range front has been observed to be approximately half of the distance between the main drainage divide and the range front (W) (Wallace, 1978; Adams, 1985; Hovius, 1996; Talling et al., 1997; Walcott and Summerfield, 2009; Sømme et al., 2010), irrespective of tectonic style and catchment size (Figure 4.2). Hovius (1996) found the average ratio of W/S to be between 1.91 and 2.23 for drainage basins with rivers reaching 90% of the distance to the main divide, while Talling et al. (1997) found a wider range of values with a higher mean of about 2.5 by using rivers that extended at least 70% of the distance to the main drainage divide. The mean spacing ratio for range-scale catchments in southeast Africa is 1.94 (Walcott and Summerfield, 2009). A similar spacing ratio has been found in other drainage basin types that are internal, commonly structurally pinned by active fault strands, or infilling 'interstitial' space between the main range-scale drainages (Walcott and Summerfield, 2009; Sømme et al., 2010). Sømme et al. (2010) tested the potential impact of geographic location, climate and other factors by plotting the spacing ratio of 594 drainage basins in 6 modern rift zones, obtaining a mean value of 2.48 (Figure 4.2). There is no significant difference in the spacing ratio between the different rift provinces (Figure 4.3). Likewise, a breakdown of the rift basin catchments into range-scale, internal/structural and interstitial, as recommended by Walcott and Summerfield (2009), revealed no significant variation in the spacing ratio (Figure 4.4). The regression of outlet spacing S against catchment length L_c from the compilation of Sømme et al. (2010) gives

$$S = 0.638L_c^{0.915} \tag{4.2}$$

The aspect ratio of W/S therefore seems to be an intrinsic product of drainage organisation. If so, the spacing ratio may be useful in estimating catchment sizes of ancient sediment routing systems where the locations of depositional apices are known from stratigraphic studies, or alternatively in estimating the outlet points of depositional systems where the range width is constrained from structural geology.

4.1.2 Bedrock Channels

The incision of rivers into bedrock controls the response of landscapes to uplift and thereby the flux of sediment into downstream basins in the sediment routing system. Incision rates can be measured at a variety of timescales. A common method is to measure the height of dated strath terraces or lava flows above the present-day river bed. In addition, erosion rates of a channel can be calculated using cosmogenic radionuclides, the concentration of radionuclides reflecting the time a sample of bedrock from the river channel has spent within the nuclide production zone at and close to the surface of the Earth. Erosion at much shorter timescales can be directly measured using calibrated pins drilled into bedrock, but

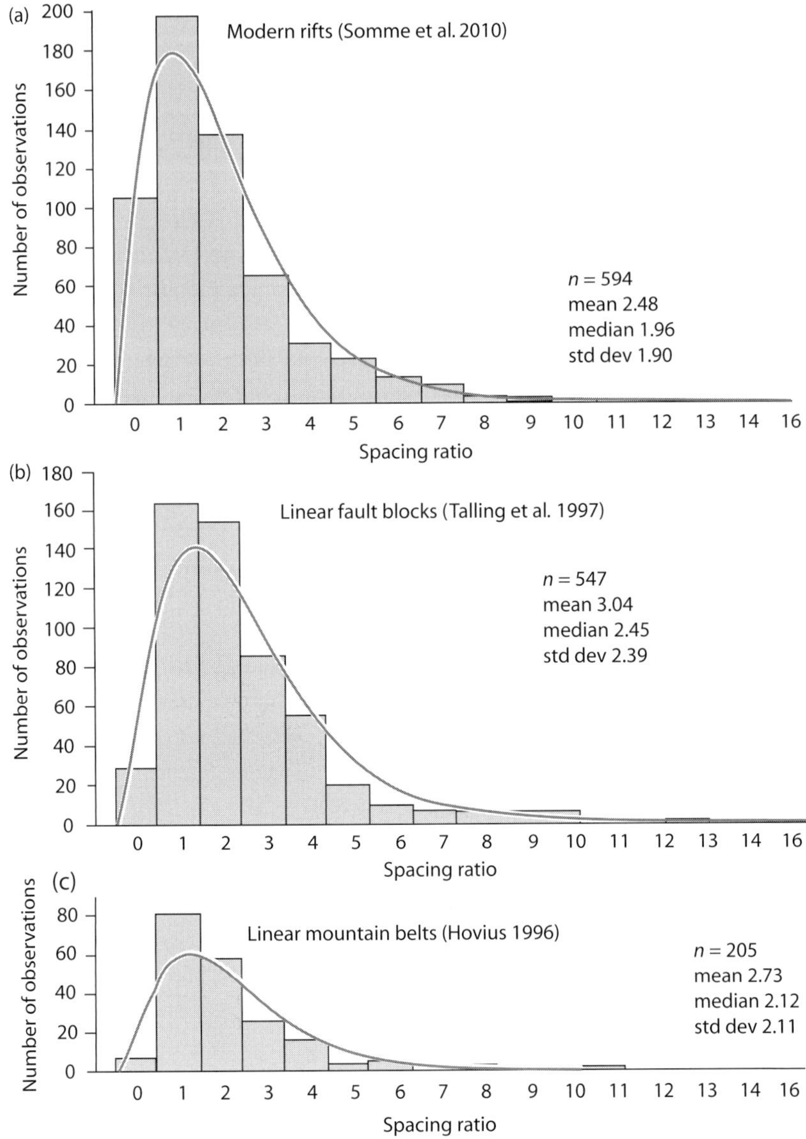

Figure 4.2 Spacing ratio for catchments in extensional (a and b), and contractional tectonic settings (c). Modified from Sømme et al. (2010), published with permission of American Association of Petroleum Geologists.

this method is restricted to bedrock channels undergoing high incision rates of greater than 1 mm yr^{-1}. Measurements in highly energetic rivers such as the gorge of the Indus River (Hancock et al., 1998) and those in tectonically very active Taiwan (Hartshorne et al., 2002) show that tens of mm of erosion can take place in single flood events.

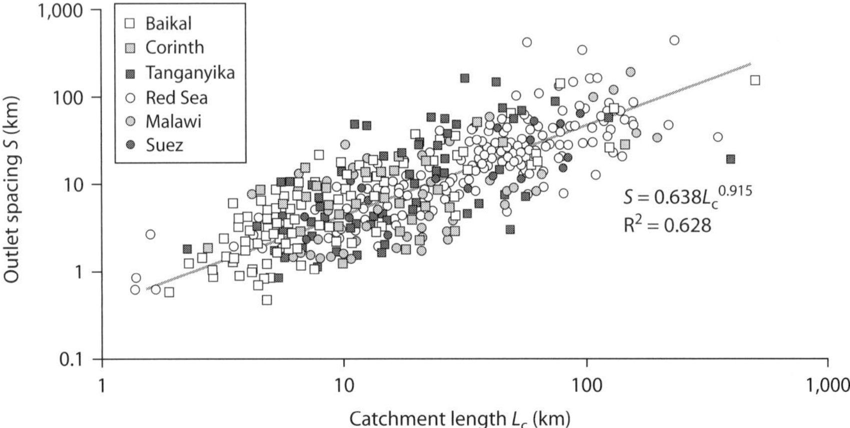

Figure 4.3 Spacing of catchment outlets versus catchment length for 6 present-day rift settings. Modified from Sømme et al. (2010), published with permission of American Association of Petroleum Geologists.

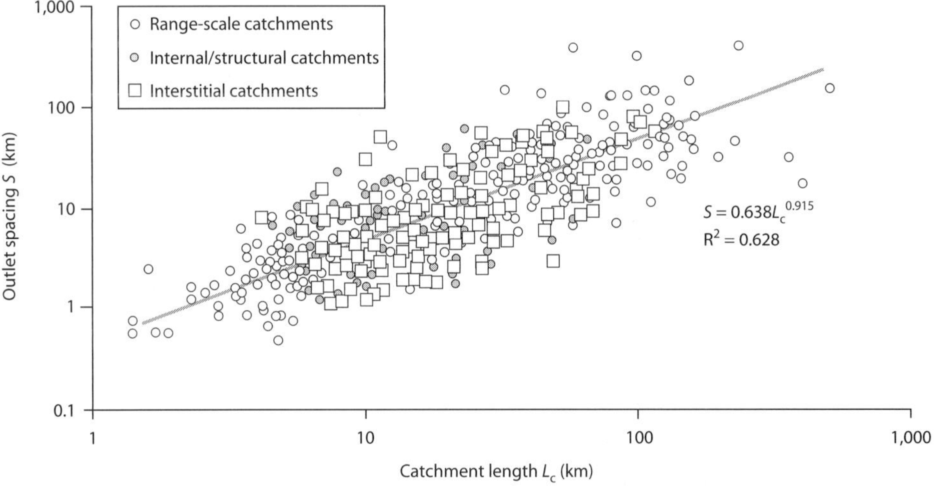

Figure 4.4 Spacing of catchment outlets versus catchment length for range-scale catchments, internal/structural catchments and interstitial catchments. Modified from Sømme et al. (2010), published with permission of American Association of Petroleum Geologists.

Hillslope processes and topographic profiles are set by the boundary condition of channel erosion, and the products of hillslope erosion are transported away by these channelised flows. Channelised flow takes place wherever a stream power threshold is locally exceeded (Montgomery and Dietrich, 1988). There is a fundamental distinction, however, between rivers incised into bedrock and alluvial rivers with beds and banks of sediment. Bedrock rivers are those that lack a coherent bed of active alluvium (Howard,

1987). Consequently, bedrock rivers have a transport capacity that is larger than that required to transport all of the available sediment. They incise by abrasion by the sediment load, plucking of blocks from the bed, and cavitation (Hancock et al., 1998). Bedrock rivers are the dominant channel type in mountainous topography.

The evolution of the profiles of bedrock channels is conventionally assessed using a framework based on *stream power*, equivalent to the rate of potential energy loss as the river flows from high to low elevations. The stream power rule is a framework that relates the elevation change in a bedrock channel to a power law of water discharge and slope. It is commonly used in landscape evolution models.

Since the discharge of a river is highly unsteady, Howard and Kerby (1983) suggested that channel incision rate is proportional to the 'dominant discharge', represented by a drainage area A_D, since water discharge must scale in some way on the product of the contributing drainage area above the channel and the average precipitation over that area. The rate of channel incision is then given by

$$\frac{\partial y}{\partial t} = kA_D^m S^n \tag{4.3}$$

where m and n are constants that vary in value depending on whether the stream power or shear stress rules are being applied. However, there is considerable uncertainty in the values of k, m and n in bedrock channels. The coefficient k is particulary hard to constrain, since it folds in a large number of variables relevant to channel incision.

The values of the exponents m and n have received a good deal of attention. In channels incising badlands topography, m and n were calculated as 0.45 and 0.70 respectively (Howard and Kerby, 1983), whereas in valleys on a basaltic shield volcano, it was found that $m = n = 1$ (Seidl, Dietrich, and Kirchner, 1994). Using a finite difference method, Stock and Montgomery (1999) optimised values of m, n and k by matching the lowering of ancient river profiles preserved as strath terraces and lava flow tops with observed profiles. Optimised values of k, m and n from 4 different settings are given in Table 4.1. Rivers with a strong area dependence on incision rate are associated with stable base levels (as in the Australian examples in Table 4.1), the stream power law taking the form

$$\frac{dy}{dt} = k^{[1]} A^{0.3-0.5} S^1 \tag{4.4}$$

Rivers with a small area dependence on incision, however, have base levels that vary strongly enough to cause the upstream propagation of knickpoints, as in the Hawaiian examples, where incision rate is dependent on locally steepened slopes. In this case the rate of incision is expressed as

$$\frac{dy}{dt} = k^{[2]} A^{0.1-0.2} S^{0-2} \tag{4.5}$$

In the simplest case, where knickpoint migration is not involved, we can assume that k is a constant determined by lithology, ranging from 10^{-4} to 10^{-2} for mudstones, from 10^{-5} to 10^{-4} for volcaniclastic rocks and from 10^{-7} to 10^{-6} for granitic and metamorphic

Table 4.1 *Optimised values of area exponent m, slope exponent n and bedrock incision coefficient k, selected from Stock and Montgomery (1999)(tabs.1, 2, p.498). Published with permission of American Geophysical Union.*

River	m	n	Lithology	Mean $k^{[1]}$	Mean $k^{[2]}$
Kauai (Hawaii)					
Laulaula	0.1–0.2	–	basalt	–	6.7×10^{-6}
Waika	0.1–0.2	–	basalt	–	6.2×10^{-6}
Waipao	0.1–0.2	–	basalt	–	7.3×10^{-6}
Paua	0.1–0.2	–	basalt	–	3.8×10^{-6}
Australia					
Tumbarumba	0.5	1	gran.-metaseds.	1.1×10^{-6}	–
Tumut	0.3	1	gran.-metaseds.	2.3×10^{-6}	–
Wheeo	0.1	0.5–0.6	gran.-metaseds.	4.4×10^{-7}	–
Lachlan	0.5	0.5	gran.-metaseds.	4.3×10^{-6}	–
California					
Cowlet	0.1–0.2	1.6–1.7	volcaniclastics	8.2×10^{-5}	8.5×10^{-5}
French	0.2	0.1	volcaniclastics	4.8×10^{-5}	2.2×10^{-5}
Japan					
Iwaki	> 2	0.1	mudstones/ volcaniclastics	7.0×10^{-3}	3.8×10^{-3}

Mean $k^{[1]}$ is calculated using the Australian end-member (stable base-level), mean $k^{[2]}$ is calculated using the Hawaiian end-member (knickpoint migration). 'Gran.-metaseds.' is granitoids and metasediments.

rocks (units of $m^{0.2}$ yr^{-1}) (Stock and Montgomery, 1999). This range is supported by the mean value of k in Nepalese rivers cutting into relatively easily erodible material at 4.3×10^{-4} (Kirby and Whipple, 2001). The wide variability of the bedrock incision coefficient with respect to lithology suggests that the landscapes of bedrock rivers and fringing hillslopes, and their response times to perturbations (Section 8.5), are strongly sensitive to bedrock geology. In the progressive exhumation of an orogen, for example, as the sedimentary cover is eroded to expose crystalline basement, there should be a slowing in the erosion rate. Likewise, erosion rates and landscape response times should vary as thrust-related anticlines grow and undergo exhumation, as in the case of the Zagros fold-thrust belt of Iran and western Pakistan (Tucker and Slingerland, 1996) (Section 8.6.3). The possibility that k may vary over 5 orders of magnitude as a result of lithological factors makes the use of the stream power rule in analysing ancient landscapes and sediment routing systems challenging.

Since k represents the proportion of stream power expended in incision as opposed to generating heat and transporting loose sediment, the coefficient should change along the river profile, since in steep reaches debris flows are likely to play an important role

Table 4.2 *Median values of power law parameters from cosmogenic nuclide studies. The normalised erodibility coefficient was calculated assuming a reference concavity index* $m/n_{ref} = 0.5$. *Data from Harel, Mudd, and Attal (2016).*

Concavity index m/n	0.51 ± 0.14
Slope exponent n	2.43 ± 0.15
Normalised erodibility coefficient k	$2.9 \times 10^{-10} \pm 1.0 \times 10^{-9}$

in scouring the channel floor. Raising the slope exponent n to between 1 and 2 would incorporate the enhanced incision due to debris flow activity.

Harel, Mudd, and Attal (2016) compiled a global data set of published basin-averaged erosion rates derived from the analysis of detrital cosmogenic ^{10}Be from 1,457 rivers in a wide range of topographic, climatic and tectonic settings (Section 7.6). At the global scale they found median values of the slope exponent n of generally greater than 1 (Table 4.2), implying that the erosion rate is a non-linear function of the channel slope. This non-linearity suggests that bedrock channel incision is a threshold-controlled process.

4.1.3 Effect of Tectonic Uplift on River Long Profiles

In Section 4.1.2, it was suggested that the longitudinal profile of a river can be modeled using the stream power rule. A tectonic displacement field causing uplift of rocks has impact on the long profile of the river, including on parameters such as m, n and k. Tectonic uplift commonly results in knickzones that propagate upstream as convex reaches. The analysis of longitudinal stream profiles is useful in assessing landscape response to forcing mechanisms such as variations in tectonic uplift rate. The stream power law can therefore be built upon by including a term for the tectonic uplift rate of rock. In steady-state conditions bedrock river erosion is balanced by the uplift field $U(x, t)$, so that

$$U(x, t) = kA^m S^n \tag{4.6}$$

U is considered initially to be spatially uniform, and m and n are constants, so equation (4.6) becomes after rearrangement

$$S = \left(\frac{U}{k}\right)^{1/n} A^{-m/n} \tag{4.7}$$

Equation (4.7) has the form of a power law (Sklar and Dietrich, 1998; Kirby and Whipple, 2001). For the uniform rock uplift case, the straight line regression of $\log S$ plotted against $\log A$ has a slope of $-m/n$, and an intercept of $(U/k)^{1/n}$, both of which can be derived from plots of field data from modern rivers (Snyder et al., 2000).

A spatially variable tectonic uplift field can be approximated by a simple uplift function $U = U_0 x^\alpha$, where U_0 is the rock uplift rate at the edge of the model, x is the horizontal coordinate and α is a constant describing the downstream gradient in tectonic uplift rate. This simple uplift function yields a power law relationship between slope and area. Assuming Hack's Law for drainage area A versus downstream distance x, $A = k_a x^h$, and steady-state conditions, this power law relationship has gradient and intercepts giving the coefficient k_s (steepness) and the exponent (concavity) θ of the stream power law as follows (Kirby and Whipple, 2001):

$$k_s = (U_0/k)^{1/n} k_a^{-(\alpha/hn)} \tag{4.8}$$

and

$$\theta = (m/n) - (\alpha/hn) \tag{4.9}$$

Equations (4.8) and (4.9) indicate that where the rock uplift rate increases downstream ($\alpha > 0$), streams should have low concavity or even convexity ($\theta < 0$), whereas where the rock uplift rate decreases downstream ($\alpha < 0$), streams should have high concavity ($\alpha > 0$). A study of the concavity of streams crossing the Siwalik Hills (Kirby and Whipple, 2001) showed strong variations in θ dependent on the gradients in tectonic uplift rate associated with differential movement across the Main Frontal Thrust (Lavé and Avouac, 2001). Channels crossing the tectonically uplifting anticline from north to south experience increasing tectonic uplift rate and have convex-up longitudinal profiles ($\theta < 0$). Tectonic uplift rates and their lateral variability can therefore potentially be inverted from stream profiles if parameter values can be well calibrated.

Deep incision by rivers causes a removal of mass from the area undergoing erosion, which drives an isostatic compensation (Molnar and England, 1990). Consequently, the high mountain peaks fringing deeply incised valleys may experience a surface uplift. This possibility can be stated in terms of a bedrock river incision rule. Assuming the isostatic uplift to be due to a local (Airy) compensation rather than to a regional flexure, the rate of uplift of the mountain peaks is

$$\frac{\partial y_p}{\partial t} = \frac{1}{2}\left(\frac{\rho_c}{\rho_m}\right) c_1 x^2 S \tag{4.10}$$

where ρ_c and ρ_m are the crustal and mantle densities respectively, y_p is the elevation of the mountain peak, x is the horizontal coordinate, and S is the slope. Using $c_1 = 0.5$ m^{-1} Myr^{-1}, profiles along the Arun and Karnali rivers in the Nepalese Himalaya show the elevation of mountain peaks some distance downstream from the drainage divides in the Tibetan plateau (Montgomery, 1984) (Figure 4.5). The greatest relief is on the edge of the Himalayan plateau, where the Karnali River has incised 4,500 m below the surrounding peaks. As much as 20 to 30% of the present elevation of the Himalayan peaks can be explained by isostatic compensation.

4.1.4 Hillslopes

Hillslopes convey both water and sediment into river channels, which are situated at the foot of the slope. As Robert and Suzanne Anderson put it (Anderson and Anderson, 2010) (p.306),

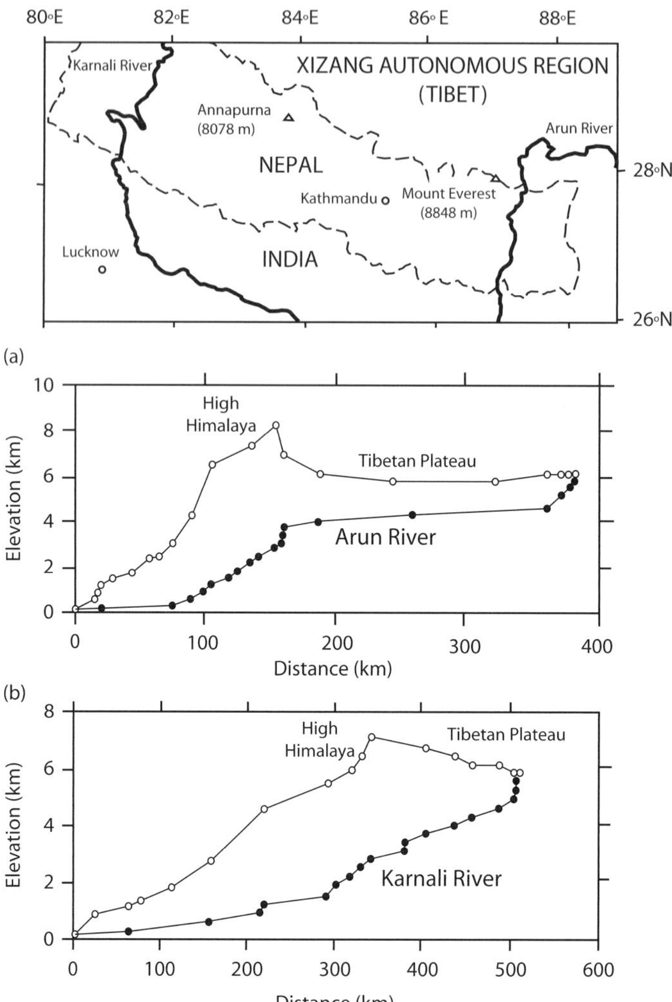

Figure 4.5 Map of the Himalayas and Tibetan plateau, showing the location of the Karnali and Arun rivers. (a), (b) Long profiles of the mountain peaks (open circles) and valley bottoms (filled circles) along the Arun and Karnali rivers, showing the high elevation of the summit about 200 km from the headwater regions, suggesting that bedrock incision causes uplift of the peaks. Modified from Montgomery (1984), with permission from American Geophysical Union.

Hillslopes . . . do not exist in isolation, but are linked to their bounding channels. To put it mathematically, channels serve as the bottom boundary conditions for the hillslopes.

Channels and hillslopes therefore act in a coupled process system. The topography generated by tectonics is degraded through the action of this coupled system. Rivers typically become entrenched in position, cutting down like cheese wires through the regionally uplifting bedrock (Burbank et al., 1996). The intervening hillslope system provides erosional material by soil creep, overland flow, gullying, debris flows, landslides and rockfalls. If the rate of hillslope erosion is slower than the rate of valley lowering, hillslopes steepen and eventually become susceptible to landsliding involving bedrock. The amplitude of hillslopes is therefore set by the effect of rock strength on landsliding. The fluvial system transports away material derived from hillslope erosion through its channel networks. Evacuation may be significantly delayed where high-magnitude trigger events such as earthquakes cause an instantaneous discharge from hillslopes, blocking valleys with landslide and debris-flow dams (Li et al., 2014).

In mountain ranges landslide activity, triggered by rainstorms and co-seismic shaking (Densmore and Hovius, 2000), is critical to the clearing of hillslopes (Densmore et al., 1997). A number of rapidly uplifting mountain belts (for example, the Southern Alps of New Zealand and the Karakorum range of the western Himalayas) are believed to exhibit a steady-state topography where the tectonic rate of uplift of rock is balanced by the landslide-dominated erosional sediment efflux. In the Southern Alps of New Zealand, the flux of material due to landsliding estimated over a 60-year period (several mm yr^{-1}) is roughly equivalent to the sediment output of the main rivers draining the region (Hovius et al., 2007).

Hillslopes involve a downslope motion of material, whether quasi-continuously by slow creep, or in discrete events. Over time, most hillslopes take on a smooth, convex-upward profile, suggesting that hillslopes might be modelled as a diffusive process. Sediment diffusion models are based on two first-order assumptions: (1) the sediment continuity equation (also known as the Exner equation), which states that the spatial variation in the sediment transport rate q_s is proportional to the vertical erosion or aggradation rate of the substrate, $\partial q_s / \partial x = -\rho_b \partial y / \partial t$, where ρ_b is the bulk density of the mobile regolith, assuming there to be no changes in the concentration of suspended sediment; and (2) that the flux is proportional to the local gradient, $q_s = -k \partial y / \partial x$, where k is a transport coefficient, which is a basic notion underpinning all diffusional problems. The change of elevation over a time step t is the familiar diffusion equation

$$\frac{\partial y}{\partial t} = \kappa \frac{\partial^2 y}{\partial x^2} \tag{4.11}$$

where $\kappa = k / \rho_b$ is the diffusion coefficient linking the change in elevation to the topographic curvature.

The derivation of solutions for the sediment discharge, slope and maximum relief of hillslopes undergoing diffusion is given in a number of sources, including appendix 42, pages 532–533, in Allen and Allen (2013); pages 30–34 in Pelletier (2008); and pages 309–313 in Anderson and Anderson (2010). The shape of diffusional hillslopes is parabolic, with a maximum slope at the river channel and a minimum slope at the hillslope crest:

$$y = -\frac{V}{2\kappa} \left(L^2 - x^2 \right) \tag{4.12}$$

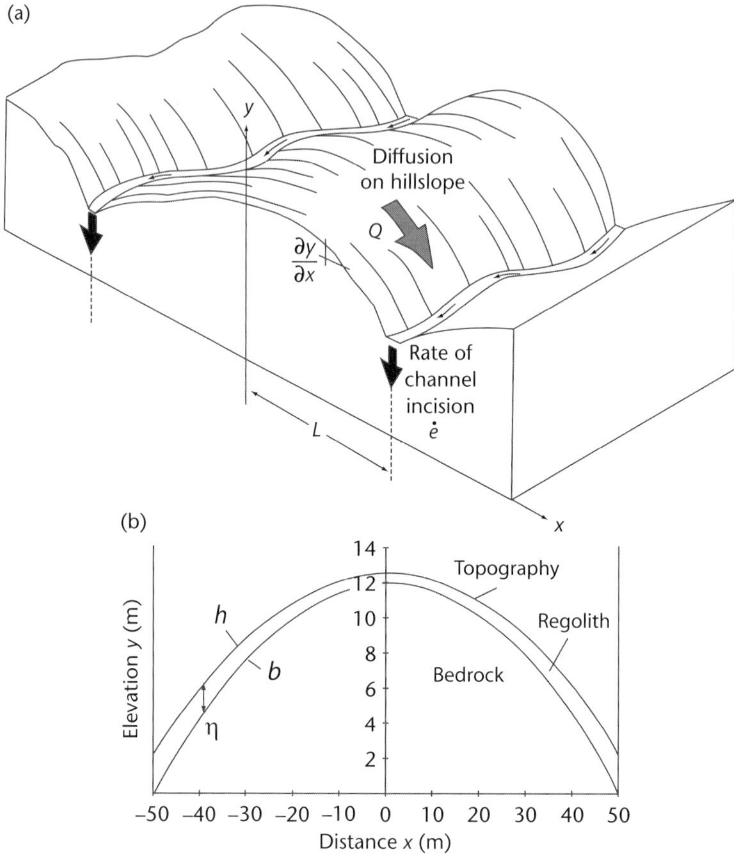

Figure 4.6 (a) Schematic diagram showing hillslopes and bedrock channels, with notation. (b) Solution (parabolic) for the hillslope profile using the parameter values given in the text. The thickness of regolith $\eta(x, t)$ is exaggerated for clarity. Part (a) from Allen and Allen (2013) (fig.7.32) with permission of John Wiley & Sons Inc.

where V is the incision rate of the channels, equivalent to the uplift rate of rock in a steady-state profile, κ is the diffusivity, L is the length of the hillslope from ridge crest to valley bottom and x is the horizontal coordinate (Figure 4.6). The characteristic timescale for diffusional hillslopes is of the familiar form $\tau = L^2/\kappa$. If two channels located 100 m apart ($L = 50$ m) incise at a rate of 0.5 mm yr^{-1}, typical of tectonic uplift rates, and a diffusion coefficient applicable to the slow processes of rainsplash and soil creep is 50×10^{-3} m^2 yr^{-1}, the maximum gradient of the hillslope is 0.5 (tan^{-1} 0.5 is 27°). The time constant is 50 kyr.

The weathering of hillslope rock results in regolith, which mantles the bedrock surface. Consequently, there are two interacting surfaces: a topographic surface with elevations given by $h(x, t)$ and a weathering front with elevations given by $b(x, t)$. The difference between these two elevations is the regolith thickness $\eta(x,t)$ (Figure 4.6). The simplest way to express the coupling of the bedrock and topographic surfaces is given by

$$\frac{\partial \eta}{\partial t} = -\frac{\rho_b}{\rho_s}\frac{\partial b}{\partial t} - \kappa \frac{\partial^2 h}{\partial x^2} \tag{4.13}$$

In other words, the change in thickness of regolith over time is the sum of the change in elevation of the bedrock surface and the change in elevation of the surface topography (see equation (4.11)). The change of elevation of the bedrock surface occurs at a rate dependent on depth below the topographic surface given by

$$\frac{\partial b}{\partial t} = -P_0 e^{-\eta/\eta_0} \tag{4.14}$$

where P_0 is the regolith production rate for bare bedrock and η_0 is the characteristic regolith depth, which expresses the rate at which regolith production reduces with increasing depth below the surface. Such a depth profile of regolith production has been verified from cosmogenic nuclide analyses (Heimsath et al., 1997).

Diffusion theory can also be used to study the degradation of fault scarps (Hanks et al., 1984), river terraces (Avouac, 1993) or independently dated wave-cut lake margin scarps, such as those of Pleistocene Lake Bonneville, Utah. The scarp profile evolves over time by an erosional smoothing of the upper portion and a depositional smoothing of the lower portion. The slope of the mid-point of the scarp decays as the square root of time. Diffusivities in the present-day arid landscapes of the American West were estimated as $5{,}100 \times 10^{-3}$ m^2 yr^{-1} (Burbank and Anderson, 2001). However, a large number of studies suggest that we cannot always assume that the diffusivity κ is constant and independent of x on hillslopes and scarps of various origins. Furthermore, the upper parts of hillslope profiles may be strongly affected by the regolith production rate, so that they are weathering-limited and therefore not strictly determined by linear diffusion (Rosenbloom and Anderson, 1994).

When mass flow such as landsliding is the dominant form of erosion on hillslopes the sediment flux increases nonlinearly with the topographic gradient. Hillslopes become linear and the drainage divide becomes knife-edge sharp instead of gently parabolic (Roering, Kirchner, and Dietrich, 1999). In a nonlinear model, the sediment flux increases rapidly as the hillslope gradient approaches a critical value. The sediment flux is given by (Howard, 1997; Roering et al., 1999):

$$\bar{q}_s = \frac{\kappa\,\partial y/\partial x}{1 - (|\partial y/\partial x|/S_c)^2} \tag{4.15}$$

where κ is the diffusivity (L^2/T), S_c is the critical hillslope gradient and $|\partial y/\partial x|$ is the absolute value of the topographic gradient. From equation (4.15), the sediment flux becomes infinite at the critical hillslope gradient S_c.

If the channel incision rate is large compared to the diffusive mass wasting of the hillslope, a critical slope will be reached after which landsliding takes place. This is why mountains with high channel incision rates, such as the Southern Alps of New Zealand and the Finisterre Range of Papua New Guinea, have hillslopes dominated by landsliding. Landslide-dominated hillslopes are straight rather than parabolic (Anderson, 1994; Densmore et al., 1997; Densmore and Hovius, 2000). The critical slope for the onset of landsliding depends on bedrock strength, vegetation cover and fluid pressures in the regolith and bedrock. It is likely to vary between about $30°$ and $60°$. For effective diffusivities in the range 10 to 100×10^{-3} m^2 yr^{-1}, the channel incision rates required to initiate landsliding at critical slopes between $40°$ and $60°$ are $0.2 - 3.5$ mm yr^{-1}. This range covers the tectonic uplift rates of recent mountain belts. The rates of rock uplift along the Alpine Fault in the Southern Alps of New Zealand is more than 5 mm yr^{-1}, with rates of 1 mm yr^{-1} along the main drainage divide of the Southern Alps (Koons, 1989). The Central Range of Taiwan and the Finisterre Range, Papua New Guinea, have similar tectonic rates of rock uplift.

On the western flank of the Southern Alps of New Zealand (Koons, 1989), where the lateral distance from interfluve to interfluve $L = 3,000$ m, the maximum topographic relief $y_{max} = 1,500$ m, and V is approximately 5 mm yr^{-1}, the effective erosional hillslope diffusivity must be 15 m^2 yr^{-1} (Figure 4.7). It is considerably higher than the value of 0.01 m^2 yr^{-1} estimated for the semi-arid to arid hillslopes in the western United States (Rosenbloom and Anderson, 1994). The effective erosional hillslope diffusivity encompasses a whole range of geomorphological processes acting on the hillslope, and therefore may lump together the effects of mean annual precipitation (more than 1 m yr^{-1}) and rainfall intensity, rock strength, vegetation and temperature-dependent weathering. The value calculated here reflects the high activity of rapid mass wasting events. The mean gradient of the hillslopes from the diffusion model is y_{max}/L, which for the Southern Alps examples is 0.5, or $27°$. This corresponds closely with measured mean values.

The transverse spacing of hillslopes must always be small (high drainage density, equation (4.1)) in order to balance high rates of tectonic uplift of rock in a steady-state landscape. On the drier eastern flank of the Southern Alps (less than 1 m yr^{-1} mean annual precipitation) the tectonic uplift rates reduce from 1 mm yr^{-1} at the main divide to zero at the east coast of South Island. On this flank, braided river valleys are broadly spaced (low drainage density, equation (4.1)) and less deeply entrenched into bedrock (Figure 4.7).

The ratio y_{max}/L can also be written as $VL/2\kappa$. This is a dimensionless ratio of a flow rate of rock across a section of the transverse drainage relative to an effective erosional diffusivity. It has the form of a Péclet number. If the erosional Péclet number is greater than about 0.6, it is unlikely that erosional processes can keep pace with tectonic uplift rates to produce a steady-state landscape. The erosional Péclet number for the wet western flank of the Southern Alps, using the parameter values provided earlier, is 0.5.

Hillslopes and bedrock channels are dynamically linked, but they have different response times (Section 8.5). The hillslope response time introduced previously is L^2/κ. The response time of the channel system must be influenced by the length of the channel and the average velocity of the knickpoints generated when the channel adjusts to its new base level. Such knickpoints migrate headwards, but slow down in their upstream velocity as the stream power decreases due to a reduction in the contributing drainage area.

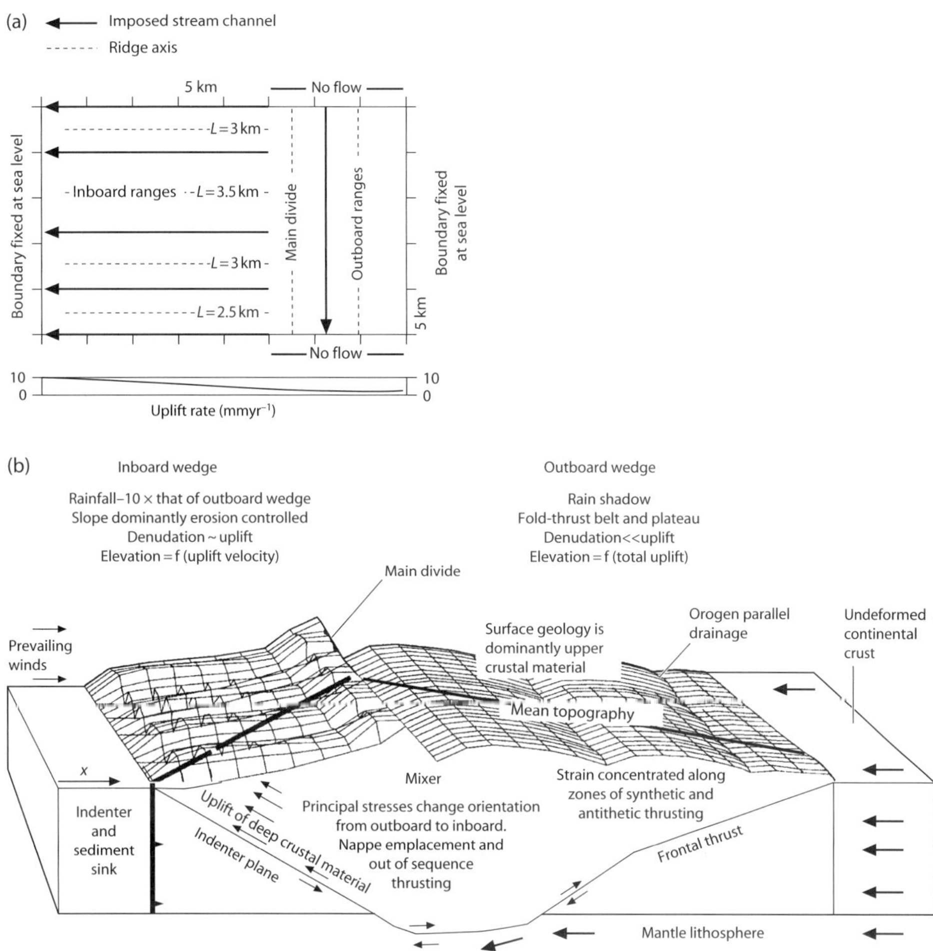

Figure 4.7 The doubly vergent orogenic wedge of the Southern Alps of New Zealand has deeply etched, narrowly spaced, transverse drainages on the wet western flank where tectonic uplift rates are high. On the drier eastern flank, rivers are much less deeply entrenched and are broadly spaced. (a) Boundary and initial conditions for the numerical model in Koons (1989). (b) Tectonic velocities and topography, with precipitation from the left. Modified from Koons (1989), published with permission of *American Journal of Science*.

Anderson (1994) calculated channel system response times assuming the stream power rule as typically 10^3 to greater than 10^4 yr. This is significantly longer than the response time for landslides, but is comparable to the hillslope response time. This suggests that hillslopes characterised by the high effective diffusivities produced by landslides are always in equilibrium with the incising channels. Less steep hillslopes dominated by slow diffusion, however, may not be in equilibrium with channel incision and may display transient rather than steady-state morphologies.

4.2 Basin-Margin Fans

Alluvial fans are Δ-shaped sediment bodies with distributive channel networks, commonly developed at basin margins where an upland catchment enters an adjacent basin. Alluvial fans therefore build up at the abrupt change in topographic slope characteristic of tectonically active basin margins. Although fans may comprise debris cones in steep bedrock channels and accumulations at tributary (wadi) junctions (Al-Farraj and Harvey, 2005), fans at mountain fronts are focussed upon here. Sediment is deposited on the fan surface by debris flow and stream flow, with localised contributions from landslides, soil creep and aeolian fall-out. They occur in all climatic zones, from arid to humid and from hot to cold. Some catchment-fans act as closed systems amenable to a balance of mass or volume, whereas others involve an export or leakage of sediment to playas and axial river systems. The progradation distance of fans away from the mountain front, because of their simple planform geometry, is an indicator of the balance between sediment supply from the catchment and the subsidence rate in the basin. The sediment budget manifests itself in a relationship between catchment area and fan area as originally proposed by Bull (1962, 1964), which is strongly influenced by the tectonic subsidence in the basin, whether spatially uniform (Allen and Hovius, 1998) or spatially variable (Whipple and Trayler, 1996).

The fan model involves a source catchment of area A_c feeding a fan of area A_f (Figure 4.8). The relation between A_c and A_f is known from geomorphic studies of modern fans (Hooke, 1968; Bull, 1977; Lecce, 1991) as having a power law form

$$A_f = cA_c^n \tag{4.16}$$

where the coefficient c and the exponent n have values reported in the literature attributed to factors such as variation in bedrock lithology, climate, rate of uplift of rock in source areas and basin subsidence. There is considerable variation in c and n in modern fans from different climatic zones (Figure 4.9). The median value for c from the global data used in Figure 4.9 is 0.55 (mean of 0.83) with a standard deviation of 0.85, whereas n shows less variation, with a median value of 0.76 (mean of 0.80) and a standard deviation of 0.40. However, fans distributed along single fault systems fall along linear trends in plots of $\log(A_f)$ against $\log(A_c)$ (Figure 4.10) indicating that some local arrays of fans experiencing similar arid climatic, lithological and tectonic forcings have $n \sim 1$.

Some material derived by hillslope erosion may be stored in valley bottoms instead of being immediately transported onto the alluvial fan. The efficiency of sediment transport from hillslopes to the fan apex (e_s) therefore affects the discharge to the fan. The storage efficiency factor e_s, which varies from zero (complete storage) to 1 (no storage), depends on factors such as the effect of climate on run-off and topography, but empirical studies suggest the effect is relatively small (Hooke, 1968; Hooke and Rohrer, 1977; Jansson, Jacobson, and Hooke, 1993). There is the possibility, especially in humid climates, that some sediment escapes the fan and is delivered to axial river systems, expressed by a second efficiency factor e_f, which varies from zero (complete bypass) to 1 (entire sediment load deposited

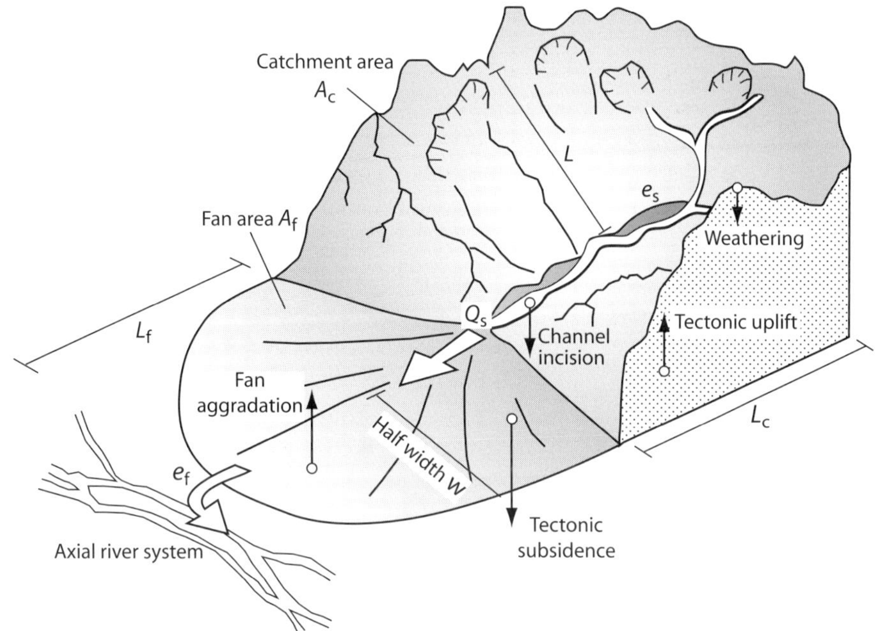

Figure 4.8 Schematic diagram of a landslide-dominated hillslope system feeding an alluvial fan. Notation is explained in the text. After Allen and Hovius (1998), with permission of John Wiley & Sons Inc.

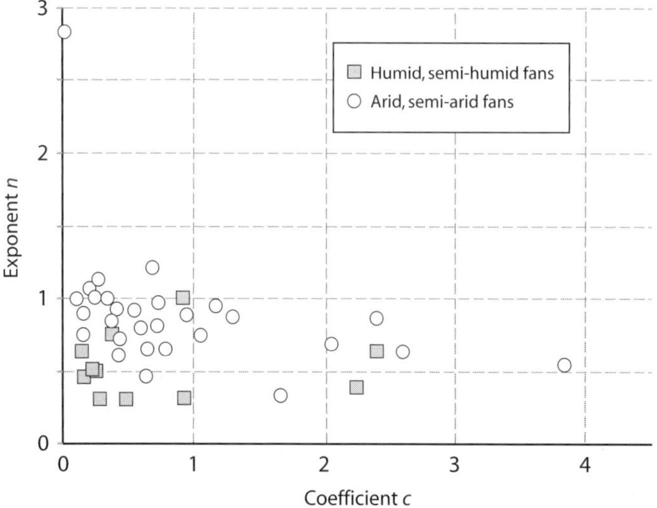

Figure 4.9 Plot of coefficient c versus exponent n in the power law of fan area and catchment area (equation (4.16)), from a wide range of climatic and tectonic settings, using data in Allen and Hovius (1998), Al-Farraj and Harvey (2005), Harvey (2005) and Jarman, Agliardi, and Crosta (2011).

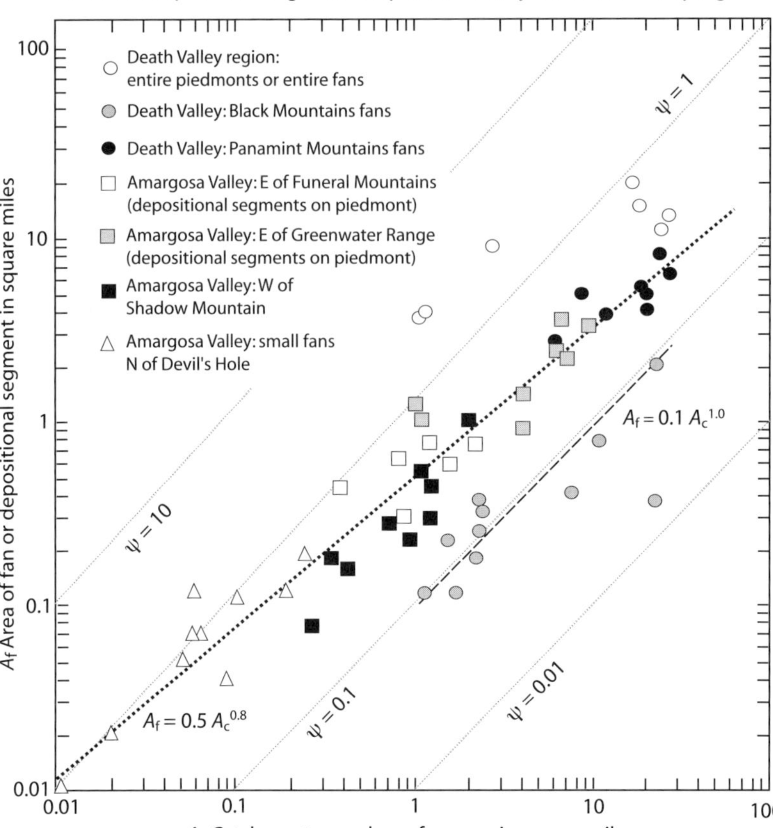

Figure 4.10 Plot of catchment area versus fan area for locations in the Death Valley region, eastern California, showing the influence of variations in tectonic subsidence rate. Data from Denny (1965). Regressions are given for fans along the eastern side of Death Valley (Black Mountains) (long dashed line) and for the full data set (short dashed line). From Allen (2008a) with permission of the Geological Society.

on fan surface). Toe-cutting of the fan by axial channels may cause e_f to become negative. Assuming topographic steady state (Willett and Brandon, 2002) and a conservation of mass, the fan area A_f becomes

$$A_f = \frac{Q_s}{\dot{y}e_f(1-\lambda_s)} = \frac{1}{\dot{y}}V\frac{(1-\lambda_r)}{(1-\lambda_s)}e_se_fA_c \tag{4.17}$$

where Q_s is the discharge of sediment onto the fan, $\dot{y}$ is the average sediment deposition rate on the fan, V is the efflux of the catchment expressed as the denudation velocity due to landsliding and is essentially a function of κ, the rate of landsliding per unit area per year, and λ_s and λ_r are the porosities of fan sediment and catchment rock respectively.

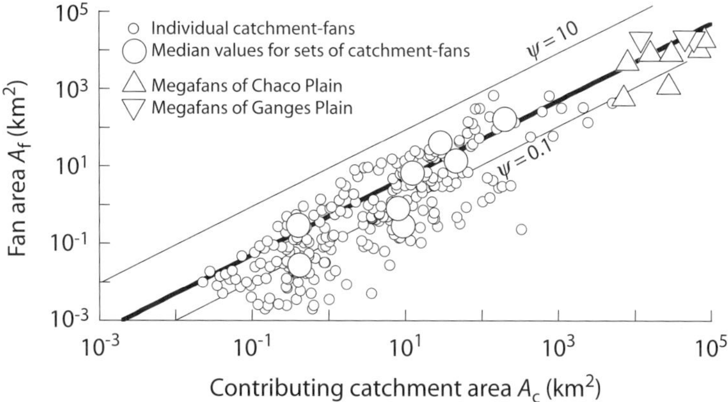

Figure 4.11 Fan area A_f versus catchment area A_c, with least squares regression for conventional fans, required to pass through origin, of the form $A_f = \psi A_c$, where $\psi = 0.5 \pm 0.35$, $R^2 = 0.95$. From Dade and Verdeyen (2007)(fig.2a) with permission of the Geological Society.

Taking the catchment-fan as an integrated system, the ratio of catchment to fan area (Figure 4.11) can be expressed in terms of tectonic, climatic and topographic parameters. As a reminder, the local erosion rate of a bedrock channel sets the boundary condition for the hillslope processes that control catchmentwide denudation and can be expressed in terms of stream power per unit width $\rho g q(x) S(x)$, where $S(x)$ is the local channel gradient at x. If the catchment has a uniform cross-valley geometry, the width of the bedrock channel at any distance from the headwaters $b(x)$ is proportional to the average width $B_c(x)$ of the catchment upstream of x (Dade and Verdeyen, 2007) so that

$$\beta = \frac{b(x)}{B_c(x)} \tag{4.18}$$

where β is the dimensionless proportionality constant. β describes the downstream variation of channel width in relation to catchment width. Assuming the catchment has a length L_c and a uniform precipitation rate P, the discharge per unit width becomes $q(x) \equiv \beta^{-1} P x$, and the average erosion rate of the catchment $\hat{E}$ is

$$\hat{E} = k_1 (1 - \lambda_r) \beta^{-1} P \bar{H} \tag{4.19}$$

where λ_r is the porosity of rock in the catchment and $\bar{H}$ is the average relief of the catchment evaluated along the main channel. A readily obtained index of catchment relief is the overall elevation difference between the headwaters and the outlet H at $x = L_c$. Introducing this substitution of H for $\bar{H}$ and assuming that the average erosion rate $\hat{E}$ equates to the uplift rate of rocks V in a steady state, equation (4.19) can be modified to

$$H = k_2 \frac{V}{P} \tag{4.20}$$

where

$$k_2 = \beta [k_1 (1 - \lambda_r)]^{-1} (H/\bar{H}) \tag{4.21}$$

and has units of [L]. Substituting $Q_s \equiv \hat{E}A_c$ into equation (4.17) gives after rearrangement an expression for the ratio of fan area to catchment area

$$\frac{A_f}{A_c} = k_3 \left(\frac{V}{PH}\right)^{-1} \tag{4.22}$$

where k_3 is a consolidated coefficient with units of $[L]^{-1}$ containing terms for bedrock resistance to erosion, cross-valley geometry and fan efficiency, given by

$$k_3 = e_f \beta^{-1} \frac{(1-\lambda_r)}{(1-\lambda_s)} k_1 (H/\bar{H}) \tag{4.23}$$

Equation (4.22) is a power law between the geometric area ratio and V/PH with a slope of -1 (Figure 4.12). It suggests that the observed variation of A_f/A_c from catchment-fan systems in different tectonic and climatic settings can be attributed to variation in V/PH, which encapsulates the effects of tectonics (through uplift rate of rocks), climate (through precipitation) and topography (in the form of catchment relief). As an example, k_1 can be calculated from equation (4.19), with erosion rate $\hat{E} = 1$ mm yr^{-1}, $\beta = 0.1$, precipitation $P = 0.5$ m yr^{-1}, average relief $\bar{H} = 1$ km and porosity of rocks in the catchment $\lambda_r = 0.05$, giving a value of $k_1 = 0.2 \times 10^{-6}$ m^{-1}. Second, k_3 can be calculated using equation (4.23), with fan efficiency factor $e_f = 1$, porosity of fan sediment $\lambda_s = 0.2$ and the overall relief of the catchment $H = 2$ km, giving $k_3 = 1.25 \times 10^{-6}$ m^{-1}. Substituting this value into equation (4.22), where the uplift rate is the same as the erosion rate in steady state, the ratio of fan area to catchment area ψ becomes 1.25. The value of V/PH for the parameter values given previously is 10^{-3} km^{-1}.

The ratio A_f/A_c is low where the tectonic velocity is high compared to precipitation, consistent with the data for the Blackwater fans in Death Valley shown in Figure 4.10, with $\psi = A_f/A_c \sim 0.1$. At the other extreme, the fans of Fresno County, California, have a high value of $\psi = A_f/A_c$ of $1-2$ due to the relatively low tectonic activity in the Great Valley.

4.3 Axial versus Transverse Drainage

A common arrangement of fluvial elements in a sedimentary basin is for basin-margin fans at the mountain range front to pass downsystem into axial river systems. In some cases, a fan may develop at the meeting of a tributary with a powerful axial river, as in the case of the San Juan River of the Argentine Andes (Colombo, 2005). A series of cycles of fan build-up, ponding of the axial river behind the sediment dam, and scarp erosion during renewed axial flow, produce segmented or 'telescopic' fans. The more general case illustrated here, however, is of an axial flow as a residue of the sediment budget of fringing alluvial fans along the mountain front.

Following Kim et al. (2011), we let the sediment discharges from hangingwall and footwall fans be q_h and q_f and the axial river discharge to be q_a. The transfers from footwall and hangingwall fans to the axial system are q_{fa} and q_{ha}. The half graben with a basin width W_b subsides by a simple rotation at a maximum rate S_{max} at $x = W_b$, reducing to zero at

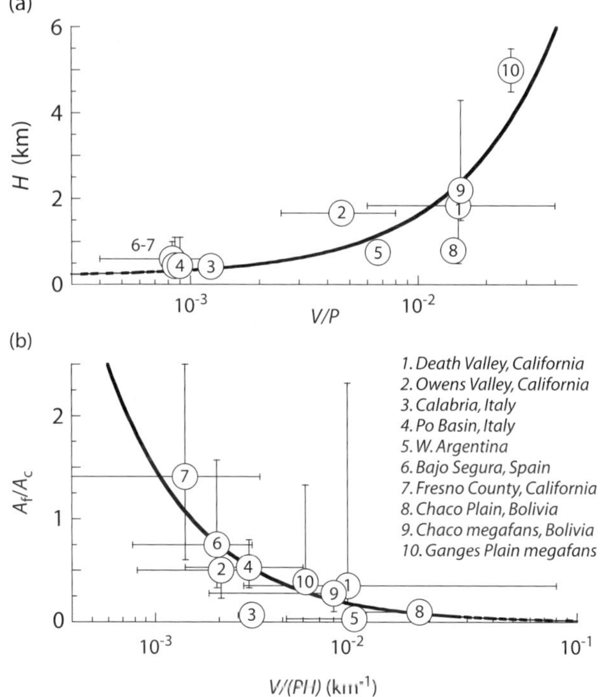

Figure 4.12 (a) Characteristic relief H of catchments serving as source areas for sediment fans as a function of the ratio of tectonic uplift and average precipitation V/P for median values of sets of catchment-fan systems. 90% confidence intervals shown by bars. Curve is least-squares regression on median values ($R^2 = 0.74$) of the form given in equation (4.20). (b) A_f/A_c ratio as a function of the ratio V/PH. Curve is least-squares regression ($R^2 = 0.71$) of the form given in equation (4.22). Data and citations of source literature in table 1, p.354 of Dade and Verdeyen (2007). Published with permission of the Geological Society.

$x = 0$ (Figure 4.13). The lengths of the fans can be calculated using a sediment mass balance, assuming that the fluvial system remains constant in surface area and there is no extension across the half graben. Taking the hangingwall fan, the sediment below the fan surface is made up of that above base level and that below base level, accommodated by tectonic subsidence. However, if the fan slope and maximum elevation do not change over time, the sediment deposited is equal to the increment in the part below base level, so that

$$w_h = \sqrt{2\frac{W_b}{S_{max}}(q_h - q_{ha})} \tag{4.24}$$

Similarly, for the footwall-derived fans, the mass balance gives

$$w_f = 2\frac{(q_f - q_{fa})}{S_{max}} \tag{4.25}$$

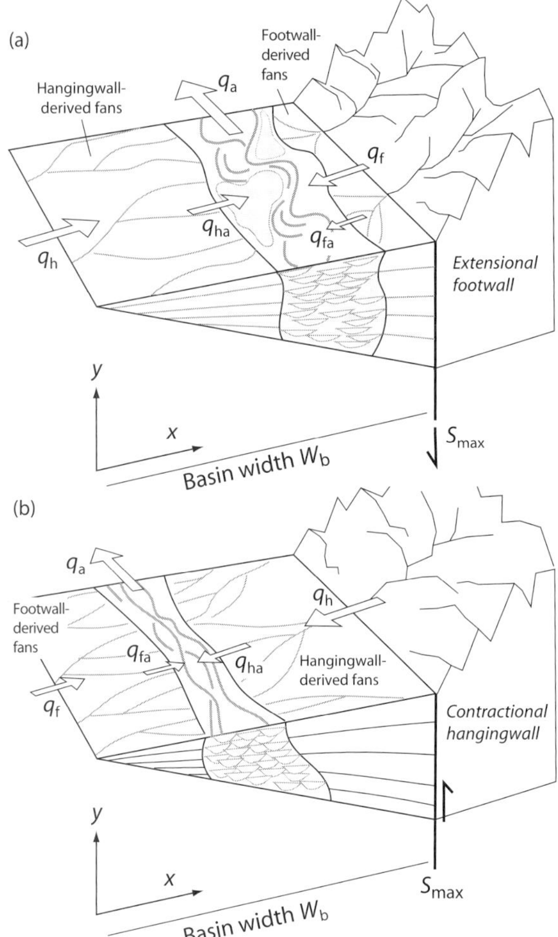

Figure 4.13 Transverse and axial sediment dispersal in a half-graben (a) or wedge-top basin (b). The position of fan toes, the width of the axial system and the axial sediment flux, depend on the two transverse supplies relative to accommodation in the basin. From Allen and Allen (2013) (fig.8.20, p.303), modified from Kim et al. (2011), published with permission of Geological Society of America.

From equations (4.24) and (4.25) it is clear that the positions of the fan toes are determined by the sediment mass balance. Inserting reasonable values for a Basin and Range half-graben, let $W_b = 20$ km, and $S_{max} = 1$ mm yr^{-1}, the widths of the fans scale on the sediment discharge terms. The total sediment budget is $q_T = W_b S_{max}/2$. The fraction delivered from each margin is therefore given by equations (4.24) and (4.25) scaled by q_T.

If the fraction of the total budget supplied from each margin remains constant, the fan toes remain stationary, despite changes in the maximum subsidence rate. Inspection of equation (4.24) shows that the length of hangingwall-derived fans remains constant when the net discharge $q_h - q_{ha}$ scales with $S_{max}/2$. In other words, when S_{max} doubles through an increase in slip rate on a bounding fault, the length of the fan stays fixed if the net discharge increases by 50%. If sediment discharge increases through a climate change, with no change in accommodation driven by fault-related subsidence, the fan length must change in response. This suggests that climatically driven movement of the fan toe should be more variable than the damped (coupled) tectonically driven changes.

The width of the axial tract should vary according to the tectonic or climatic forcing. For the coupled tectonic model, the tract remains almost stationary, but an increase in the net discharge forced by climate results in fan toe progradation from each side of the basin and the eventual meeting of the oppositely derived fans.

The preceding concept can be extended to the case of an array of folds bordering wedge-top basins. The basement surface is assumed to subside in a sinusoidal pattern and we neglect horizontal advection of rock during tectonic shortening. Accommodation is created by an increase in the amplitude of the sine curve, keeping wavelength constant. The new accommodation generated up to a transverse fan length from the proximal margin is

$$h_0 S_{max} \int_0^{w_f} \sin\left(\pi + \frac{2\pi x}{\lambda}\right) dx \tag{4.26}$$

which when evaluated gives

$$w_f = \frac{\lambda}{2\pi}\left[\cos^{-1}\left(-\frac{(q_f - q_{fa})}{h_0 \lambda S_{max}} + 1\right) - \pi\right] \tag{4.27}$$

For example, with a wavelength λ of 40 km (basin width of 20 km), a maximum subsidence rate S_{max} of 0.1 mm yr^{-1}, a net sediment discharge to the fan $q_f - q_{fa}$ of 400 m^2 yr^{-1}, an amplitude h_0 of 1 km, the proximal fan length is approximately 12 km. If using Excel^{TM}, equation (4.27) needs to be solved by taking arccos(x)=cos^{-1}(x) and using an on-line calculator, or by making a look-up table of cos(x) in Excel^{TM}, but the latter is slow.

We can make further use of the idea of an axial sediment discharge. For a closed budget, the axial discharge is that remaining after sediment extraction to build fringing fans, that is, $q_T = q_h + q_f$, and $q_{fa} + q_{ha} = q_a$ (Figure 4.14). The axial budget is broken down into that deposited and that exported downslope q_e. The fraction of the total discharge exported through the axial system is a key parameter in understanding the dynamics of sediment routing systems in tectonically active regions. For $q_e/q_T \sim 0$, we anticipate perfectly filled basins with little far-field transport. For $q_e/q_T >> 0$, we anticipate major longitudinal downslope export of sediment guided by the tectonic grain. When there is marked asymmetry in the sediment supply combined with a high total discharge, we anticipate extensive systems that bury underlying tectonic structures and follow the regional transverse palaeoslope. These various outcomes have been identified in the Tertiary alluvial fans of the fold-thrust belt and foreland basin of the central Pyrenees in northern Spain (Allen et al., 2013) (Figure 4.15).

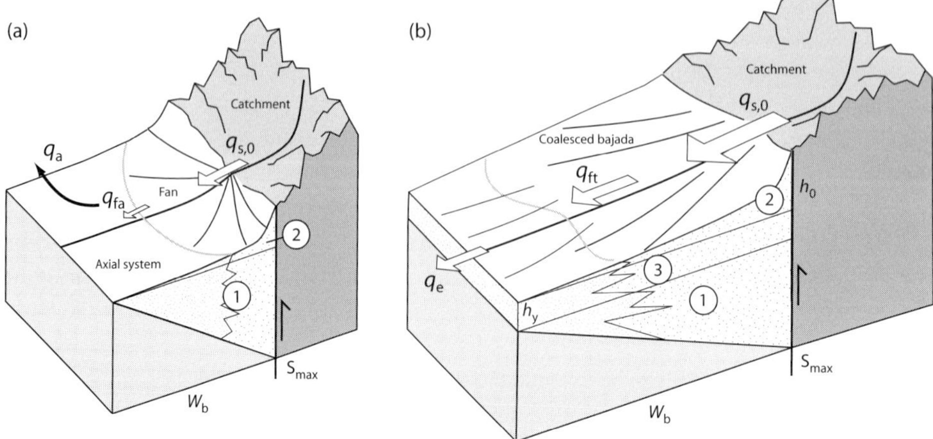

Figure 4.14 Block diagrams illustrating contrasting behaviours determined by the sediment budget. In both cases, the backshedding of sediment into the basin is set to zero. (a) A steep catchment feeds a basin-margin fan at a rate $q_{s,0}$, which fluxes sediment to an axial system at a rate q_{fa}. The sediment supply is less than or approximately equal to the accommodation generation. The cumulative volume of accommodation is denoted '1', which increases over time by slip on the border fault at the rate S_{max}. The volume above base level, denoted '2', can be assumed to not vary over time. (b) A larger catchment feeds a higher sediment supply into the basin, causing the progradation and coalescence of large transverse fans and potentially the burial of the distal basin margin. A large portion of the sediment supply is discharged through the fan to downstream transverse systems, q_{ft}, and exported to distal foredeep depocentres, q_e. The volume denoted '3' may accumulate over time as base level rises and the distal margin of the wedge-top basin is buried in sediment. From Allen et al. (2013) (fig.7), with permission of John Wiley & Sons Inc.

4.4 Alluvial Rivers

4.4.1 Fluvial Geomorphic Elements in Sedimentary Basins

It has previously been suggested that a survey of many (724) sedimentary basins in a wide range of climatic and tectonic settings (Weissmann et al., 2010), and the results of physical experiments (Sheets et al., 2002), show that terrestrial sedimentation is dominated by distributive fluvial systems (Figure 4.1). The dominance of distributive systems in net aggradational settings contrasts with the dominantly contributive patterns observed in modern degradational settings. The occurrence of distributive rather than contributive patterns in the rock record illustrates the role of accommodation generation, or 'preservation space' (Blum and Törnquist, 2000), in controlling fluvial form. Distributive fluvial systems take the form commonly attributed to fans, occurring where a river becomes unconfined on entering a sedimentary basin from an upland catchment or incised valley, spreading in a radial pattern from the depositional apex. Within this general fluvial form, individual channels may be straight, braided or meandering. In large-scale distributive fluvial systems there is a downstream decrease in channel size caused by bifurcation, infiltration

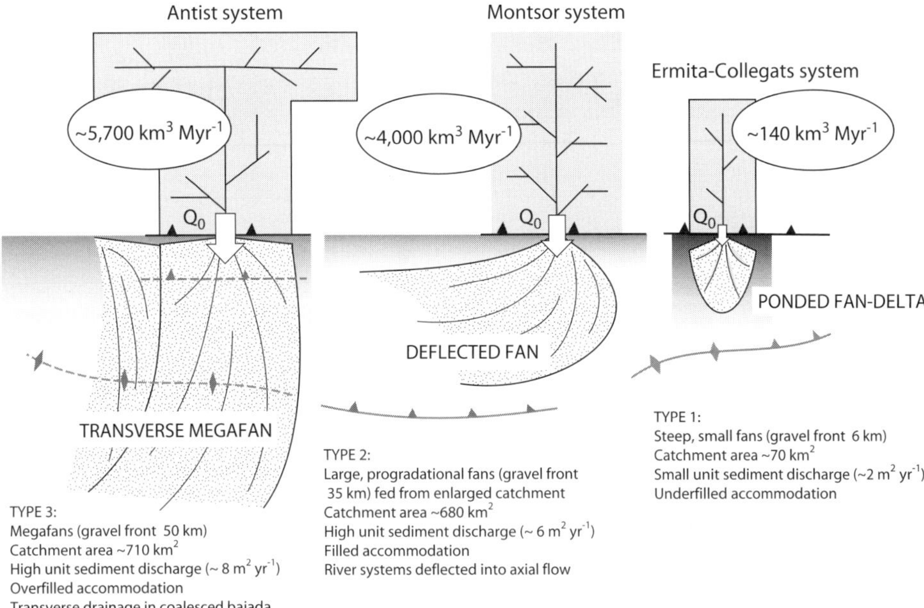

Figure 4.15 Summary of three sediment routing systems from the mid Eocene to Oligocene of the south-central Pyrenees. Numbers in ellipses are the approximate total sediment discharges of catchments (based on the unit sediment discharges and catchment sizes in table 2, Allen et al. (2013)). From Allen and Allen (2013) (fig.7.58), published with permission of John Wiley & Sons Inc.

and evapotranspiration, and an increase in the ratio of floodplain area to channel area. Switching of channel position within the floodplain is commonly by nodal avulsion near the apex.

Sedimentary basins contain a range of fluvial elements (Table 4.3) (Figure 4.16). The largest are megafans, typical of the Himalayan and Andean foreland basins, with surface areas up to ca. 700,000 km². Megafans have large catchment areas stretching far into the mountain sediment source area. The apices of megafans are separated by inter-megafan rivers, which drain the foothills region, and distally pass into an axial, trunk system (DeCelles and Cavazza, 1999). Smaller fans typically join axial systems in extensional basins (Gawthorpe and Leeder, 2000). Coalesced alluvial fans along a range front produce bajadas. The areas occupied by distributive systems strongly exceed the areas covered by tributive fluvial systems, such as axial tributary rivers held in an axial position or focussed in interfan positions.

The large megafans of the Himalayan and Andean foreland basins have upstream catchments stretching far into the mountain belt, but the ratio of DFS surface area to catchment area ($A_f/A_c = \psi$) is highly variable (Figure 4.17). Taking all data combined the value of ψ is approximately unity, which is consistent with the global compilation of Dade and Verdeyen (2007).

Table 4.3 *Fluvial geomorphic elements in selected sedimentary basins. Typical surface areas, terminology and categorisation are from Weissmann et al. (2015).*

Fluvial element	Surface area $(10^3 km^2)$	Example
Distributive systems		
Megafans	5–700	Magdalena River, Colombia
		Himalayan foreland
		Chaco Plain (Andean foreland)
Fluvial fans and alluvial fans	0.1–12	Brahmaputra Valley, Assam
		Death Valley, United States
Bajada or piedmont	0.1–80	Mongolia
Incised distributive systems	0.2–40	Taquari, Brazil
Tributive systems		
Axial tributary	0.03–25	Paraná River, Argentina
Interfan rivers	0.005–2	Between Kosi-Baghmati, India

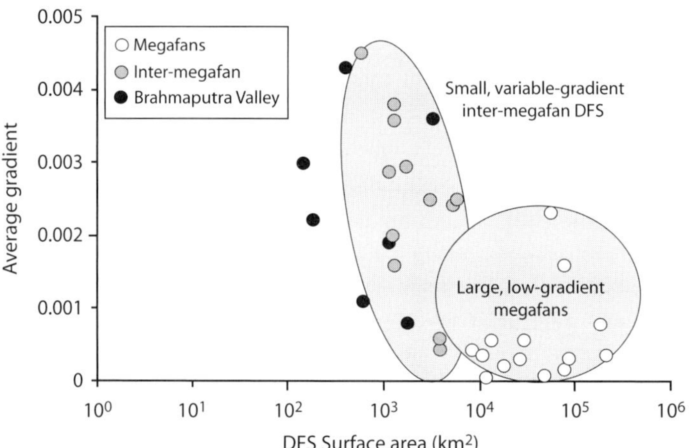

Figure 4.16 Surface area of distributary fluvial system (DFS) versus average gradient for a range of examples in foreland basin settings. Data in Weissmann et al. (2015).

4.4.2 *Long-Range Sediment Transport and Deposition*

Alluvial rivers (those with beds and banks of sediment) cause long-range sediment transport from the erosional engine to inland depocentres or the ocean. The mechanics of sediment transport by rivers is complex. In this section, the broad aspects of far-field sediment transport are tackled broadly in terms of long-term behaviour rather than in terms of individual

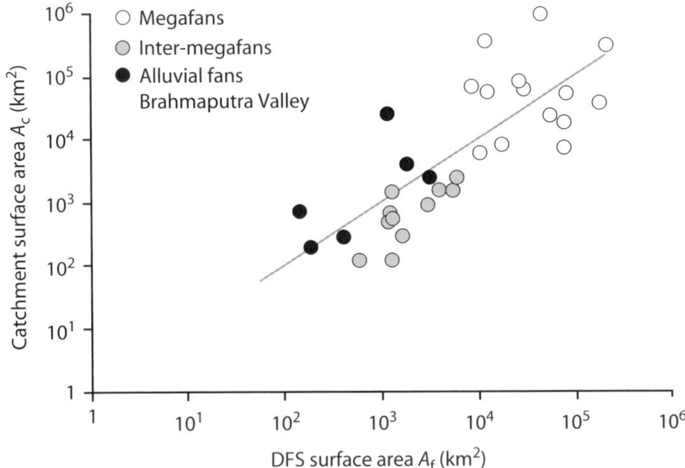

Figure 4.17 Plot of DFS surface area versus catchment area for a range of distributive fluvial systems, using data in Weissmann et al. (2015).

flood events. More comprehensive treatments of long-range sediment transport by rivers are found in Parker (1978a), Parker (1978b), Paola et al. (1992) and Dade and Friend (1998). Bridge (2003) and Bridge and Demicco (2008) provide much useful information on rivers and fluvial deposits.

A useful starting point for considering sediment transport and deposition in alluvial rivers is to treat the system as overall diffusive in character. That is, the sediment flux scales on the topographic gradient, and the elevation change of the river bed over time scales on the topographic curvature.

Various forms of diffusion equation have been applied to long-range fluvial transport. Here we concentrate on an assessment of long-range fluvial transport of a mixture of gravel and sand. The quantitative development of this group of models can be found in Paola et al. (1992) and Marr et al. (2000). It is well known that in modern mixed gravel-sand fluvial systems there is a relatively steep gravelly proximal zone with an abrupt change to a lower gradient sandy zone (Sambrook Smith and Ferguson, 1996). To be able to predict the movement of the *gravel front*, or *gravel-sand transition*, would be of considerable benefit in basin analysis (Section 8.1).

Most models of sediment transport in rivers make the same set of assumptions (Paola, 2000):

(1) It is assumed that the long-term flow of water in a river approximates that of a steady uniform flow down an inclined plane. For a flow of depth h and density ρ_f on a slope $\partial y/\partial x$, the downslope component of the fluid weight on a unit area of the river bed is $\rho_f g h \partial y/\partial x$. This downslope acting force must be opposed by an equal and opposite drag force exerted on the fluid by the unit area of bed. This is the shear stress τ_0. Consequently, the force balance gives

$$\tau_0 = -\rho_f g h \frac{\partial y}{\partial x} \tag{4.28}$$

Rivers are anything but steady and uniform. However, equation (4.28) can be used as an approximation for long-term river behaviour. It works best for shallow high-gradient streams and worst for deep, low-gradient streams. If the flow depth is large compared to the channel width, h should be replaced by the hydraulic radius R in equation (4.28), where the hydraulic radius is the cross-sectional area divided by the wetted perimeter, $R = wh/(w + 2h)$, and w is the channel width.

(2) A fluid moving over its bed and banks experiences frictional losses of energy known as flow resistance. Where the bed of the river is rough, as is always the case in natural rivers, the energy losses should in some way be related to the length scale of the roughness of the bed. This roughness can be expressed in a number of ways. A common method is to use the Darcy–Weisbach friction factor f

$$f = \frac{8\tau_0}{\rho_f u^2} \tag{4.29}$$

where u is the flow velocity. The friction factor (or similar forms such as the Chézy coefficient C and Manning's n) varies strongly according to the grain size of the sediment on the river bed and is especially affected by the presence of bedforms such as ripples and dunes and of macroforms such as bars, chutes and pools.

(3) Third, the discharge of water through the system is conserved. Consider a slice of width B of an alluvial basin of length L, which has active channels on its surface with cumulative width b, of flow depth h and containing flows of velocity u (Figure 4.18). Let β be the fraction of the section width B occupied by channels, so $\beta = b/B$. The discharge of water Q_w in the channels occupying the width of floodplain B is bhu, or averaged across the floodplain is βhu. Consequently,

$$\beta = \frac{Q_w}{hu} \tag{4.30}$$

(4) Finally, use is made of the sediment continuity equation, modified by the use of a Shields-type dimensionless shear stress τ^*,

$$\tau^* = \frac{\tau_0}{\left(\rho_s - \rho_f\right) g D} \tag{4.31}$$

where D is the median grain size and ρ_s and ρ_f are the sediment and fluid densities respectively. There appear to be strong limits on the value of the dimensionless shear stress, so that it can be treated as a constant, at about 1.4 times the critical shear stress at the threshold of particle motion in coarse-grained braided rivers, and between 1 and 2 in alluvial sand-bed rivers (Paola and Seal, 1995; Dade and Friend, 1998; Parker et al., 1998). It is probably a constant because if it is too high, the stream erodes its banks and widens its bed, thereby reducing the shear stress per unit area of stream bed (Parker, 1978a,b). There is a sharp jump in τ^* at the gravel front in many rivers.

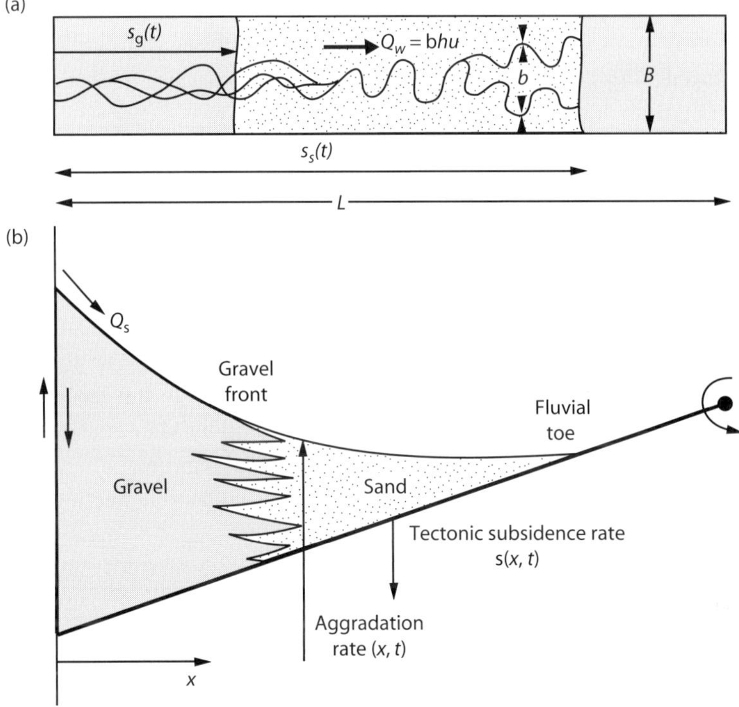

Figure 4.18 Map view (a) and cross-section (b) of a slice of alluvial basin of length L, width B, containing channels of cumulative width b undergoing tectonic subsidence $\sigma(x)$ approximated by a linear tilt from a distant hinge. After Marr et al. (2000) modified by Allen and Allen (2013) (fig.7.39, p.262) with permission of John Wiley & Sons Inc.

The modified sediment continuity equation where rivers are flowing over a template experiencing tectonic subsidence $\sigma(x)$ has the form

$$\frac{\partial y}{\partial t} = \frac{\partial}{\partial x}\left(\kappa \frac{\partial y}{\partial x}\right) - \sigma(x) \tag{4.32}$$

where the effective diffusivity or transport coefficient κ of the alluvial system is:

$$\kappa = I_f \left(\frac{q_s^*}{\tau^{*3/2}}\right) \frac{C_d^{1/2} q_w}{C_0} \left(\frac{1}{s-1}\right) \tag{4.33}$$

where s is the submerged specific weight of the sediment, equal to $(\rho_s - \rho_f)g$, C_d is a drag coefficient, q_w is the average water discharge per unit width, q_s^* is a dimensionless sediment discharge per unit width determined by τ^* (the dimensionless shear stress) and a sediment transport formula, C_0 is the volume sediment concentration in the deposit, proportional to porosity, and I_f is a term to represent transport rate fluctuations termed intermittency (Paola, 2000). Equation (4.33) is a useful formulation of the effective fluvial diffusivity for present-day rivers, but has limited applicability in geological studies. An alternative formulation (Smith and Bretherton, 1972; Simpson and Schlunegger, 2003), where sediment-hydrodynamic constraints are lacking is:

$$\kappa = \kappa_l + c_f (\alpha x)^m \tag{4.34}$$

where κ_l is the linear hillslope diffusivity, α is the mean annual precipitation, x is the horizontal coordinate along the direction of concentrative flow, which causes the water discharge q_w to increase downstream, m is an exponent, which is normally assumed to equal 2, and c_f is a fluvial transport coefficient, assumed to be equal to 10^{-6}.

Equations (4.33) and (4.34) are functionally similar, since q_w approximates αx, the drag coefficient, submerged specific weight of the sediment and the dimensionless shear stress can be assumed to be constant, and κ_l is negligible compared to the second term on the right hand side of equation (4.34). Typical values for κ used by Marr et al. (2000) are 0.01 km^2 yr^{-1} in the gravel regime and 0.1 km^2 yr^{-1} in the sand regime.

A discussion of the response times of alluvial systems is found in Section 8.5.1.

4.4.3 River Planform Patterns and Long Profiles of Alluvial Rivers

Viewed from the air, rivers display a variety of patterns, but the main characteristics can be simply summarised as comprising two forms: a single, sinuous channel and interwoven multiple channels, known conventionally as meandering and braided respectively. Meandering and braided patterns are associated with combinations of discharge characteristics, grain size of the bed and banks and channel slope. The two planform patterns can be inferred from the architecture of fluvial stratigraphy, very occasionally from exhumed planform surfaces in outcrops, and increasingly from very high resolution seismic reflection data sets.

Braided patterns are favoured by river banks that lack cohesion, sediment transport mostly as bedload, and highly fluctuating flow discharge, which results in multiple threads of shallow channels. Braided rivers are typical of glacial outwash and basin-margin fans. Braided systems can be generated in numerical models and in laboratory experiments (Murray and Paola, 1994). Meandering patterns depend on the existence of cohesive bank material. The meanders are self-similar, with wavelengths of the loops roughly 11 times the river width (Leopold and Wolman, 1960). Meandering rivers have also been generated in numerical models (Sun, Meakin, and Jossang, 1996, 2001; Lancaster and Bras, 2002), but are more difficult to reproduce in experiments. The alluvial architecture resulting from the preservation of the deposits of braided and meandering rivers is found in innumerable sedimentological studies. In essence, braided rivers produce sheets of sand and gravel with little preservation of fine sediment. Meandering systems produce plugs where lateral migration was limited, and sheets of sand encased in floodplain fine sediments where the meander migrated laterally. Meander migration lengthens the river over time, but eventually the meander is cut off at the neck of the loop, leaving behing an ox-bow lake.

Ox-bow lakes are storage sites of sediment (Section 4.5), with lengths L that are log-normally distributed given by

$$L = 3.0e^{0.82S} \tag{4.35}$$

where S is the sinuosity (Constantine and Dunne, 2008).

Alluvial rivers are concave up in longitudinal profile, the elevation of the mouth of the river acting as a base level. Base level might be the ocean, an inland lake, a playa or alluvial plain. It acts as a bottom boundary condition and may move dynamically as a result of tectonic processes, climate change or the periodic tidal variation of the ocean. The stream profile is a logarithmic function of the distance from the drainage divide. This can be demonstrated by introducing a stream gradient index (Hack, 1973), which describes how the local slope $\partial y/\partial x$ depends on the downstream distance x. The elevation of the river is then

$$y = -\left[\frac{\partial y}{\partial x}x\right]\ln(x/L) \tag{4.36}$$

where the term in square brackets is the stream gradient index and L is the length of the system at which the stream elevation is at base level. The alluvial river therefore has a concave up profile, with the river discharge increasing downstream as the local slope reduces. The downstream reduction in slope is accompanied by a reduction in grain size in the river bed. This grain-size reduction is achieved by a combination of attrition or comminution during transport, weathering during periods when clasts are at rest, and selective deposition (Section 9.4).

River long profiles can also be matched closely with curves representing the solution for thermal diffusion following instantaneous heating (Allen and Allen, 2013) (appendix 48, p.541). This approach, originally proposed by Culling (1960), is widely applied to the cooling of the ocean lithosphere. Like the logarithmic profile discussed previously, diffusive profiles are also concave-up.

We assume a steady sediment discharge to solve for the elevation of the depositional surface h (Turcotte and Schubert, 2002) (p.193), which has the form of a complementary error function (erfc):

$$h = \frac{2q_{s,0}}{\kappa}\left[\left(\frac{\kappa t}{\pi}\right)^{1/2}\exp\left(-\frac{x^2}{4\kappa t}\right) - \frac{x}{2}\mathrm{erfc}\left(\frac{x}{2\sqrt{\kappa t}}\right)\right] \tag{4.37}$$

where x is the horizontal coordinate, $q_{s,0}$ is the unit width sediment discharge at the sediment source ($x = 0$), t is the time elapsed, and κ is the diffusivity. The quantity $\sqrt{\kappa t}$ is a diffusion distance and is an index, in this application, of the rate of downstream decay of elevation. The sediment discharge at the source is a function of the slope at $x = 0$,

$$\left(\frac{\partial h}{\partial x}\right)_{x=0} = -\frac{q_{s,0}}{\kappa} \tag{4.38}$$

The elevation (at $x = 0$) relative to base level at the distal limit of the river profile is given by

$$h_0 = 2q_{s,0}\left(\frac{t}{\pi\kappa}\right)^{1/2} \tag{4.39}$$

for $t > 0$, and the unit width sediment discharge at the source can be written

$$q_{s,0} = -\frac{\pi h_0^2}{4t(\partial h/\partial x)} \tag{4.40}$$

For example, let the proximal slope be -0.075 and the elevation at the depositional apex be $h_0 = 400$ m. The unit width sediment discharge at $x = 0$ is then 1.68 m^2 yr^{-1} for $t = 10^6$ yr, from which κ must be 7.1 m^2 yr^{-1}.

The heat equation can be used to model delta clinoforms and alluvial fans where there is a single sediment source, but also entire fluvial systems. The long profile of the alluvial section of the Magdalena River, for example, can be closely matched by a thermal diffusion model, with a slope at $x = 0$ of 0.333×10^{-3} and an elevation h_0 of 100 m. It should be emphasised, however, that real river systems, such as the Magdalena, have many tributaries (Restrepo et al., 2006a,b; Kettner, Restrepo, and Syvitski, 2010) and the value of $q_{s,0}$ cannot be expected to compare closely with the sediment discharge through an upstream gauging station.

4.5 Floodplains as Sediment Stores

Floodplains, as the name suggests, are inundated during river flood events when channel banks are overtopped. Floodwaters transport sediment into the floodplain, but only the finest grain sizes succeed in escaping the channel and reaching the floodplain where they settle out. Flood events are stochastic and river discharges vary strongly in magnitude and frequency. As a result, there is much interest in the response of the landscape to small, frequent floods relative to large, rare floods. The magnitude-frequency characteristics of floods affect river incision, bedload sediment transport and the inundation of floodplains.

The distribution of the largest flood discharges can be fitted by a power law (Turcotte and Greene, 1993; Lague, Hovius, and Davy, 2005; Malamud and Turcotte, 2006):

$$N(Q) \propto Q^{-\alpha} \tag{4.41}$$

where $N(Q)$ is the number of floods with discharge greater than Q and α is an empirical constant with a value between 1 and 6 (Figures 4.19, 4.20). In arid climates, there are relatively more high magnitude floods compared to low magnitude floods, so the slope of the power law α is smaller than in rivers in temperate climates. Rivers might therefore do more geomorphic work in arid climates than in more humid or temperate climates because of the effect of discharge variability, parameterised by α. The relationship between the power law exponent α and the annual precipitation represented by $\bar{P}$ shows the relation

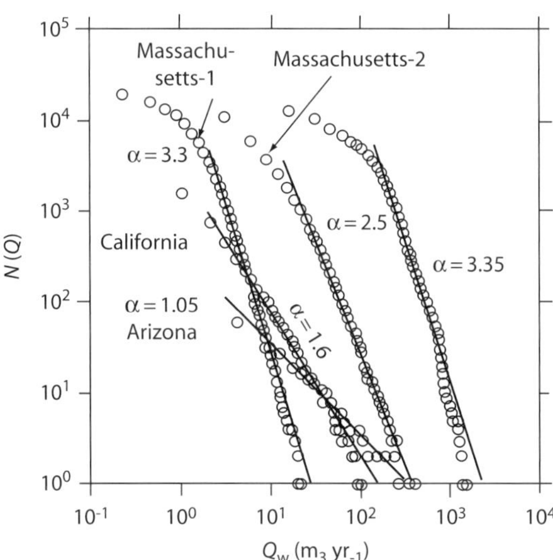

Figure 4.19 Examples of plots of $N(Q)$ versus Q for daily discharge data collected at gauging stations in the United States. After Molnar et al. (2006) (fig.2) with permission of American Geophysical Union.

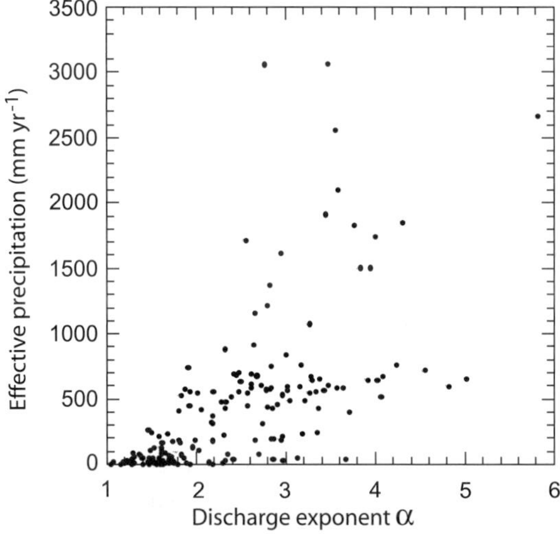

Figure 4.20 Plot of measured values of effective precipitation $\bar{P}$ versus estimated values of discharge exponent α from discharge data at 155 gauging stations. After Molnar et al. (2006) (fig.3), published with permission of American Geophysical Union.

between climate differences and discharge variability, where the *effective* mean annual precipitation $\bar{P}$ is the average annual discharge divided by the total contributing drainage area above the gauging station. The exponent α increases with $\bar{P}$, with a relation

$$\alpha - 1 = \bar{P}^n \tag{4.42}$$

where $n \sim 1.6$. Despite the scatter, the trend in Figure 4.20 supports the view that large floods occur more frequently relative to small floods in arid climates than in more humid or temperate climates. The exponent α therefore varies from catchment to catchment depending on precipitation and precipitation variability. The impact of precipitation variability on erosion and sediment yield is further developed in Section 9.2.

For the floodplain to be inundated, the river must exceed its bankfull depth, which can be expressed as $Q_w > Q_b$, the condition that the river discharge exceeds the bankfull discharge. Since the size of particulate sediment in the channel depends on depth within the flow, only the finer grains are able to be advected into the floodplain. As the shallow floodwaters decelerate by the effects of the frictional resistance of the substrate, sediment is dropped from suspension onto the floodplain surface. The sediment concentration profile of an open channel flow is given by

$$C = C_a \left[\left(\frac{h-y}{y} \right) \left(\frac{a}{h-a} \right) \right]^{-w/ku^*} \tag{4.43}$$

where C is the suspended sediment concentration, C_a is the concentration at a reference level close to the bed a, which might be the concentration in the moving bedload layer predicted by a bedload sediment transport equation, w is the settling velocity, u^* is the shear velocity, h is the total flow depth and k is von Karman's constant, generally taken as 0.4. The dimensionless group w/ku^* is termed the Rouse Number. A Rouse Number of 2.5 corresponds to the case $w = u^*$, a rough criterion for suspension. If the settling velocity exceeds the shear velocity, the Rouse Number should exceed 2.5 and bedload transport rather than suspension should take place. Equation (4.43) can be used to plot the concentration profile for a given grain size with a single value of settling velocity w. An illustration of the use of equation (4.43) is given below for the Mississippi River at St Louis, Missouri. The river at St Louis is 8.5 m deep and the velocity profile measured on 24 April 1956 is as given in Table 4.4 (Colby, 1963). A reference depth of $a = 0.5$ m is taken. From the velocity profile, the shear velocity u^* can be calculated from

$$u^* = k(u_1 - u_2)/2.3(\log y_1 - \log y_2) \tag{4.44}$$

which gives $u^* = 0.245$ m s^{-1}. We take five grain-size classes, with average settling velocities as given in Table 4.5. Clearly, the vertical profile of sediment concentration depends on grain size, with fine sizes distributed uniformly in the flow, and coarse grain sizes concentrated near the bed (Figure 4.21), Consequently, during flood events when the river overtops its banks, the fine sediment is decanted into the floodplain while the coarse sediment is retained in the channel. A similar process in subaqueous turbiditic settings is known as 'flow stripping'.

The deposition rate of a floodplain may vary spatially due to proximity to the channel, presence of vegetation and the topographic effects of abandoned channel belts. On overtopping the channel bank, the coarsest grains are deposited in levees, while the finest grains accumulate in flood basins. The deposition rate varies as well as the grain size

Table 4.4 *Velocity profile measured on 24 April 1956 in the Mississippi River at St Louis, using data in Colby (1963).*

Distance above bed y (m)	Velocity $\bar{u}$ (m s^{-1})
7.7	1.25
5.2	1.20
3.2	1.05
1.8	1.00
0.9	0.90
0.6	0.80

Table 4.5 *Suspended sediment concentrations at the reference height $y = a$ and average settling velocity w for five grain-size classes.*

Grain size (mm)	0.002–0.016	0.016–0.062	0.062–0.125	0.125–0.25	> 0.25
Concentration at $y = a$ (ppm)	145	100	60	270	100
Settling velocity w (mm s^{-1})	0.6	6	11	30	70

(Figure 4.22). The time averaged deposition rate r as a function of distance from the river channel x can be calculated using

$$\frac{r}{a} = (1 + x/x_m)^{-b} \tag{4.45}$$

or

$$\frac{r}{a} = e^{-bx/x_m} \tag{4.46}$$

where a is the mean channel-belt deposition rate, x_m is the maximum floodplain distance from the edge of the channel-belt and b is an exponent describing the lateral decrease of floodplain deposition rate (Bridge and Mackey, 1993). In the case of a single flood of the Mississippi b was between 5 and 10 for both power and exponential functions (Kesel et al., 1974) and a similar value of 5 was found for the Rhine-Meuse delta (Middlekoop and Asselman, 1998). However, calculations based on longer term rates of floodplain accumulation give a value of b of $0.5-1.8$ for the power function and $0.35-1.4$ for the exponential function (Pizzuto, 1987), showing that b values decrease with increasing timescale of observation.

Floodplain deposition is an important part of the overall sediment budget of alluvial rivers (Figure 4.23). Using the Amazon River as an example of the exchange of sediment

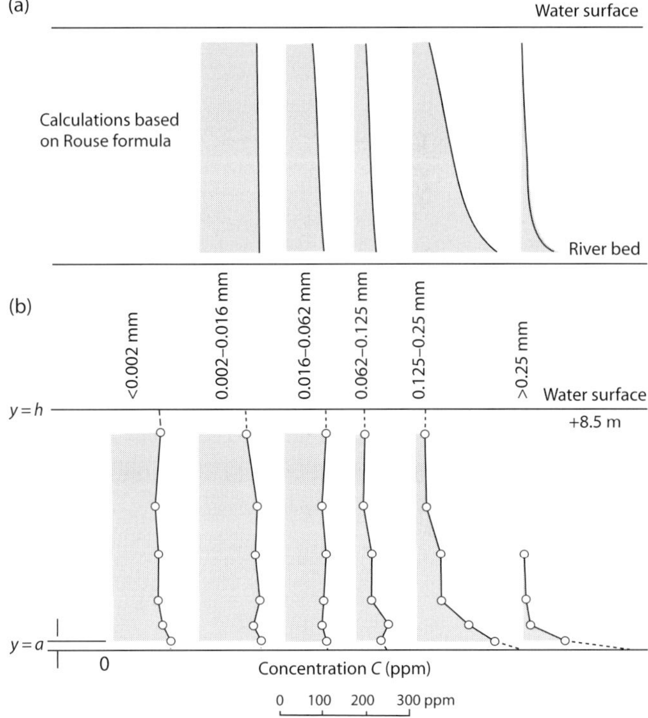

Figure 4.21 Measured suspended sediment concentration profiles for the Mississippi River at St Louis (b), compared to the predictions using equation (4.43) (a). Note that very little sand-grade sediment is measured or predicted in the near-surface flow, implying that floodplain sedimentation is predominantly composed of silt and clay. Part (b) from Colby (1963), published with permission of U.S. Geological Survey.

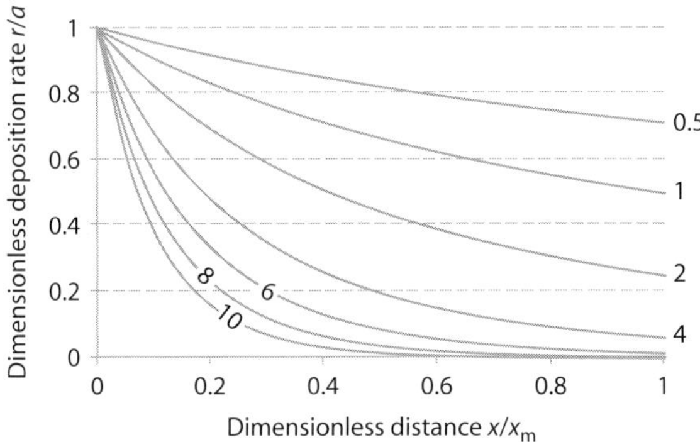

Figure 4.22 Plot of dimensionless distance from the edge of the channel belt versus dimensionless floodplain deposition rate using equation (4.45).

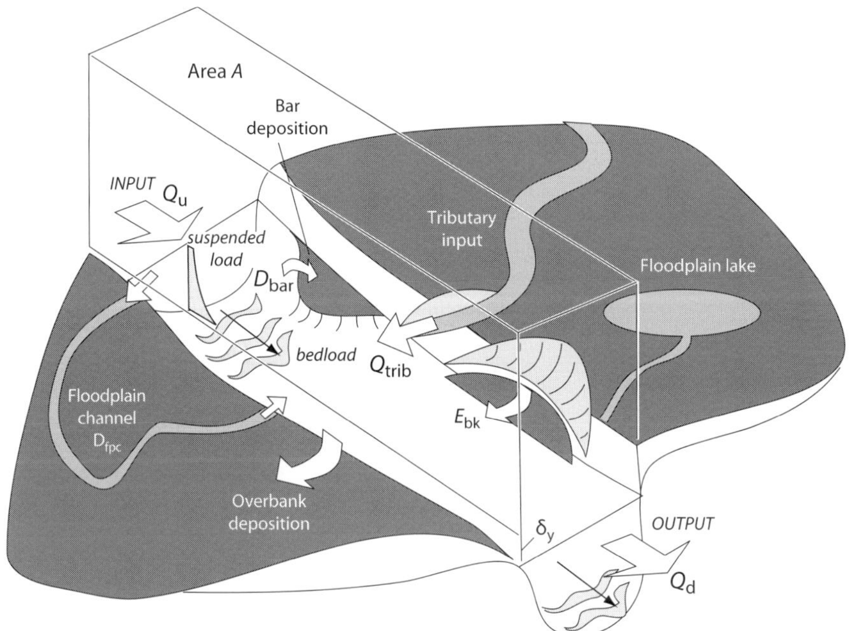

Figure 4.23 Sediment budget of a floodplain-channel reach. Sediment gains to the reach are the suspended and bedload sediment supplies from the upstream channel, from tributary inputs and from bank erosion. Sediment losses from the reach are by sediment transport in the channel, deposition on bars, overbank deposition including in floodplain lakes, and deposition in floodplain channels. Notation explained in main text. Modified from Dunne et al. (1998) (fig.1) with permission of Geological Society of America.

between channels and floodplains (Dunne et al., 1998), the sediment budget for a given reach of the river i can be written

$$Q_u + \Sigma_i Q_{trib\,i} + E_{bk} = Q_d + D_{bar} + D_{ob} + D_{fpc} + A_c \rho_b \frac{\partial y}{\partial t} \tag{4.47}$$

where Q_u and Q_d are the total (bedload plus suspended) annual sediment fluxes at the upstream and downstream ends of each reach, $\Sigma_i Q_{trib}$ is the total sediment flux from tributaries entering the reach, E_{bk} is bank erosion, D_{bar} is deposition on bars within and adjacent to the channel, D_{ob} is overbank (floodplain) deposition, D_{fpc} is deposition in floodplain channels, A_c is the surface area of the main channel bed and banks within the reach, ∂y is the average elevation change of the channel and banks within the reach, ∂t is the time interval of the calculation, and ρ_b is the bulk density of the bed material (1.7×10^{-6} Mt m^{-3}). Fluxes and deposition and erosion rates are in Mt yr^{-1}. The last term on the right is the change of sediment mass on the bed of the channel reach of area A_c over a time interval ∂t, that is, the rate of change of channel storage.

Delivery of sediment to floodplains in episodic. Individual floods have left a sediment record of packages 20–80 cm thick, taking place every 8 years on average, in the floodplain

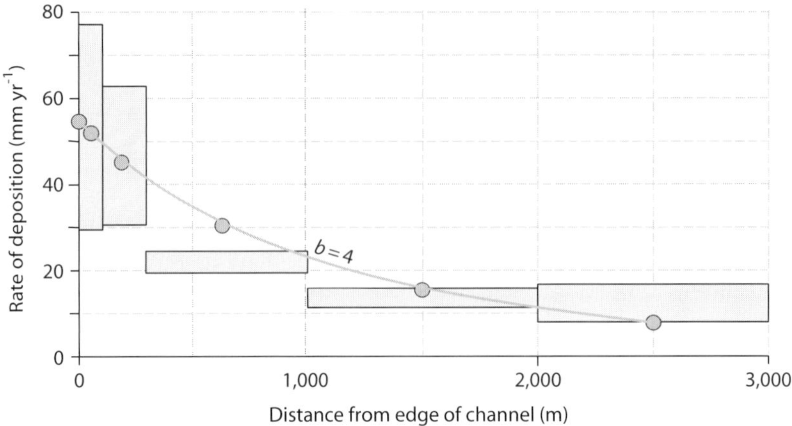

Figure 4.24 Lateral distribution of rate of deposition of flood deposits of the Beni River, with curve fitted from equation (4.45) with $b = 4$, $x_m = 4$ km and $a = 55$ mm yr^{-1}. Modified from Aalto et al. (2003) (fig.3) with permission of Nature Publishing Group.

of the Beni River, a tributary of the Amazon (Aalto et al., 2003). Such major floods occur at times of enhanced precipitation in the Andean catchments related to La Niña events. The best-fitting value of b in equation (4.45) is 4 (Figure 4.24), consistent with the values found in the floodplains of other major rivers.

The sediment budget of the Amazon channel-floodplain system can be assessed for the 2,010 km-long stretch between São Paulo de Olivença and Óbidos (Dunne et al., 1998) (Table 4.6). In broad terms, 616 ± 44 Mt yr^{-1} of sediment enters the channel at the upper end of the system at São Paulo de Olivença, which is added to by contributions from lowland tributaries (117 ± 8 Mt yr^{-1}) and from the sediment-laden River Madeira (715 ± 94 Mt yr^{-1}), making a total input of $1,448 \pm 146$ Mt yr^{-1}. The annual sediment discharge through the most downstream gauging station at Óbidos averages $1,239 \pm 130$ Mt yr^{-1}, so 209 Mt of sediment must accumulate within the channel-floodplain system each year, presenting 14% of the influx. Most of the stored sediment is silt and clay, the remainder being sand. A further $300-400$ Mt yr^{-1} of sediment is deposited on the delta plain downstream of Óbidos.

Analysis of the exchanges within the channel-floodplain system indicates that the combination of deposition on bars and in the floodplain exceed erosion of channel banks by $\sim$500 Mt yr^{-1}, in contrast to the estimate of $\sim$200 Mt yr^{-1} for the entire 2,010 km-long stretch, but the general magnitude and sign are in agreement. Exchanges within the channel-floodplain system exceed the values for channel transport. If the average sediment discharge past the Óbidos gauging station (1,240 Mt yr^{-1}) is expressed as a value of 1, the input at São Paulo de Olivença is 0.5, the contribution of the Madeira River is 0.6, that of other tributaries is 0.1, bank erosion is 1.3, while deposition in channel bars (0.3), overbank areas (1.0) and floodplain channels (0.4) together amount to 1.7. These figures

Table 4.6 *Summary of the channel-floodplain sediment budget of the Amazon River valley. From Dunne et al. (1998) (tab.3, p.465), with permission of Geological Society of America.*

Channel sediment transport (Mt yr^{-1})	
Input from São Paulo de Olivença Q_u	616 ($\pm$44)
Input from lowland tributaries Q_{trib}	117 ($\pm$8)
Input from River Madeira Q_{trib}	715 ($\pm$94)
Output at Óbidos Q_d	1,239 ($\pm$130)
Accumulation	209 ($\pm$167)
Channel-floodplain exchanges (Mt yr^{-1})	
Bank erosion E_{bk}	1,570
Bar deposition D_{bar}	380
Diffuse overbank sedimentation D_{ob}	1,230
Channelised floodplain sedimentation D_{fpc}	460
Net transfer to floodplain	500

imply that most or all of the sediment entering the ocean downstream of Óbidos has spent time stored in floodplain locations, where it is subject to chemical weathering.

It is clear that an important part of the sediment budget of many drainage basins is the sequestration of sediment in river floodplains. Alluvial sequestration not only reduces the potential downstream discharge of sediment into the coastal ocean, but also serves to buffer the transmission of signals through the sediment routing system (Romans et al., 2015). The potential for sequestration can be appreciated by applying the sediment load model BQART (Section 7.3) to an idealised catchment with run-off (m yr^{-1}) decreasing towards the coast, but discharge (km^3 yr^{-1}) increasing towards the coast (Syvitski and Kettner, 2008) (Table 4.7). The volume of sediment stored in the alluvial system can be expressed as a conveyance loss (%), which increases with downstream distance, reaching 38.2% at the most downstream station in the example given in Table 4.7.

Linked catchment-fluvial segments vary in the proportion of the drainage basin area occupied by net depositional alluvial systems compared to net erosional areas (referred to in Section 2.3.2 as the alluvial fraction, denoted by F_a). On the basis of an analysis of 28 submodern source-to-sink systems, Sømme et al. (2009) found that the proportion of the segment with topographic gradients of less than 0.01 increased with drainage basin area A with a power law regression of

$$F_a = 43.4A^{0.258} \tag{4.48}$$

where F_a is the alluvial fraction (approximated as having an slope of less than 0.01) and A is in units of $10^6 \times$ km^2 (Figure 4.25). Consequently, it can be argued that large drainage basins act as important storage zones for sediment. These large drainage basins are typical

Table 4.7 *Application of the BQART predictor to an idealised temperate drainage basin with total area 400,000 km² and maximum relief 3.55 km. From Syvitski and Kettner (2008) (tab.2, p.6) with permission of International Association of Hydrological Sciences (IAHS).*

1	2	3	4	5	6	7	8
Run-off m/yr	Area 10^3 km²	Relief km	Q_w km³/yr	BQART Q_s Mt/yr	BQART Yield t/km²/yr	Est. load Mt/yr	Conveyance loss %
2.50	50	2.55	125	32	643	32	0
1.50	100	2.70	150	51	509	64	20.8
1.00	200	3.00	200	88	437	129	31.9
0.75	300	3.30	225	122	407	193	36.6
0.65	400	3.55	260	159	397	257	38.2

Parameters: 1, Run-off from precipitation concentrated in upland area; 2, Upstream contributing area; 3, Topographic relief; 4, Water discharge; 5, BQART prediction for sediment discharge; 6, BQART prediction for sediment yield, decreasing downstream as commonly observed; 7, Estimate of sediment load assuming a constant sediment yield within the entire basin set at the upper catchment value of 643 Mt km^{-2} yr^{-1}; 8, Percentage of total load sequestered in alluvial system, representing the difference between (5) and (7).

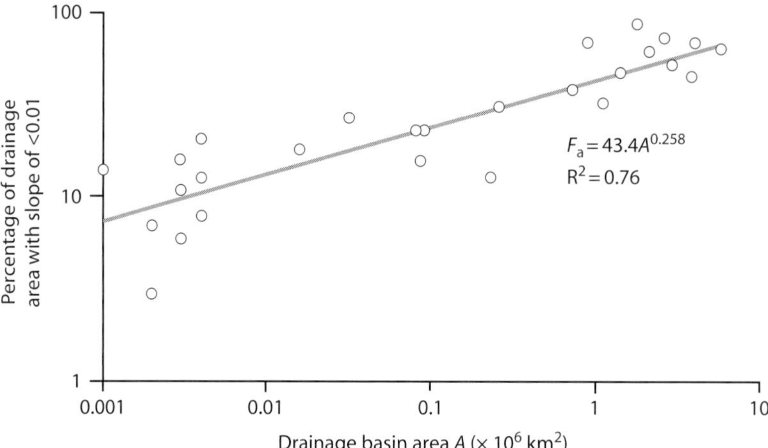

Figure 4.25 Plot of the percentage of the drainage basin with a slope of less than 0.01 versus drainage basin area, using data in table 1 of Sømme et al. (2009).

of passive margins, whereas small catchment areas with little alluvial storage are typical of active margins.

The tendency for large drainage basins to have high proportions of low-gradient topography, which is used here as a proxy for the alluvial fraction, is also seen in the

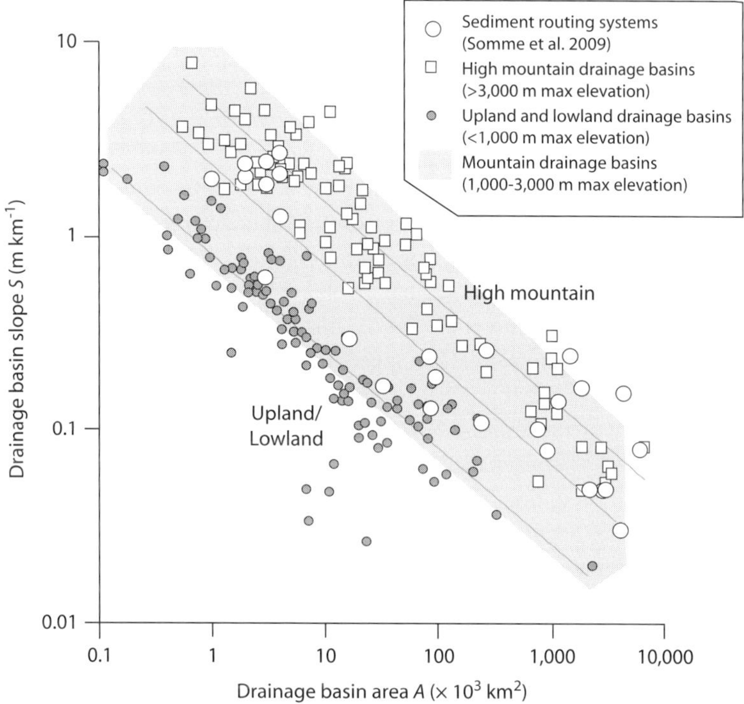

Figure 4.26 Average gradient versus area, categorised using elevation class of the drainage basin, from Milliman and Farnsworth (2011), and average channel gradient versus area for 28 source-to-sink systems from Sømme et al. (2009) (open circles). Grey lines are regressions for high mountain, mountain and upland/lowland categories. Data points for sediment routing systems fall within the 'mountain' elevation class. Modified from Milliman and Farnsworth (2011) (fig.2.16), published with permission of Cambridge University Press.

plot of the average slope $\hat{S}$ of 1340 drainage basins versus drainage basin area (fig.2.16, p.32 in Milliman and Farnsworth (2011)). The power law regression gives

$$\hat{S} = 0.39A^{-0.468} \tag{4.49}$$

The variation about this regression is partly accounted for by the elevation class of the drainage basin (Figure 4.26). High mountain basins (maximum elevation greater than 3,000 m) plot with high average gradients for a given drainage basin area, mountain basins (maximum elevation 1,000–3,000 m) plot in an intermediate position, and upland and lowland basins (maximum elevation less than 1,000 m) plot with low average gradients for a given drainage basin area. The 28 source-to-sink systems analysed by Sømme et al. (2009) have a power law regression of $\hat{S} = 0.45A^{-0.441}$, where A is in km² and slope is in m m⁻¹, which has a similar exponent (slope) to that in equation (4.49) (0.441 compared to 0.468). There is a similarly close correspondence for the coefficient (0.45 compared to

0.39), suggesting that the 'average' source-to-sink system has slope attributes similar to those of drainage basins in the mountain elevation class. Like the alluvial fraction, mean channel gradient is a useful indicator of sediment storage potential.

The storage potential of alluvial floodplains can also be demonstrated from the sediment discharges calculated at gauging stations and from isotopic geochemistry of sediment sampled in river floodplains. Galy and France-Lanord (2001) and Goodbred and Kuehl (1998), for example, estimated that about half of the sediment liberated by erosion in the Himalayan mountain chain was deposited in the Ganges–Brahmaputra floodplains before entering the ocean. Similarly, about half of the sediment released by the Andes is deposited in the floodplains of the Amazon (Aalto et al., 2006). It is estimated that 14% of the total annual sediment load of the Magdalena River is trapped in the Mompox Depression, which reduces the export to the ocean to 144 Mt yr^{-1} (Restrepo and Kjerfve, 2000). Before construction of the Three Gorges Dam on the Yangtze River, there was a downstream reduction of sediment discharge of 100 Mt yr^{-1} between the Yichang and Datong gauging stations. This reduction is attributed to deposition in floodplains, channels and lakes (Xu et al., 2007).

Cosmogenic nuclides accumulate in erosional source areas and their concentrations can be used to estimate denudation rates (Section 7.6). However, changes to nuclide concentrations may also take place when the quartz grains bearing the cosmogenic dose are transported through and stored within alluvial systems. The short timescale of transport events prevents significant change to cosmogenic nuclide concentrations, whereas storage near the surface of the floodplain and channel banks and bars allows further accumulation of nuclides, and burial at greater depths shields the quartz grains, allowing concentrations to decrease over time by radioactive decay. Burial at depths below the production zone (of order a metre) is likely in old, static floodplains, where storage times may be in excess of a million years. Shallow and short-lived burial is typical of dynamic floodplains characterised by rapid channel migration.

The total nuclide concentration of floodplain sediment C_{total} is made up of a concentration inherited from the source and supplied through an upstream cross-section of the channel-floodplain system C_{up} plus a concentration C_{bank} generated during floodplain storage of duration Δt, which is calculated following Schaller et al. (2004)

$$C_{total} = C_{up}e^{(-\lambda \Delta t)} + C_{bank} \qquad (4.50)$$

where λ is the decay constant (yr^{-1}) for the nuclide being measured (^{26}Al, ^{10}Be, ^{14}C). In each following compartment of the channel-floodplain system, C_{up} is mixed with sediment and nuclide fluxes derived by bank erosion as the river shifts in its position across the floodplain. The nuclide flux derived from bank erosion C_{bank} is mainly a function of the surface production rate P_0, the storage depth in the river bank and the nuclide production profile with depth. The resulting mixture therefore has a concentration

$$C_{mix} = \frac{(Q_{up}C_{up}) + (Q_{be}C_{mix(y)})}{Q_{total}} \qquad (4.51)$$

Table 4.8 *Parameters used in the floodplain cosmogenic nuclide model of Wittmann and von Blanckenburg (2009)(tab.1, p.248). Published with permission of Elsevier.*

River/ country	Channel migration rate (m yr^{-1})	Channel depth (m)	Channel belt width (km)	Q_{up} sediment flux (Mt yr^{-1})	Sediment residence time (kyr)	Cumulative duration (kyr)
Amazon/ Brazil	5	23	35	700	7.0	14
Mississippi/ USA	25	6	100	140	4.0	16
Beni/ Bolivia	15	20	6	210	0.4	4
Rhine/ Netherlands	5	4	3	3.5	0.6	9
Pearl/ USA	1	6	4	2.0	4.0	16
Vermillion/ USA	0.4	1	0.2	0.1	0.5	16

Sediment residence time is channel belt width divided by channel migration rate. Cumulative duration is age of alluvial sediment from radiometric dating and sediment budget studies.

where the total sediment flux Q_{total} is the sum of the sediment discharged into the channel-floodplain system, such as that measured at a gauging station Q_{up}, and the flux released by bank erosion Q_{be}, and $C_{mix(y)}$ is the concentration tapped by bank erosion integrated over the depth y. This procedure is repeated for every compartment of the channel-floodplain system (Wittmann and von Blanckenburg, 2009).

The age of sediment in river floodplains can be measured using U-series dating (Dosseto et al., 2006) or by the calculation of sediment budgets (Hoffmann et al., 2007). On this basis, residence times of sediment are typically 10^3-10^4 years (Table 4.8). Active, dynamic floodplains limit the residence time of sediment, so ^{10}Be and ^{26}Al do not decay and concentrations may show a slight increase relative to the inherited dose from the source region as a result of shallow burial and surface exposure. This means that transport of sediment through the channel-floodplain system does not change the ^{10}Be concentration of river-borne sediment and that the signal from the source region is preserved. In sediment-laden rivers in dynamic settings, the upstream flux overwhelms the signal generated by bank erosion, so that even in systems with long residence times, the upstream signal is preserved. This is supported by measurements of ^{10}Be along about 500 km of the floodplain of the Beni River tributary of the Amazon (Wittmann and von Blanckenburg, 2009), which show no significant downstream variation. Consequently, cosmogenic nuclide concentrations of long half-life isotopes such as ^{10}Be can be used to estimate denudation rates in sediment source regions despite having undergone a protracted history of floodplain storage (Wittmann et al., 2009).

In contrast, the results for in situ-produced ^{14}C, which has a short half-life, show a significant decrease in concentrations during residence in the channel-floodplain system. This isotope might be used in future to evaluate floodplain residence times.

4.6 Palaeohydrology of Rivers

The mass balance of a sediment routing system can be estimated by approximating the total sediment mass passing through a chosen cross-section, or fulcrum, of an axial fluvial system (Holbrook and Wanas, 2014). The discharge of sediment across the fulcrum can be estimated using standard palaeohydrological methods. Characteristics of the rivers responsible for the transport and deposition of sediment can be estimated by 'inversion' of properties shown by alluvial stratigraphy such as channel storey dimensions and constituent grain sizes. Such inversion techniques are based on the fundamental principle that the driving force for the movement of water is the downslope component of gravity acting on the mass of water. The downslope component is balanced against the upslope-acting frictional force at the bed. The fulcrum approach has the advantage over other methods in not requiring the catchment area to be defined and in not requiring basin volumes to be known. However, the fulcrum method involves uncertainties in the algorithms used to calculate water and sediment discharges and in the measurement of channel bodies that are representative of the alluvial architecture over the time span of interest (Hajek and Wolinsky, 2012).

Palaeohydrological analysis is fundamentally based on the equation for steady uniform flow down an inclined plane (equation 4.28) (Section 4.4.2) combined with an expression for flow resistance (equation 4.29) such as the friction factor f, Manning's roughness coefficient n or the Chezy coefficient C.

The key steps in a palaeohydrological analysis of former rivers, whether recently active Quaternary channels (Maizels, 1986) or of channel bodies preserved in outcropping or subsurface stratigraphy (Davidson and Hartley, 2010; Hajek and Wolinsky, 2012; Holbrook and Wanas, 2014) rely on the measurement of channel body dimensions and grain size. Bankfull depth of main trunk rivers can be estimated on the basis of the thickness of fining-upward cycles (storeys) preserved in outcrops, cores and well logs. Bankfull channel width is measured in outcrops perpendicular to the palaeoflow direction, and where flow-perpendicular sections are not available, empirical relationships can be used, dependent on fluvial style (for example, meandering single-thread channel versus braided channels) (Gibling, 2006; Blum et al., 2013). Flow resistance and palaeoflow velocity are estimated from measurements of bedform geometry and grain size (Rubin and McCullough, 1980; LeClair and Bridge, 2001) combined with formulations of the resistance coefficient (Van Rijn, 1984). In summary, Holbrook and Wanas (2014) give a suite of equations for the bankfull palaeodischarge of water Q_w, the discharge of bedload sediment Q_b, the discharge of suspended sediment Q_{ss} and the total sediment discharge Q_t.

A compilation of data on the dimensions of fluvial channel bodies shows that the mobile channel belts of braided and low-sinuosity rivers make up the bulk of the fluvial

(a) Braided and low-sinuosity rivers

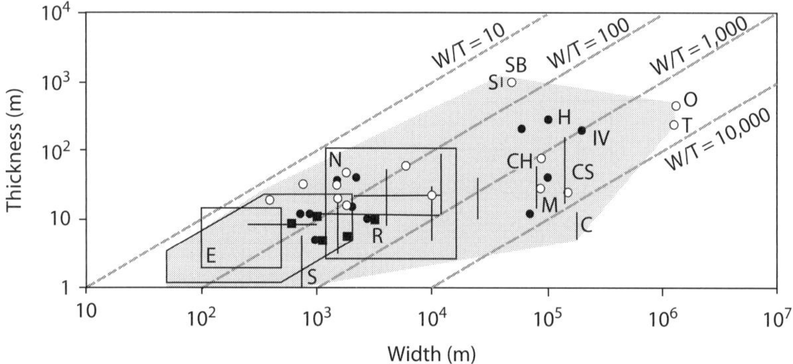

(b) Meandering rivers

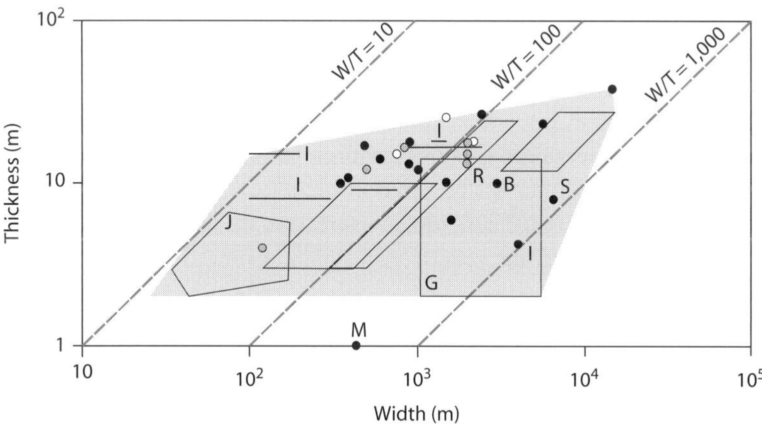

Figure 4.27 Compilation of width/thickness ratio for mobile channel belts of braided and low-sinuosity rivers (a) and meandering rivers (b). Open circles, minimum W at known T; black circles, W, T for single channel bodies (bedrock); black squares, W, T (Quaternary); grey shaded area, envelope in W/T space. In (a), C, Cadomin Formation; CH, Cypress Hills Formation; CS, Castlegate Sandstone; E, Escanilla Group; H, Hawkesbury Sandstone; I, Ivashak Sandstone; M, Mesa Rica Formation; N, Newcastle Coal Meaures (38 bodies); O, Ogalalla Group; R, Quaternary, Riverina, Australia; S, Siwalik Group; SB, South Bar Formation; T, Tuscarora Formation. In (b), B, Beaufort Group; G, German Creek Formation; J, Joggins Formation; M, Miocene, Spain (Murillo el Fruto); I, Indonesian Cenozoic; R, Rangal Coal Measures (grey circles); S, Scalby Formation. Redrawn from Gibling (2006) (fig.6, p.741) with permission of Society of Economic Paleontologists and Mineralogists (SEPM).

stratigraphic record, whereas meandering systems do not on the whole build basin-scale deposits (Gibling, 2006). The width/thickness ratios of mobile channel belts (Figure 4.27) includes very large examples of braided and low-sinuosity rivers with widths greater than 40 km and thicknesses less than 1, 200 m, as in the case of the Castlegate Sandstone of Utah and the Hawkesbury Sandstone of eastern Australia. Meandering river deposits form a coherent group with moderate width and thickness of channel bodies (Figure 4.27b). Although W/T data overlap with those of low-sinuosity and braided rivers, the field for meandering rivers is more restricted. Maximum thicknesses for meandering types is 38 m and widths are greater than 15 m and typically less than 3 km. Gibling (2006) provides further W/T plots, together with analysis, of distributary systems and channel bodies in a range of other settings. Of these other types, valley fills on bedrock surfaces show a very wide range of dimensions, from 12–1400 m thick and 75–5200 m wide. A fine example is the Eocene Sis palaeovalley of the Spanish Pyrenees (Vincent, 2001). In the Cretaceous Dunvegan Formation of western Canada, an empirical relation was used for single-thread, meandering channels (Plint and Wadsworth, 2003; Lin and Bhattacharya, 2017):

$$W_c = 9.8T_c^{1.82} \tag{4.52}$$

where W_c is channel body width and T_c is mean bankfull channel depth estimated from deposit thickness.

The use of the results from palaeohydrological analysis in making estimates of the mass balance of stratigraphic systems is problematical because the relationship between sediment transport rates in short-term flood events and longer-term discharge averages is poorly known. Holbrook and Wanas (2014) expressed this relationship in terms of a dimensionless year-averaged bankfull duration $\hat{t}_b$ (the bankfull event duration divided by the recurrence interval of the bankfull event) and a multiplier expressing the inverse proportion of the total sediment load transported over the year-averaged bankfull duration b:

$$Q_{mas} = Q_{bs}(\hat{t}_b)b \tag{4.53}$$

where Q_{mas} is the total sediment volume discharged per year on average and Q_{bs} is the averaged bankfull sediment discharge rate. Average values of mean annual sediment discharge of individual channels can then be used to calculate the total mass or volume of sediment discharged across the river fulcrum cross-section over the time span of interest. Long-term averaged sediment discharges calculated in this way can be compared with 'regional curves' from different climates using modern drainage basin averages (Davidson and North, 2009).

Examples of the mass balance approach using river palaeohydrology include the Cretaceous Bahariya Formation of Egypt (Holbrook and Wanas, 2014) and the Cretaceous Dunvegan Formation of western Canada (Lin and Bhattacharya, 2017). In addition to providing important information on sediment fluxes and budgets in palaeo-rivers, these examples allow the size of sediment source areas to be estimated and compared with estimates derived from other methods.

5

The Continental Shelf Segment

5.1 Dynamics at River Mouths

The sediment routing system is strongly impacted by deposition at the coast in subaerial and subqueous deltas. The coast is the most dynamic of transition zones and moving boundaries in sediment routing systems. Deltas comprise the inclined and commonly sigmoidal sediment bodies, or *clinoforms*, that are the building blocks for much of the stratigraphic record of continental margins. The inclined depositional surfaces migrate by progradation driven by sediment supply. Deposition in deltas is strongly influenced by the sediment discharge of rivers as well as the physical oceanography and bathymetry of the marine basin into which the river debouches (Wright, 1977; Allen, 1997). The clinoforms in the stratigraphic record, caused by the migration of the front of deltas, are commonly recognised on seismic reflection profiles. The shape of the clinoforms reflects the interplay of river discharge and basin energy (Swenson et al., 2005).

The processes at river mouths fall broadly into the phenomena known as *mixing*. Mixing is encountered in a variety of natural environmental settings, such as the meeting of two rivers with different sediment loads at a confluence, the flow into a lake of a meltwater stream, the mixing of fresh and salty water in an estuary, and the eruption into the atmosphere of a column of ash from a volcano. The primary concern here is the dynamics of rivers entering the ocean, which can be divided into two broad categories:

1. Buoyancy-driven flow resulting from density differences between fresh and salty water, which gives rise to a low-salinity plume moving over a lower salty layer, and
2. Inertia-driven flow of river water into a reservoir such as a lake or sea, known as a jet.

Instabilities commonly develop at the interface between different fluids as a result of interfacial shear. These instabilities may grow into vortices that draw in ambient fluid, causing mixing of the two fluid masses. Alternatively, instabilities may be damped by the density contrast of two fluid masses. Consequently, mixing is fundamentally important to sediment deposition since mixing leads to a reduction of the density differences that drive fluid motion and keep sediment grains suspended, mixing leads to flow expansion by the entrainment of one fluid mass into the other, causing deceleration, and mixing may also cause geochemical changes leading to precipitation and flocculation.

The simplest case of mixing is when a narrow jet of fluid enters a large, deep reservoir. The jet expands into a conical shape and mixes with the ambient fluid. If the reservoir, such as a lake, is deep, the jet is unaffected by solid boundaries and is axisymmetric. If there is no entrainment across the boundaries of the jet, fluid in the jet decelerates rapidly with distance travelled x, so that the mean velocity $\hat{U}$ varies inversely with the square of x. If ambient fluid is drawn in, the mean velocity of the fluid in the jet decelerates less rapidly, with $\hat{U}$ varying simply inversely with x. Jets become progressively diluted with ambient fluid as they spread away from the outlet.

However, although the simple axisymmetric jet model might apply to an eruption column from a volcano, it is unlikely to apply to the conditions at river mouths except for rivers entering deep, steep-sided lakes, such as the Rhein River effluent in Lake Constance (Müller, 1966; Allen, 1997). Instead, it is more common for the jet to be unable to spread freely in all directions due to the presence of a shallow bottom boundary comprising the sea bed. Incorporation of sea bed friction leads to marked lateral expansion and rapid deceleration of the plane jet, causing sediment to be spread laterally in a narrow zone close to the shoreline.

The mixing of fresh and salty water where rivers enter the sea results in buoyancy-driven outflows and generally takes place in estuaries that deeply indent the shoreline. Estuaries are not passive funnels through which sediment is passed en route for the ocean. They are dynamic entities in which complex physical and chemical changes take place. Their mixing behaviour has a profound influence on the seaward extrusion of low-salinity plumes onto the continental shelf. The presence of a salinity and therefore density gradient along the estuary results in a pressure difference that drives a seaward flow of low-salinity water over a deeper landward flow of saline water. Tidal flows cause a mixing of the water masses of different salinity, so estuaries may be of 'salt wedge' or 'stratified' type characterised by a seaward low-salinity plume over an intruding salt water wedge, or 'mixed', or 'partially mixed' where the stratification in completely or incompletely broken down.

If estuaries are wide enough, the seaward and landward flows may be separated by the topography of channels and shoals, with the Coriolis force causing a deflection to the right of low-salinity outflowing water and landward flowing salty water in the northern hemisphere, when looking down the flow direction.

Salt wedge intrusion may extend considerable distances up river channels. During low-discharge conditions of the Mississippi River, a salt wedge extends more than 300 km upstream from its mouth. Under high discharge conditions, however, the salt wedge is pushed entirely out of the estuary and sediment-laden river water flows into the Gulf of Mexico as a plume that is recognised as far as the continental edge 50 km from shore. In the case of the Amazon River, low-salinity water is found as much as 1,000 km from the river mouth. Flow of the plume is caused by lateral pressure gradients set up by the density contrast with the underlying water, so mixing reduces the effectiveness of the buoyancy-driven flow.

A wide range of marine processes potentially modify the outflow dynamics of turbulent jet inertia and buoyancy-driven flow, including internal waves, coastal and nearshore currents, wind-driven ocean currents, thermal stratification and the presence of sea ice, but chief of which are:

1. Tidal currents: an index of the strength of tidal currents is the tidal range of the sea into which the rivers enters. Powerful currents are found in macrotidal seas (range in excess of 4 m) and commonly exceed the speed of the river, as in the delta of the Ganges–Brahmaputra in the Bay of Bengal, the Klang of Malaysia and the Ord of northern Australia.
2. Surface gravity waves: regions of small fetch and with shallow offshore slopes that attenuate wave energy by friction tend to have low wave energy at the coast. However, outflows that enter large seas with steep offshore slopes, such as that of the Senegal delta, western Africa, may be strongly modified by waves, enhancing mixing of plumes and redistributing deposited sediment shorewards and alongshore. The strength of the inertial flow relative to wave energy is expressed as a *discharge effectiveness index* (Wright and Coleman, 1973).

Delta morphology and sediment budgets are therefore strongly affected by outflow dynamics. In particular, wave action redistributes sediment provided by river discharge. Low river discharges relative to wave power causes sediment reworking into a wave-dominated configuration of linear shorelines with barriers, spits and aeolian dunes, as is typical of examples such as the Sao Francisco and Senegal deltas (Wright and Coleman, 1972). Where the river discharge is large compared to wave power, effluent-dominated configurations result, forming prograding digitate distributaries, and indented shorelines with marshes and bays, as in the Mississippi delta. Oblique wave approach intensifies longshore drift and forces the delta to be asymmetrical, with pronounced differences in morphology and sedimentary facies on the updrift and downdrift sides of the river effluent (Komar, 1973; Bhattacharya and Giosan, 2003; Weiguo, Bhattacharya, and Yingmin, 2011). An asymmetry index is defined as

$$A = \frac{Q_L}{Q_R} \tag{5.1}$$

where Q_L is the net longshore transport rate in m^3 yr^{-1} and Q_R is the river discharge in 10^6 m^3 yr^{-1}. Symmetrical deltas, such as the Tiber in Italy and Ebro in Spain, have values of $A < 200$, whereas asymmetrical deltas such as the Danube, Black Sea and Brazos, Gulf Coast United States, have values of $A > 200$.

The strength of the river effluent can be assumed to be given by the river discharge per unit width of river mouth, which can be estimated from gauging station records. The strength of waves in the receiving basin can be assumed to be related to the wave power per unit width of wave crest, but its measurement is problematical. At what water depth should wave power be calculated, and is sufficient intrumentation deployed off the world's major deltas? To circumvent these problems, deep water (water motion unaffected by the

Table 5.1 *Mean annual river discharge and wave power data for 7 of the world's major deltas. From Wright and Coleman (1972) (tab.2, p.283) with permission of the American Association for the Advancement of Science.*

River	Deep-water wave power (erg s^{-1})	Nearshore wave power (erg s^{-1})	River discharge (10^3 m^3 s^{-1})	Discharge effectiveness index	Attenuation ratio
Mississippi	1.06×10^{10}	1.34×10^6	17.69	1.00	7,913.3
Danube	2.30×10^9	1.40×10^6	6.29	0.214	2,585.0
Ebro	7.28×10^9	5.09×10^6	0.55	0.0487	12,99.5
Niger	6.76×10^9	6.59×10^7	10.90	0.000803	102.8
Nile	1.36×10^{10}	3.21×10^8	1.47	0.000586	42.46
Sao Francisco	3.71×10^{10}	9.97×10^8	3.12	0.000237	37.2
Senegal	1.56×10^{10}	3.77×10^9	0.77	0.0000475	4.16

Wave power is per unit metre of wave crest. The *DEI* is normalised to the maximum *DEI* so that the *DEI* of the Mississippi is 1.0.

friction of the sea bed) wave characteristics can be calculated from published records of wind data. These deep water wave conditions are then used to compute the effects of shoaling, refraction and frictional attenuation. Wave power is calculated at decrements of water depth from deep water to the shoreline. The discharge effectiveness index is then

$$DEI = \frac{Q_R}{P_W} \tag{5.2}$$

where Q_R is the average river discharge per unit channel width with dimensions of $[L^3T^{-1}]$ and P_W is the average nearshore power per unit width of wave crest with dimensions of $[ML^2T^{-3}]$. Consequently *DEI* has dimensions of $[M^{-1}LT^{-2}]$ and absolute values have no physical meaning. Nevertheless, modern deltas can be ranked according to their value of *DEI* (Table 5.1). Values for modern deltas range from the highly river-dominated Mississippi delta (*DEI* > 10^5) to the wave-dominated Senegal delta (*DEI* < 1) (Figure 5.1). Homewood and Allen (1981) carried out an analysis of a lower Miocene shallow marine succession in the Molasse of western Switzerland that enabled them to compare their Miocene example with modern deltas in terms of outflow dynamics.

Delta morphology is correlated with the value of *DEI* rather than with absolute values of river discharge or deep water wave power. Deep-water waves lose their power due to frictional attenuation on the sea bed, which depends on the subaqueous slope (Figure 5.1a). Shallow slopes cause greater frictional attenuation of wave power, whereas steep subaqueous profiles allow strong waves to impact the nearshore zone. Consequently, the shallow sloping linear or slightly convex seabed profile of the Mississippi delta causes low wave

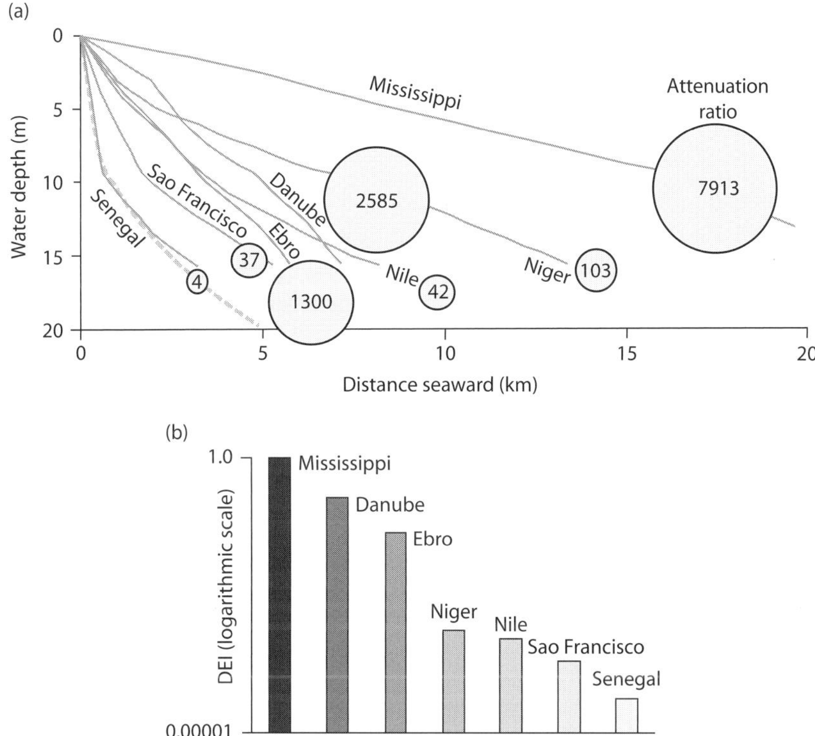

Figure 5.1 (a) Subaqueous profiles for 7 of the world's major deltas, with Attenuation Ratio, using data in Wright and Coleman (1972). Dashed grey line is shoreface profile assuming the Bruun relationship $h = Ax^n$, where h is water depth (m), x is the horizontal coordinate of seaward distance (m), A is a scale coefficient taken as 0.67, and the exponent n is taken to be 2/5. (b) Major deltas ranked according to Discharge Effectiveness Index on a logarithmic scale. The slope of the subaqueous profile scales more closely on *DEI* than on Attenuation Ratio *AR*.

energy at the coast and a high *DEI*. In contrast, the steep and concave seabed slopes of the Senegal and Sao Francisco deltas allows high wave energy to reach the coast, leading to low values of *DEI*. Wave-built profiles tend to be concave.

The extent of the frictional reduction of wave power is seen in the Attenuation Ratio (AR), which is given by

$$AR = \frac{P_0 r_s^2}{P_s} \tag{5.3}$$

where P_0 and P_s are the deep-water and shoreline wave power respectively, and r_s is a refraction coefficient, such that a value of 1 indicates complete conservation of power, and values greater than 1 indicate the degree to which power is lost. Frictional attenuation increases in shallower water depths, especially shorewards of the 10 m isobath. Seabed slope and delta morphology therefore depend on the supply of sediment by rivers relative to the ability of waves to redistribute it. Redistribution of sediment by waves results in sediment transport in littoral cells parallel to the coast and therefore influences the pathway followed by sediment en route to the deep sea (Section 6.2).

Syvitski and Saito (2007) used tidal range and wave height as a proxy for marine power P_m at delta fronts

$$P_m = T_i^2 + W_a^2 \tag{5.4}$$

where T_i is the tidal range and W_a is the maximum monthly wave height. A proxy for river power P_r is

$$P_r = 11 Q_w S_{dp} \tag{5.5}$$

where Q_w is the average water discharge and S_{dp} is the gradient of the delta plain, and the coefficient 11 is chosen so that $\Sigma P_m = \Sigma P_r$. The effectiveness of the fluvial sediment supply in building a constructional delta against the erosive forces of the marine basin is indexed by a supply-to-dispersal parameter φ (Table 5.2) (Figure 5.2)

$$\varphi = \frac{10^{-3}(Q_s + Q_b)}{P_m} \tag{5.6}$$

Table 5.2 *Marine versus fluvial power and the supply-to-dispersal parameter for 51 of the world's major deltas. Data from Syvitski and Saito (2007) (tab.1, pp.264–265). Published with permission of Elsevier.*

Delta	Wave height W_a (m)	Tidal range T_i (m)	Marine power P_m	Fluvial power P_f	Supply-to dispersal φ
Amazon	2.0	6.0	40	21.9	0.97
Arno	2.0	0.7	4.49	0.48	0.02
Brazos	1.5	0.7	2.74	0.79	0.11
Ceyan	1.5	0.7	6.25	0.32	0.07
Chao Phraya	1.5	2.0	25.25	3.05	0.06
Colorado (CA)	0.5	5.0	2.74	0.23	0.15
Colorado (TX)	1.5	0.7	20.56	8.18	0.02
Copper	3.0	3.4	2.25	4.24	0.12
Danube	1.5	0	2.29	3.23	0.99
Ebro	1.5	0.2	21.25	2.59	0.28
Eel	3.5	3.0	18.25	5.25	0.03
Fly	1.5	4.0	22.5	2.35	0.13
Fraser	1.5	4.5	13.96	13	0.03
Ganges–Brahmaputra	1.0	3.6	13	7.87	2.55
Godavari	2.0	3.0	13.96	0.61	0.43

Table 5.2. *(cont.)*

Delta	Wave height W_a (m)	Tidal range T_i (m)	Marine power P_m	Fluvial power P_f	Supply-to dispersal φ
Homathko	1.0	3.6	2.89	16.28	0.01
Huanghe	1.5	0.8	2	2.86	12.0
Inidigirka	1.0	1.0	37.25	4.19	0.21
Indus	3.5	5.0	19.89	11.93	0.21
Irrawaddy	1.5	4.2	21.25	0.68	0.42
Klamath	3.5	3.0	17	1.20	0.02
Klinaklini	1.0	4.0	3.25	3.33	0.01
Kolmya	1.5	1.0	13	9.77	0.08
Krishna	2.0	3.0	2	1.79	0.17
Lena	1.0	1.0	22.25	4.60	0.33
Limpopo	2.5	4.0	22.25	4.96	2.8
MacKenzie	1.2	0.2	1.48	5.36	0.5
Magdalena	1.5	1.5	4.5	8.28	0.9
Mahanadi	2.0	3.5	16.25	2.32	0.6
Mekong	1.5	2.6	9.01	5.72	0.4
Mississippi	0.5	0.4	0.41	3.40	0.8
Niger	1.4	3.0	10.96	4.72	0.4
Nile	1.5	0.4	2.41	3.45	0.3
Orange	2.5	1.5	8.5	4.86	93.3
Orinoco	1.3	1.9	5.3	7.59	0.3
Parana	0.5	4.0	16.25	9.57	1.1
Pechora	2.0	3.0	13	0.45	0.7
Pescara	1.5	0.6	2.61	0.27	50.9
Po	1.5	0.7	2.74	0.67	3.4
Rhone	2.0	0.5	4.25	2.06	3.4
Song Hong (Red)	1.5	3.5	14.5	2.08	1.2
Squamish	1.0	3.0	10	0.33	4.1
Tigris–Euphrates	1.0	3.0	10	3.63	1.8
Var	2.0	0.5	4.25	0.37	4.3
Vistula	2.0	1.0	5	0.23	1.1
Volga	1.0	0.0	1	1.80	0.8
Waipaoa	3.0	1.2	10.44	2.10	87.5
Yana	1.5	1.0	3.25	0.33	0.5
Yangtze	1.5	4.5	22.5	18.66	1.7
Yukon	2.0	1.5	6.25	8.01	1.4
Zhujiang (Pearl)	1.5	5.0	27.25	5.41	1.0

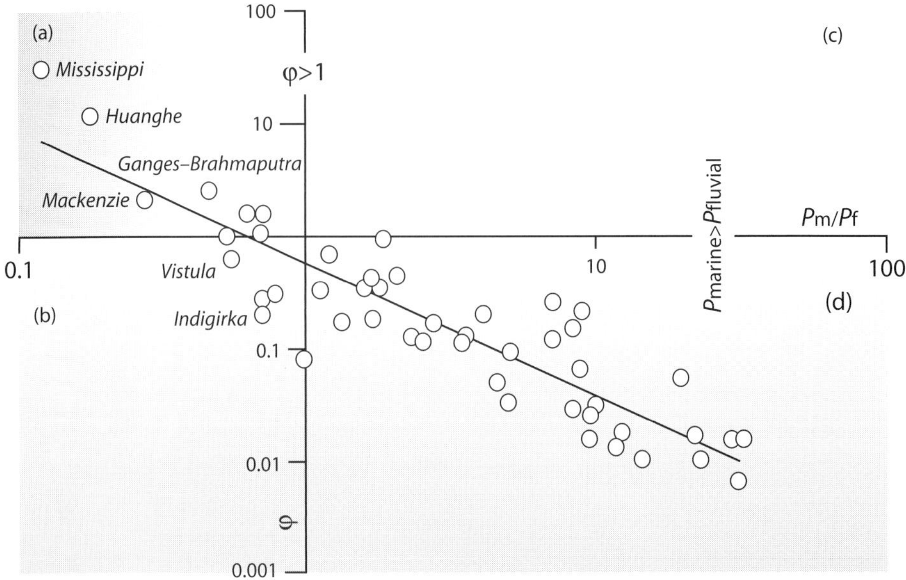

Figure 5.2 Log-log plot of the supply-to-dispersal parameter φ versus the ratio of marine to fluvial power P_m/P_f. Quadrant (a) is characterised by deltas with relatively high supply rates and relatively low marine power, such as the Mississippi and Huanghe. Quadrant (b) has relatively low supply rates and relatively low marine power, such as the Vistula and Indigirka. Quadrant (c) has relatively high supply rates and high marine power and is unpopulated. Quadrant (d) is characterised by relatively low sediment supply rates and high marine power, and contains the bulk of the world's major deltas.

The grain size of the topset (overbank) deposits is calculated for 19 of the 51 deltas in the Syvitski and Saito (2007) database. A power law relationship is found between river mainstem length L in km and the mean size D in mm of the suspended load (Table 5.3)

$$D = 18.47L^{-0.681} \tag{5.7}$$

with a considerable amount of scatter ($R^2 = 0.584$) (Figure 5.3). In essence, long rivers feed fine sediment to delta topsets (as in the Amazon, Orinoco and Mekong), and short rivers feed relatively coarse sediment to the delta topset (as in the Waipaoa and Po). Downstream grain-size fining is controlled in part by the extraction from the sediment load by selective deposition (Section 9.4). Consequently, some of the variation seen in Figure 5.3 is undoubtedly caused by variations in the tectonic subsidence generating accommodation in the catchment-fluvial segment.

Table 5.3 *Mean grain size of delta topsets in relation to the mainstem length of some of the world's major rivers. Data from Syvitski and Saito (2007) (tab.1, pp.264–265). Published with permission of Elsevier.*

Delta	Mainstem length L (km)	Grain size D (mm)
Amazon	6,516	0.03
Copper	40	0.25
Ebro	930	0.2
Fraser	1,370	0.25
Ganges–Brahmaputra	2,840	0.16
Homathko	115	0.14
Huanghe	4,845	0.06
Irrawaddy	2,150	0.05
Klinaklini	150	0.3
MacKenzie	4,250	0.2
Mekong	4,425	0.1
Mississippi	6,020	0.01
Niger	4,170	0.15
Nile	6,669	0.03
Po	652	0.4
Rhone	820	0.2
Squamish	110	0.3
Waipaoa	47	10
Yangtze	4,670	0.05

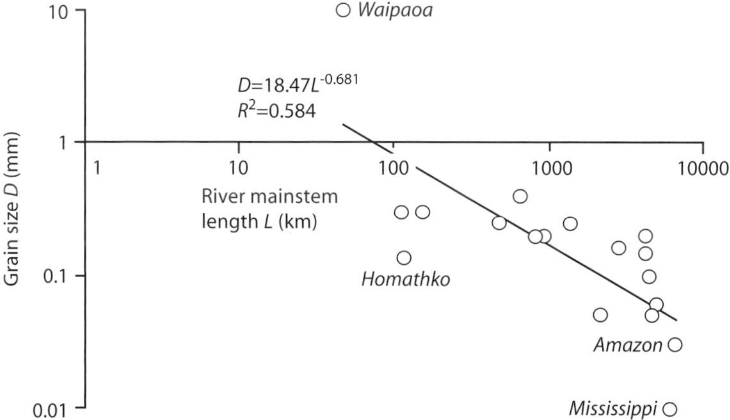

Figure 5.3 Log-log plot of mean grain size (mm) in the delta topset against the mainstem length of the river (km) for 19 of the world's major deltas, using data from Syvitski and Saito (2007) (tab.1).

5.2 Natural Range of Deltaic and Subaqueous Clinoforms

Sediment delivered to the ocean by rivers is commonly preserved in basinward-dipping architectures known as clinoforms (Patruno, Hampson, and Jackson, 2015). Clinoforms occur over a range of vertical scales, from tens to thousands of metres, and comprise, in a proximal to distal direction, (1) subaerial deltas, (2) subaqueous deltas, (3) shelf prisms and (4) continental margins (Helland-Hansen and Gjelberg, 2012). Subaerial and subaqueous clinoforms have vertical relief of tens of metres, whereas shelf-prisms consist of clinoforms of 100–500 m vertical relief, and the clinoforms that build continental margins may be several thousands of metres in relief (Figure 5.4).

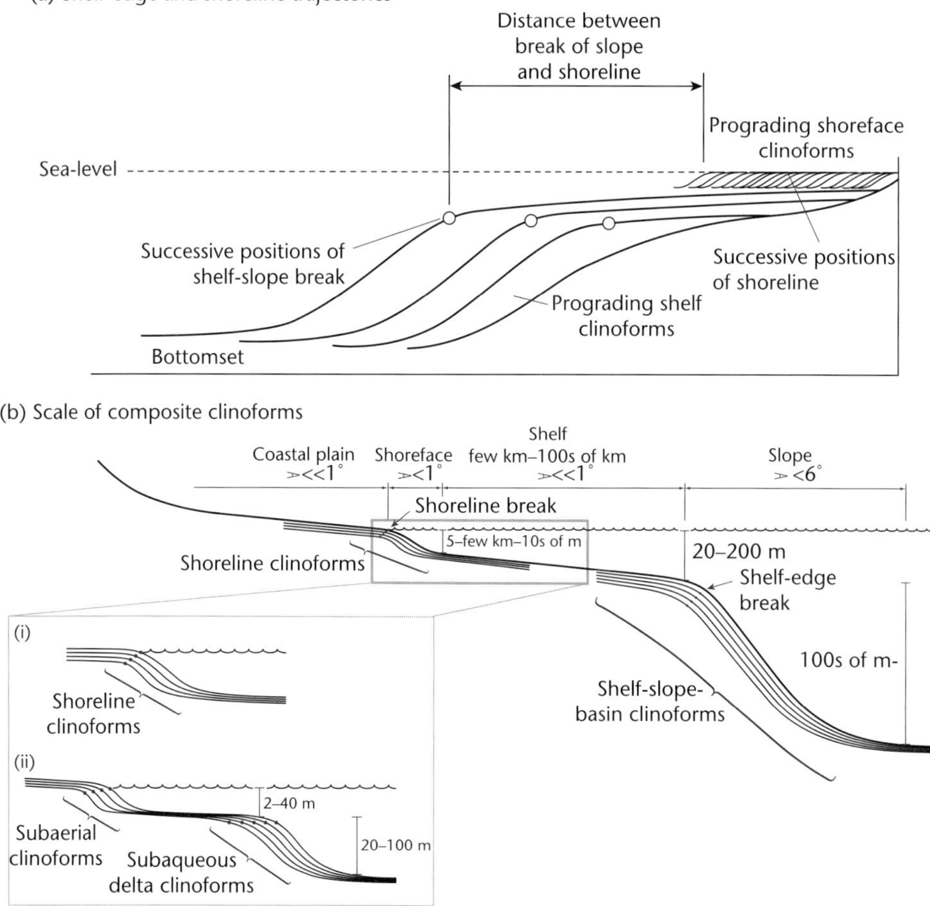

Figure 5.4 (a) Schematic diagram illustrating shoreline and shelf-edge trajectories for the case of prograding shoreface and shelf-slope clinoforms. (b) Dimensions of clinoforms associated with the shoreface, subaerial deltas, subaqueous deltas and the shelf slope. After Allen and Allen (2013) (fig.8.30, p.313) with permission of John Wiley & Sons Inc.

Delta-scale clinoforms may be subaerial (with shorelines behind which are delta top and coastal plain deposits) or subaqueous (with submarine topsets dominated by sediment bypass across the inner shelf). Subaerial deltas are affected by river discharge, and are therefore oriented normally or radially from the coastline, whereas sediment transport in subaqueous clinoforms in shallow marine environments is dominated by ocean basin dynamics (waves, tides, currents), which orients them parallel to the alongshore advective transport direction. Subaerial and subaqueous delta-scale clinoforms commonly occur together and take part in the same sediment routing system, as in the western Adriatic and western Yellow Sea. Such instances are known as compound clinoforms (Nittrouer et al., 1996).

Each delta has a clinoform rollover, marking the point of gradient change from topset to foreset, which corresponds to the shoreline in subaerial clinoforms and the shelf-slope break in subaqueous clinoforms (Figure 5.4). The trajectory of the shoreline and shelf-slope break is an indicator of the dynamics of delta growth. Progradation rates of subaerial and subaqueous clinoforms are highly variable. In the Ganges–Brahmaputra system, the subaerial delta is prograding very slowly, whereas the subaqueous delta is prograding rapidly at 10 m yr^{-1} (Kuehl et al., 1997). In general, rivers entering marine basins with high energy (tidal range and wave energy) have well-developed subaqueous deltas, as in the Bay of Bengal, Papua New Guinea and the Amazon shelf, whereas those entering low-energy marine basins, such as the Gulf of Mexico, have poorly developed subaqueous deltas.

5.3 Simple Models of Delta Progradation

Kenyon and Turcotte (1985) developed a theoretical 2-dimensional model for the progradation of a river delta where the delta front moves by bulk transport, as in creep induced by wave loading, debris flow and shallow landsliding. Deposition on the delta slope is treated as diffusive, analogously to the creep of soil, in which case the change in the elevation of the sea bed, ignoring the porosity of sea bed sediments, is given by the familiar law of conservation of mass (Fick's law):

$$\frac{\partial h}{\partial t} = K\frac{\partial^2 h}{\partial x^2} \tag{5.8}$$

where h is the elevation above base level, such as the sea bed in front of the delta, x is the horizontal distance, t is time, and K is a sediment transport coefficient equivalent to the diffusivity κ. Equation (5.8)) can be solved for the case of steady progradation using a coordinate system that moves with the landward edge of the delta front. The length coordinate becomes

$$x^* = x - Ut \tag{5.9}$$

where U is the progradation speed, and t is the time coordinate. If the shape of the delta front is constant as it progrades, equation (5.8) can be solved for the new coordinate system to give

$$h = A\exp\left(-\frac{Ux^*}{K}\right) + B \tag{5.10}$$

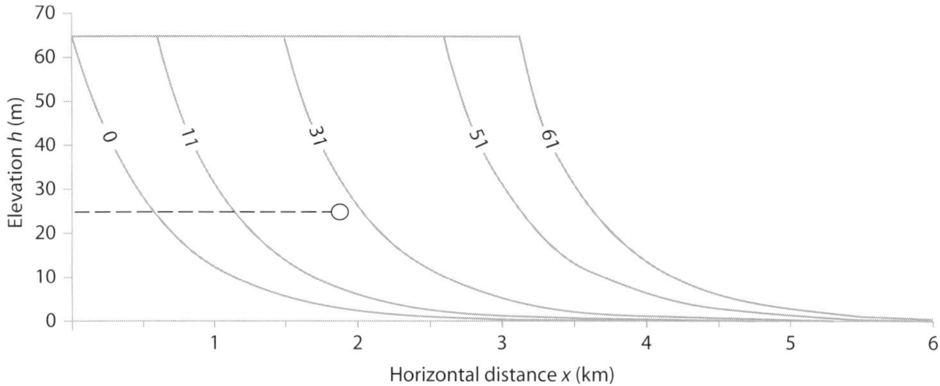

Figure 5.5 Delta profiles using a diffusive bulk transport model for the Rhein delta in Lake Constance (Bodensee). Profiles are shown for years after the construction of a new outlet in 1900. The depth of the lake basin h_0 is set at 65 m, the progradation rate U is 50 m yr^{-1} and the sediment transport coefficient K is 3×10^4 m^2 yr^{-1} (see Table 5.4). The slope of the delta front at a water depth of 40 m ($h = 25$ m) is -0.050, compared to an observed slope in 1929 ($t = 29$ yr) of -0.059.

where A and B are constants. Since $h = h_0$ at $x^* = 0$, and $h \to 0$ as $x^* \to \infty$, A must be equal to h_0 and B must be equal to 0. Consequently

$$h = h_0 \exp\left(-\frac{Ux^*}{K}\right)$$
(5.11)

or equivalently

$$h = h_0 \exp\left[-\frac{U}{K}(x - Ut)\right]$$
(5.12)

The sea-bed elevation of the delta front therefore decreases exponentially with distance from the shoreline (Figure 5.5). In order to allow comparison with observations on modern and ancient clinoform stratigraphic geometries, we need to obtain solutions for the slope or the increment of volume of sediment on the delta front per unit time, which is the sediment supply rate. Equation (5.12) can be differentiated to give a solution for the slope

$$\frac{\partial h}{\partial x} = -\frac{Uh}{K}$$
(5.13)

whereas the sediment supply rate per unit width q_s is found from

$$U = \frac{q_s}{h_0}$$
(5.14)

For example, if the progradation rate of the delta front is 10 m yr^{-1} and the height of the clinoforms is 100 m, the sediment transport rate per unit width of delta front from

Table 5.4 *Observations from three deltas and calculated sediment transport coefficient K using equation (5.15). From Kenyon and Turcotte (1985)(tab.1, p.1460) with permission of Geological Society of America.*

Delta	Slope $(\partial h/\partial x)$	Basin depth h_0 (m)	Progradation rate U(m yr^{-1})	Sediment supply Q(m^2 yr^{-1})	Transport coefficient K(m^2 yr^{-1})
Fraser[a]	−0.0612 (1959) −0.0699 (1929) (at $h = 279$m)	347	2.3 (shore) 8.5 (91m)	1,850 (1959) 2,130 (1929)	2.4×10^4
Rhine	−0.0588 (at $h = 35$m)	65	50	1,450	3.0×10^4
Mississippi River	−0.00955 (at $h = 70$m)	107	76	9,480	5.6×10^5

a, For the Fraser River, British Columbia, data on slope and sediment supply are given for the years 1929 and 1959. Slopes are given at values of water depth h.

equation (5.14) is 1000 m^2 yr^{-1}. Combining equation (5.13) and equation (5.14) provides a solution for the sediment transport coefficient K:

$$K = -\frac{q_s h}{h_0\left(\frac{\partial h}{\partial x}\right)} \tag{5.15}$$

Consequently, information on the shape and progradation rate of deltaic clinoforms can be used to solve for the sediment transport coefficient K.

Kenyon and Turcotte (1985) compared their theory with observations on three major deltas: the Fraser River delta in British Columbia; the Rhein delta in Lake Constance (Bodensee), Switzerland-Germany; and the Mississippi delta at Southwest Pass, Gulf of Mexico (Table 5.4). In each case, the wave and tidal energy in the receiving basin are small and bathymetric slopes are steep, so the river effluents act primarily as inertial jets.

An alternative way to explain the subaqueous slope of deltas is to continue to treat the transport and deposition as diffusive, that is, the sediment flux is proportional to the bathymetric gradient, but to use a solution identical to that of the instantaneous cooling of a semi-infinite half-space (Allen and Allen, 2013)(appendix 48, p.541). Assuming a steady sediment discharge, the elevation of the sea bed is given by

$$h = \frac{2q_{s,0}}{\kappa}\left\{\left(\frac{\kappa t}{\pi}\right)^{1/2}\exp\left(-\frac{x^2}{4\kappa t}\right) - \frac{x}{2}\text{erfc}\left(\frac{x}{2\sqrt{\kappa t}}\right)\right\} \tag{5.16}$$

where $q_{s,0}$ is the unit width sediment discharge at the sediment source (the river effluent) ($x = 0$), t is the time elapsed, κ is the diffusivity and erfc is the complementary error function. The elevation h_0 above the base level over which the clinoforms migrate is given by

$$h_0 = 2q_{s,0}\left(\frac{t}{\pi\kappa}\right)^{1/2} \tag{5.17}$$

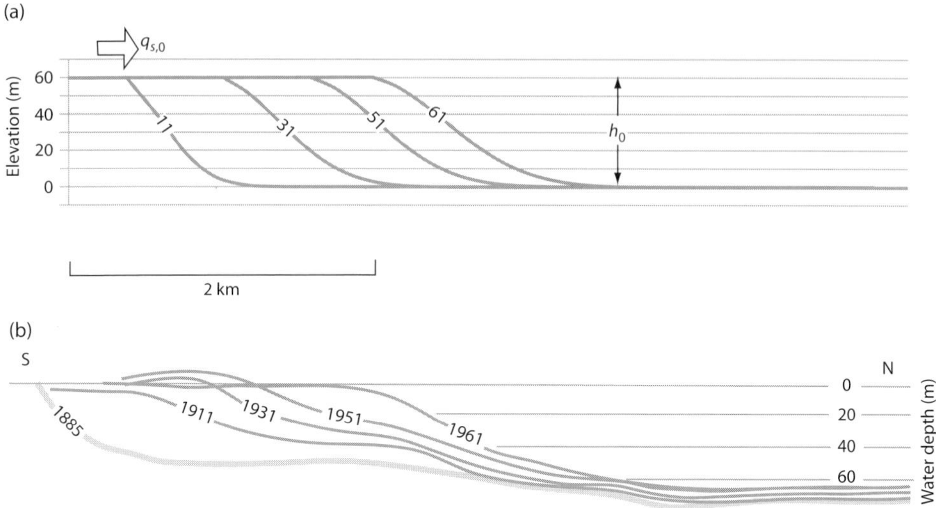

Figure 5.6 (a) Delta profiles using a heat transport equation with $h_0 = 60$ m and a sediment discharge $q_{s,0}$ of 1.8 Mt yr^{-1} (derived from the Bundesanstalt für Gewässerkunde, Sedimentbilanz Rhein) from a 200 m-wide outlet. The sediment discharge is held constant. (b) N-S cross-section of the delta of the Rhein River where it enters Lake Constance (Bodensee) close to Fussach, showing steady progradation of the delta between the years 1885–1961 (33 m yr^{-1}). A new outlet was constructed in 1900. Bathymetric profiles in (b) after Müller (1966) and Allen (1997) (p.250).

We take the example of the Rhein delta in Lake Constance (Bodensee). The entry point of the Rhein River into the lake was artificially changed in the year 1900. The outlet is 200 m wide with a water depth of 4 m. The present-day average annual sediment discharge is 1.8 Mt, giving an average annual discharge per unit width of approximately 3330 m^2 yr^{-1}. Since 1900 the delta has advanced steadily into the lake. There is little riverine sediment beyond 3 km from the present-position of the river mouth. This is in good agreement with the simulation of delta profiles using the heat transport equation with an annual steady advance of the outlet of 33 m, which results in an outlet position of $x = 2$ km after 61 years of progradation.

The relationship of clinoform geometry to the controlling factors of subsidence and sea level change, as well as to the magnitude of the sediment supply, can be explored using a kinematic wave approximation (Petter et al., 2013). Shelf-margin accretion in clinoform wedges has been used as a proxy for sediment flux (Carvajal, Steel, and Petter, 2009) by using the horizontal and vertical components of the shelf-edge trajectory (Figure 5.7). The progradation rate of the shelf edge, and the delivery of sediment to the deep sea, are both related to modern river loads (Wetzel, 1993; Carvajal, Steel, and Petter, 2009). Shelf-margin clinoforms are envisaged as comprising sigmoidal sediment bodies with topsets, foresets and bottomsets fed from a riverine or littoral sediment source. Sediment is transported across the foreset by shallow marine processes such as waves, tides and storms, or by

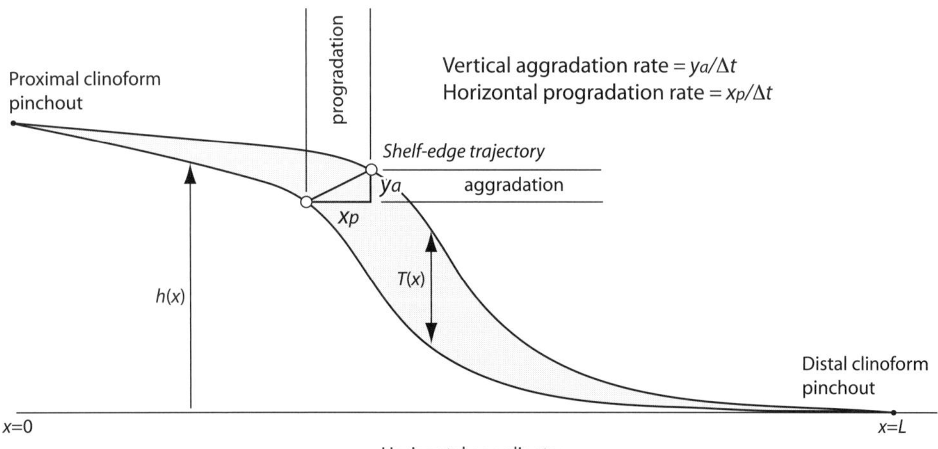

Figure 5.7 Schematic cross-section showing the migration of clinoform profiles and the shelf-edge trajectory. The horizontal and vertical components of the shelf edge trajectory are x_p and y_a respectively, which can be used to calculate the rates of progradation and aggradation. The clinoform set has a thickness $T(x)$ and each clinoform has an elevation above the basal datum of $h(x)$. Modified from Petter et al. (2013) (fig.1), with permission of Geological Society of America.

progradation of the deltaic shoreline to the shelf edge (Burgess and Hovius, 1998). The very large volumes of sediment required to drive shelf-margin growth, which are evident from study of ancient continental margins, indicate that deltas must directly feed the upper part of the continental margin slope (Steel et al., 2009).

The kinematic wave equation expresses the change in elevation over time as proportional to the local topographic (bathymetric) slope by the speed of the wave form, or celerity C:

$$\frac{\partial h}{\partial t} = -C\left[\frac{\partial h}{\partial x}\right] \tag{5.18}$$

where C is equivalent to the progradation rate of the shelf-margin. Equation (5.18) requires that an increase in deposition rate is necessary to maintain a constant celerity on clinoforms with steeper slopes. Since we know from the Exner equation (Paola and Voller, 2005) that the change in elevation of the seabed depends on the spatial variation in the sediment transport rate

$$\frac{\partial h}{\partial t} = -\frac{1}{1-\theta}\left[\frac{\partial q_s}{\partial x}\right] \tag{5.19}$$

where θ is the porosity of the sea-bed sediment, it follows by combining equations (5.19) and (5.18) that

$$P\left[\frac{\partial h}{\partial x}\right] - A(x) = \frac{1}{1-\theta}\left[\frac{\partial q_s}{\partial x}\right] \tag{5.20}$$

Table 5.5 *Parameter values used in the calculation of shelf-edge sediment flux, following the method developed by Petter et al. (2013). Porosity is taken as 0.2. After Petter et al. (2013) (tab.2, p.585) with permission of Geological Society of America.*

	Lewis margin	North Slope
Progradation rate P (m yr^{-1})	0.048	0.024
Aggradation rate A (m yr^{-1})	0.00027	0.0001
Clinoform height (m)	452	1,000
Clinoform length (m)	140,000	50,000
Sediment flux (m^2 yr^{-1})	47.6	23.2

where A is the vertical aggradation rate, which varies with changes in relative sea level, and P is the progradation rate. The integration of equation (5.20) with respect to x gives the sediment flux at any point x up to the distal clinoform pinchout at $x = L$:

$$P(h_L - h_x) - \int_x^L A(x)dx = \frac{1}{1 - \theta}[q_s(L) - q_s(x)] \tag{5.21}$$

If sediment is entirely conserved within the clinoform set, that is, $q_s = 0$ at $x = L$, meaning there is no downdip export of sediment, equation (5.21) simplifies to

$$q_s(x) = (1 - \theta)\left[Ph(x) + \int_x^L A(x)dx\right] \tag{5.22}$$

The term $A(x)dx$ is a sink term representing the sediment aggradation necessary to keep pace with subsidence and sea level change while maintaining clinoform geometry. It is therefore the threshold flux necessary for progradation.

If the shelf margin progrades at some rate P_i into a basin with a gently dipping sea floor and with rising relative sea level, the clinoforms are forced to increase their height and length. In order for the progradation rate to stay constant, the sediment flux must increase to satisfy the sediment volume requirements of the new clinoform (Petter et al., 2013). Alternatively, if the sediment flux stays constant, the progradation rate will decrease as the shelf margin progrades into deeper water. Taking clinoform data from clinoform 8 of the Cretaceous Lewis margin, Washakie Basin, Wyoming, United States and from the Cretaceous Colville Trough, North Slope, Alaska (Carvajal and Steel, 2006; Petter et al., 2013), the shelf-edge unit width sediment flux can be simply calculated (Table 5.5). The relationship between sediment flux and clinoform profile for clinothem 8 of the Lewis margin is shown in Figure 5.8.

The large-scale curvature of clinoforms comprising the continental slope (linear/planar, exponential/concave, sigmoidal, convex) can be related to the slope angle, sediment composition, transport mechanisms and hydrodynamic regime (Adams and Schlager, 2000).

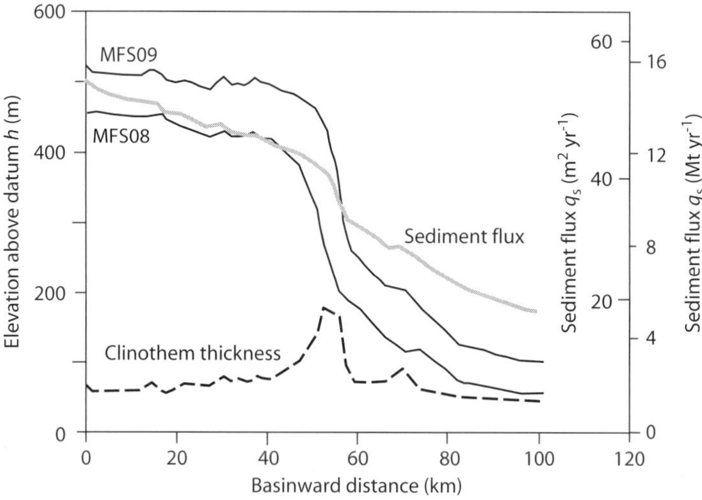

Figure 5.8 Cross-section of clinothem 8 from the Lewis margin, with clinothem thickness, and the calculated value of the sediment flux showing a basinward reduction. Sediment flux is shown as a unit width volumetric discharge (m^2 yr^{-1}) and as a discharge of mass (Mt yr^{-1}) using a basin width length scale of 100 km. Modified from Petter et al. (2013) (fig.6) with permission of Geological Society of America.

Linear profiles are thought to be caused by sediments resting at their angle of repose, whereas exponential profiles in siliciclastic systems are generated by an exponential decay with increasing distance of the capacity of the gravity-controlled transport mechanisms. Sigmoidal profiles result from modification of exponential forms by extrinsic factors such as basel level change and action of ocean currents. In the case of the Ebro clinoforms (Figure 5.9), sigmoidal profiles correspond to unmodified slopes while exponential geometries result from a modification of the original shape of the slope by erosional processes (Kertznus and Kneller, 2009).

The submarine Ebro continental margin is characterised by a series of progradational clinoforms in the 1–1.5 km-thick Pliocene-Pleistocene succession, deposited since the end of the Messinian salinity crisis (Figure 5.9). The margin is fed with sediment from the Ebro Basin, which since the Tortonian (late Miocene) has been connected to the Mediterranean Sea (Garcia-Castellanos et al., 2003). The drainage area of the modern Ebro Basin is 85,820 km^2, whereas the modern Ebro delta has a total area of just 2,170 km^2, of which 320 km^2 is subaerial. Much of the sediment is delivered by the Ebro River ('natural' or 'pristine' values of 18 Mt yr^{-1}, 0.15 Mt yr^{-1} for 'anthropogenic' values). Subaerial exposure of the margins surrounding the Mediterranean Sea during the Messinian caused deep incision by river systems, generating a distinctive unconformity. Global sea level rose in the early Pliocene at 5.3 Ma re-establishing marine conditions in the Mediterranean basin, but Plio-Pleistocene sedimentation was strongly influenced by the relief generated in the Messinian desiccation event. The Plio-Pleistocene succession is divided into three packages separated

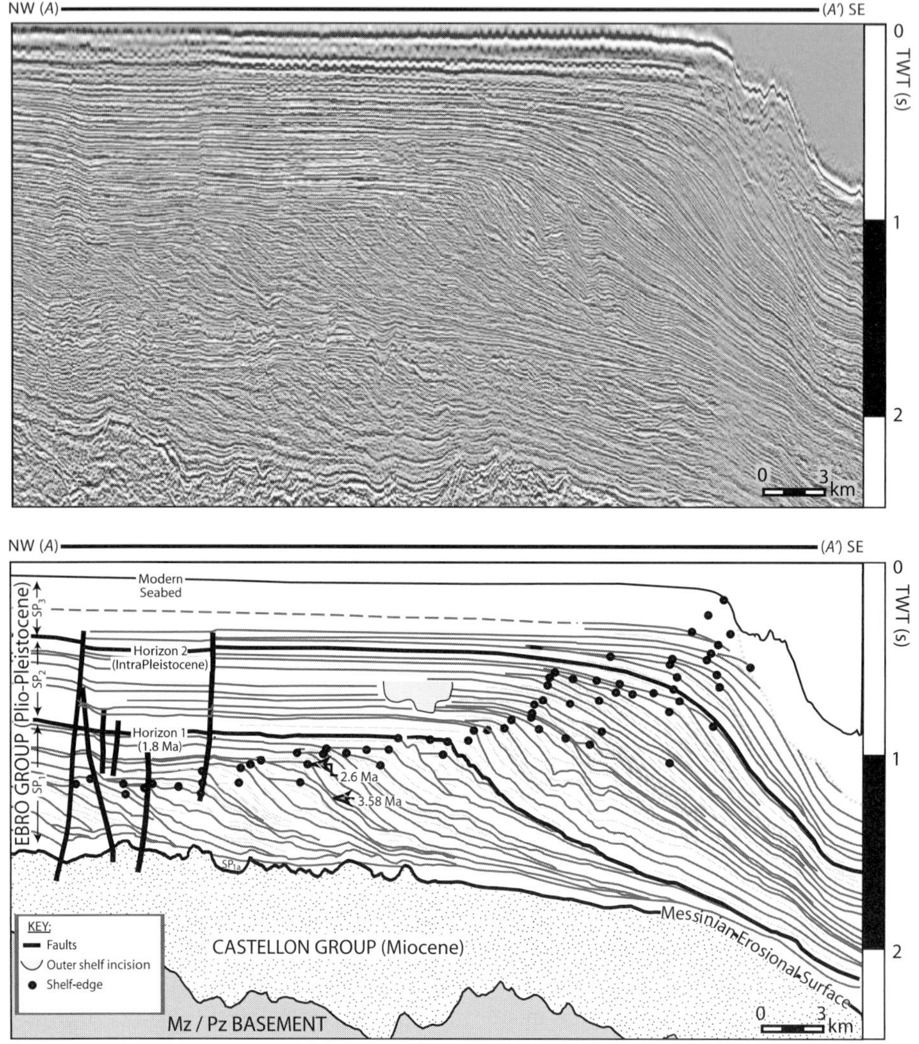

Figure 5.9 Uninterpreted (top) and interpreted (bottom) seismic profile along the depositional dip direction of the Ebro continental margin showing the three seismic packages comprising the Plio-Pleistocene, from Kertznus and Kneller (2009) (fig.4, p.743) with permission of John Wiley & Sons Inc.

by regionally extensive horizons dated 1.8 Ma and approximately 1.0 Ma. Clinoform size and geometry in the three seismic packages are given in Table 5.6.

The sediment fluxes at the shelf edge $q_s(x_{SE})$ required to fill the clinoform packages on the Ebro continental margin using the data in Table 5.6 can be calculated from a simplified version of equation (5.22)

$$q_s(x_{SE}) = (1 - \theta)(Ph + AL) \tag{5.23}$$

Table 5.6 *Measurements of clinoform growth parameters for 3 seismic reflection packages spanning the Plio-Pleistocene of the Ebro continental margin. Sediment fluxes are calculated using equation (5.23) from Carvajal and Steel (2006) and Petter et al. (2013). From Kertznus and Kneller (2009) (tab.2, p.748) with permission of John Wiley & Sons Inc.*

Parameter	Units	Seismic package 1	Seismic package 2	Seismic package 3
Progradation amount				
Min-max	km	n/a	7–11	1.2–9
Average	km	40	7.9	6.8
Progradation rate				
Min-max	km Myr^{-1}	n/a	8.7–13.7	1.2–9
Average	km Myr^{-1}	11.4	9.8	6.8
Aggradation amount				
Min-max	TWT (ms)	680–1,060	290–350	270–340
	m	647–1,014	282–335	261–324
Average	TWT (ms)	850	320	320
	m	810	307	304
Aggradation rate				
Min-max	m Myr^{-1}	185–290	352–419	261–324
Average	m Myr^{-1}	231	383	304
Clinoform height				
Min-max	TWT (ms)	150–757	757–1,500	1,500–1,894
	m	145–720	720–1,430	1,430–1,800
Slope				
Min-max	degrees	1–3	3–6	5–8
Calculated sediment flux				
Average	m^2 yr^{-1}	14.0	30.0	30.5

Sediment fluxes at the shelf edge increase from the Pliocene seismic package 1 through the early Pleistocene seismic package 2 to the late Pleistocene seismic package 3. The increase in sediment supply at the end of the Pliocene, combined with a wider shelf in the Pleistocene, which allowed more effective redistribution of sediment, and accommodation created by compaction of the underlying sediment prism, resulted in increasing clinoform height, average slope and shelf margin relief. The sediment flux values for the Plio-Pleistocene can also be compared with the sediment load of the Ebro delta. Taking the pristine value of 18 Mt yr^{-1}, and assuming that the Ebro River is the sole source of sediment along a 150 km stretch of shelf margin, we obtain a unit width sediment delivery of 45 m^2 yr^{-1}. This indicates that the Ebro River at its current sediment discharge is capable of building the clinoforms observed on the continental margin.

The method of calculating depositional sediment fluxes proposed by Petter et al. (2013) has the advantage that it requires minimal data from 2D data sets and full information on the geometry of the clinoform is not needed. The depositional sediment flux can be calculated as a function of basinward distance, as has been carried out on the Maastrichtian Lewis margin of the Washakie Basin, Wyoming (Carvajal and Steel, 2006). The average depositional flux at the shelf edge is 8.8 Mt yr^{-1} based on analysis of clinoforms, compared to a range of $4-16$ Mt yr^{-1} based on estimates of the volume of basin stratigraphy. Depositional fluxes decrease from the rollover basinwards, principally due to upstream deposition in the sediment routing system.

Present-day continental margins are thought to export from less than 10% to 90% of their sediment supply off the shelf into deeper water (Walsh and Nittrouer, 2003), with the wide variation dependent mostly on shelf width. If rivers enter the ocean a short distance from the shelf edge, a high proportion of the sediment supply is discharged into deeper water. Petter et al. (2013) (fig.12, p.591) estimated the fraction of the total sediment supply deposited beyond the shelf edge for three Neogene continental margins (Ebro, Zambezi and New Jersey). They found that approximately $60-70\%$ of the sediment supply reached the shelf edge over the timescale represented by these three examples, supporting the estimates of the partitioning of sediment into topset, slope and basin-floor of 1:1:1 in the Washakie Basin (Carvajal and Steel, 2012). If these findings are of generic value, it would represent a useful, simple tool for estimating sediment export to the deep ocean.

Pirmez, Pratson, and Steckler (1998) related clinoform growth to the distribution of shear stresses on the sea bed seaward of the river mouth coupled to suspended sediment transport. The distribution of sedimentation rate varies from low in the erosive topset region, to high on the sloping foresets, and low again in the distant bottomsets. This variability of sedimentation rate is captured in the model by shear stresses in shallow water being too high to allow deposition, so sediment bypasses the topset region. With increasing water depth, shear stresses on the sea bed reduce, allowing foreset deposition, with decreasing rates of accumulation as water depth increases.

Swenson et al. (2005) related clinoform development to the relative strength of river floods and basin hydrodynamics represented by tidal and storm wave power. They were able to simulate compound clinoform geometries involving both subaerial and subaqueous progradation. Increasing the frequency or magnitude of coastal storms, decreasing flood frequency or river discharge, and reducing grain size all have the effect of increasing the fraction of the sediment supply delivered to the shallow marine environment, which in turn controls the extent of subaqueous delta progradation relative to subaerial delta development. Swenson et al. (2005) modelled compound clinoform development using an impulse function with characteristic events of certain amplitude and intermittency to approximate the unsteady river discharge, and the wave height of breakers and the strength of downwelling currents caused by coastal set-up to approximate the energy of the receiving marine basin. The magnitude and frequency of floods on land and storms at sea control the partitioning of sediment between the subaerial and subaqueous deltas, which in turn controls their growth or progradation rates. Linked subaerial and subaqueous deltas therefore vary in growth rates, posing problems for the interpretation of relative sea level from clinoform geometry. When river floods dominate over coastal storms, clinoforms

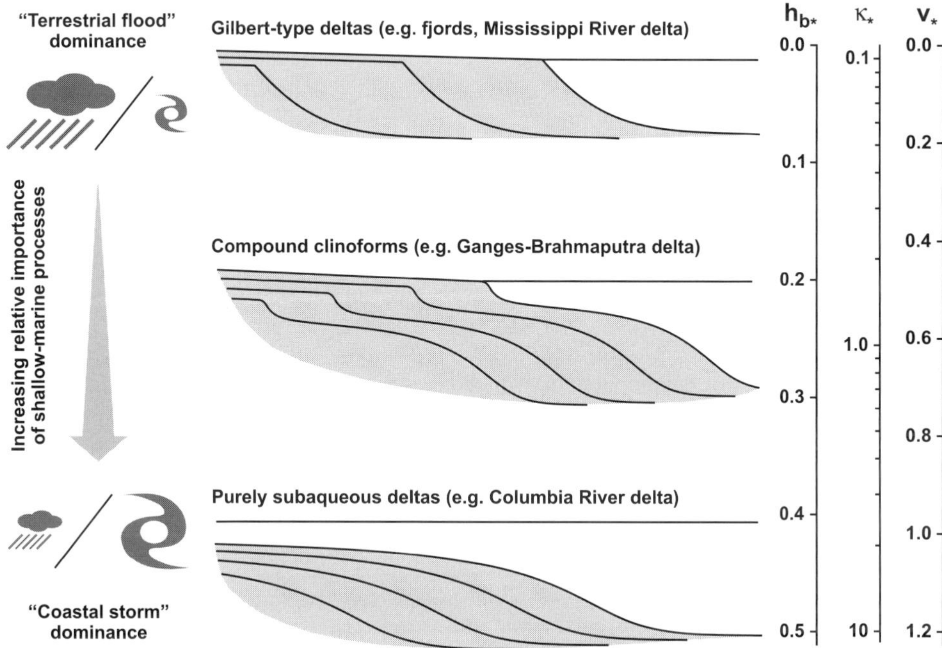

Figure 5.10 Clinoform geometries based on a spectrum from dominance of terrestrial (river) floods to dominance by coastal storms. River dominance favours Gilbert-type deltas, compound clinoforms are found in intermediate positions, and purely subaqueous sigmoidal deltas are found in regions of coastal storm dominance. h_b^*, κ^* and v^* are dimensionless parameters. h_b^* is the fraction of the basin depth affected by high wave energies based on the water depth of breaking waves (high for shallow basins and low for deep), and κ^* and v^* are the slope and current-driven diffusivities controlling the shallow marine sediment flux. These dimensionless parameters collectively quantify how flood and storm parameters and basin geometry combine to control clinoform behaviour. From Swenson et al. (2005) (fig.14) with permission of American Geophysical Union.

are of the simple Gilbert-type, as in the case of the Mississippi River delta (Figure 5.10). Compound clinoforms are found at intermediate positions, and purely subaqueous deltas are found in regions of coastal storm dominance, such as the Columbia River delta in the northeastern United States.

5.4 Sediment Transport on the Shelf

Of the total discharge of sediment generated by erosion in source area catchments, only a portion reaches the deep sea. In order to reach the deep abyssal plains of the oceans, sediment must be transported from river mouths across the continental shelf. The continental shelf has an average gradient of 0.1°, with an oceanward termination in water depths of 130–200 m. This means that during the LGM, when sea levels were lower than today's by approximately 130 m, many of the world's continental shelves were subaerially exposed. On passive (stretched) continental margins, the continental slope, which has an average gradient of $3-6°$, passes distally into the flat (average slope of less than $0.1°$) abyssal plains at water depths of greater than $4,000$ m. On active (convergent) margins, the continental shelf is narrower and sediment supply rates from steep mountainous catchments are higher than on passive margins.

Sediment transport on the continental shelf involves a complex set of physical oceanographic processes set up by a combination of plume buoyancy, tidal currents, wave action and gravity flow. The fate of sediment entering the ocean depends on a number of factors (Ma, 2009):

1. Deposition of a progradational delta, which takes place where high sediment discharges enter a wide, gently dipping continental shelf of low energy as in the Mississippi River (Coleman, Roberts, and Stone, 1998) and Yellow River (Huang He) deltas (Bornhold et al., 1986).

2. Bypass of the inner shelf and deposition on the middle shelf is favoured by high riverine sediment supply and a highly energetic coastal zone. Sediment accumulates as subaqueous clinoforms offshore of the river mouth. Sediment bodies are elongated along the shelf by vigorous coastal currents, while sediment is also transported across-shelf down clinoform surfaces by gravity flows (Friedrichs and Wright, 1986). Sediment dispersal of this type is common, including in the regions offshore the Amazon, Yangtze and Fly rivers.

3. Wide dispersal of sediment to the mid-outer shelf by energetic waves and tidal currents acting on fresh-water plumes. Fine sediments settling out from plumes are commonly resuspended and driven across the shelf as gravity-driven turbid layers (Harris, Traykovski, and Geyer, 2005). Examples are the continental shelves offshore the Eel River of the United States Pacific margin, and the Waiapu and Waiapoa rivers of New Zealand.

4. Escape of sediment from the shelf system through capture by submarine canyons incised into the continental slope and shelf. Canyons may extend across the shelf to the river mouth, as is the case of the Sepik River, Papua New Guinea (Walsh and Nittrouer, 2003). Capture of the sediment cascade by submarine canyons is favoured by tectonically active, narrow shelves, such as of the California Borderland, but capture may also take place where relative sea level fall has caused strong incision at lowstands when continental shelves were subaerially exposed. Canyons are important in funnelling sediment to the deep sea off the Eel River, Congo River and Ganga-Brahmaputra (Section 6.2).

These possibilities of sediment dispersal can be further illustrated by a selection of examples.

- In the case of the Ganges–Brahmaputra system (Goodbred and Kuehl, 1999) (Figure 5.11), 30% of the total sediment supply from the Ganges and Brahmaputra rivers ($1,060$ Mt yr^{-1}) is deposited in the floodplains and delta-plain areas. Of the sediment that enters

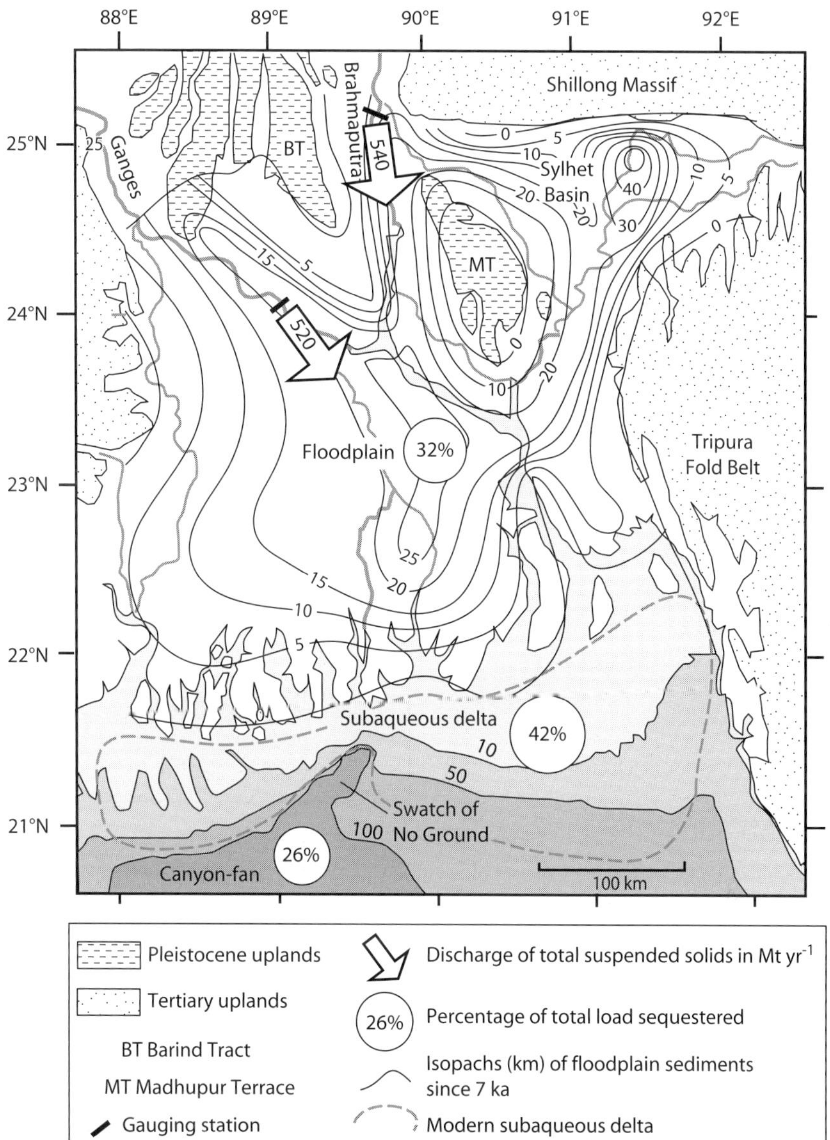

Figure 5.11 Sediment budget of the Ganges–Brahmaputra system. Isopachs are for floodplain and delta-plain sediments deposited since the middle Holocene slow-down of sea-level rise at about 7000 yr BP. The volume deposited is approximately 30% of the annual discharge of the combined Ganges–Brahmaputra rivers per year. Approximately 40% of the sediment discharge since 7 ka is sequestered in the subaqueous delta. Modified from Goodbred and Kuehl (1999) (fig.1) with permission of Geological Society of America.

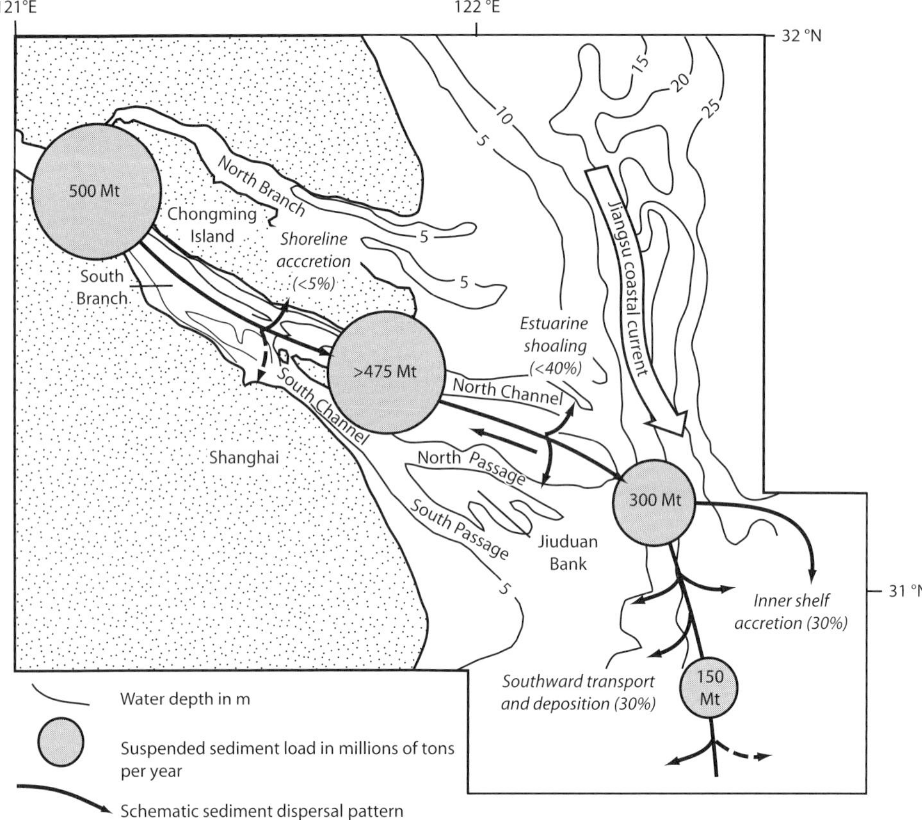

Figure 5.12 Sediment budget and dispersal patterns for the Changjiang River estuary, China, and adjacent shelf. Modified from Milliman et al. (1985) (figs.2, 7) with permission of Pergamon Press.

the Bay of Bengal, a further 42% is deposited in the subaqueous delta, leaving 26% of the total budget that is transported to the deep-water Bengal Fan through the Swatch of No Ground Canyon.

- The Changjiang (Yangtze) River delivers approximately 500 Mt of sediment (mostly silt and finer) annually into the South China Sea (Milliman et al., 1985) (Figure 5.12). Over the last 2 to 3 thousand years, 5% of this sediment load has been sequestered in progradation of the shoreline and a further up to 40% has been deposited in the river estuary, chiefly in South Channel. About half of the riverine sediment supply therefore escapes from the delta region and is entrained in the north-south flowing Jiangsu Coastal Current. The sediment load of the coastal current causes accretion of the inner shelf, leaving ∼30% of the riverine supply that is exported further southward.
- The Rhone River delivers 7 to 20 × 10^6 m^3 of sediment each year into the western Mediterranean Sea, through two outlets, the Grand Rhone and the Petit Rhone (Sabatier et al., 2006) (Figure 5.13). Sediment escaping from the river mouths is entrained in

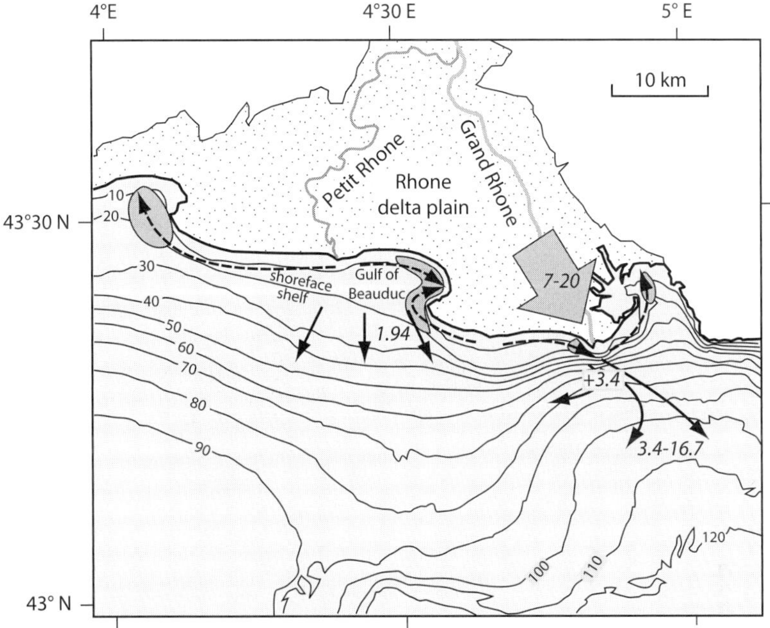

Figure 5.13 Pattern of littoral drift off the Rhône delta, southern France, showing bathymetry at the end of the twentieth century, and sediment budget in 10^6 m^3 yr^{-1} during the twentieth century. Dark grey areas of shoreface are areas of net deposition at the termini of littoral sediment transport vectors (dashed arrows). Solid black arrows are fluxes to deep water. Modified from Sabatier et al. (2006) (fig.1) with permission of Elsevier.

coastal currents, causing deposition on the shoreface (water depths of less than 20 m) at the termini of sediment transport vectors influenced by the physiography of the coastline. Part of the riverine sediment supply is advected into deeper water depths. A similar picture of ocean sediment dynamics is found along the Catalonian (Durán et al., 2012) and Languedoc-Roussillon (Brunel et al., 2014) coasts to the west of the Rhone delta. The Llobregat River is the main supplier of sediment to the Catalan coast northeast of Barcelona, with an annual discharge of 100×10^3 yr^{-1}. A portion of the sediment from the Llobregat catchment and the other smaller catchments draining the Coastal Catalan Chain (Figure 5.14) descends into submarine canyons that incise deeply into the continental shelf. Other sediment participates in littoral cells close to the shore, whereas the Northern Current flows along the shelf edge approximately 20 km offshore.

5.5 River Plumes and Dispersal Scaling

Large rivers commonly produce buoyant hypopycnal outflows into the coastal ocean, since fresh water is less dense than saline. In buoyant outflows, fresh water flows oceanward above a salt water wedge. The landward limit of the salt intrusion is commonly a turbidity

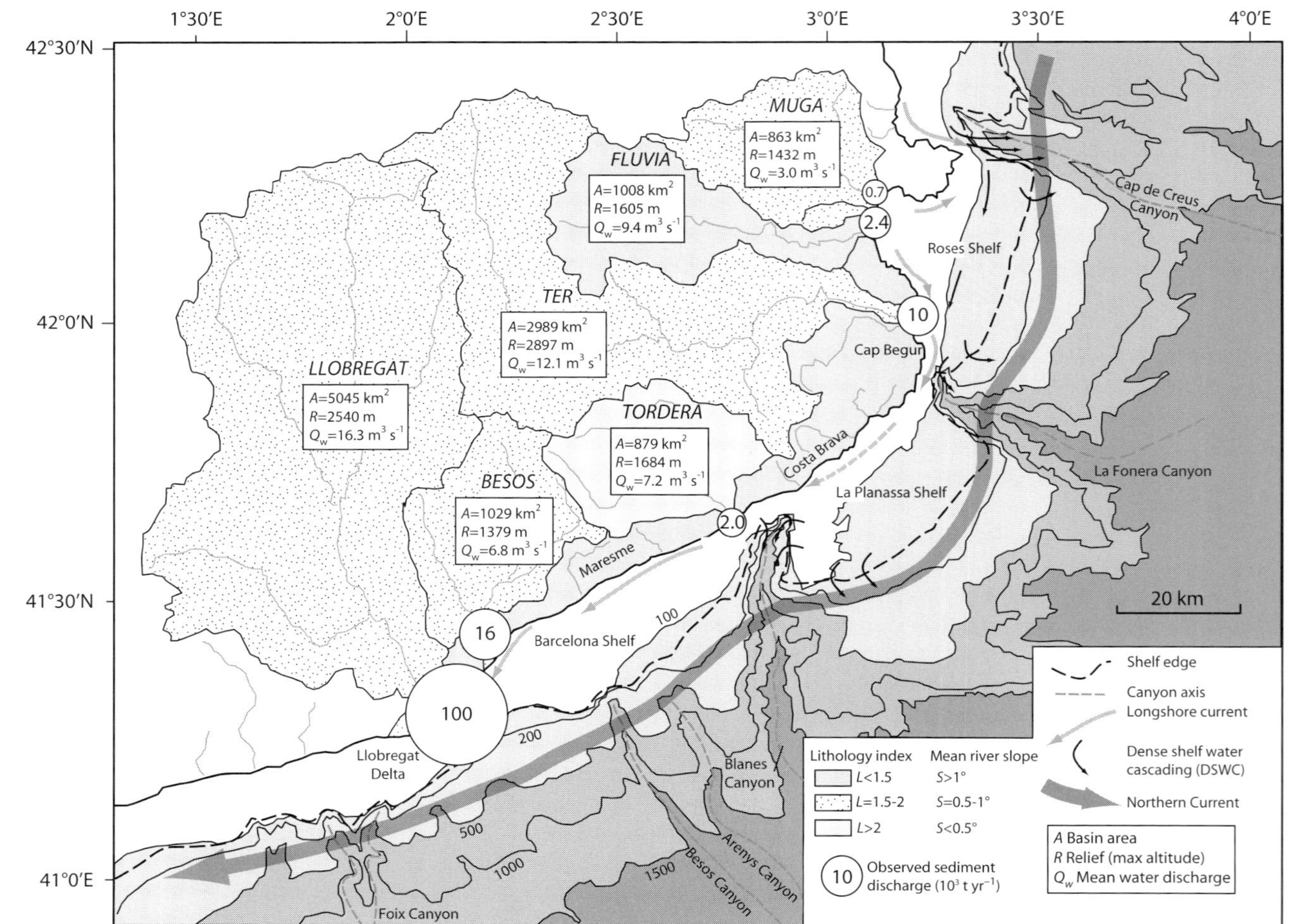

Figure 5.14 Fluvial and oceanographic setting of the Catalan Shelf, northeastern Spain. Modified from Durán et al. (2012) (fig.3) with permission of Elsevier.

maximum and the site of sediment deposition. In some cases, however, such as the Yellow River (Wang et al., 2006), and a large number of steep, mountainous rivers, the high sediment concentration of the effluent produces hyperpycnal outflows that hug the sea bed. The mode of delivery of sediment to the coastal ocean is therefore influenced by the buoyancy of outflows.

The balance between the river effluent and the salt water intrusion (the balance between turbulent jet diffusion and buoyant outflow) can be expressed by the densimetric Froude Number (F')

$$F' = \frac{U}{\sqrt{g'h'}} \qquad (5.24)$$

where g prime is a 'reduced' or density-modified gravity, U is the mean effluent outflow velocity and h' is the thickness of the plume at the river mouth, so is the depth to the density interface. When inertial forces are dominant (F' high) the outflow approximates the turbulent jet model, whereas buoyancy dominates when F' is small. For the same river water density, buoyancy is favoured by cold seawater and concentrated water such as the brines of an enclosed sea or lake, whereas turbulent jets are favoured by warm seawater.

Sediment is trapped near the salinity front defining the leading edge of the plume. The position of the salinity front depends on outflow velocity and geometry, and the strength of wave activity and tidal currents in the receiving marine basin. Outflows with high velocities can push the salinity front far offshore, as in the case of the Amazon outflow. The frontal zone is sharp and the plume is well stratified where the receiving water body is deep and winds, waves and tides are weak (as in the case of the Sepik River), but where the receiving basin is shallow, with strong winds, waves and tides, enhanced mixing results in a broad frontal zone extending far from the river mouth (as in the case of the Amazon and Changjiang (Yangtze) rivers. The efficiency of sediment trapping is illustrated by the length scale L

$$L = \frac{U}{w_s}h' \qquad (5.25)$$

where w_s is the sediment settling velocity. Large amounts of sediment are trapped when the width of the frontal zone is greater than L, but when the frontal zone is sharper and less than L, coarse sediment is deposited, whereas fine sediment bypasses the frontal zone and is advected by ocean currents, typically in an along-shelf direction.

River plumes are affected by the rotation of the Earth. The river plume beyond the salinity front broadens and slows down, with a width that scales on the inverse of the Coriolis parameter f

$$R' = \frac{\sqrt{g'h'}}{f} \qquad (5.26)$$

where R' is the Rossby deformation radius and the Coriolis force per unit mass is the product of the Coriolis parameter and the horizontal velocity of the parcel of fluid. At scales greater than R' the river plume is deflected to the right or left in the northern and southern hemispheres respectively. If the plume velocity is in the same direction as the ambient oceanic current, the plume extends for a long distance away from the river mouth and may create a coastal mud wedge, as in the case of the Amazon plume (Geyer, Hill, and Kineke, 2004), the Changjiang (Yangtze) plume in winter (Milliman et al., 1985) and the western Adriatic supplied principally by the Po River (Brommer et al., 2009).

Although plumes generated by some of the world's major rivers, such as the Amazon and Mississippi, are well known, a significant contribution to the coastal and shelfal sediment budget, and therefore also to the biogeochemistry and productivity of coastal waters, is from many small mountainous rivers. River plumes during high-discharge flood events initially have enough inertia to disperse as a thin (1–10 m), extensive (from 1 to thousands of km^2) hypopycnal jet. Since rivers have self-similar scaling of their geomorphological and hydrological properties, a first-order assumption is that there will be power law (fractal) relationships between river catchment attributes and plume dispersal (Warrick and Fong, 2004). In other words, if river plume surface area is A_p and catchment area is A_c, we can write

$$A_p = cA_c^b \tag{5.27}$$

and the number of plumes with area greater than $A_c = i$ is

$$N_i = aA_p^{-\beta} \tag{5.28}$$

where the coefficients a and c, and the exponents b and β are derived from regional or global compilations.

Plume areas can be assessed from satellite imagery of turbidity fronts following river flooding. For example, river plumes off the Moroccan coast varied in area from 21 to 1,100 km^2 and when compared to catchment areas gave a power law slope b of 0.6 (Figure 5.15). This is similar to the exponent (b=0.63) derived from small Californian catchments and plumes (A_p of 1–230 km^2). Incorporating data from the floods of major world rivers gives $b = 0.68$. The Californian size-frequency data give a power law scale β of very close to 1 (Figure 5.15).

The dispersal of river plumes can also be approached from laboratory experiments (Fischer et al., 1979), which show that plume dispersal is a function of the fluid fluxes of mass, momentum and buoyancy. The plume can be described by a mass flux length scale (l_Q), which characterises the zone of flow establishment near the river mouth and a momentum length scale (l_M), which characterises the offshore transition from momentum-dominated to buoyancy-dominated forcing. Making the approximation that the plume area is the square of the length scale gives

$$A_Q = Q/V \tag{5.29}$$

for the mass flux length scale, and

$$A_M = Q^{1/2}V^{3/2}(g')^{-1} \tag{5.30}$$

for the momentum length scale, where A_Q and A_M are the plume areas associated with the mass flux and momentum length scales, Q and V are the discharge and average velocity

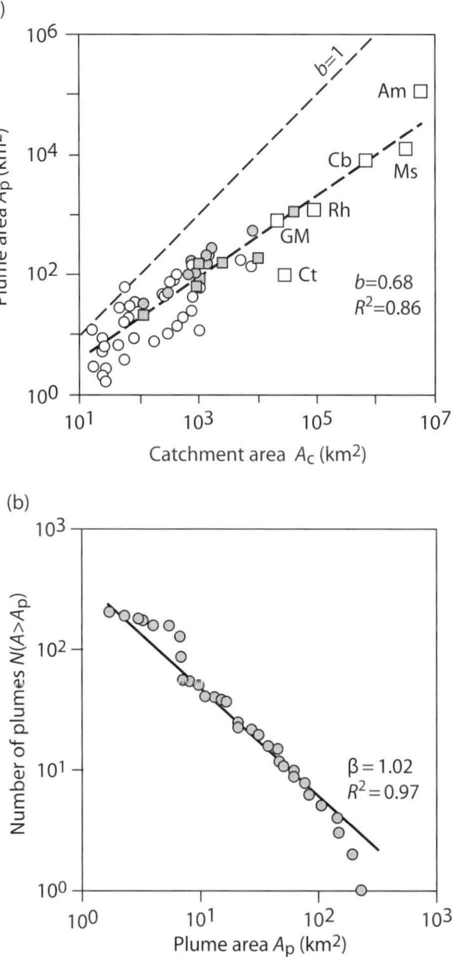

Figure 5.15 (a) River plume area versus catchment area from analyses of remote sensing for coastal Morocco (small filled squares) and coastal California (unfilled circles), maximum river plume areas based on numerical model results (filled circles) and maximum plume areas for exceptional floods of large rivers (large unfilled squares). Am, Amazon; Ms, Mississippi; Cb, Columbia; Rh, Rhône; GM, Gulf of Maine, Ct, Connecticut. Redrawn from Warrick and Fong (2004) (fig.2f). (b) Size frequency relationships of coastal California plumes. Data from Mertes and Warrick (2001). Part (a) published with permission of American Geophysical Union.

of the river, and $g' = (\Delta\rho/\rho)g$ is the 'reduced gravity' buoyancy anomaly of the river water compared to ocean water, which is normally a few percent of g. Within a region of relatively uniform geomorphology and hydrology, river discharge and velocity can be related to catchment area using well-known scaling relationships. Dunne and Leopold (1978) suggested that Q is roughly proportional to $A_c^{0.85}$. Assuming g' is constant regionally, A_Q is proportional to $A_c^{0.85}$ and A_M is proportional to $A_c^{0.43}$, so b is less than 1, as found using the satellite data. The value of b for the area associated with the inshore, mass flux length scale is twice the value of b for the area associated with the offshore, momentum length scale. The slope of the power law relation between A_c and A_p therefore decreases with distance from the river mouth on the basis of laboratory experiments. A seaward decrease in b is also found in numerical models of plume dispersal in the absence of tides, waves and ocean currents.

The significance of $b < 1$ is that the combination of many small rivers spread their fresh water and suspended sediment over a wider cumulative area than does one large river with the same total drainage area. This is due to the fact that small, mountainous rivers have higher rates of fluid mass flux and momentum per unit area of catchment. Small, mountainous rivers have been shown to have relatively high sediment yields (Milliman and Syvitski, 1992). Since the inclusion of data from some of the world's largest plumes, which have different forcing mechanisms, still gives $b < 1$, the patterns of plume size are assumed to be global.

5.6 The Bottom Boundary Layer

The boundary layer is the region where the velocity is affected by the frictional resistance of the boundary and is therefore characterised by a high velocity gradient. Waves are particularly effective at stirring up sediment because of the high velocity gradients generated in their reversing, thin boundary layer. The continental shelf boundary layer is affected by a combination of wave oscillation and quasi-steady currents (Grant and Madsen, 1979) driven by tides or wind stress.

Some continental shelves have seaward-fining sands graded by the action of shoaling waves, passing into mid-shelf muds below wave base, at water depths of $20-30$ m. The Roussillon-Languedoc shelf of the western Golfe du Lion (France) is an example (Jago and Barusseau, 1981). The shelf is incised by submarine canyons to a depth of about 100 m, and has a nearshore sand prism extending to a water depth of 20 m that merges gradually into a mid-shelf mud blanket extending between water depths of 30 m and 80 m. The shelf is microtidal. Wind-driven waves have a mean wave period of 4.2 s and a mean height of 0.5 m. The wave power exerted on a unit area of sea bed by the tangential stress of waves is

$$\omega = C_D \rho \left(\frac{4}{3} u_m^3 \right) \tag{5.31}$$

where C_D is a drag coefficient, u_m is the maximum near-bed orbital velocity due to waves and ρ is the density of sea water. By summing wave power estimates for all waves at any depth, and repeating the computation at successive depths, a wave power profile is obtained that can be compared with the seaward fining of the grain-size profile (Figure 5.16). The two data sets show a similar nearshore-offshore profile, suggesting that waves are primarily responsible for the textural changes in the nearshore sand prism.

For grain sizes of less than 0.5mm, the threshold of sediment motion under waves using linear (Airy) wave theory is given by (Komar and Miller, 1973):

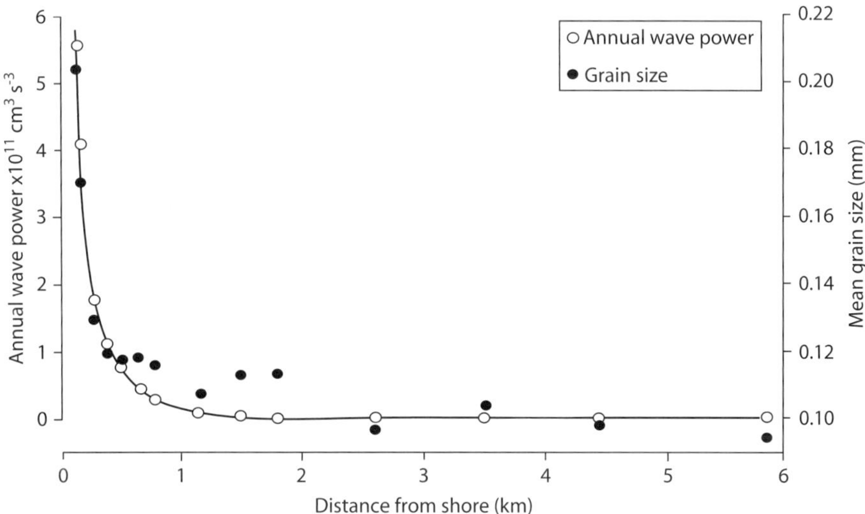

Figure 5.16 Profile for annual wave power and seaward-fining grain size as a function of distance from shore, Languedoc-Roussillon shelf. After Jago and Barusseau (1981) (fig.5) with permission of Elsevier.

$$\frac{\rho u_m^2}{(\rho_s - \rho)gD_t} = 0.21 \left(\frac{d_0}{D_t}\right)^{1/2} \tag{5.32}$$

where d_0 is the near-bed orbital diameter of the wave motion, u_m is the maximum orbital velocity and D_t is the grain diameter at the threshold of sediment motion. d_0 and u_m can be found from information on the wave climate using

$$u_m = \frac{\pi d_0}{T} = \frac{\pi H}{T\sinh(2\pi h/L)} \tag{5.33}$$

where H, L and T are the height, wavelength and period of the surface waves respectively, and sinh is the hyperbolic sine function. Using equations (5.32) and (5.33), the grain size at the threshold of sediment motion for typical wave conditions can be plotted against water depth and compared with the observed shore-normal grain-size data (Figure 5.17).

The oscillatory motion of surface waves is transferred to the sea bed above wave base, causing sediment to be agitated by each half cycle of wave motion (Figure 5.18). The wave base therefore depends on wave conditions and sediment grain size and extends to the outer shelf for small storms (wind speed of 8 m s⁻¹, wave height of 2 m and wave period of 10 s) and to the upper continental slope for large storms (wind speed 20 m s⁻¹, wave height of 6 m and wave period of 15 s). Waves interact with the steady geostrophic currents set up by wind stress and acted upon by the Coriolis force, causing the current to veer to the right in the northern hemisphere. The onshore component of the resulting Ekman layer causes coastal set-up. The gradient of the water surface drives an offshore 'slope' or 'gradient' current that also veers to the right in the northern hemisphere, leading to a geostrophic flow

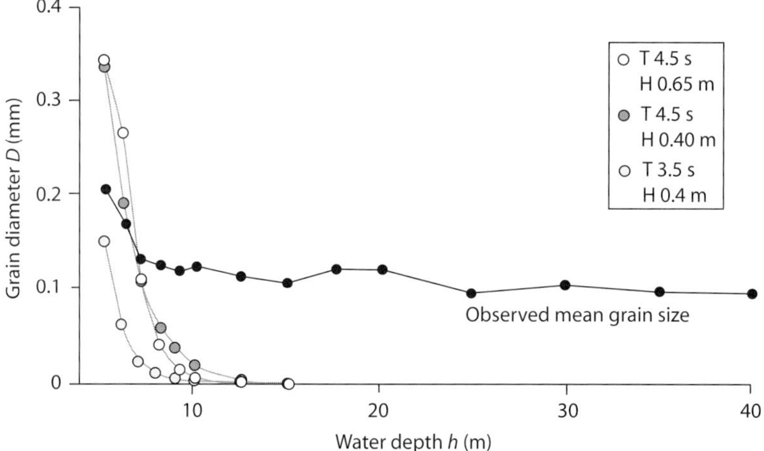

Figure 5.17 Grain-size profiles for various wave conditions typical of the Languedoc-Roussillon continental shelf, using linear wave theory and the threshold condition, compared with field observations. After Jago and Barusseau (1981) (fig.6) with permission of Elsevier.

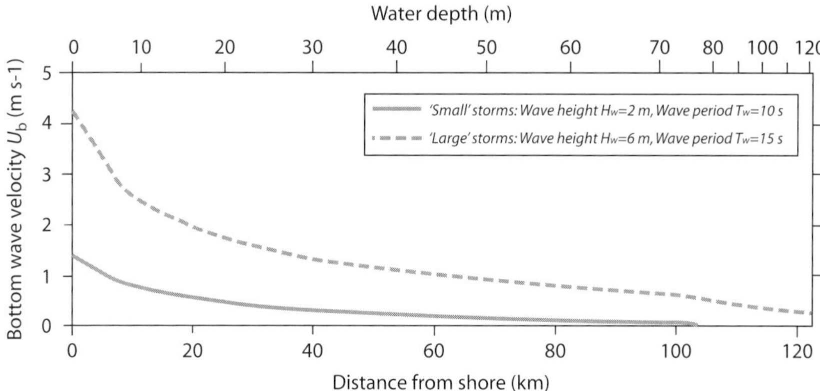

Figure 5.18 Maximum orbital velocity near the sea bed as a function of water depth and distance from shore for 'small' and 'large' storms. Wave base is half of the surface wavelength of the waves, equal to 78 m for small storms and 176 m for large storms. The sea bed influenced by overlying wave motion includes most of the shelf for small storms and extends down the upper slope for large waves. From Cookman and Flemings (2001) (fig.3) with permission of Pergamon Press.

oriented parallel to the shore, or, where the geostrophic flow 'feels' the frictional resistance of the sea bed, a flow oriented obliquely offshore.

Although the predicted threshold grain size from linear wave theory shows a similar onshore-offshore trend to the observed mean grain size on the Languedoc-Roussillon continental shelf, the calculated textural gradient is steeper than those observed in the field. This may be due to a number of factors, including the effect of variable wave conditions in causing sediment transport. Similar rates of seaward fining in sands have been found on other wave-graded continental shelves, such as that of the Adriatic Sea (Passega, Rizzini, and Borghetti, 1967) and the southeastern Australian shelf (Griffin, Hemer, and Jones, 2008).

Shoaling waves transform as they enter shallowing water so that linear wave theory no longer correctly describes the wave motion. Under shoaling waves the orbits of water particles become asymmetrical, resulting in a shoreward mass transport or 'Stokes drift'. However, comparison of the simple linear theory with Stokes theory indicates that the effects of shoaling transformations on grain size are relatively small.

The sediment transport on the continental shelf by wind and waves can be predicted by solving horizontal momentum equations for along-shelf and cross-shelf transport (Cookman and Flemings, 2001; Harris and Wiberg, 2001; Syvitski and Hutton, 2001). The sediment transport on an idealised, two-dimensional continental shelf can be solved for steady wind and wave forcing of the flow field using, for example, the STORMSED1.0 program of Cookman and Flemings (2001). Such a sediment transport model helps the geologist to understand better the formation of marine sedimentary successions, but also aids the assessment of the sediment routing from the coast to the absorbing states of the continental margin and deep sea. The model links three components: (1) a circulation model driven by steady wind conditions that apply a wind stress to the water surface, (2) a wave model, and (3) a sediment transport model.

The combined wave-current boundary layer is used to solve for the erosion, transport and deposition of a single-sized sediment, initially with a wind that causes downwelling currents. For the small and large storms modelled, fine sand is transported in shallow water depths as bedload, whereas medium and coarse silt is transported as suspended load. The zone of increasing sediment flux, which corresponds to a zone of erosion, includes almost the entire shelf for the case of large storms (Figure 5.19).

With multiple grain sizes and small storms, the locations of sediment flux peaks, which correspond to the change from erosion to deposition, shift from a few kilometers offshore for medium sand to 30 km offshore for medium silt. At the same time, the flux of silt increases by three orders of magnitude relative to sand as suspended sediment concentrations increase over the finer grained sea bed. Where the sea bed comprises a mixture of grain sizes, finer grain sizes may be eroded and transported away while coarser grain sizes are being deposited in a winnowed lag.

When the direction of the wind is reversed, so that the surface Ekman layer flows offshore, there is an onshore flow above the sea bed, producing upwelling. Suspended and bedload sediment fluxes of fine sand are directed shorewards, with a peak at a water depth of approximately 20 m. As a result, there is erosion in deeper water depths and deposition in shallow-intermediate water depths. This illustrates the different sedimentation and erosion patterns generated by a change in the orientation of the wind stress. Such a change in orientation may take place between storm events or even within the course of a single storm. From the perspective of the sediment routing system, erosional and sedimentation patterns on storm-dominated continental shelves are complex, but it is clear that cross-shelf fluxes do not allow widespread export of sand-grade sediment beyond the shelf edge to the deep sea.

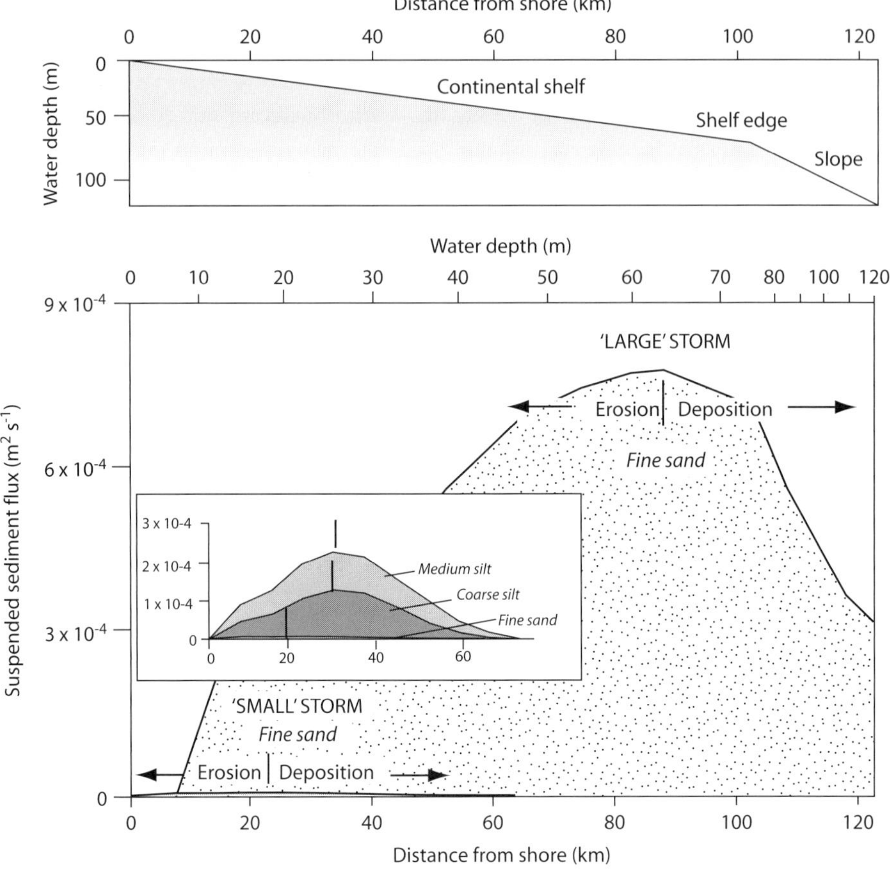

Figure 5.19 Cross-shelf suspended sediment flux on a model continental shelf based on the Eastern Seaboard of the United States. During large storm events, fine sand (0.1 mm) is resuspended and transported to a water depth of 70 m on the outer shelf, and is deposited in deeper water of the shelf edge and upper slope. During small storms, very little sand grade sediment is transported, but significant fluxes of medium (0.01 mm) and coarse (0.04 mm) silt are eroded from shallow water depths of about 25 m and deposited in the mid-shelf region (inset). The deposition/erosion rate is proportional to the gradient in the cross-shelf sediment flux. Modified from Cookman and Flemings (2001) (fig.13) with permission of Pergamon Press.

A two-dimensional numerical code known as 2D SEDFLUX 1.0C also simulates sediment movement on the continental shelf, but includes a multi-sized sediment load and continually evolving boundary conditions (fluvial sediment supply, seafloor bathymetry, sea level etc.) to build a picture of sedimentary architecture (Syvitski and Hutton, 2001). It is part of a full interactive model that takes a source-to-sink approach called EARTHWORKS (Syvitski, Pratson, and Morehead, 1997). Sediment is supplied by a single river and enters the marine basin as hyperpycnal or hypopycnal (plume) flows. In momentum-driven plumes, sediment is fractionated by grain size as it drops out of the overlying plume (Syvitski and Alcott, 1993). The plume may be deflected by entering a crossflow such as a coastal current. Although the spreading angle about the centreline position is narrow (approximately $20°$) for fully turbulent jets, deposited sediment is commonly resuspended and advected by wave energy, bottom currents and bioturbation. These processes are lumped together and specified by a single diffusivity $k(t, y)$, where t is time and y is the vertical coordinate, which reflects the maximum energy of the coastal ocean available to move sediment. The amount of sediment that is capable of being resuspended is then

$$\frac{\partial h}{\partial t} = \frac{\partial}{\partial x}\left(k(t, y)\frac{\partial h}{\partial x}\right) \tag{5.34}$$

where h is the bathymetry. For wave processes, $k(t, y)$ may be assumed to reduce exponentially with water depth. The amount of sediment of each grain size moved on the sea bed is given by a further diffusivity term $\bar{k}$:

$$\frac{\partial \bar{k}}{\partial t} = \bar{k}_i \frac{\partial h}{\partial t} \tag{5.35}$$

where the subscript i denotes the ith grain-size class. In other words, the change in sea bed elevation (water depth) attributable to the removal of sediment grains of size i is a fraction of the total change in the water depth by an index h_i, which varies from 0 to 1.

In 2D SEDFLUX, sediment may also be moved into deeper water by gravitational failure. Failure of the sea bed, for example of the delta front, may generate debris flows or turbidity currents . Turbidity currents may also issue from the river mouth as hyperpycnal flows. As the turbidity current flows down the sea bed it entrains ambient water, thus increasing the flow volume. Sediment may be lost from the flow by deposition or gained by erosion of the sea bed. The balance between these two processes gives the change of the flux of the ith grain-size class of the suspended load in the transport direction. The concentrations and fluxes across the shelf of suspended sediment of mixed grain size has also been modelled by Zhang et al. (1999) and Harris and Wiberg (2001).

The 35 km-wide Valencia Shelf is situated south of Barcelona in the western Mediterranean and experiences a north-to-south flowing coastal current (Figure 5.20). It also has a storm-dominated nearshore-inner shelf belt of winnowed terrigenous sand and gravel, extending to 20 m water depth, beyond which calcareous mud is accumulating (Maldonado, 1972). Rates of accumulation based on ^{210}Pb dating of piston core sediment are 2.6 mm y^{-1} near the Ebro delta, north of the Valencia Shelf, reducing to half this rate on the southern portion of the Valencia Shelf (Maldonado et al., 1983). A circulation model of water movement on the shelf in response to northeasterly storms shows a three-layer geostrophic flow: a surface drift caused by the Coriolis deflection of a wind-driven current towards the shore; a pressure gradient current caused by the coastal set-up (less than 1 m), which is turned geostrophically to flow along the shelf to the southwest; and a bottom flow retarded by friction on the sea bed, flowing obliquely offfshore to the southeast. Sediment is therefore advected along-shore to the southwest, with a bottom flow drifting mud obliquely offshore into an inner shelf mud belt. The source of the mud is probably a combination of the Ebro River, shoreface erosion of muddy alluvial fans during sea level rise, and small contributions from local streams.

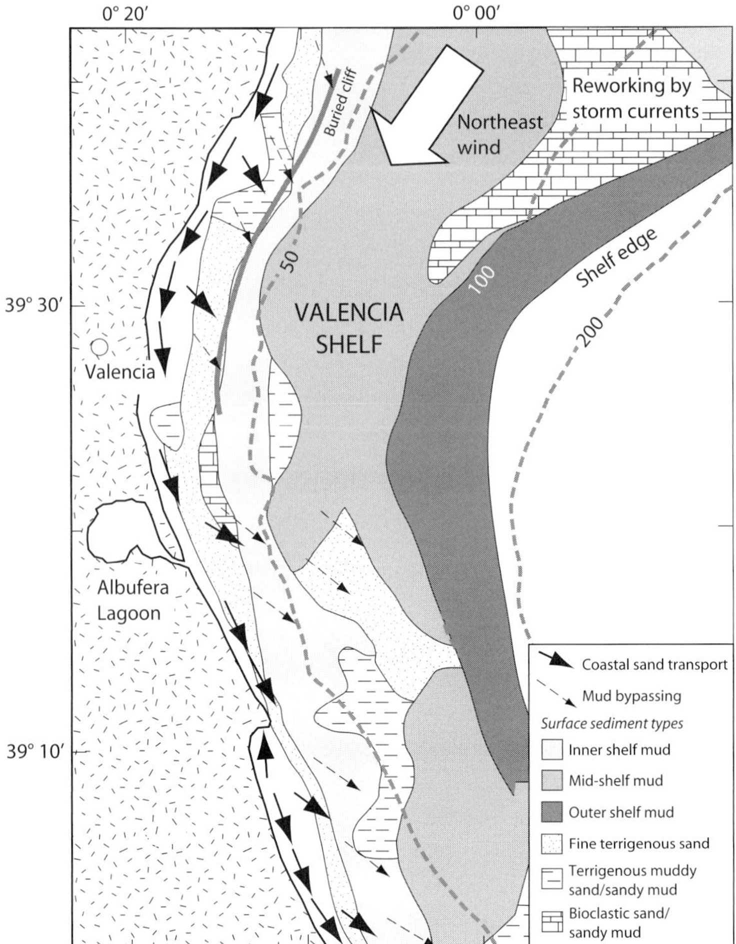

Figure 5.20 Distribution of surface sediment types on the Valencia Shelf based on grab samples and uppermost 5 cm of piston cores, with main coastal current transporting sand driven by a prevailing (winter) northeast wind, and direction of bypassing of mud. Isobaths are in m. After Maldonado et al. (1983) (figs.4, 12) with permission of Pergamon Press.

Cross-shelf transport can also be affected by tidal action, but the extent of seaward flux rather than balanced reversing flow is not clear. The seaward fining of grain size under tidal flows, as occurs in the Celtic Sea, suggests the fractionating effect of net transport. The interaction of storm flows with tidal sand ridge topography may cause across-shelf transport, as on the sandy shelves of eastern Canada (Amos and Judge, 1991).

5.7 Interaction between Ocean Currents and Coastal Waters

Ocean currents intrude onto the continental shelf in less than 3% of the world's oceans. Where they do, they exert a strong influence on sediment transport and therefore participate in the marine part of the sediment routing system. There are a number of ways in which ocean currents interact with coastal waters:

- It is common for a shallow water countercurrent to the offshore deep ocean current to form. An example is the western Atlantic Ocean north of Cape Hatteras, where a nearshore coastal current flows to the southwest in the opposite direction to the Gulf Stream.
- Upwelling along the eastern sides of continents, such as the Peru Current in the south Pacific Ocean and the Benguela Current in the south Atlantic Ocean. Upwelling is a result of the interaction of a wind drift current and a geostrophic flow, and brings cold, deep, nutrient-rich water to the surface, thereby strongly influencing biological productivity and the cycling of carbon.
- Oceanic waters can be driven by strong transient currents during cyclones, moving roughly parallel to the coast.
- Ocean currents may intrude onto continental shelves. The Agulhas Current off the southeast African coast is the best example, which is able to intrude because of the narrowness and steepness of the shelf. Bottom currents that exceed 0.2 m s^{-1} for more than 50% of time (Cronin et al., 2013) are found beneath the core of the ocean current at approximately the shelf-edge and cause significant transport of carbonate sand and reworked gravels (Figure 5.21).

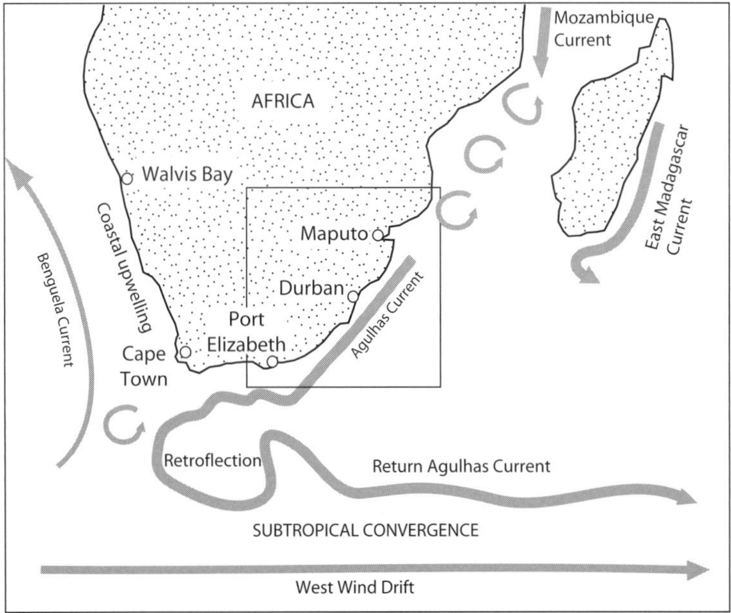

Figure 5.21 Ocean currents surrounding the southern African continent. Box indicates detail of the sediment routing systems along the coast from Port Elizabeth to Maputo shown in Figure 5.22. Modified from Lutjeharms (2006) (fig.1, p.420) with permission of South African Journal of Science.

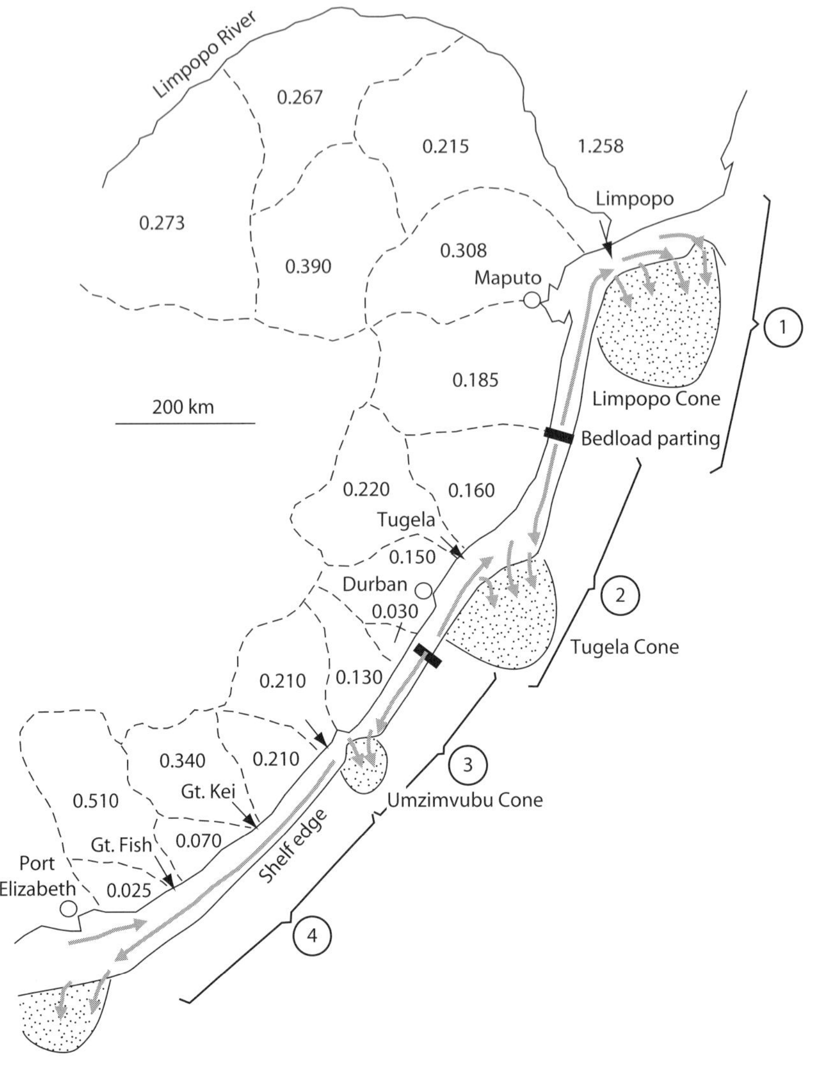

Figure 5.22 Sedimentary model of Agulhas sediment routing system, showing transport direction of bedload and spilling over shelf edge of suspended sediment to build deep-water cones. Catchments show sediment discharge in 10^6 m^3 yr^{-1}. Source-to-sink compartments are labelled 1 to 4. Map covers rectangular box in Figure 5.21. Adapted from Flemming (1981) (fig.14, p.275) with permission of Elsevier.

The continental shelf along the southeast African continental margin is very narrow (3–40 km) and the adjacent continental slope is very steep (maximum values of greater than 12°), which allows the intrusion onto the shelf of the powerful Agulhas Current. The nearshore sand prism is dominated by the wind-generated swell wave regime. Some sediment escapes the nearshore zone and is fed into a sand stream on the central shelf driven southwestwards by the Agulhas Current (Flemming, 1981; Green, 2009). At points along the flow path, structural offsets guide sediment dispersal oceanwards, causing suspended sediment to spill over the shelf-slope break. Clockwise gyres are developed in the lee of the structural offsets, producing countercurrents transporting sand northeastwards (Figure 5.22).

The southeast African continental margin is incised by regularly spaced submarine canyons. Where canyon heads are far from shore because of the existence of a relatively wide shelf, as off Durban, their effect on sediment dispersal is minimal. However, elsewhere the shelf is very narrow, as north of Cape St Lucia, and canyons intersect sediment pathways. The continental shelf between Port Elizabeth and the mouth of the Limpopo River north of Maputo is divided into four compartments (Flemming, 1981). The presence of bedload partings and countercurrent eddies restricts the mixing of sediment between compartments.

The sediment routing system is fed by sediment originating from a number of sources (Figure 5.22). The predominant source is river supply to the coast. About 75% of riverine sediment supplied to the shelf comes from the 5 largest rivers. In addition, local minor contributions are made by modern and relict biogenic products, erosion of the coast and shallow marine seabed and authigenesis at the sea bed.

The total sediment input from 18 catchments entering the coastal ocean is estimated at 95×10^6 m^3 yr^{-1} (Flemming, 1981). Bedload is estimated to be about 5% of the total sediment load of rivers, and the biogenic production is thought to be about 15% of the bedload figure. These total sediment supply rates exceed the volume preserved on the shelf, requiring some sediment to be expelled from the shelf segment into the deep sea.

6

The Deep Marine Segment

The deep marine segment, comprising the continental slope and deep basin floor, is the final destination in the sediment cascade from source to sink. High volumes of sediment are deposited in the deep sea, as demonstrated by the surface areas of large submarine fans such as the Indus Fan ($1{,}100{,}000$ km^2) and the Bengal Fan (3,000 km long and 1,000 km wide). The Bengal Fan has a maximum sediment thickness of 16.5 km (Curray, Emmel, and Moore, 2002). Of the sediment volume released from the combined catchments of the Ganges–Brahmaputra river systems, about 25% is delivered to the Bengal Fan (Goodbred and Kuehl, 1999). The fraction of the sediment supply delivered to the deep sea varies from one sediment routing system to another (Table 6.1). Negligible sediment in the Amazon system currently reaches the deep sea, whereas 75% of the sediment supply of the Congo/Zaire system reaches the deep basin floor at the present day.

The morphology of the continental slope is strongly influenced by the sediment supply of associated rivers (O'Grady et al., 2000), so slope length shows a positive correlation with river length (Sømme et al., 2009). Small, tectonically active systems generally have steep slopes compared to large systems nourished with high sediment supply rates, which have lower gradients and pass transitionally into the basin floor. Slope length L_s, therefore, can be used as a proxy for the longest associated river catchment L_r in studies of ancient sediment routing systems, using $L_s = 1.08L_r^{0.54}$ (Sømme et al., 2009). Delivery of sediment to the shelf edge to build the continental slope depends on the across-shelf sediment transport rate. The remaining sediment is transported to the basin floor through submarine canyons. The sediment discharges across the shelf and down submarine canyons, however, may be highly variable over time (Covault and Graham, 2010). Sediment storage on the shelf must also vary over time in a reciprocal fashion. The potential for shelf storage depends on shelf area and on the presence of waves, tides and geostrophic flows capable of widely distributing the mud, silt and sand delivered to the coast by rivers (Chapter 5).

6.1 Slope Morphology

The residence of sediment in slope accommodation represents a significant storage in the global budget. Sediment fills accommodation on the continental slope in a number of different settings. The accommodation on the continental slope is governed by the topography

Table 6.1 *Sediment delivery to the deep sea as a percentage of the sediment supply, for a number of submodern sediment routing systems. From compilation in Sømme et al. (2009) (tab.1, p.364; tab.2, p.375) with permission of John Wiley & Sons Inc.*

Sediment routing system	Delivery to deep sea segment (%)	Deep sea fan volume ($\times 10^3$ km^3)	Long-term fan deposition ($\times 10^6$ t yr^{-1})
Bengal[a]	40	12,500	325
Congo	75	500	5.7
Amazon	0	700	70
Rhône[a]	25	12	14.5
Valencia	2	6.2	2.5
Ebro	31	1.7	0.6
Cap Ferret	3	1.3	12.3
Danube	0	30	19
Var	91	8	3.2
Monterey	50	50	6.8
Astoria	5	27	39.7
Redondo	53	0.2	0.12
Hueneme	59	0.4	0.34

[a]in these cases the percentage values of sediment bypassing the shelf to the deep sea have been corrected to exclude floodplain storage.

of the depositional surface and its graded (steady-state) profile, the graded profile being an idealised profile representing the long-term equilibrium between deposition and erosion. Accommodation may be categorised as ponded, healed slope, incised submarine valley and slope accommodation (Figure 6.1).

Ponded accommodation is found in three-dimensionally closed topographic lows forming intra-slope basins. Ponded accommodation is generated by localised withdrawal of mobile substrates (typically salt or shale) and less commonly by structural deformation resulting in footwall synclines in contractional settings and hangingwall basins in extensional settings. Healed-slope accommodation occurs in mid-slope regions and is the space between a lower, stepped profile and a higher-angle equilibrium profile, tapering both landward and seaward. Accommodation in incised submarine valleys and canyons is generated by submarine erosion, and is commonly filled with channel deposits. Slope accommodation is the space between the highest stable graded-slope profile and the top of healed slope accommodation. The graded slope angle is dependent on pore pressure within slope muds, high pore pressure reducing shear strength and promoting mass movement.

Slopes evolve through time depending on the sediment flux from the continental shelf, but also depending on the mobility of the substrate. Slopes can be classified topographically in a continuum on the basis of the attainment of grade (Figure 6.2) (Prather, 2003):

1. Above-grade slopes with conspicuous ponded accommodation and large amounts of mid- to upper-slope healed-slope accommodation, as in the Gulf of Mexico. These

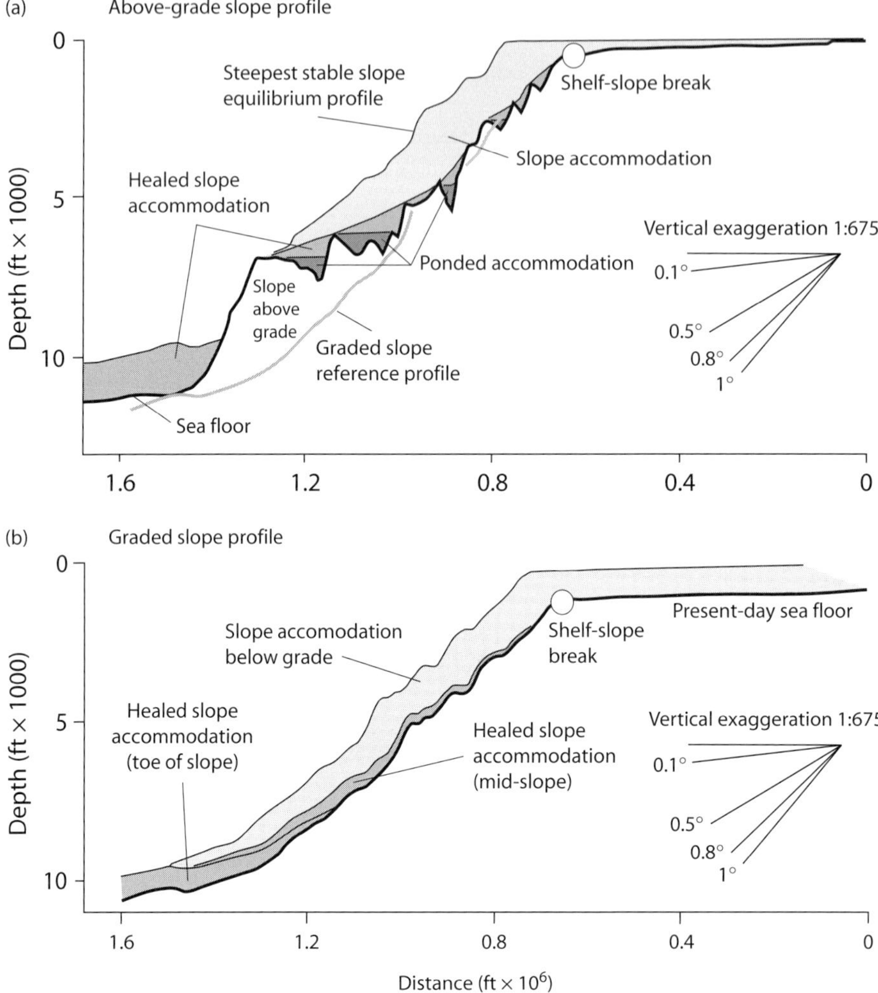

Figure 6.1 (a) Seafloor profile showing the distribution of accommodation on an end-member above-grade slope profile with ponded (perched) basins, based on the central Gulf of Mexico. (b) Seafloor profile showing the distribution of accommodation on a graded slope profile, based on the eastern Gulf of Mexico. Modified from Prather (2003) (figs.3, 4, p.531) with permission of Elsevier.

slopes are characterised by highly mobile substrata and episodic or relatively low sediment flux, allowing ponded intra-slope basins to form. Slope ponded topography is smoothed out by 'fill and spill' processes.

2. Above-grade slopes typified by stepped or terraced profiles lacking well-developed ponded accommodation, as in the Niger delta slope, the Lower Congo slope and the NW Borneo slope. Stepped slopes are associated with less mobile substrates and relatively high rates of sediment flux that causes the slope topography to be smothered.

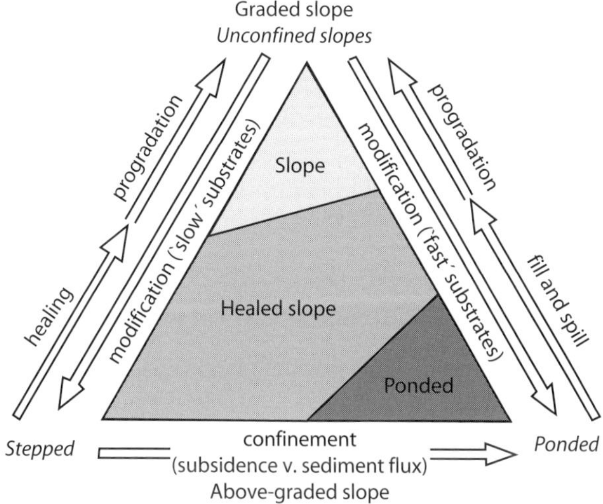

Figure 6.2 Slope type end-members and key processes. From Prather (2003) (fig.21, p.542) with permission of Elsevier.

3. Graded slopes that lack significant topography, as in the eastern Gulf of Mexico. They are associated with low rates of substrate mobility. The dominant sediment transport processes are bypass and erosion, leading to aggradational turbidite sequences at the toe of slope and deep marine basin. The resulting fan apron forms a platform for slope progradation.

Slopes may also be classified in tems of the grain size of the sediment supply (gravel-rich, sand-rich, mud/sand-rich, mud-rich) and the configuration of the feeder system (point-sourced, line-sourced and multiple-sourced) (Reading and Richards, 1994). However, slope and basin floor systems may change radically in their character over time, in response to changing sea level, source area climate and tectonics. This makes any static classification system vulnerable.

The storage of sediment on the continental slope, as a fraction of the delivery to the coastal ocean by rivers, depends on the availability of slope accommodation combined with the existence of a sediment transport system. For example, the Waipaoa River delivers 15 Mt yr^{-1} of sediment to the coastal ocean along the active margin of eastern New Zealand. The slope traps 15–18% of the riverine discharge at the present day. Although some of the sediment accumulating on the proximal slope is derived from turbidity currents, the bulk of it appears to be due to hemipelagic deposition (Alexander et al., 2006). In contrast, the Gulf of Mexico is situated on an extensional margin and has been supplied with sediment throughout the Cenozoic. Sediment supply to the ocean has over the long term exceeded the available accommodation of the coastal plain and shelf created by tectonic subsidence, compaction and sediment loading, so more than two-thirds of the deposited sediment volume occurs beyond the contemporary shelf edge on the continental slope and abyssal plain (Galloway, Ganey-Curry, and Whiteaker, 2009). Sediment supply was by a suite of large rivers whose loads were affected by tectonic uplift and climate changes in the source areas during the Cenozoic (Galloway, 2005). Each progradational cycle caused coastal plain deposits to accumulate over an apron of continental slope sediments, which accumulated by the filling of accommodation generated by extension and salt withdrawal superimosed on broader background patterns of compaction and flexural subsidence.

6.2 Sediment Transfer to the Deep Sea: Critical Role of Submarine Canyons

Submarine canyons are incisions into continental margins (Shepard and Dill, 1966) that act as conduits of sediment en route to the deep sea (Table 6.1) and play an important role in the exchange between oceanic and shelf water masses. Sediment transport in submarine canyons may be triggered by a number of physical processes (Puig, Palanques, and Martín, 2014):

1. Initiation of turbidity currents by the surface waves of storms. Current meters deployed close to the sea bed have recorded flow speeds during storm events of greater than $0.5 \ \text{m s}^{-1}$ in submarine canyons such as Scripps and La Jolla off the Californian coast. Considerably higher speeds lasting several hours were recorded in nearby Monterey Canyon using acoustic laser doppler profilers (Xu, Noble, and Rosenfeld, 2004). All these high velocity flows are thought to have been turbidity currents, captured by canyons whose heads were incised across the shelf to within a short distance of the coast. Sediments delivered by along-shelf currents and river outflows are deposited temporarily near the canyon head before being flushed oceanwards during storms. Canyons that incise to positions farther from the coastline at mid-shelf to shelf-break depths can also funnel fine grained sediment stirred up by high wave orbital velocities during storms.
2. Advection of resuspended sediment from adjacent shelves. Increased sediment fluxes through submarine canyons has been attributed to resusupension of shelf sediment during storms as *nepheloid layers*. The advection of these layers of resuspended sediment is favoured by the presence of cold, dense waters over the shelf, as in Cap de Creus Canyon, Gulf of Lions, France (Palanques et al., 2008) (Figure 5.14). Cascading of dense shelf waters is caused by cooling, evaporation or freezing and may reach speeds in excess of $0.8 \ \text{m s}^{-1}$ in the Cap de Creus Canyon. Similar processes operate in the northern Adriatic shelf, where dense water cascades into the Bari Canyon.
3. Negatively buoyant plumes from river discharges. Offshore moving turbidity currents are well known to be associated with large river discharges. They move due to the excess density of suspended sediment in a sea bed-hugging plume, but some may be triggered by the failure of oversteepened, recently deposited sediment near the canyon head. Var Canyon (off Nice, France) carries turbidity flows of this type during times of approximately yearly floods of the Var River (Mulder et al., 1998). However, larger turbidity currents are generated with decadal frequency and by slope failures.

4. Large slope failures evolving into turbidity currents. Large submarine slope failures generate debris flows and very fast turbidity currents with long run-out distances. The Grand Banks event of 1929, which was triggered by an earthquake, is best known (Heezen and Ewing, 1952). The volume of sediment deposited in the Sohm abyssal plain (greater than 150 km^3) suggests that much of the sediment was mobilised from the upper canyon rather than representing continental slope sources (Hughes Clark et al., 1990).
5. Internal waves oscillating along canyon axes. Oscillating flows with a semi-diurnal or diurnal frequency may reach speeds of 0.3 m s^{-1}, high enough to transport sand-grade sediment (Shepard, Marshall, and McLoughlin, 1974).

The abundance of data documenting the vigorous transport of sediment and water down submarine canyons at the present day demonstrates that sediment is exported to the deep sea during conditions of sea level highstand. Sediment transport down submarine canyons depends on both the tapping of a sediment pathway such as a river outlet or a littoral drift and the oceanographic (and to some extent atmospheric) conditions that drive sediment transport and resuspension on the continental shelf. Submarine canyons therefore play a vital role in transporting sediment to the absorbing state of the deep sea, but they also serve as sediment storage areas. Sediment advected down a canyon may settle onto the sea bed and remain there for hours to years, centuries and millennia before renewed transport by infrequent, large flows triggered by seismically generated slope failures or exceptional river floods.

Submarine canyons may be incised into bedrock or into unconsolidated or partially consolidated marine sediments deposited during previous cycles of shoreface and shelf progradation and retrogradation. Canyons incised into mechanically weak sediments, which is typical of large passive margins, can form relatively rapidly ($10^3 - 10^5$ yr), allowing basin floor fan systems to change position. Active margins, in contrast, are commonly associated with bedrock canyons that form over protracted periods of time and are less mobile.

Larger shelf-slope systems where the shelf is greater than 50 km in width, as is typical of passive margins, generally have canyon heads located more than 10 km from the river mouth in relatively deep water ($\sim$100 m), as in the case of the Amazon, Cap Ferret and Danube systems. Smaller systems typically associated with active margins have canyon heads located less than 10 km from the river mouth in relatively shallow water ($\sim$50 m), as is the case in the Tyrrhenian Sea, Delgado and Golo examples (Sømme et al., 2009).

The heads of submarine canyons must intercept coastal and shelf sediment-transporting currents in order to funnel large quantities of sediment into the deep sea (Figure 6.3). It is estimated that 200 to 250 × 10^3 m^3 yr^{-1} of sand enters the upper part of Monterey Canyon from the Santa Cruz littoral cell in the northern half of the bay (Best and Griggs, 1991) and a similar amount may enter the canyon by littoral transport from the southern half. Sand accumulates continuously in the upper part of the canyon and is advected episodically to the deep sea fan (Paull et al., 2005).

Figure 6.3 This aerial view of Monterey Bay from the south was created by combining computer-generated topographic and bathymetric data. Vertical relief has been exaggerated to better show the Monterey Canyon and mountains on either side of the bay. The arrows illustrate some sources of sand (rivers and streams), as well as the generalised pattern of sand flow along the beaches of Monterey Bay. Canyon image in colour from David Fierstein ©2000 MBARI, mbay fierstein-sand-jpg.

6.3 Basin Plains and Deep Sea Fans

Deep sea fans comprise some of the greatest accumulations of sediment on Earth. There is a relationship between fan area or fan length and long-term volumetric deposition rate (Wetzel, 1993) (Figure 6.4), with presently active fans, modern fans and fossil fans all plotting along the same powerlaw regression. Sømme et al. (2009) expressed this relationship as

$$S = 0.0055 L_f^{1.354} \tag{6.1}$$

where S is a sediment deposition rate by mass ($\times 10^6$ Mt yr^{-1}) and L_f is fan length in km. When fan length is plotted against the suspended sediment discharge of associated rivers

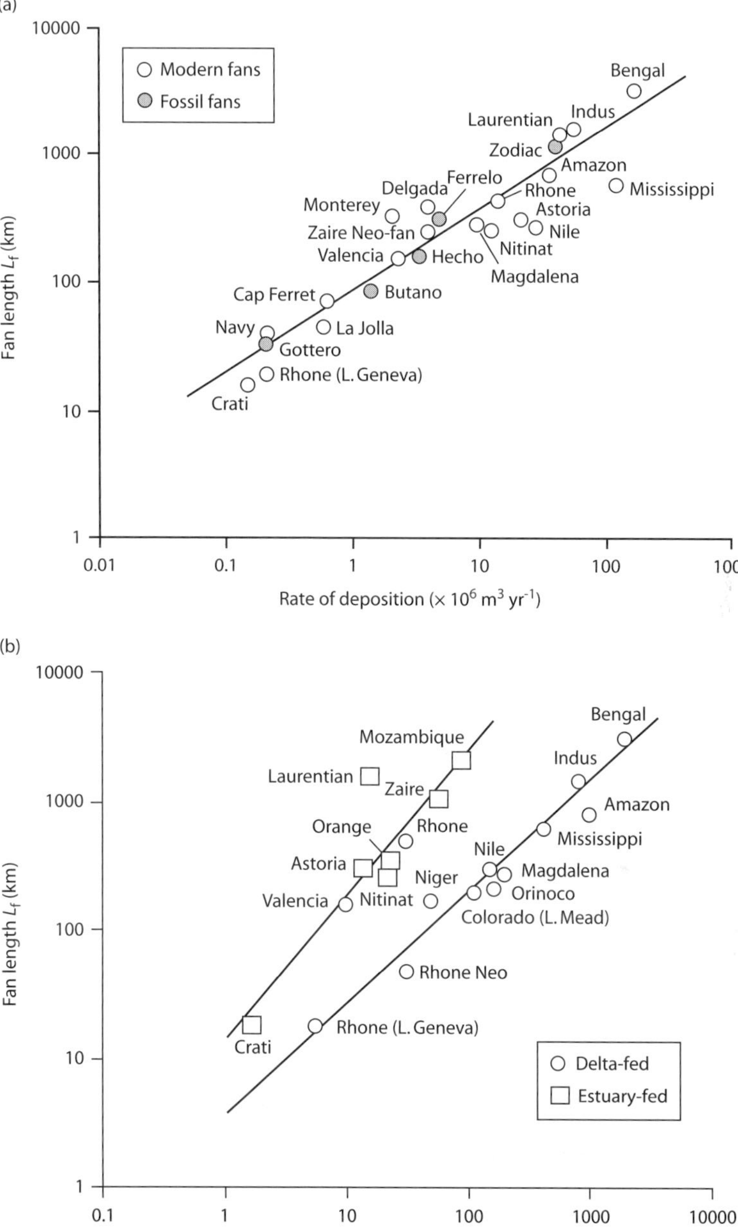

Figure 6.4 Scale of submarine fans. (a) Fan length versus fan deposition rate, with good power law fit for both presently active, modern and fossil fans. (b) Fan length versus discharge of riverine suspended sediment, showing separate power laws for estuarine- and delta-fed morphologies. Modified from Wetzel (1993) (figs.3, 6) with permission of American Association of Petroleum Geologists.

($\times 10^6$ Mt yr^{-1}), fans with deltaic sources have smaller lengths than those with estuarine sources for the same riverine suspended sediment discharge (Figure 6.4b).

The area of basin-floor fans scales closely on the catchment area of the river basin feeding the fan with sediment:

$$A_f = 176.3 A_c^{0.786} \tag{6.2}$$

where A_f is fan area $\times 10^3$ km^2 and A_c is catchment area $\times 10^6$ km^2 (Sømme et al., 2009). Equation (6.2) can be compared with equation (4.16) to show the difference between the geometry of terrestrial and submarine fans. Note that the submarine regression is for A_c in 10^6 km^2 and fan area is in 10^3 km^2, whereas the scales for terrestrial mountain front fans are both km^2.

Small fans at active margins are approximately two orders of magnitude smaller in fan area than fans at passive margins, and about three orders of magnitude smaller in their associated catchment areas. Large fans at passive margins, linked to major river systems draining to the stretched plate margin, therefore dominate the sequestering of sediment in the deep ocean from a global perspective.

Most deepsea fans are located at the base of the continental slope and may extend onto the basin floor for thousands of kilometers. The water depth above the distal part of submarine fans ranges from hundreds of metres to more than 6,000 m. The distal parts of small fans at active margins are in significantly shallower water depths (centred on 1 km) than fans at passive margins and large fans at active margins, suggesting that small fans occupy basin compartments perched on the continental slope or fault-bounded troughs in the continental shelf. Sediment may be routed through intra-slope basins at active margins and bypass the fan on its journey to locations deeper on the continental slope. The distal parts of large active margin deepsea fans and those at passive margins are in water depths of several thousand metres, typical of the ocean-continent crustal transition.

Early models of stratigraphy at continental margins made a strong connection between submarine fan deposition and periods of sea level fall and lowstand (Chapter 10). The argument was that during sea level fall the continental shelf is subaerially exposed and incised by canyons, so that the sediment supply from the land surface bypassed the continental shelf and was routed directly to the deep sea (Vail, Mitchum, and Thompson, 1977). If this were so, it would imply that there is negligible sediment transport to the deep sea at the present day highstand of sea level. Yet, turbidite deposition has continued in the deep sea during the Holocene transgression in, for example, the Mississippi, Amazon and Bengal fans as well as the California Borderland. Subsequently, a compilation of data from 22 deep sea fans has enabled fan deposition rate to be related to sea level since 35 ka – a time period over which sea level change is well calibrated (Lambeck and Chappell, 2001). From this analysis, it is clear that deposition on deep sea fans can take place at any stage of sea level (Covault and Graham, 2010) (Figure 6.5), particularly if submarine canyons incise the continental shelf to intersect littoral sediment pathways. It has been argued that at highstands, deltas may prograde to the edges of continental shelves and deliver sediment to the deep sea when riverine sediment supply is sufficient (Burgess and Hovius, 1998).

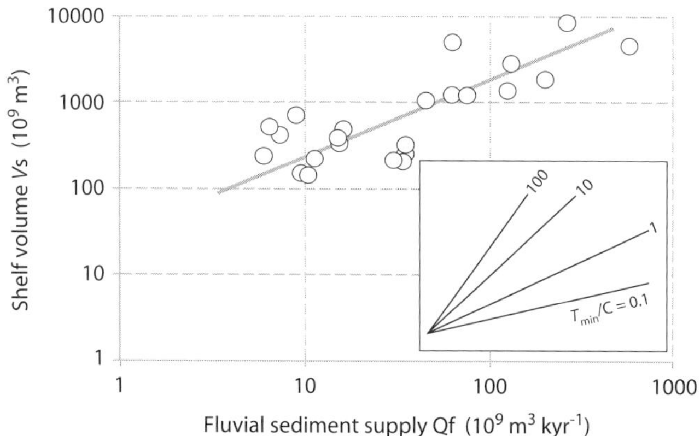

Figure 6.5 Logarithmic plot of fluvial sediment supply versus shelf volume using data in Burgess and Hovius (1998). Calculations of minimum times for delta progradation to the shelf edge T_{min} assume no gain or loss from sources or sinks on the continental shelf. Inset shows a log-log plot of V_s versus Q_f for different values of the exponent T_{min}/C in equation (6.4).

The minimum time required for the transit of a delta to the shelf edge T_{min}, assuming that there is no storage elsewhere on the shelf, is simply

$$T_{min} = C\left(\frac{V_s}{Q_f}\right) \tag{6.3}$$

where V_s is the shelf volume, Q_f is the volumetric flux of fluvial sediment, and C is a dimensionless constant that incorporates the effects of sediment loading, thermal subsidence and compaction. Equation (6.3) can be rearranged to give

$$\log Q_f = \log V_s^{(T_{min}/C)} \tag{6.4}$$

so that T_{min}/C is the slope of a log-log plot of Q_f versus V_s. C is likely to vary from margin to margin, but was assumed to be a constant ~ 1.4 for the 24 deltas studied by Burgess and Hovius (1998). Consequently, T_{min} ranges from less than 10 kyr to greater than 100 kyr in Figure 6.6. These estimates of progradation time to reach the shelf edge are, however, potentially strongly affected by loss of sediment in shelf storage. Much sediment from the Amazon delta, for example, is transported along-shelf by wind-driven ocean currents, thereby reducing shore-normal rates of progradation (Nittrouer et al., 1986). The actual delta progradation time is therefore better expressed as

$$T = C\left(\frac{V_s}{Q_f - Q_m}\right) \tag{6.5}$$

where Q_m is the marine flux of sediment on the shelf. For 16 of the 24 examples given in Burgess and Hovius (1998), the estimated marine flux is greater than the riverine input, indicating that sediment on the shelf is gained from sources other than adjacent rivers. Of the remaining 8 deltas, the actual progradation times vary between 112 and 582 kyr. These

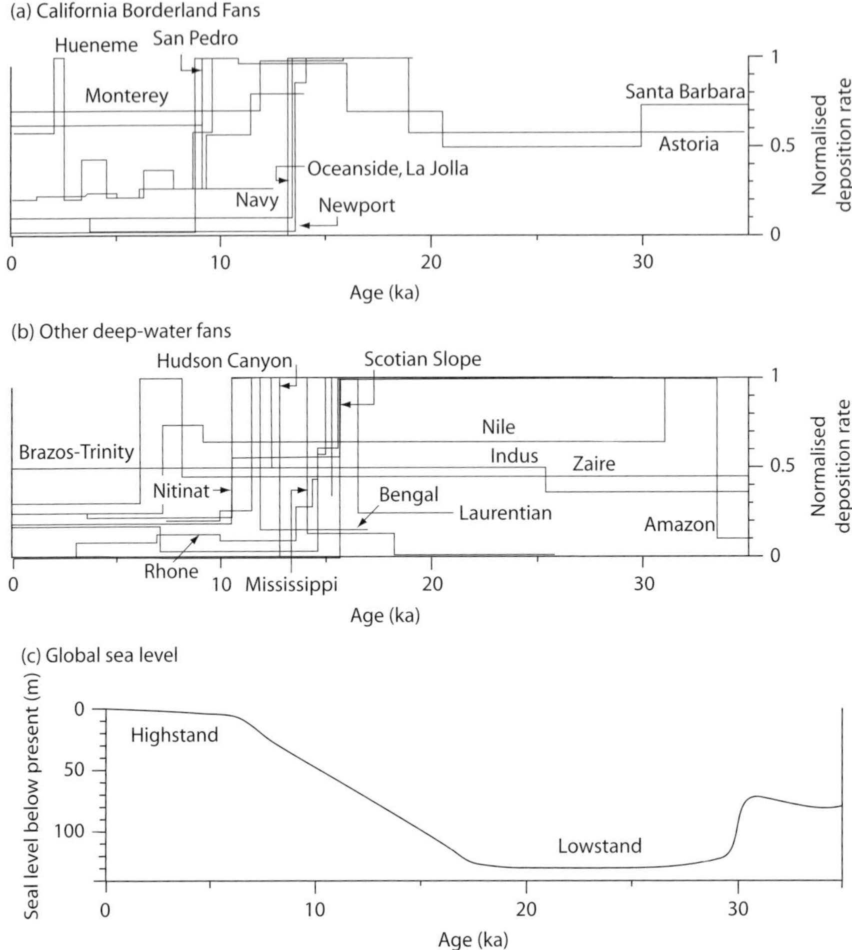

Figure 6.6 Normalised deposition rates over the last 35 kyr for (a) fans on the California Borderland and (b) other major deep-water fans. (c) Global sea level curve from Lambeck and Chappell (2001). Data from Covault and Graham (2010) (fig.2), GSA Data Repository 2010259. Published with permission of Geological Society of America.

progradation times are rapid enough to ensure progradation to the shelf edge during highstands of sea level cycles with periodicity 1 Myr, but inadequate to allow progradation to the shelf-edge at the 100 kyr periodicity of the glacial-interglacial cycles of the Pleistocene.

A question that needs addressing is whether delivery of sediment to the deep sea from highstand deltas includes significant fractions of sand. The present-day sand fraction of rivers ranges widely from 98% for the Copper River to 2% for the Mississippi, Indus and Zaire rivers (Table 6.2). If these present-day values can be extrapolated to times when the shelf-edge delta directly feeds the deep sea, there would be significant sand delivery at

Table 6.2 *Grain-size fractions in present-day rivers and supply rate of sand to selected deep sea fans, from Burgess and Hovius (1998) (tab.2, p.220) with permission of the Geological Society.*

River	%clay	% silt	% sand	Sand supply rate to fan $(10^9 \times \mathrm{m}^3\ \mathrm{kyr}^{-1})$
Brahmaputra	5	45	45	117
Amazon	45	28	16	92
Copper	0	2	98	34.3
Krishna	23	35	42	13.7
Nile	35	45	20	12.5
Orinoco	26	67	7	5.25
Mississippi	50	48	2	4.00
Mackenzie	45	50	5	3.13
Indus	20	78	2	2.50
Po	7	70	23	2.07
Mahakam	35	35	30	1.80
Brazos	63	27	10	1.55
Colorado	?70	?25	?15	0.98
Rio Grande	75	21	4	0.60
Zaire	36	52	2	0.33

highstands. If this sand were evenly spread over the whole fan surface, the deposition rate would vary from $\sim 0.002-0.013$ mm yr^{-1} for the sand-poor cases such as the Indus and Mississippi respectively, to $\sim 0.18-0.28$ mm yr^{-1} for the more sand-rich Nile and Amazon examples, respectively (Table 6.2).

A fall in sea level towards a glacio-eustatic lowstand would cause the merging of rivers on the subaerially exposed continental shelf (Mulder and Syvitski, 1996) (Section 10.2.1), as seen on the continental shelf of the Gulf of Mexico (Figure 10.9). As an example, if sea level fell by 200 m, all of the rivers currently discharging into the Adriatic Sea would become tributaries of the Po River. The merging into larger river systems with lower gradients would cause a drop in sediment concentration and the reduction in hyperpycnal outflows, despite the global increase in sediment delivery to the ocean. Consequently, the delivery of sand at times of lowstand may not be significantly greater than during highstand, but the spatial distribution of sand may differ markedly. Whereas at lowstands a small number of megarivers would deposit sand-rich sediments at widely spaced canyon mouths, at highstands, deltas that have prograded close to the shelf edge in the absence of submarine canyons would represent more of a line source of sediment.

Canyons incised into continental shelves to very shallow water depths maintain connections between the erosional source area and deep-sea depositional systems, so that the sediment routing system is sustained. This situation is particularly found along tectonically active continental margins where fault systems create a narrow shelf and constrain the

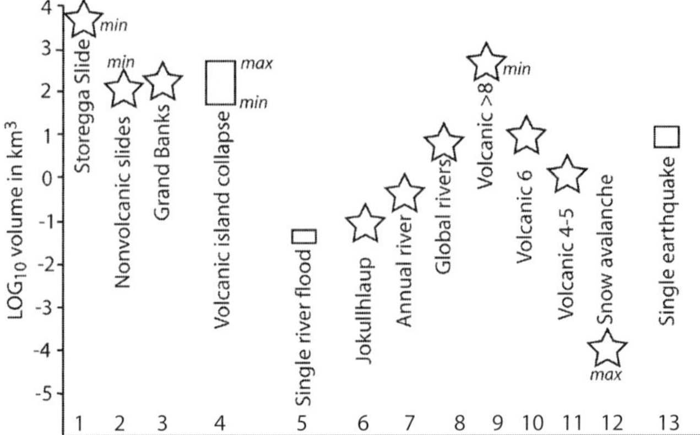

Figure 6.7 Volumes of different types of events involving transport of particulates and mass flows, data from Talling et al. (2014). (1) Storegga Slide, offshore Norway; (2) 15 non-volcanic slides in the last 36,000 years greater than 100 km³ in volume; (3) 1929 Grand Banks landslide-turbidite event; (4) volcanic island flank collapses of the western Canary Islands; (5) largest floods of single rivers; (6) Jokullhlaup, Iceland, 1996; (7) largest annual sediment discharge from a single river, Amazon; (8) sediment discharge from all the rivers in the world for a year; (9) Largest explosive volcanic eruptions magnitude greater than 8, e.g. Toba, 74,000 ka; (10) magnitude 6 explosive eruptions, e.g. Kratatau, 1883 AD; (11) Magnitude 4-5 explosive eruptions, e.g. Mount St. Helens, 1980 AD; (12) snow avalanches; (13) sediment mobilised on land during a single major earthquake.

location of submarine canyons, as in the California Borderland. Sediment deposition during transgression and highstand is most likely due to the increased contribution of littoral cells at times when river systems are unable to reach canyon heads across drowned shelves.

Canyon heads that are detached from terrestrial sediment sources are mostly active during times of sea level lowstand and transgression, when river systems are able to cross the subaerially exposed shelf and deliver sediment to canyon heads. Increased sediment deposition on deep sea fans may also take place during transgression due to the connection between mountain belt uplift, the onset of monsoon rains and the delivery of large sediment supplies to deep sea fans even during transgression, as in the Himalayan-Bengal-Indus systems (Clift et al., 2008a). At the present-day, the Bengal Fan is supplied with 412 km³ Myr⁻¹ from the river catchments of the Eastern Ghats, 520 km³ Myr⁻¹ from the Ganges, 540 km³ Myr⁻¹ from the Brahmaputra and 193 km³ Myr⁻¹ from the Salween, making a total sediment delivery to the deep-water Bengal Fan west of Ninety East Ridge of 1,686 km³ Myr⁻¹. The Irrawaddy supplies 533 km³ Myr⁻¹ to the Andaman sea bed east of the Ninety East Ridge.

Whereas many turbidites are fed from river discharges into the coastal ocean, continental margins are also the site of debris flow and gravity slide processes, where the source is the shelf edge and continental slope. A similar situation is found on the slopes of volcanic islands and seamounts in the ocean. Gravity-driven slides potentially transport enormous volumes of sediment to the deep sea (Figure 6.7). Peter Talling and co-workers state in Talling et al. (2014) (p.33):

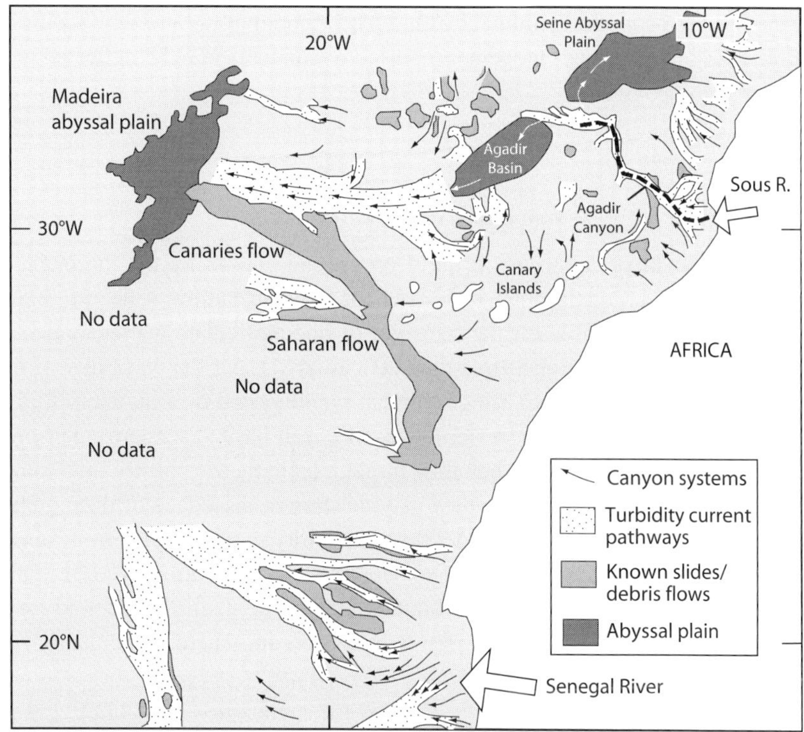

Figure 6.8 Deep-water sediment routing systems offshore NW Africa, showing the transport paths of turbidity currents and giant debris flows. The Canaries debris flow originated by the loading of the sea bed by a rock avalanche caused by catastrophic collapse of a nearby volcanic seamount. The Saharan debris flow was due to failure of the continental margin. After Weaver et al. (1992) (figs.1, 15) with permission of Elsevier and Wynn et al. (2002) (fig.1) with permission of John Wiley & Sons Inc.

Submarine landslides on open continental slopes can be exceptionally large, with volumes that far exceed those of any terrestrial landslide. For instance, the Storegga landslide that occurred 8,200 years ago offshore Norway covers an area of sea floor that is larger than Scotland, and it contains over 3,000 km^3 of sediment. This volume is more than 300 times the annual sediment flux transported to the ocean by all of the world's rivers.

Gravity flows are important agents in transporting sediment to the basin floors of the world's oceans. The sea bed off northwest Africa is a good example of the prevalence of mass flows (Weaver et al., 1992; Gee et al., 1999; Talling et al., 2007) (Figure 6.8). Catastrophic collapse of volcanic islands in the Canaries has repeatedly produced debris avalanches that load the adjacent sea bed and trigger large debris flows. Some of these debris flows have yielded turbidity currents that travelled downslope to the Madeira abyssal plain at water depths in excess of 5 km. Other flows originate from the Agadir Canyon and are routed for thousands of kilometres to the deep ocean (Wynn et al., 2002). Turbidites en route to the deep sea commonly interact with sea bed topography caused by sedimentation, gravitational tectonics and salt movement. Turbidite channels may be deflected or guided by active sea bed topography.

6.4 Deep-Water Circulation

By far the greatest fluxes of water in the deep sea are caused by the thermohaline circulation. The global circulation pattern is vital for the distribution of heat on the Earth's surface, but is also responsible for the transport of fine particulate matter. In general, the deep ocean circulation is sluggish (less than 20 mm yr^{-1}), but the capability of the deep thermohaline circulation in transporting sediment is influenced by a number of processes: (1) currents are intensified on the western sides of ocean basins (*western intensification*), (2) current speeds are increased as they flow through bathymetric constrictions and reduce where they expand, (3) large-scale eddies may develop at the edges of deep geostrophic flows, resulting in enhanced mixing and a reduced ability to hold sediment in suspension. The more vigorous bottom currents are able to maintain concentrations of 0.01–0.5 mg l^{-1} of very fine-grained particulate matter (about 12 μm mean size) in nepheloid layers that advect material beyond the shelf edge into the deep ocean, where it eventually rains out onto the sea bed.

The density forcing of currents in the deep sea is maintained by cooling, evaporation or a combination of both, in which case they are said to be thermohaline. These density currents of cold or salty water enter the larger ocean through a narrow or shallow strait and veer to the right (in the northern hemisphere) or left (in the southern hemisphere) by the Earth's rotation. Once in the unconfined ocean, the speed of the large scale current is determined by the slope of the ocean bed and the density difference between the current and the ambient overlying sea water. The speed can reach much higher values, for example where it encounters topographic features such as submarine canyons or seamounts, or in western boundary currents. Faster currents carry higher sediment loads, which potentially accrete in sediment drifts upon deceleration.

Bottom currents deposit large sediment drifts, particularly at the edges of deep bottom water pathways. The sediment contained in them cannot be linked easily with a cascade from source to sink. Sediment drifts occur where secondary gyres are shed from the main current, or where the bottom current runs along the contours of the continental slope. Fine-grained beds within deep water sediment drifts are consequently commonly known as *contourites* (McCave and Tucholke, 1986). Contourites are defined as 'sediments deposited or substantially reworked by the persistent action of bottom currents' (Rebesco et al., 2014). Contourite drifts typically have a mounded character, generally elongated parallel or slightly oblique to the strike of the deep-water basin margin, with an adjacent concave moat. Contourite drifts can be more than 100 km wide, several hundreds of kilometres long and up to 2 km thick, with a submarine relief of up to 1.5 km. Small patch drifts are roughly equivalent to individual turbidite lobes ($\sim$100 km^2), and giant elongated drifts match roughly the size of the largest deep sea fans (greater than 10^8 km^2). The largest contourite drifts are found on the western side of the ocean basins and around the Antarctic and Arctic Oceans.

The sand, silt and mud of contourite drift deposits is therefore an important absorbing state in the ocean, representing a storage zone of global significance. Bearing in mind the difficulty of sampling modern contourite drifts, and the nascent state of research on fossil contourite depositional systems, it is not surprising that little is known of the provenance of their sediment at present. Neglecting the reworking of previously deposited ocean sediment, the bulk of the sediment comprising contourite depositional systems presumably originates in the suspended and wash loads of river outflows that become incorporated in the oceanic circulation.

Although contourites were intially discovered in the North and South Atlantic Oceans, they have since been found in every major ocean basin and even in lakes. One hundred and sixteen major contourite areas were known at the time of writing the review by Rebesco et al. (2014), associated with a wide range of oceanographic processes and the circulation of surface, intermediate and deep water masses. The definition of the word term 'contourite' has therefore broadened from an association with thermohaline currents to include a variety of other forms of ocean circulation. *Contourites* are distinct from turbidites deposited by downslope density currents, and *pelagites*, deposited by vertical settling, but may be interstratified with them.

The large wind-driven and thermohaline circuation of the oceans also affects the slow settling of primarily biogenic and organic debris of calcareous and silicious plankton living within the photic zone of the ocean. These pelagic sediments may be contaminated with other fine sediment derived from volcanic and wind-blown dust and a fine rain of particles from the cosmos. Although these contaminants may be important sources of nutrients, affect sea surface temperature and are a significant component of sea bed sediment, particularly where winds blow sand, silt and dust from the world's deserts such as the Sahara (Foltz and McPhaden, 2008; Evan et al., 2011), they are not major contributors to the budgets of sediment routing systems.

6.5 Record of Glaciation in the Deep Sea

On orbital (Milankovitch) timescales (10^4–10^5 yr), the growth and decay of ice can be recognised globally (Patterson et al., 2014). But at higher frequencies, the terrestrial and marine records contain a range of local and regional influences, including those relating to ice dynamics (e.g. polar versus temperate systems), drainage basin topography and size, climate change, sea level, tectonics (active versus passive continental margins) and oceanographic setting (Jaeger and Koppes, 2015). These influences are potentially forcings that may be decipherable in terms of the transmission of sediment flux signals through the sediment routing system. For example, a climatic forcing may cause a glacial system to increase bed erosion and consequently result in the transmission of a spike in sediment flux from the glaciated catchment to the transfer zone of fjords, the continental shelf and continental slope, and to eventual deposition in the deep ocean. This spike in sediment flux is most likely reduced in amplitude and lagged in time compared to the initial signal as it is propagated through the sediment routing system (Romans et al., 2015). Glaciated systems therefore behave like non-glaciated sediment routing systems, albeit with different dynamical forcings.

Glacial conditions have changed periodically over the Quaternary era, modifying landscapes in mid- to high latitudes and producing sedimentary signals that may be global in extent. Intensification of glaciation beginning 2–4 Myr ago is held responsible for the deposition of thick (up to 5 km) clastic wedges at continental margins. Moving ice, the primary agent of erosion and deposition in glaciated sediment routing systems, can occur throughout almost the entire system, from high mountains to the continental shelf edge. The most complete depositional record of glaciation is found in the deep sea, where the glacial signal is convolved with oceanographic effects and tectonic processes.

The sediments liberated by glacial erosion are dispersed into other segments of the sediment routing system en route to the absorbing state of the deep sea, but in contrast to fluvial systems, glaciers come and go, so may disappear from the landscape while glacigenic sediment is still being dispersed. The processes of sediment dispersal are varied, including transport on, within and at the base of moving ice, and more importantly within a subglacial hydrological system. Subglacial transport is greatly aided by large volumes of meltwater, so is relatively important in temperate glacial systems. As in fluvial systems, ice (and water) may be stored, thereby changing the discharge to down-system locations in ice-proximal regions.

The sediment evacuated from glaciers undergoes a number of transport processes in glacilacustrine and marine settings. Most sediment enters temperate fjords as meltwater discharge and direct melting from the calving ice front is less important. Sediment accumulation rates may therefore be extremely high near the grounding line of temperate glaciers. Rates then decay exponentially with distance from the ice front, due to the settling of suspended sediment (Syvitski, 1989). For hyperpycnal flows to take place, sediment concentrations in the freshwater outflow must exceed the density of fjord seawater, requiring concentrations of at least 30 mg l^{-1}. At smaller concentrations, sediment moves in a surface plume, whereas at higher values, sediment moves along the sea bed as a gravity-driven flow.

In contrast, sediment transfer to polar fjords is primarily by calving and ice rafting, most of the sediment accumulating close to the ice front. Variations in the flux of ice-rafted debris (IRD) depends on variations in the amount of sediment entrained in the ice, changes in the ice flux at the terminus, the time spent in the zone of iceberg transit and the melting rate, as well as interaction with ocean currents (Andrews, 2000).

Continental shelves are dominated by ice streams flowing from the main ice mass. Deep (100–2,000 m) troughs crossing the shelf link with major sediment depocentres and dispersal systems on the continental slope. Between the major ice streams, the shelf is marked by nested arcuate moraines marking the episodic positions of former ice margins (Dowdeswell et al., 2008; Ottesen and Dowdeswell, 2009) (Figure 6.9a). Glacigenic

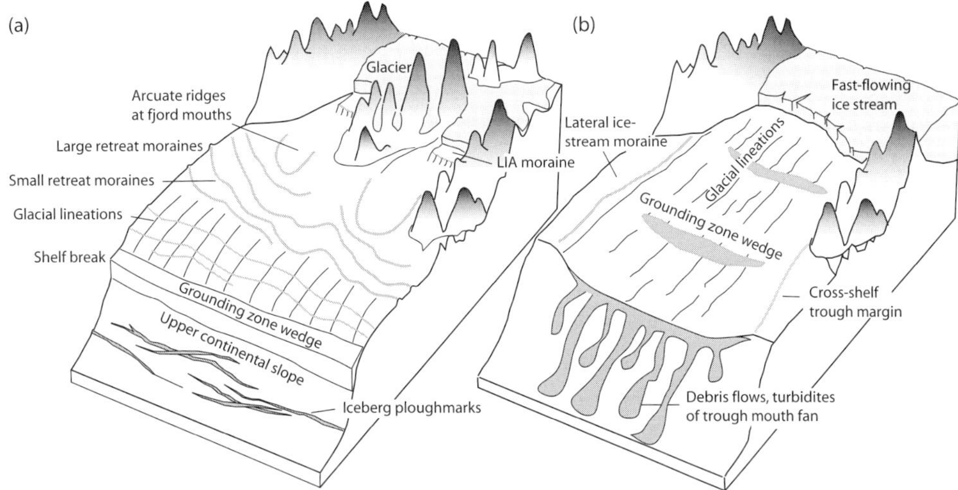

Figure 6.9 Schematic diagram of submarine landforms on the continental shelf typical of inter-ice-stream (a) and palaeo ice-stream (b) sedimentary environments. LIA, Little Ice Age. Adapted from Dowdeswell et al. (2008), Ottesen and Dowdeswell (2009) and Cofaigh et al. (2013) in Jaeger and Koppes (2015) (fig.10) with permission of Elsevier.

sediment is transported down the continental slope and into the deep sea by the familiar processes of debris flows and turbidites and by the settling of fine hemipelagic sediments. Debris flows are common at the distal ends of ice-stream troughs, forming 'trough-mouth fans' (Cofaigh et al., 2013) (Figure 6.9b).

Cross-margin troughs and associated trough mouth fans are common on high-latitude continental margins, suggesting that ice streams are the principal means by which sediment is dispersed in glaciated sediment routing systems. For example, the sediment flux at the 100 km-wide outlet of the Norwegian Channel ice stream reveals a very high sediment yield of 1.1 Gt yr^{-1} for the period 20−19 ka during the last glacial stage, equivalent to 8,000 m^3 yr^{-1} per metre width of ice stream front. Such values are similar to the present-day yields of the largest of modern rivers (Nygård et al., 2007). The sediment flux comprises debris flows on the trough mouth fan of the palaeo-ice stream of the Norwegian Channel, which transported more than 3.2×10^4 km^3 of sediment across its grounding line in a period of 0.5 Myr. The debris flows appear to represent subglacial till that moved downslope above a low strength layer. The delivery of sediment to the North Atlantic between 20 − 19 ka was punctuated, as a result of the Nowegian Channel ice stream switching on and off. Such switching might have been driven by high-frequency climate cycles such as the Dansgaard-Oeschger events prominently recorded in Greenland ice cores.

The sediment flux must scale in some way with the ice velocity, with the form

$$q_s = kU_s \tag{6.6}$$

where q_s is the sediment flux per unit width (m^2 yr^{-1}), U_s is the ice stream velocity (m yr^{-1}) and k is a proportionality coefficient that is conceptually the effective thickness of mobile debris (m). If q_s is 8,000 m^3 yr^{-1}, and k ranges between 0.1 and 10 m, the streaming velocity of the ice varies between 80 and 0.8 km yr^{-1}. The value of the thickness of mobile debris in the ice stream k is uncertain, but comparisons with other ice streams, the thickness of transparent units is high-resolution seismic profiles and the state of consolidation of equivalent tills suggest that k is most likely in the range 3−6 m, giving ice velocities of 1.3−2.7 km yr^{-1}.

At the timescale of the period since the LGM, the fast retreat of the termini of modern tidewater glaciers indicates that the sediment effluxes should be elevated during deglaciation. This trend is found in the sedimentary records of polar and temperate glaciers. Sediment effluxes are reflected in sediment isopachs, which reveal much-increased fluxes during retreat following the LGM compared to the present day (Hogan, Dowdeswell, and Cofaigh, 2012). In central western Greenland, the sediment flux from meltwater plumes and melting icebergs coming from the palaeo-Jakobshavn ice sheet while the ice margin was quasi-stable is estimated to be 1.6 to 3.6×10^7 m^3 yr^{-1}. Rapid retreat of ice margins in western Greenland during deglaciation is characterised by a rapid transition to ice-distal sedimentation of rainout from iceberg-rafted debris and hemipelagic material, accompanied by a rapid reduction of sediment accumulation rate to about 2 mm yr^{-1}.

At the orbital timescale of Quaternary glacial-interglacial cycles (10^5 yr), sediment fluxes through the sediment routing system are difficult to assess landward of the continental shelf break because of reworking by successive ice advances. Beyond the continental shelf break on the continental slope and in the deep sea, however, gravity-flow deposits and IRD record the dynamics of ice extent. In Baffin Bay, for example, glacigenic sediment production was from the northeastern Laurentide, southern Innuitian and western Greenland ice margins over the last 115 kyr (Simon et al., 2014) (Figure 6.10). These different sources are marked by different contents of detrital grains and clay minerals.

Knowledge of the provenance of detrital material allows stratigraphic trends in the sediments of Baffin Bay to be used to assess the sensitivity of the different ice masses to forcing mechanisms. The northeastern Laurentide and southern Innuitian ice masses, for example, appear to have been sensitive to high frequency climate fluctuations (such as Dansgaard-Oeschger events with a recurrence time of a multiple of 1,470 years), whereas the larger western Greenland margins were more sensitive to large-scale reorganisations of climate and oceanography. Such reorganisations might have involved changes of relative sea level or the advection of warmer Atlantic waters into the bay (Simon et al., 2014). Whereas the conventional view that low sediment fluxes are associated with glacial advance and high sediment fluxes characterise glacial retreat over a full glacial-interglacial cycle, the highest sediment accumulation rates were found during the post-LGM transition associated with elevated input from the relatively small ice streams from the northeastern Laurentian and southern Innuitian ice sheets, whereas the sediment delivered from the larger and fast-flowing Uummannaq ice stream on the western Greenland margin dominated the period of the LGM (32 − 16 ka).

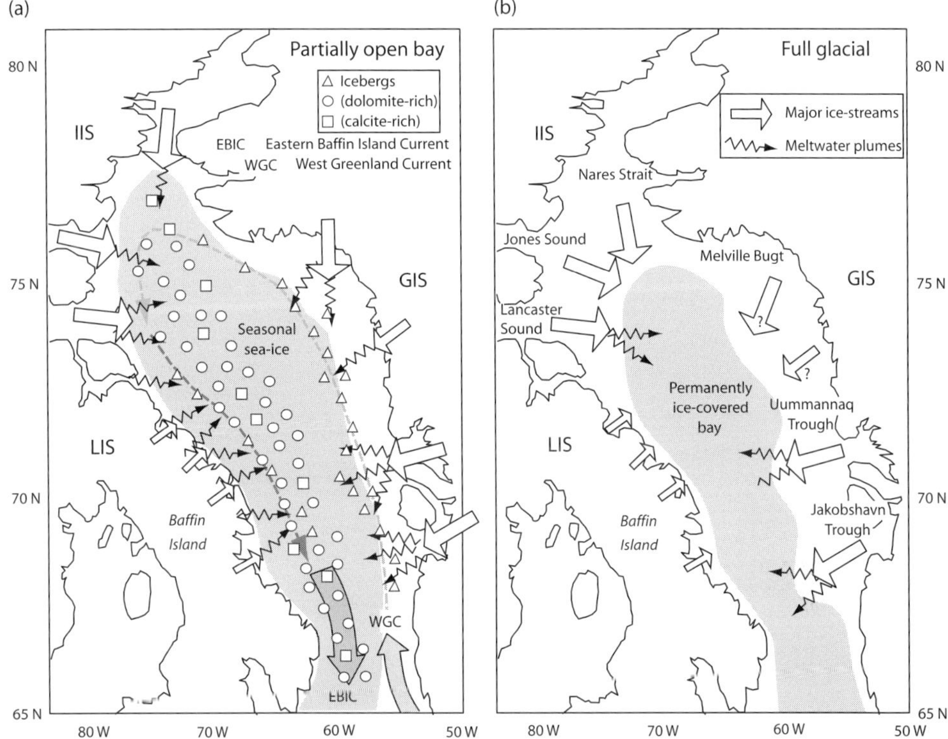

Figure 6.10 Simplified palaeogeography of Baffin Bay during the last glacial cycle. (a) Partially open bay with ice-rafted debris (IRD) delivered by the northern ice-streams and by meltwater plumes from Baffin Island. The West Greenland Current traps IRD close to the coast. (b) Full glacial mode dominated by diamicts (glacial flour sediments) from Greenland and Baffin Island in a permanently ice-covered bay. Fast flowing ice-streams (e.g. Lancaster and Uummannaq) feed an ice-shelf in the bay. Main ice masses are Laurentide Ice Sheet (LIS), Innuitian ice Sheet (IIS) and Greenland Ice Sheet (GIS). Redrafted from Simon et al. (2014) (fig.7), with permission of John Wiley & Sons Inc.

Changes in the volume of sediment deposited in the deep sea revealed by isopachs imaged in high-resolution seismic profiles may not directly reflect ice dynamics. Although isopachs may record sediment dispersal in relation to changing ice extent and ice flow, delivery to the deep sea may be dominated by large-scale submarine mass failures eroding pre-Quaternary strata, as in the Norwegian and Lofoten deep sea basins (Dowdeswell et al., 1996). Although pelagic, hemipelagic and glacimarine processes were active during the LGM, they were responsible for only 10% of the total sediment volume deposited in deep water.

Part III

The Functioning of Sediment Routing Systems

7

Denudation and Sedimentation

7.1 Range of Techniques

The denudation of continental landscapes varies spatially from individual hillslopes to vast drainage basins, and temporally from the scale of individual sediment transport events to the growth and decay of mountain belts. There are consequently many techniques available to capture aspects of the fluxes, budgets and rates of sedimentation and denudation of sediment routing systems, chief of which are the measurement of suspended loads at river gauging stations, dams and reservoirs, the estimation of sediment volumes in the depositional part of sediment routing systems, the use of thermochronometers such as (U-Th)/He diffusion, fission track analysis and other short half-life isotopic systems, and cosmogenic radionuclide analysis.

In present-day sediment routing systems, denudation can be assessed from the fluxes of particulate and dissolved matter measured in the rivers draining certain source areas (Section 2.3). A common approach is to measure suspended sediment loads using historic records (Milliman and Syvitski, 1992; Summerfield and Hulton, 1994; Syvitski and Morehead, 1999; Syvitski et al., 2003, 2005; Syvitski and Milliman, 2007; Milliman and Farnsworth, 2011). Historic records from gauging stations are prone to a number of biases: sediment fluxes may be strongly affected by anthropogenic influences (Syvitski et al., 2005; Wilkinson and McElroy, 2007), may fail to register large magnitude-low frequency events that dominate long-term behaviour, and neglect mass transport as bedload (Turowski, Rickenmann, and Dadson, 2010) and as solute load (Meybeck, 1976). Short-period historic measurements from gauging stations may not, therefore, be representative of longer timescales.

Whereas suspended sediment loads have been measured at gauging stations, commonly at intervals of several days, in some countries for more than a century, cosmogenic radionuclides reveal catchment-averaged denudation rates at timescales of $10^2 - 10^5$ years (von Blanckenburg, 2005) (Section 7.6). Cosmogenic radionuclides, such as ^{10}Be and ^{26}Al, are produced in situ by the bombardment of cosmic rays. Since cosmic rays penetrate a certain distance into a host rock, slower denudation results in longer exposure and vice versa. In this way, the erosion rate of static individual boulders can be analysed. However, spatially averaged denudation rates can be evaluated by taking samples from the bedload of streams, which has the effect of averaging denudation over the contributing drainage area for the stream (Brown et al., 1995; Bierman and Steig, 1996; Granger, Kirchner, and Finkel, 1996). The estimation of catchment-averaged denudation rates from stream samples ideally requires sediment to be derived without intermittent storage in upstream areas, and for the contributing catchment to be defined precisely. Consequently, the use of cosmogenic radionuclides to estimate denudation rates applies only to present-day or Recent landscapes.

In contrast, low-temperature thermochronometric methods are very widely used to estimate erosion rates in geological contexts. The dating technique is based on the fading or annealing of metastable tracks caused by the spontaneous fission of ^{238}U, which takes place below a critical temperature. The distribution of track lengths reveals information about thermal history below this critical temperature. For apatite the closure temperature is 110°C ± 10°C. Apatite fission track analysis has been used in a wide range of tectonic contexts, including convergent mountain belts, passive margin escarpments and rift flanks (summary in Gallagher, Brown, and Johnson (1998)). The rate of cooling derived from fission track thermochronometry is very valuable in estimating the long-term sediment effluxes of erosional hinterlands, and in the timing of distinct exhumation events.

The accumulation of ^{4}He from the decay of uranium and thorium to stable lead is increasingly widely used as a thermochronological tool (Wolf, Farley, and Silver, 1996). The helium produced diffuses from the crystal over time, at a rate dependent on the temperature. As in fission track analysis, there is a critical temperature below which helium is retained, which is approximately 80°C (Zeitler et al., 1987) or 68 ± 5°C (Reiners, 2002). Consequently, (U-Th)/He methods extend the temperature range of sensitivity to lower temperatures compared to apatite fission track analysis. The drawbacks, however, are that erosion rates derived from the (U-Th)/He method may be complicated by the curvature of isotherms beneath surface topography (Braun, 2005; Reiners, Ehlers, and Zeitler, 2005), helium concentrations may be affected by thermal events taking place at the surface of the Earth such as forest fires (Mitchell and Reiners, 2003; Heffern et al., 2008), and helium ages in detrital apatite grains are prone to being reset during even small amounts of burial in sedimentary basins.

Erosion rates derived from the time taken for an apatite crystal to cool from its critical temperature to its surface temperature during exhumation are pointwise estimates. The erosion rate of upland catchments serving as source regions for sediment can also be calculated from a cross-plot of age and elevation at which the apatite sample was collected (Brewer, Burbank, and Hodges, 2003; Stock, Ehlers, and Farley, 2006). As in the case of cosmogenic radionuclides, we can 'let Nature do the averaging' and collect sediment samples from streams at the mouths of catchments. The probability distribution function of helium ages of detrital apatites gives spatially averaged information on erosion rate in the catchment.

Erosion estimates from low temperature thermochronometry can be compared with sediment volumes in the depositional part of sediment routing systems. Sediment volumes may be calculated using seismic reflection observations in the ocean allowing average erosion rates in contributing continental catchments to be assessed and compared with historic records of sediment discharge to the ocean (Métivier et al., 1999; Clift, 2006). Alternatively, sediment volumes may be estimated by mapping of palaeo-sediment routing systems from surface outcrops. Michael et al. (2014b) for example, compared preserved sediment volumes in the mid-upper Eocene Escanilla sediment routing system in the foreland of the Pyrenean orogen in northern Spain with apatite fission track and (U-Th)/He ages from bedrock in the contributing catchments in the Pyrenees defined from provenance studies, and from detrital grains in the basin-fill. Estimates from the different techniques were consistent with catchment-averaged erosion rates of 0.2–0.3 mm yr^{-1} over a time period of nearly 8 million years.

It is clear that erosion rate can be estimated at different temporal resolutions. As a result, erosion rates derived from a very short time resolution cannot easily be compared with an erosion rate derived from a long time resolution. In general, short term erosion rates obtained from gauging stations and from the infill rate of reservoirs are considerably smaller than the rates obtained from cosmogenic nuclide dating and from low-temperature thermochronometers such as apatite fission track and (U-Th)/He (Kirchner et al., 2001). This is presumably because the longer timescale methods have a greater chance of capturing large magnitude events. In the Central Range of Taiwan, however, erosion rates at the decadal timescale derived from suspended sediment concentrations at gauging stations and the infilling of reservoirs, have been compared with rates obtained from uplifted Holocene river terraces (kyr timescale), and from apatite fission track thermochronometry (Myr timescale) (Dadson et al., 2003). Despite differences in timescale, all these sources of erosion rate data gave values between 3 and 6 mm yr^{-1}. Global compilations of erosion rate estimated from cosmogenic radionuclide dating using ^{10}Be ($n = 1252$) can be compared with estimates from suspended sediment concentrations at gauging stations (Covault et al., 2013). The authors found that cosmogenic-derived rates and gauging station-derived rates were broadly similar for locations in close proximity.

A number of other tools are available for dating of sediments and thereby allow the assessment of accumulation rates. These methods include the use of short half-life chronometers such as radiocarbon, ^{210}Pb, ^{137}Cs, ^{55}Fe and ^{32}Si. Modern accumulation rates on the Eel shelf, western United States, were calculated using ^{210}Pb and ^{137}Cs as geochronometers (Sommerfield and Nittrouer, 1999; Wheatcroft and Sommerfield, 2005) and ^{7}Be as a tracer of the dispersal of fine-grained river sediment on the continental shelf (Sommerfield, Nittrouer, and Alexander, 1999). The spatial pattern of radioactive fall-out of ^{137}Cs, caused by nuclear weapons tests, can be used to measure rates of soil erosion and deposition. Owing to their short half-lives, cesium and lead isotopes are used to measure young geomorphic processes. ^{210}Pb was used initially to measure the accumulation of snow, but has been extended to the dating of sediment deposition in marine and lacustrine environments (Krishnaswamy et al., 1971; Koide, Soutar, and Goldberg, 1972; Koide, Bruland, and Goldberg, 1973). ^{210}Pb is useful for dating sediments up to 100 years in age, and ^{32}Si is useful for dating sediments up to 2,000 years in age. ^{32}Si and ^{210}Pb are both produced naturally in the atmosphere.

7.2 Controls on Sediment Yield and Erosion Rate

Numerous attempts have been made to correlate sediment yield with topographic and climatic factors. The resulting relationships invariably work reasonably well at restricted spatial scales, or in certain prescribed geomorphic settings, but involve a large amount of scatter at the global scale.

A number of authors have investigated the link between sediment yield and mean annual precipitation, rainfall variability and specific run-off (water discharge per square kilometer of drainage basin, in mm yr^{-1}). Any relationship between precipitation and run-off variables and sediment yield must be strongly mediated by the effects of vegetation (Leeder, Harris, and Kirkby, 1998) (Section 9.2). Erosion rate can be linked to a variety of topographic parameters such as mean or maximum elevation, large-scale or local relief, slope and drainage basin area (Ahnert, 1970; Schröder and Theune, 1984; Milliman and Syvitski, 1992; Summerfield and Hulton, 1994; Pazzaglia and Brandon, 1996; Allen, 1997; Hovius, 1998). The bewildering range of correlations with topographic and climatic parameters proposed by different authors suggests that erosion rate data cannot be collapsed onto one simple plot. A multivariate analysis of sediment yield data reveals that a combination of environmental and topographic factors only explains about half of the variance in global sediment yield data (Hovius, 1998). However, when plotted against a proxy for tectonic uplift rate, tectonically inactive and tectonically active settings are discriminated (Figure 2.18). For example, tectonically inactive cratonic settings are characterised by very low sediment yields of less than 100 t km^{-2} yr^{-1}, whereas currently or recently tectonically active contractional mountain belts have sediment yields of 100 to 10,000 t km^{-2} yr^{-1}. Pinet and Souriau (1988) similarly noted different relationships for young, tectonically active orogens and old, tectonically inactive landscapes. This suggests that physical insights based on the different sets of processes dominating in low-relief and high-relief areas must be included in the analysis. Montgomery and Brandon (2002) suggested a non-linear regional scale relation between erosion and mean slope derived from a digital elevation model (DEM):

$$E = E_0 + \frac{KS}{\left[1 - \left(\frac{S}{S_c}\right)^2\right]} \tag{7.1}$$

where E is the erosion rate, E_0 is a background erosion rate due to chemical weathering (approximately 0.01 mm yr^{-1} based on the mean chemical denudation rate for the world's 35 largest drainage basins (Summerfield, 1991)), S is the mean slope, S_c is a limiting hillslope gradient for landsliding and K is a rate constant. Here, mean slope is derived from the range of values within a 10 km-diameter analysis window around points spaced every 2 km, using a 10 m-resolution DEM. In the Olympic Mountains of Washington (United States), $E_0 = 0.016-0.059$ mm yr^{-1}, $K = 0.6$ mm yr^{-1}, and $S_c = 40°$. Erosion rate can be expressed in terms of a mean local relief, derived from the same 10 m-resolution DEM with a 10 km-diameter analysis window, giving

$$E = E_0 + \frac{KR_y}{\left[1 - \left(\frac{R_y}{R_c}\right)^2\right]} \tag{7.2}$$

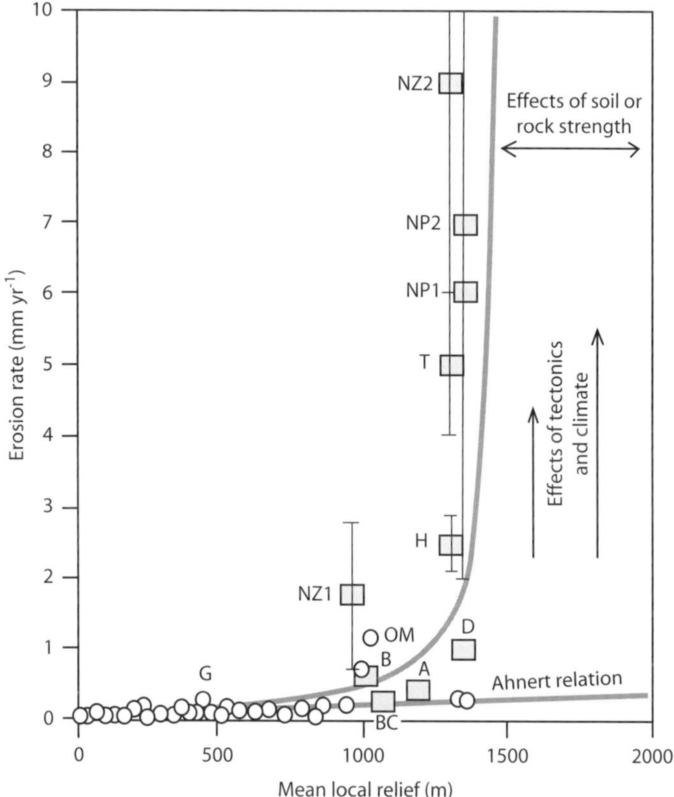

Figure 7.1 Plot of erosion rate versus mean local relief using data compiled from Ahnert (1970), Summerfield and Hulton (1994) and Pazzaglia and Brandon (1996). Solid line is model fit using equation (7.2) with $E_0 = 0.01$ mm yr^{-1}, $R_c = 1,500$ m, and $K = 2.5 \times 10^{-4}$ mm yr^{-1}. Solid squares are from tectonically active regions: NZ1 and NZ2, Southern Alps of New Zealand; H, Central Himalaya; NP1 and NP2, the Himalayan portion of the Indus Basin; T, Taiwan; BC, British Columbia Coast Range; OM, Olympic Mountains; D, Denali portion of the Alaska Range; A, European Alps. G, Ganges and B, Brahmaputra are also shown for reference. From Montgomery and Brandon (2002) (fig.4), redrawn by Allen and Allen (2013), with permission of Elsevier.

where R_y is the mean local relief, and R_c is the limiting local relief. Data from low-relief, tectonically inactive areas show a linear relation between erosion and mean local relief, as previously suggested by Ahnert (1970) (Figure 7.1). However, this linear relation does not hold for high-relief, tectonically active areas. In such cases, equation (7.2) gives a good fit with $E_0 = 0.01$ mm yr^{-1}, $R_c = 1,500$ m, and $K = 2.5 \times 10^{-4}$ mm yr^{-1}. The mean local relief R_y appears to reach a maximum of about 1,500 m ($R_c = 1,500$ m), and is very rarely greater than 2,000 m, suggesting that rock strength may control the maximum relief attainable in a mountain belt.

There are therefore essentially two different sorts of landscape in terms of erosion rate. In low-relief landscapes, such as low-lying shield areas and alluvial plains, thick regoliths may develop. The rate of removal of regolith is determined mainly by the erosivity of the transport processes, rather than by the availability of loose, easily transportable material. This may also be the case in regions of extreme aridity where streamflow processes are negligible. In these transport-limited circumstances, parameters such as mean annual rainfall and specific run-off may offer the best correlation with sediment yield. Where the removal of loose material by streams is efficient, the rate of removal of sediment may, however, be limited by the rate at which loose material is supplied by hillslope weathering. These circumstances are therefore weathering-limited, and hillslope erosion processes determine the erosion rate, which depends linearly on mean slope or local relief. These conditions may be typical of low-relief, mid-latitude, temperate zone landscapes. In high-relief landscapes, however, physical transport processes of hillslope erosion by landsliding and channelised flow are almost always capable of removing regolith above a critical rate of rock uplift. Erosion rates then stabilise at a certain mean local relief determined by rock and soil strength. It is estimated that rock uplift rates of 1 mm yr^{-1} are required to sustain combined chemical and physical erosion in mountainous areas (Koons, 1995). In high-relief landscapes, therefore, the key process is the landsliding of critically steep hillslopes and efficient removal of debris by streams.

7.2.1 Erosion Rate in Glaciated Basins

A key question in geomorphology is the effectiveness of erosion and sediment efflux in glacierised relative to non-glacierised landscapes (Hallet et al., 1996; Elverhoi et al., 1998). The sediment yield from a glaciated system depends largely on the erosion rate at the base and sides of the glacier or ice stream, which in turn is determined by the ice mass balance, basal temperature, availability of meltwater and variations in ice flow in space and over time (Hallet et al., 1996; Koppes et al., 2015). Erosion rate estimates are affected by the timescale over which the rate is measured, which depends on the technique employed. At a short timescale of less than 10^2 yr, sediment yields calculated from annual discharges of glaciers provide the highest values of erosion rate. With increasing timescale, radiocarbon dating (timescale of $10^2 - 10^4$ yr), cosmogenic surface exposure ages (timescale $10^4 - 10^5$ yr) and low-temperature thermochronology (timescale $10^6 - 10^8$ yr) provide progressively lower erosion rates (Koppes and Montgomery, 2009) (Figure 7.2). The 100-fold decrease in glacial erosion rate from decadal timescales to the multi-million year timescale for the growth and decay of orogenic topography suggests that today's high glacial erosion rates are a transient response as landscapes are reshaped following the Pleistocene glaciations. The time-dependence of glacial erosion rates suggests that the erosional impact of glaciers is high during the warming at the end of a glacial cycle, but are lower when averaged over a full glacial-interglacial cycle.

Erosion rates associated with glaciated basins can be compared with equivalent rates associated with rivers (Koppes and Montgomery, 2009) (Figure 7.3). Erosion rates can be assessed with a range of geochronometric tools that allow fluvial and glacial erosion rates to be compared over a range of timescales ($1-10^7$ yr) and with a range of forcings, such

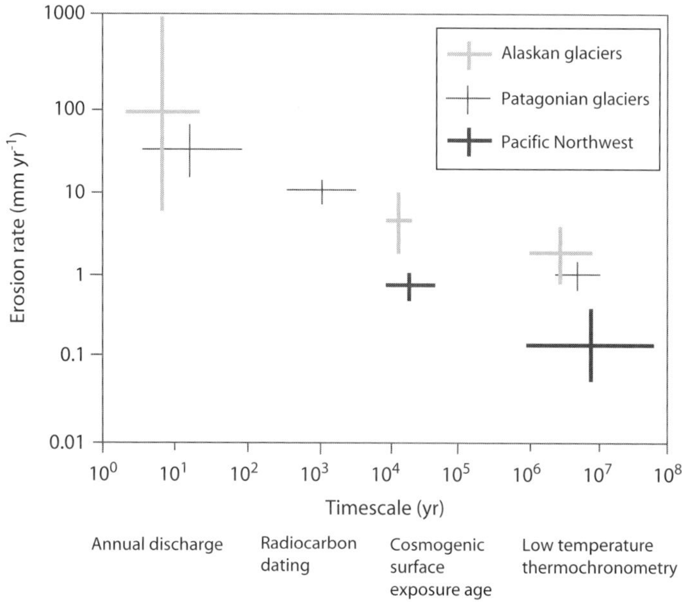

Figure 7.2 Erosion rates from glaciated basins in Alaska, Patagonia and the Pacific Northwest (Washington and British Columbia) as a function of time resolution using a range of chronometric techniques. Modified from Koppes and Montgomery (2009) (fig.2) with permission of Nature Publishing Group.

as tectonic uplift rate, volcanic activity, climate-driven ice dynamics and modern intensive agriculture. Contemporary sediment yields calculated over the last 20 years plotted against catchment area show no distinction between fluvial and glaciated basins (Figure 7.3). The highest erosion rates have been measured from retreating tidewater glaciers and volcanically disturbed rivers. The tectonic uplift rate, rather than specific geomorphic agents, appears to set the pace for erosion. Whereas erosion rate shows a good diminishing trend with timescale of chronometric resolution (Figure 7.2), fluvial basins show little variability between erosion rates at different timescales (Figure 7.4). However, the efficacy of erosion varies over two orders of magnitude between tectonically active and tectonically quiescent landscapes. The uniformity of fluvial erosion rates across timescales indicates that these landscapes have been in dynamic equilibrium with tectonic uplift rates throughout the late Cenozoic, with the exception of rivers draining active volcanoes. Rivers draining active volcanic landscapes show extremely high rates of erosion following major eruptions (Gran and Montgomery, 2005), long after the phase of eruption and emplacement of debris flows and lahars.

Glaciated landscape evolution models that simulate ice flux and glacial erosion allow the geomorphic evolution of topography and sediment production to be visualised and better understood. It is currently recognised that large, fast-moving, temperate tidewater glaciers erode rapidly, and that polar glaciers and small, high-altitude alpine glaciers erode much

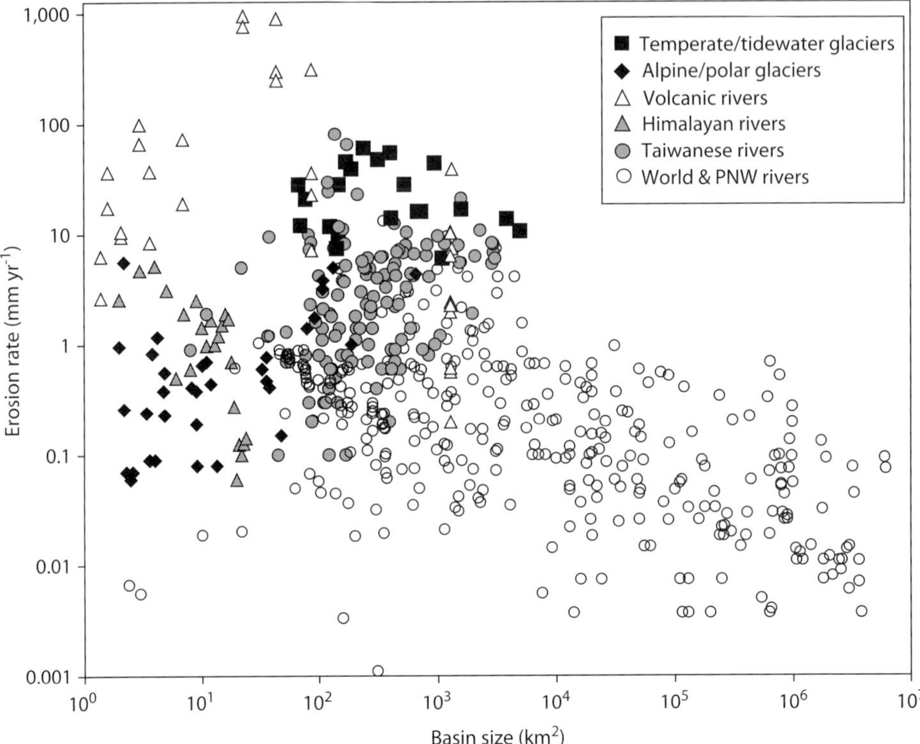

Figure 7.3 Comparison between erosion rates associated with fluvial, glacial and composite land-scapes derived from sediment yield data collected over 1–20 years. PNW is river basins in the Pacific Northwest. From Koppes and Montgomery (2009) (fig.1, p.645) with permission of Nature Publishing Group.

more slowly. The large meltwater discharge of temperate glaciers flushes large amounts of debris from under the ice, resulting in major sediment accumulations at the ice terminus, whereas in colder climates, little water reaches the bed to facilitate sliding and the evacuation of sediment. The most common way to link sediment production to glaciological or climatic variables is to assume that erosion rates are proportional to the ice sliding velocity at the bed (Egholm et al., 2012). The bedrock erosion rate E is then given by

$$E = K_g u_s^n \tag{7.3}$$

where u_s is the glacier sliding speed, K_g is a constant describing the susceptibility of the bedrock to erosion and n is a constant that is commonly taken as unity, but which may vary over a broader range of 1–3. The sliding rate most likely varies strongly with the thermal regime of the glacier.

A quantitative study of data from the last 50–100 years from glaciers occurring over a range of latitude reveals the impact of thermal regime on erosion rate (Koppes et al., 2015). For each glacier catchment, sediment yields are calculated from the sediment

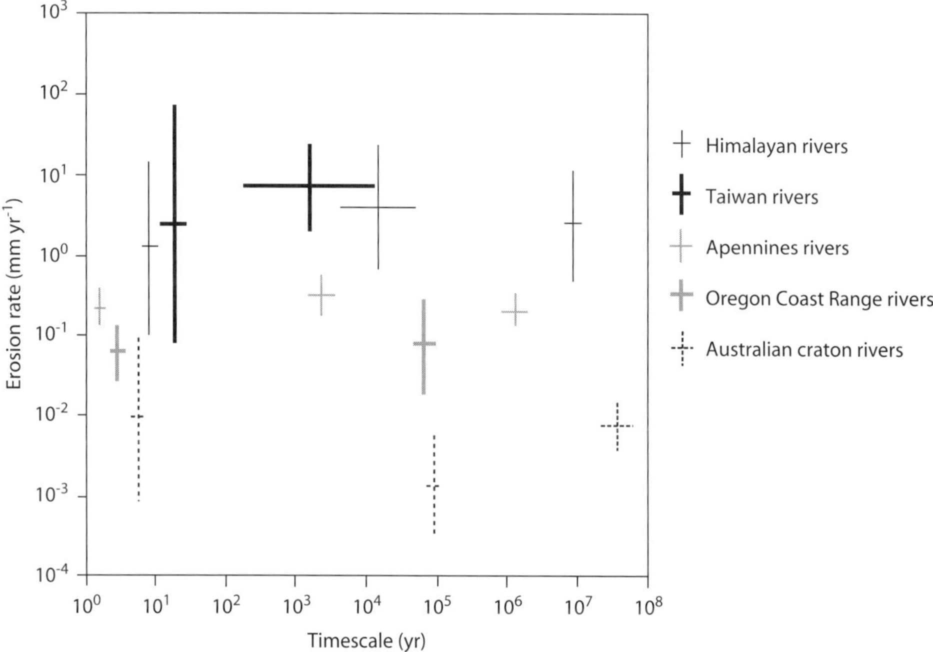

Figure 7.4 Comparison of short-term and long-term erosion rates in fluvial basins. From Koppes and Montgomery (2009) (fig.2b, p.646) with permission of Nature Publishing Group.

volumes imaged in recently deglaciated fjords, and using published ^{210}Pb analyses of cores to estimate sediment accumulation rates combined with the spatial extent of sediment depocentres measured from multibean swath bathymetry. Erosion rates are then calculated as the ratio of the sediment volume to the area of the glaciated basin. Erosion rates decrease with increasing latitude, from high rates in the temperate glaciers of Patagonia to the polar glaciers of the Antarctic Peninsula. Likewise, erosion rate increases with increasing mean annual temperature. Along the N-S transect, erosion rates also increase non-linearly with basal sliding speed, though the correlation is not strong ($R^2 = 0.39$), suggesting that the power law erosion 'rule' in equation (7.3) is broadly applicable.

Modern erosion rates for the western Antarctic Peninula (0.01 to less than 0.1 mm yr^{-1}), which are similar to those of polar fjords in the Arctic, are two orders of magnitude lower than the rates for Patagonian glaciers (1 to greater than 10 mm yr^{-1}). A similar range of two orders of magnitude in the erosion rate between temperate and polar settings is also found when comparing rates at short to long timescales, as noted previously.

Outlet glaciers in Patagonia, the South Shetland Islands and the Antarctic Peninsula have basal sliding velocities of between about 100 and 3,600 m yr^{-1} calculated from maximum surface velocities adjusted for the velocity depth profile, and basal erosion rates of 0.01 to 12 mm yr^{-1} calculated by dividing the centennial sediment yield by the glacier catchment area (Koppes et al., 2015). The best fit on a log-log plot of erosion rate (mm yr^{-1}) versus basal sliding velocity (m yr^{-1}) yields an exponent $n = 2.34$ and an intercept $K_g = 5.2 \times 10^{-8}$ using 13 outlet glaciers ($R^2 = 0.39$). The fit improves by removing the two outliers (Fourcade Glacier and Tyndall Glacier), giving $n = 2.62$ and $K_g = 5.3 \times 10^{-9}$ ($R^2 = 0.62$).

7.3 The BQART Predictor

Sediment discharge to the ocean is controlled by a range of geomorphic and geologic factors, but is also increasingly conditioned by the activities of man (Syvitski et al., 2005; Syvitski and Milliman, 2007). Consequently, there has been a concerted effort to incorporate anthropogenic factors into an equation that would explain sediment discharge at the present day. This global predictor is called BQART.

The evolution of the BQART predictor begins with the recognition that the sediment discharge of rivers Q_s relates to their maximum relief R and basin area A (Milliman and Syvitski, 1992; Mulder and Syvitski, 1996). A global regression analysis gives

$$Q_s = \alpha A^m R^n \tag{7.4}$$

where m and n are area and relief exponents, and α is a coefficient whose value is dependent on climatic and lithologic effects. When Q_s is in Mt yr^{-1}, area A is in km$^2 \times 10^6$ and relief R is in km, the global regression gives $m = 0.412$ and $n = 1.3$. α is likely to vary markedly but the median from 488 rivers gives $\alpha = 17$. Although the regression in equation (7.4) does not explicitly take into account precipitation, climate, vegetation and bedrock geology, it was assumed by Syvitski and Morehead (1999) that climatic factors, including temperature, were embedded into the coefficient and exponents of equation (7.4). Taking basin-averaged temperature as a separate parameter, they modified equation (7.4) to derive an estimate of the sediment discharge as a function of area A, relief R and temperature T in the form of the so-called ART equation:

$$Q_s = c_1 A^{c_2} R^{c_3} e^{kT} \tag{7.5}$$

where c_1, c_2 and c_3 are empirically derived coefficients based on climate characteristics, T is temperature, and k is another coefficient related to the strength of physical versus chemical weathering. The ART equation applies to pristine conditions, unaffected by the activities of man. However, equation (7.5) accounts for very little of the variance of a global compilation of data from nearly 500 rivers, suggesting that anthropogenic factors need to be considered. Consequently, sediment discharge is expressed in a modified equation that explicitly accounts for human factors, the BQART equation of Syvitski and Milliman (2007):

$$Q_s = 0.02 B Q_w^{0.31} A^{0.5} R T \tag{7.6}$$

for basin-averaged temperatures of greater than 2°C, and

$$Q_s = 0.04 B Q_w^{0.31} A^{0.5} R \tag{7.7}$$

for basin-averaged temperatures of less than 2°C, Q_s is in units of kg s^{-1}, Q_w is in km^3 yr^{-1}, A is in km^2, R is in km and T is in °C, and climatic and human factors are embedded in the new variable B. This variable incorporates the effect of temperature on weathering rates and the effects of glaciation, bedrock lithology in the catchment and also human activities affecting river sediment loads. It is given by

$$B = IL(1 - T_E)E_h \tag{7.8}$$

where I is a glacial erosion factor, L is a basinwide lithology factor, T_E is a trapping efficiency in lakes and man-made reservoirs, and E_h is a human-influenced soil erosion factor. B can be calculated for individual catchments (Table 7.1) or as a global average (Allen and Allen, 2013) (appendix 41, p.530).

The accuracy of equations (7.6) and (7.7) is tested on 20 rivers from different topographic, tectonic and climatic settings in Table 7.1. The sediment discharge is calculated using the BQART predictor and compared with the observed discharges at gauging stations using the global database of Syvitski and Milliman (2007) (Figure 7.5).

The sediment discharge of the BQART equation can be divided by drainage basin area to derive the average yield for suspended particulate matter (Table 7.1), since measured riverine sediment discharges do not account for dissolved loads or bedload. Consequently

$$Y_s = 0.02BQ_w^{0.31}A^{-0.5}RT \tag{7.9}$$

for basin-averaged temperatures of greater than 2°C.

There are several parameters controlling sediment yield, so the topic is amenable to dimensional analysis. Sediment discharge [MT^{-1}], drainage basin area [L^2] and relief [L] are clearly important to include in a dimensional analysis. Since rivers transport sediment at the surface of the Earth under the influence of gravity, we should also expect sediment density ρ_s [ML^{-3}] and gravity g [L T^{-2}] to occur in the analysis. This gives 3 dimensions and 5 parameters, so there should be 2 dimensionless groups. We can make the sediment mass discharge dimensionless by dividing by $R^{5/2}\rho_s\sqrt{g}$, so one of the dimensionless groups is

$$\left[\frac{Q_s}{R^{5/2}\rho_s\sqrt{g}}\right] \tag{7.10}$$

and the other group must be a dimensionless drainage area:

$$\left[\frac{A}{R^2}\right] \tag{7.11}$$

Equation (7.4) can now be expressed in terms of these dimensionless groups as follows:

$$\left[\frac{Q_s}{R^{5/2}\rho_s\sqrt{g}}\right] = \alpha\left[\frac{A}{R^2}\right]\left(\frac{A^{1-m}R^{n-0.5}}{\rho_s\sqrt{g}}\right) \tag{7.12}$$

A plot of dimensionless mass discharge versus dimensionless area using the dimensionless groups in equation (7.12) is shown in Figure 7.6. This plot incorporates all of the parameters thought to influence sediment yield.

Table 7.1 *Comparison of sediment discharge Q_s in Mt yr^{-1} calculated using the BQART equations with the discharge of sediment derived from the suspended loads entering the coastal ocean, using data in the global database of Syvitski and Milliman (2007). Equivalent yield of particulate sediment in mm yr^{-1} is also given.*

River	Drainage basin area A (km^2)	Anthropogenic factor B	Max relief R (km)	Temp. T(°C)	Q_w (km^3 yr^{-1})	Q_s BQART (Mt yr^{-1})	Q_s observed (Mt yr^{-1})	Yield Y_s (mm yr^{-1})
Amazon	5,853,804	1	5.5	22	6,300	2,781	1,200	0.176
Dneipr	495,839	1	0.330	7	42.9	3.3	2.3	0.002
Fly	61,413	1.5	4.0	25	179	120	80	0.690
Fraser	217,504	2.286	4.0	2.2	112.3	25.6	20.0	0.044
Ganges	1,628,405	3.052	7.0	17	996.5	2,485	1,100	0.565
Godavari	312,575	3	1.6	27	83.6	180	170	0.214
Irrawaddy	405,963	1.09	3.0	22	427.6	189	259	0.173
Magdalena	252,743	1.5	3.3	20	230	169	140	0.248
Niger	1,240,019	0.5	0.820	28	41.2	40.0	41.2	0.012
Nile	2,026,122	0.5	3.8	25	109.9	183	120	0.033
Orinoco	939,362	0.7	6.0	25	230	169	140	0.222
Po	72,183	0.354	4.8	11.5	46	10.9	15	0.056
Song Hong (Red)	150,000	1.2	3.1	22	119.3	88.0	77.5	0.217
Tigris–Euphrates	1,050,000	1	2.5	10.8	47.3	57.7	53	0.020
Susquehanna	72,147	0.6	0.96	8	31.8	0.57	1.8	0.003
Yenisei	2,578,730	0.75	3.5	-10	616.6	39.0	13.0	0.006
Yukon	855,715	1.905	6.0	-7	208.8	69.9	53.7	0.030
Zambezi	1,400,000	0.15	2.35	24	100	26.3	48.0	0.007
Zhujiang (Pearl)	370,000	0.75	2.0	18	258.6	58.0	77.5	0.058

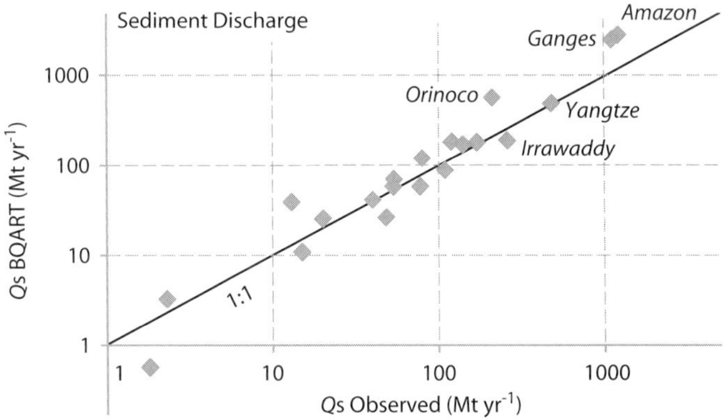

Figure 7.5 Comparison of sediment discharge in Mt yr^{-1} calculated using the BQART equations with observed values from the suspended sediment loads at gauging stations, for 20 selected river drainage basins identified in Table 7.1.

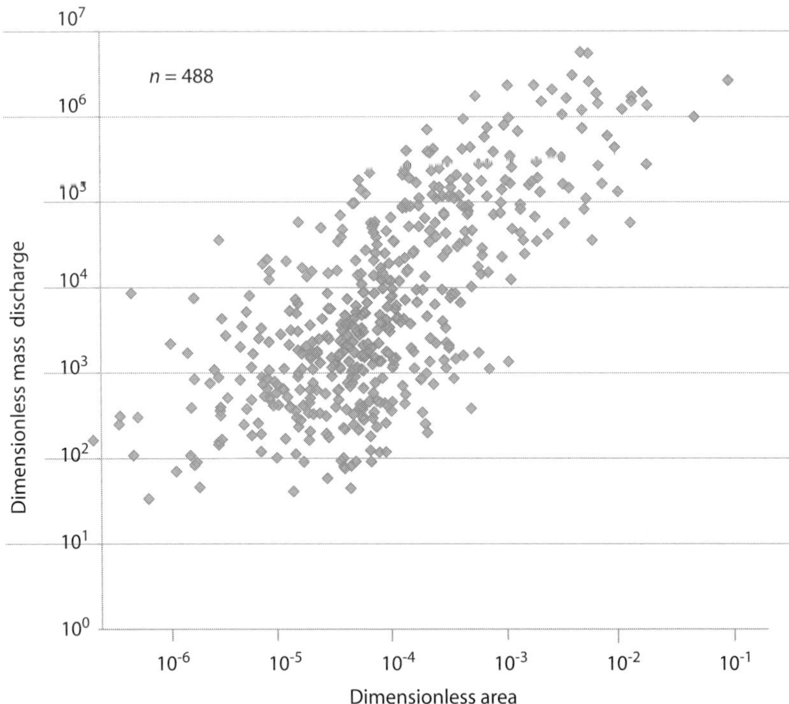

Figure 7.6 Bilogarithmic plot using 488 rivers of dimensionless suspended sediment discharge versus dimensionless catchment area, using the BQART equations and database of Syvitski and Milliman (2007). For consistency with Syvitski and Milliman (2007), Q_s is in Mt yr^{-1}, area A is in Mkm^2, and relief R is in km. There is clearly a correlation between dimensionless discharge and dimensionless area scaled by relief, but there is considerable scatter (note that Figure 7.6 is a log-log chart).

The BQART equation allows sediment discharges to the ocean, and the equivalent sediment yield of river drainage basins, to be estimated from best guesses at the parameters involved. It is worthwhile to pause to consider whether these parameters can be constrained in the case of ancient sediment routing systems. River water discharge would need to be estimated from the catchment area and average precipitation rate modified by a factor that incorporates the effects of evapotranspiration and losses to groundwater and soil soak-in. Alternatively, water discharge might be estimated by a palaeohydrological analysis of channel geometries and dimensions (Section 4.6). Average catchment temperature and the possibility of glaciated catchments would need to be estimated from palaeoclimatic reconstructions. Basin-area would need to be estimated from provenance analysis.

7.4　Estimates of Erosion from Strontium Isotope Ratios

Sr and Nd isotopes have been widely used as provenance indicators of terrestrial and marine sediment since the basin sediment retains the isotope composition of the various sediment sources (Clift et al., 2008a). Temporal variations in Sr and Nd isotopes measured from stratigraphy of known ages are used to infer climatic and tectonic changes controlling physical erosion in the past.

There is a continuous and robust record of marine $^{87}Sr/^{86}Sr$ through much of the geological past and particularly in the Cenozoic (Richter, Rowley, and DePaolo, 1992). Variation in the ratio of the radiogenic isotope ^{87}Sr relative to the non-radiogenic isotope ^{86}Sr is generally attributed to changes in the rate of continental weathering. Increases in the ratio in the Cenozoic are commonly linked to tectonic uplift of the Himalayan mountains driving enhanced rates of chemical weathering. A key test of the idea is to measure the $^{87}Sr/^{86}Sr$ ratios in river water draining from areas undergoing erosion (Krishnaswami et al., 1992; Tripathy, Singh, and Krishnasami, 2011).

The main source of dissolved Sr to rivers is chemical weathering of silicate and carbonate rocks cropping out in their drainage basins. Each mineral has its own $^{87}Sr/^{86}Sr$ ratio. Sr isotopes from these various minerals are released into overlying soils during weathering of fresh rock surfaces, which is enhanced by high rates of physical erosion driven by tectonic uplift, seismic shaking and climate change. Sr is released from rocks in a mineral dissolution sequence over time, with the most easily weatherable trace phases (calcites, apatites) being released quickly (the initial few hundred years), followed by less easily weatherable phases such as biotite ($10^2 - 10^4$ years), and finally by plagioclase releasing Sr from surfaces that are about 10^5 years old (Blum and Erel, 1995; Erel et al., 2004). The presence of different mineral weathering rates suggests that the new exposure of fresh weathering surfaces should make the $^{87}Sr/^{86}Sr$ of rivers significantly more radiogenic for periods of approximately 20 kyr following their exposure. Strontium isotope signals are transferred to the ocean by rivers, so high-frequency climatic events such as Quaternary glaciations are potentially registered by the $^{87}Sr/^{86}Sr$ of marine carbonate (Zachos et al., 1999).

The impact of mineral weathering kinetics can be seen in the seasonal variations of $^{87}Sr/^{86}Sr$ in rivers. During the monsoon, the $^{87}Sr/^{86}Sr$ of headwater tributaries of the Marsyandi River and the Ganga are less radiogenic than during non-monsoon periods (Bickle et al., 2003; Tipper et al., 2006). This is explained by the Sr isotopic composition of rivers in the Himalaya being determined by two contributions: (1) marginally radiogenic sedimentary carbonates, and (2) more highly radiogenic silicates (Krishnaswami, Singh, and Dalai, 1999). The weathering rates of the carbonates are much higher than those of silicate sources, which enhances the contribution of rivers draining carbonates in periods of high run-off. For example, it is estimated that in the headwaters of the Ganga River system about 70% of the dissolved Sr in the river water is derived from carbonates and only 20% is derived from silicates (Tripathy and Singh, 2010). Since carbonates are less radiogenic than silicates, the higher proportion of the contribution from carbonate sources makes the riverine Sr lower in $^{87}Sr/^{86}Sr$. The reverse is true for dry periods, when the relative contribution of silicate sources is increased by the long time available for water-rock interactions. The fact that $^{87}Sr/^{86}Sr$ ratios in rivers may be significantly different to the $^{87}Sr/^{86}Sr$ ratios of the parent material in the eroding hinterland suggests some caution in the use of strontium isotope ratios in evaluating climatic and tectonic events in catchment areas.

The concentration of dissolved Sr and its isotopic composition have been measured in many of the world's major rivers (Table 7.2). Dissolved Sr concentrations vary over an order of magnitude. The Huang He (Yellow River) has the highest concentration, and the Orinoco River the lowest. The Chang Jiang (Yangtze River) delivers the highest amount of Sr to the ocean -7% compared to about 2.5% in terms of water discharge. The global annual flux is estimated to be $36.5-47 \times 10^9$ mol yr^{-1} depending on the methods used (Tripathy et al., 2011).

The $^{87}Sr/^{86}Sr$ ratio of major rivers varies from 0.7089 to 0.7291, with the highest values in the Ganga and the lowest in the Danube. The Sr isotopic composition of carbonates varies little and is generally unradiogenic (0.705–0.709), but metamorphic alteration of carbonates can make them much more radiogenic. Silicates, however, show a wide range of Sr isotopic composition from unradiogenic volcanics to highly radiogenic granites and gneisses. The lithological make-up of different silicate rocks in the catchment therefore has an important impact on the Sr isotopic composition of river water. The Ganga-Brahmaputra rivers, with headwaters in the Himalaya, have relatively high $^{87}Sr/^{86}Sr$ ratios (0.7118) compared to others, and account for approximately 21% of the global riverine flux compared to their contribution to global water discharge of about 10%. Increased chemical weathering in the Himalayan source area has therefore been invoked to explain the trend in $^{87}Sr/^{86}Sr$ in the global oceanic reservoir in the Cenozoic.

Changes of rates of silicate weathering over the million year timescale are thought to be an important regulator of global climate through the CO_2 budget. The present day Sr concentration and its isotopic composition of rivers draining the Himalayan region suggests that marine $^{87}Sr/^{86}Sr$ can act as a proxy for silicate weathering on the continents. Contemporary rates of silicate weathering calculated for headwaters of tributaries in the

Table 7.2 *Strontium concentration, flux and* $^{87}Sr/^{86}Sr$. *Sr concentration (nM) is discharge-weighted. From Tripathy et al. (2011) (tab.26.1, p.530) with permission of Springer.*

River	Water discharge ($km^3\ yr^{-1}$)	Drainage area ($10^6\ km^2$)	Sr flux ($10^9\ mol\ yr^{-1}$)	$^{97}Sr/^{86}Sr$ ratio	Sr concentration (nM)
Global rivers					
Amazon	6,590	6.112	2.04	0.71165	310
Zaire (Congo)	1,200	3.698	0.38	0.7155	313
Orinoco	1,135	1.1	0.24	0.7183	210
Mississippi	580	2.98	1.24	0.70957	2,130
Parana	568	2.783	0.3	0.7139	520
Lena	525	2.49	0.58	0.71048	1,100
Tocantins	372	0.757	0.14	0.72067	380
Amur	344	1.855	0.17	0.70923	500
St Lawrence	337	1.02	0.4	0.70962	1,200
Mackenzie	308	1.786	0.84	0.71138	2,740
Columbia	236	0.669	0.23	0.7121	982
Danube	207	0.817	0.57	0.7089	2,760
Yukon	200	0.849	0.32	0.7137	1,590
Niger	154	1.2	0.04	0.714	913
Rhine	69	0.224	0.43	0.7092	6,227
Himalaya-Tibet					
Chang Jiang	928	2	2.63	0.71032	2,830
Brahmaputra	510	0.58	0.37	0.7192	730
Ganga	493	1.05	0.28	0.7291	560
Irrawaddy	486	0.41	-	0.7101	-
Mekong	467	0.795	1.44	0.71025	3,080
Pearl	363	0.437	0.28	0.7119	767
Indus	238	0.47	0.88	0.7111	3,689
Salween	211	0.325	0.28	0.71405	1,330
Hong (Red)	123	0.12	0.14	0.71284	1,158
Huang He	41	0.752	0.31	0.7111	7,458
Peninsular India					
Godavari	105	0.313	0.14	0.7152	1,375
Mahanadi	66	0.132	0.05	0.7193	704
Narmada	39	0.102	0.08	0.7114	2,156
Krishna	30	0.259	0.11	0.7142	3,742

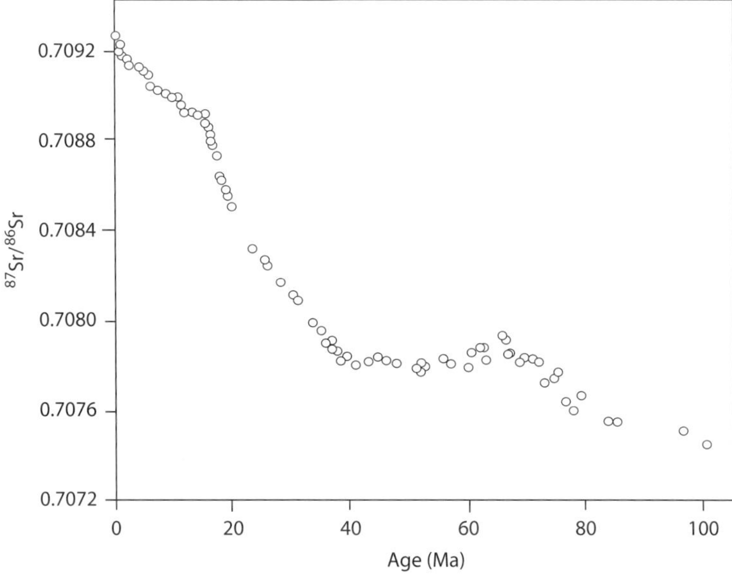

Figure 7.7 Evolution of seawater Sr isotopic data for the last 100 Myr. Note the increasing trend of the strontium isotope values at 40 Ma. Compiled by Richter et al. (1992) (fig.1, p.12). Published with permission of Elsevier.

Himalayas range between 10 and 15 tons km^{-2} yr^{-1}, which is two to three times the global average of 5.5 tons km^{-2} yr^{-1} (Gaillardet et al., 1999).

The Sr isotopic evolution of the oceans shown by the $^{87}Sr/^{86}Sr$ in marine carbonate (Veizer, 1989) (Figure 7.7) involves a marine budget where the Sr is derived from river inflows, hydrothermal sources, diagenetic alteration/dissolution of carbonates in marine biota and the removal of calcareous skeletons by burial on the sea bed (Richter et al., 1992). The Sr flux from diagenetic alteration/dissolution is much smaller than the other fluxes and is commonly neglected. The $^{87}Sr/^{86}Sr$ of sea water is consequently governed essentially by the balance between the radiogenic river supply and the unradiogenic hydrothermal supply of mantle-derived Sr, the latter being commonly assumed to be proportional to the production rate of new ocean floor.

However, there is uncertainty in linking observed strontium isotopic composition of marine carbonates with changes in rates of silicate weathering, because the Sr riverine flux is not well known over the Cenozoic, the possible important impact of different source compositions on the Sr concentration and isotopic composition of river water, and uncertainties in the magnitude of the hydrothermal flux, which may originate from volcanic islands and island arcs as well as from newly created ocean floor.

Calculation of the erosion rate from strontium data requires a number of approximations. The Sr river flux over time is calculated with a model that reproduces the observed seawater Sr isotopic evolution over the last 50 Myr, with the $^{87}Sr/^{86}Sr$ of river water held at its

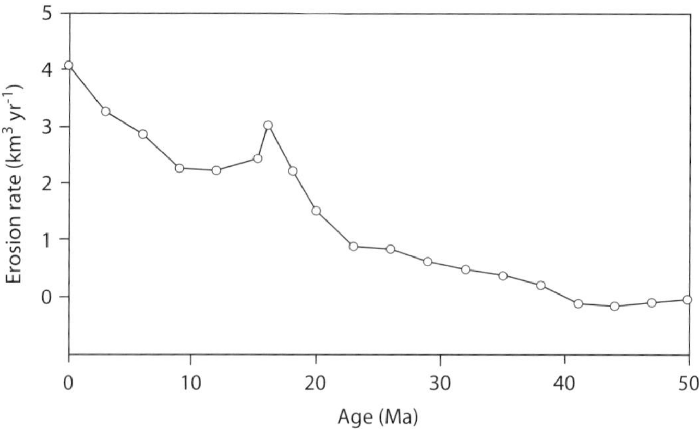

Figure 7.8 The erosion rate of the Himalaya-Tibet region during the Cenozoic required to provide the amount of dissolved Sr needed to account for the rise in seawater ^{87}Sr/^{86}Sr shown in Figure 7.7. From Richter et al. (1992) (fig.8) with permission of Elsevier.

modern value of 0.711, and a hydrothermal flux (with ^{87}Sr/^{86}Sr of 0.703) that changes in proportion to the rate of generation of new sea floor. This disslolved Sr flux is converted back to a source rock volume eroded per unit time assuming that 25% of the Sr is dissolved and 75% is detrital, the density of rocks acting as sources is 3,000 kg m^{-3} and the concentration of Sr in the source is 350 ppm. Consequently, uncertainties are at least 30% in Figure 7.8.

The erosion rate in the Himalaya-Tibet region that provides the dissolved Sr necessary to explain the variation of ^{87}Sr/^{86}Sr over the last 50 Myr (Figure 7.8) shows an overall increasing trend towards the presentday, but with a peak at 16 Ma corresponding to the rapid unroofing of the Tibetan Plateau at 15−20 Ma. The event at approximately 16 Ma also corresponds to an acceleration of the sediment mass flux delivered to the Arabian Sea (Clift and Gaedicke, 2002). Richter et al. (1992) suggested that there was a time lag of 10 Myr between the start of the increasing erosion trend and the collision of the Indian indenter with Asia, which could be explained by the time taken to generate high topography capable of sustaining high rates of chemical weathering or by the need to remove a sedimentary cover of carbonates before exposing the more radiogenic basement rocks. The presence of an accelerating erosion rate, rather than strontium isotope ratios reaching a plateau at the achievement of topographic steady state in the mountain belt, is an interesting puzzle.

The weathering of the Deccan Traps (emplaced at 65 Ma) may also have contributed unradiogenic Sr from the chemical weathering of basalts (Taylor and Lasaga, 1999), leading to a decline in seawater ^{87}Sr/^{86}Sr during the Early Tertiary (Das et al., 2005; Das, Krishnaswami, and Kumar, 2006), without invoking changes in the rate of silicate weathering on the continents.

7.5 Denudation from Low-Temperature Thermochronometry

There is a wide range of techniques used to estimate pointwise and spatially averaged erosion rates. Any individual technique is useful for a particular time range, using particular materials. A large number of techniques are applicable only to Quaternary materials (see Noller, Sowers, and Lettis (2000) for a compilation). For example, dendrochronology (tree rings) can only be used on materials up to about 10 ka in age, and radiocarbon dating (^{14}C) has a useful time range of only less than 35 ka. U-Th radioisotopic dating, amino acid racemisation, thermoluminescence and optically stimulated luminescence (OSL) all have upper age limits of approximately 300 ka. The time range is significantly extended by the use of in situ cosmogenic radionuclides (for example, ^{10}Be and ^{26}Al), and even further by the use of thermochronometric methods such as He diffusion during U-Th decay and apatite/zircon fission track analysis. Dating techniques making use of radiometric clocks with higher closure temperatures and longer half-lives, such as ^{40}Ar-^{39}Ar (half-life of 1.25×10^9 yr), ^{87}Rb-^{87}Sr (half-life 1.4×10^9 yr), ^{40}K-^{40}Ar (half-life 4.89×10^9 yr), ^{235}U-^{207}Pb (half-life 7.04×10^8 yr) and ^{147}Sm-^{143}Nd (half-life of 1.06×10^{11} yr), provide estimates of very long term ($10^6 - 10^7$ Myr) erosion rates. It needs to be stressed that the numerical values for erosion rate derived from different techniques with different temporal sensitivities cannot be directly compared. Use of a number of radiometric clocks with different closure temperatures allows cooling histories to be estimated. Combined with assumed or estimated geothermal gradients, these cooling histories can be used to calculate denudation rates over geological time periods.

7.5.1 Fission Track Analysis

Fission track analysis is a relatively modern (since the 1960s, Naeser 1967) thermochronological technique based on the atomic damage caused to minerals by their spontaneous decay, almost exclusively by the fission of ^{238}U. The fission of ^{238}U produces a trail of damage to the lattice known as a fission track. Fission tracks become visible when the crystal is chemically etched and viewed at high magnification. It is safe to assume that statistically fission takes place at a constant rate. The concentration of the parent isotope ^{238}U in minerals such as apatite ($Ca_5(PO_4)_3(F,Cl,OH)$), zircon ($ZrSiO_4$) and sphene ($CaTiO(SiO_4)$) is high enough to produce plenty of tracks, but low enough so that the crystal is not completely criss-crossed by tracks, making measurement difficult. Details of the technique of using fission tracks to obtain a sample age are found in a number of reviews including Brown, Summerfield, and Gleadow (1994) and Gallagher, Brown, and Johnson (1998). The calibration method is given in the seminal paper by Hurford and Green (1983).

Importantly, fission tracks are metastable features that fade or anneal, which causes track lengths to shorten. Annealing is mostly controlled by temperature (and also time), so track length distributions reveal information about thermal history. The higher the temperature experienced by the apatite, zircon or sphene crystal, the greater the annealing.

Individual minerals have a closure temperature below which fission tracks are preserved. For apatite, the closure temperature is $110°C \pm 10°C$, but at geological timescales there is a range of temperature of at least $60°C$ below the closure temperature where significant annealing takes place: this is termed the partial annealing zone (PAZ). For zircon, the PAZ is between 200 and $350°C$, but the annealing behaviour of zircon is not as well known as for apatite. Because of the different closure temperatures of apatite and zircon, apatite fission track analysis is particularly useful in thermal studies of the upper few km of the crust, whereas zircon fission track analysis is more informative for deeper levels of burial.

Apatite fission track distributions are sensitive to different thermal histories in a wide range of tectonic situations including convergent mountain belts, rift flanks and passive margin escarpments (Gallagher et al., 1998), each involving a different trajectory and speed of heating and cooling. Commonly, the fission track age bears no relation to any specific thermal or tectonic event. Only where rapid cooling takes place, such as caused by rapid exhumation of a rift flank or orogenic belt, are many long fission tracks recorded, and the fission track age is then a good indication of the timing of the cooling event. Thermal histories can be illuminated by use of 3D heat equation solver software packages such as Pecube (Braun, van der Beek, and Batt, 2006).

A fundamental tool in the assessment of erosion is the relationship between the cooling age of a sample and its present-day elevation. Samples might be recovered from a vertical drill hole, but samples from field campaigns are generally obtained from sloping valley sides. The isotherms in these regions are likely to be curved by the effect of irregular topography and geotherms may also be affected by high rates of advection at high Péclet Number Pe given by

$$Pe = \frac{EL}{\kappa} \tag{7.13}$$

where E is the exhumation or erosion rate, L is a length scale (such as crustal thickness), and κ is the diffusivity, which is approximately 10^{-6} m^2 s^{-1} for typical crustal rocks. At values of $Pe > 1$ the geotherm in a piece of exhuming crust is significantly curved, corresponding to an erosion rate of greater than 1 mm yr^{-1} (see Allen and Allen (2013), pp.37–39, fig.2.25). These rates are commonly achieved in tectonically active settings. Ignoring these complications for the moment, the different ages of differently elevated samples are assumed to result from the time since passing through the same characteristic isotherm. Consequently, the expectation is that the highest elevated samples have the oldest cooling ages (Figure 7.9), and the slope of the age-elevation relationship is indicative of the rate of denudation (Brewer et al., 2003; Stock et al., 2006) (Figure 7.10). The vertical spacing of the highest and lowest samples and the difference in the cooling ages therefore defines the erosion rate. The long-term erosion rate E is then given by

$$E = R/\Delta t \tag{7.14}$$

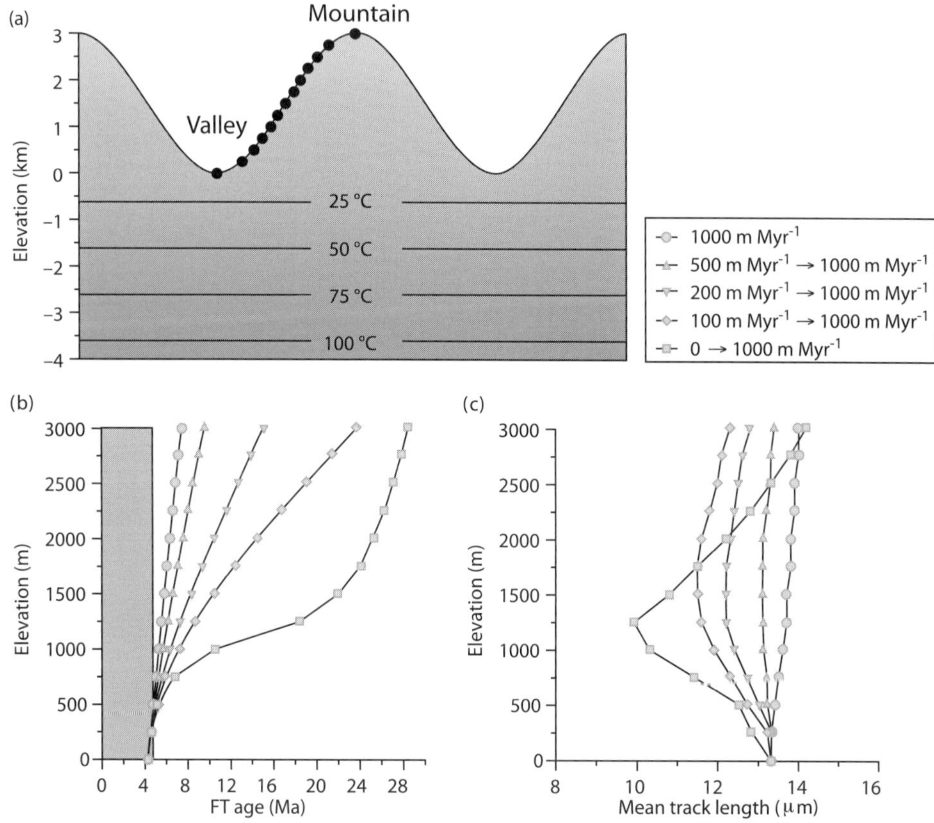

Figure 7.9 Predicted apatite fission track age-elevation relationships for different denudation histories. (a) Simple model of thermal structure beneath a varying surface topography, with geothermal gradient 25°C km^{-1} and a surface temperature at sea level of 10°C. Black dots are surface samples for which age and mean track length are predicted. Isotherms close to the topographic surface are increasingly unrealistic. (b) and (c) Predicted fission track ages and mean track lengths as a function of elevation for 6 different denudation histories with up to 1 km Myr^{-1} (values in box) prior to 4.8 Ma followed by rapid exhumation for all samples since 4.8 Ma (shaded box in (b)). From Braun et al. (2006) (fig.1.5, p.10) with permission of Cambridge University Press.

where Δt is the difference between the youngest and the oldest cooling ages and R is the relief between mountain and valley.

The age-elevation relationship may be approximately linear, but may also show a break in slope reflecting the non-linear increase in annealing rate of fission tracks, or diffusion rate of helium, with temperature (Figure 7.11). Thermochronological ages therefore rapidly decrease in the PAZ (or partial retention zone (PRZ) for U-Th/He). Rock uplift of the crust may preserve a fossil PAZ at a shallower depth or higher elevation, thereby revealing the amount of denudation that has taken place.

 Denudation and Sedimentation

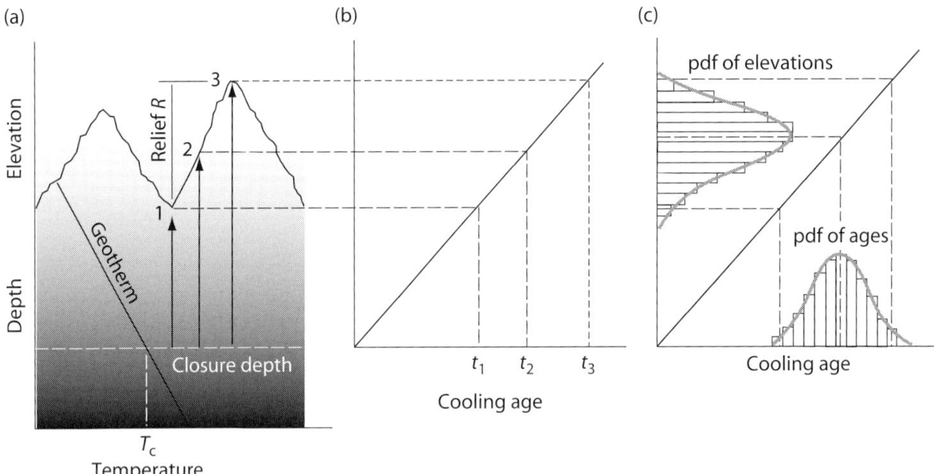

Figure 7.10 Use of elevation profiles to estimate exhumation rate in a steady-state landscape where the erosion rate of the ridge and valley floor are the same. (a) Three samples, labelled 1 to 3, from a valley side, span a range of topographic elevation R, giving three cooling ages $t1$ to $t3$. (b) The long-term erosion rate is $R/\Delta t$. (c) Samples of detrital grains from a stream bed draining this upland area have a pdf of cooling ages, which can be mapped onto a pdf of elevations in the catchment, assuming that the geothermal gradient and the catchment hypsometry are known. After Brewer et al. (2003) (fig.2) and Anderson and Anderson (2010) (p.150, fig.6.36), published with permission of John Wiley & Sons Inc.

Conversion of a cooling history to a denudation history is not necessarily straightforward. The geothermal gradient is rarely well constrained, though a near-surface value of $25–30°C$ km^{-1} is often assumed. This geothermal gradient may be affected by changes in the basal heat flow, the internal radiogenic heat production, the advection of heat from magmas and hot fluids, and the bending of isotherms by the upward exhumation of rock towards an erosional surface of high relief. This latter effect is especially important to account for with low temperature chronometers. Rapid erosion rates (greater than 5 mm yr^{-1}), associated with high thermal Péclet Numbers, may cause very high geothermal gradients ($60−100°C$ km^{-1}) to exist in the exhuming rock body (Stüwe, White, and Brown, 1994; Mancktelow and Grasemann, 1997), and geothermal gradients may be very different between mountain peaks and valley floors.

Spatially distributed thermochronometric data can be used to estimate the total denudational efflux of a mountain range or piece of continental crust. For example, the compilation of apatite fission track cooling ages across onshore Ireland was used to calculate the denudational flux over the last 200 Myr and compared with sediment volumes in offshore basins (Allen et al., 2002). Erosion rates were shown to decrease from a high in the Mesozoic to a low at $100 − 140$ Ma and then to increase irregularly towards the present-day. However, the total volume of sediment exported from the area now occupied by the Irish landmass (1 to 20 $\times 10^3$ km^3) is an order of magnitude smaller than the preserved solid volume of Cenozoic sediment in the Porcupine–Rockall basin system, suggesting

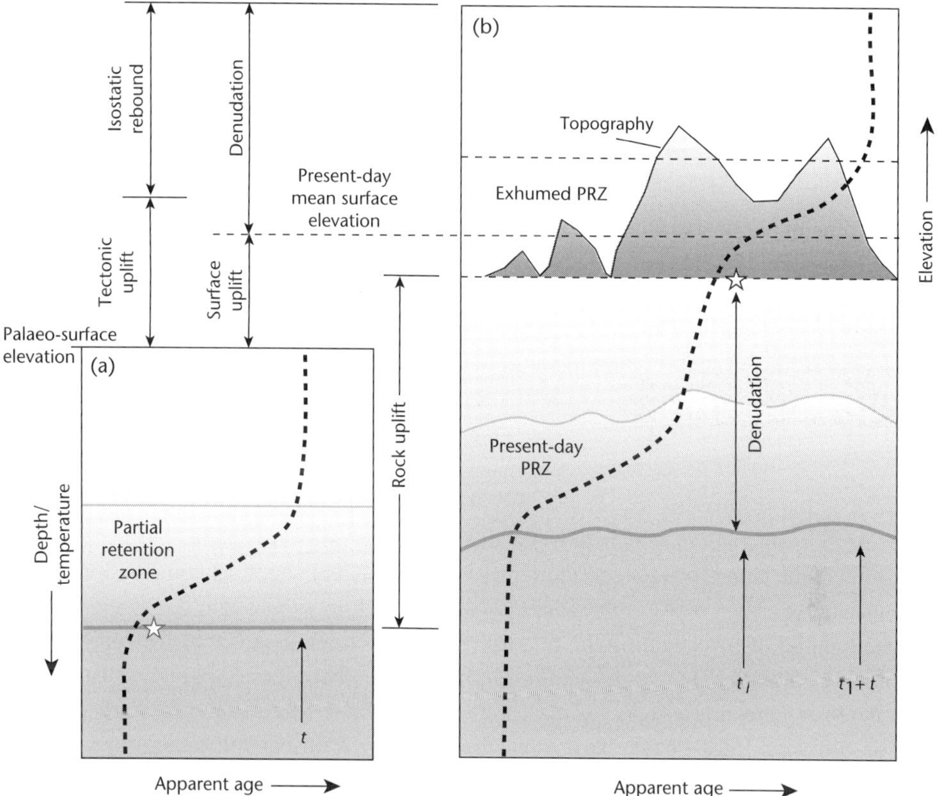

Figure 7.11 Different types of uplift, denudation and age-elevation relationship, showing the position of the partial retention zone (PRZ) for U-Th/He or partial annealing zone (PAZ) for apatite fission track. (a) Initial thermochronological age-depth profile developed under conditions of tectonic stability of duration t. Thermochronological ages decrease through the PRZ. (b) At time t_1, the crust starts to be uplifted and partially eroded causing the old PRZ to be exhumed. A sample (star) originally near the base of the pre-existing PRZ will record the age of the onset of erosion t_1, and its elevation compared to the original depth will give the amount of rock uplift. From Braun et al. (2006) (p.11, fig.1.6) modified from Fitzgerald et al. (1995) (fig.2). Published with permission of American Geophysical Union.

the operation of other sediment routing systems beyond the present-day Irish landmass. Cogné, Chew, and Stuart (2014) combined AFT and (U-Th)/He data derived from samples along the Irish Atlantic margin that were collected along pseudo-vertical profiles. Inverse modelling of the profiles using thousands of iterations (Gallagher, 2012) showed there to be a cooling event at $125-180$ Ma that was diachronous from north to south, interpreted as linked to rift-shoulder uplift and erosion of approximately 2.5 km. The rift-shoulder uplift was followed by slow heating during burial, followed by renewed exhumation, possibly in the Palaeogene and more certainly in the Neogene. Neogene uplift and erosion is a common feature of the northwest European continental margin.

The spatial distribution of thermochronometric data in present-day mountain belts, such as the European Alps, can also be used to estimate the timing and location of tectonic uplift events and sediment efflux history of the orogen (Vernon et al., 2009).

7.5.2 *Helium Diffusion: (U-Th)/He*

The accumulation of ^{4}He from the decay of uranium and thorium to stable lead can also be used as a thermochronological tool (Wolf et al., 1996). The basis for (U-Th)/He dating is that ^{4}He nuclei (α particles) are produced by the decay of ^{238}U, ^{235}U and ^{232}Th. The helium produced diffuses from the crystal over time, escaping through its surface faces. The rate of diffusion scales on temperature with an Arrhenius-type relationship

$$\frac{\kappa}{a^2} = \frac{\kappa_\infty}{a^2}\exp^{(-E_a/RT)} \tag{7.15}$$

where κ is the diffusivity, κ_∞ is the diffusivity at an infinite temperature, and a is the diffusion domain radius, which may be the physical grain dimension. Diffusivities are measured in the laboratory for certain minerals with different chemical compositions, grain size and shape.

One method of investigating the relationship between temperature history and He diffusion is to study the He age-distribution in boreholes where the temperature distribution with depth is known. He ages decrease rapidly downhole due to the effect of temperature on diffusion rates (Wolf, Farley, and Kass, 1998). Consequently, there is a Helium partial retention zone (HePRZ), which extends from approximately 40°C to 70°C. This is conceptually similar to the PAZ in fission track analysis. The existence of a HePRZ was confirmed by Stöckli, Farley, and Dumitru (2000) who found a HePRZ overlying an apatite PAZ in samples taken from the White Mountains of California. Stöckli et al. (2000) recognised a distinct break in slope at 12 Ma when apparent He ages were plotted against depth, signifying the onset of rapid exhumation caused by extensional faulting in the White Mountains. This event is barely recognisable from the distribution of apatite fission track apparent ages.

The age-elevation relationship of thermochronometers can also be used to estimate where sediment comes from in an upland catchment (Stock et al., 2006) (Figure 7.12). The (U-Th)/He ages of detrital sediment grains collected from the mouth of a catchment can be presented as a pdf (probability density function), and assumed to contain spatially averaged information on erosion rate. The simplest model for spatial averaging is that the sediment is derived uniformly across the catchment. The pdf of measured (U-Th)/He in detrital apatites can be compared with the pdf predicted by a convolution of the observed catchment hypsometry with the local age-elevation relationship. In this way, each elevation in the catchment is assigned an apatite He age. If the measured pdf matches the predicted pdf perfectly, then erosion is viewed as uniform (Brewer et al., 2003; Ruhl and Hodges, 2005), but mismatches of the pdfs indicate localised erosion, glacial modification of the

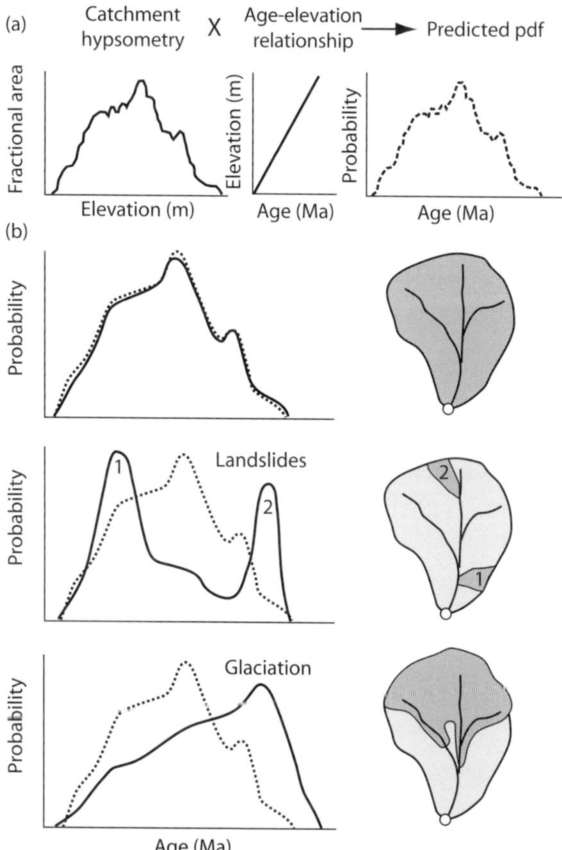

Figure 7.12 Cooling age pdfs can be used to determine where sediment comes from in an upland catchment. (a) A predicted pdf is made from a catchment hypsometry and age-elevation relationship. (b) The predicted pdf will be closely mimicked by the measured pdf derived from low-temperature thermochronological methods such as (U-Th)/He if sediment is derived uniformly from the catchment. Local derivation of grains from discrete, localised events such as landslides will cause strong departures from the predicted pdf, whereas a systematic non-uniform derivation, for example from a glaciated upper catchment, will cause a skew in the measured pdf towards higher elevations and therefore towards older ages (Brocklehurst and Whipple, 2004). After Stock et al. (2006) (fig.1, p.726) with permission of Geological Society of America.

catchment landscape (Brocklehurst and Whipple, 2004) or sediment storage (Figure 7.12). Results from two catchments in the Sierra Nevada of eastern California show that one (Inyo Creek) is eroding uniformly, whereas the other (Lone Pine Creek) has a strong mismatch of the measured and predicted pdf. However, identification of the precise factors causing the mismatch of pdfs is challenging.

7.6 Denudation from Analysis of Cosmogenic Nuclides

Erosion rates are increasingly estimated from measurements of the cosmogenic radionu-clide ^{10}Be in quartz. The half life of cosmogenic nuclides is of the order of 1 Myr, so they are particularly useful in dating materials throughout the Quaternary. The concentration of cosmogenic radionuclides measured with Accelerator Mass Spectrometry reflects the time spent by the sample close to the surface of the Earth. Consequently, cosmogenic radionuclide concentrations can be used to measure rates of surface processes such as erosion rate (Bierman and Nichols, 2004), since the production rate P is known to decrease exponentially with depth y:

$$P(y) = P_0 e^{-y/y^*} \tag{7.16}$$

where y^* is a depth decay constant (also known as the absorption mean free path Λ) equal to 0.5 to 0.7 m for most common lithologies (Figure 7.13a). If we assume that the surface production rate is 6 atoms per gram of quartz per year and that the depth decay constant is 0.7 m, the production profile reduces to less than one atom per gram of quartz per year at a depth of 1.25 m and is negligible (less than 0.1 atoms) below a depth of approximately 3 m (Figure 7.13a). The surface production rate varies with altitude, since the atmosphere reduces the incoming cosmic flux. At an altitude of 1.5 km, the surface production rate decreases to 1/e of that at sea level. Production rates are also affected by latitude, since the

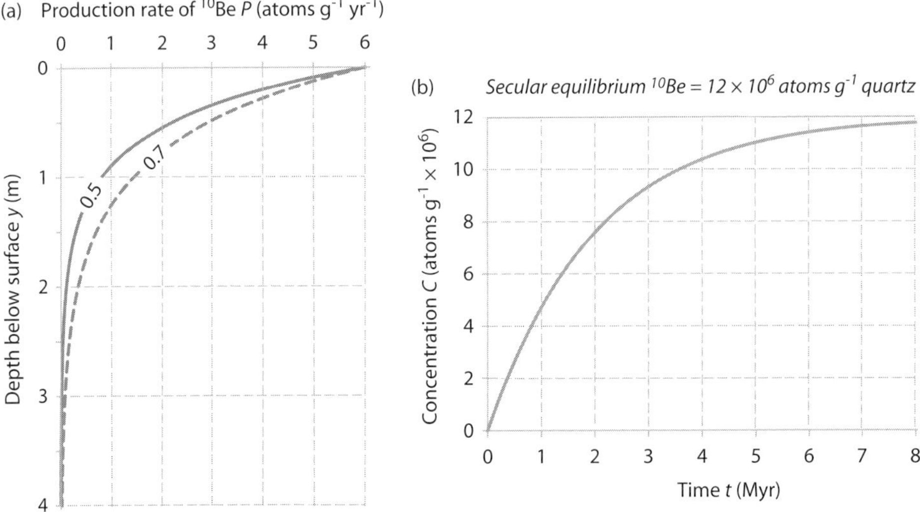

Figure 7.13 (a) Production rate of the ^{10}Be nuclide in atoms per gram of quartz per year as a function of depth below the Earth's surface y in m, for two values of the depth decay constant y^*. (b) Build up over time of concentration C of the ^{10}Be nuclide in atoms per gram of quartz towards an asymptotic value known as secular equilibrium, where the surface production rate P_0 is 6 atoms g^{-1} yr^{-1} and the decay rate of the ^{10}Be nuclide λ is 5.0 × 10^{-7} yr^{-1}.

Earth's predominantly dipole magnetic field impedes and deflects cosmic charged particles towards the poles. To obtain reliable cosmogenic dates therefore requires the altitude of the sample, and its altitude history where this has changed over time, and the past variation of the magnetic field to be known. Cosmic rays can also be shielded by topography, especially in mountainous areas.

Radionuclides such as ^{10}Be are produced in situ in the rock, sediment or soil under consideration, but nuclides also decay. The rate of change of nuclide concentration C per unit mass of material is therefore a trade-off between production and decay and is given by

$$\frac{dC}{dt} = P - \lambda C \tag{7.17}$$

where P is the production rate of new nuclides per unit mass of material, λ is the decay rate for the nuclide (5.0×10^{-7} yr^{-1} for ^{10}Be, equivalent to a half-life of 1.387×10^6 years) and C is the number of atoms of the nuclide per unit mass of material. Equation (7.17) can be solved for C to give

$$C = \frac{P}{\lambda}(1 - e^{-\lambda t}) \tag{7.18}$$

The rate of increase of concentration of nuclide over time therefore decreases, reaching an asymptotic concentration or 'secular equilibrium' at a value of $C = P/\lambda$, since dC/dt is set to zero in equation (7.17). We can define a timescale representing the time taken to reach 95% of the secular equilibrium level τ_{95}

$$\tau_{95} = -\frac{1}{\lambda}\ln\left(1 - \frac{\lambda C_{95}}{P_0}\right) \tag{7.19}$$

which simplifies to

$$\tau_{95} = -\frac{1}{\lambda}\ln(0.05) \tag{7.20}$$

since $C = P/\lambda$ at steady-state concentrations (secular equilibrium). For the example given in Figure 7.13b, this timescale is 6 Myr.

Using the correct parameter values for ^{10}Be in quartz, with $P = 6$ atoms per gram of quartz per year, the secular equilibrium for concentration of ^{10}Be is 12×10^6 atoms per gram of quartz (Figure 7.13b). After a certain number of years, the concentration of nuclide can no longer be used to reliably measure time or age. This limiting value is a small multiple of $1/\lambda$ years, known as the 'mean life', equal to 2 Myr for ^{10}Be. This sets the resolution of the dating technique in terms of time or age at less than about 4 Myr.

Cosmogenic radionuclides have been used to date bedrock surfaces, depositional surfaces such as fluvial and marine terraces and cave deposits. They have been used to estimate pointwise rates of erosion of boulders and bedrock straths, and rates of erosion averaged over contributing catchment areas can be calculated from detrital stream samples (Section 7.6.1).

A bedrock surface, once abandoned, or a bedrock boulder after deposition, will accumulate cosmogenic nuclides at a rate dictated by the surface production rate. Equation (7.18) can be solved for age t to give:

$$t = -\frac{1}{\lambda}\ln\left(1 - \frac{\lambda C}{P_0}\right) \tag{7.21}$$

If $\lambda = 5 \times 10^{-7}$ yr^{-1} and $P_0 = 6$ atoms per gram of quartz per year, the age of an abandoned bedrock surface or boulder is 16.7 kyr if the measured concentration $C = 10^5$ atoms per gram of quartz. However, equation (7.21) applies only to the case of no erosion.

The calculation of ages of depositional surfaces is complicated by the process of 'inheritance' of nuclide doses from a detrital grain's previous history of exposure to cosmic rays (Anderson, Repka, and Dick, 1996). This inheritance may differ strongly from grain to grain, so several samples are commonly taken, each representing an amalgamation of inherited ^{10}Be, from a vertical profile below the depositional surface to reveal a 'mean inheritance' (Repka, Anderson, Finkel, 1997).

The accumulation of cosmogenic radionuclides in rocks at the Earth's surface therefore depends on latitude, altitude and depth below the surface. If a rock is brought closer to the surface by erosion, it experiences a greater amount of cosmic-ray bombardment per unit time interval, thereby increasing the rate of nuclide production, but a relatively short duration of production as the grain or crystal is advected towards the Earth's surface. The cosmogenic radionuclide abundance is therefore partly a function of the rate of erosion. To calculate the rate of erosion, imagine a rock or sediment surface that is undergoing steady erosion that brings a buried quartz grain or crystal closer to the surface over time. As the grain gets closer to the surface, it will experience a higher production rate of nuclides as described by equation (7.18). We initially neglect nuclide decay. By the time the grain or crystal arrives at the surface, it will have a concentration that is the integration of its production rate history:

$$C = \int_0^\infty P_0 e^{-\epsilon t/y^*}\,dt = P_0 y^*/\epsilon \tag{7.22}$$

where $t = y^*/\epsilon$ is the time it takes to pass through a depth zone y^* thick at a velocity ϵ.

If nuclides decay during this process of exhumation, equation (7.22) becomes

$$C = \frac{P_0}{(\epsilon/y^*) + \lambda} \tag{7.23}$$

where λ is the inverse half-life of the nuclide, equal to 5×10^{-7} yr^{-1} for ^{10}Be. Erosion rates calculated in this way have been obtained from a wide variety of tectonic and climatic settings (Bierman, 1994; Small et al., 1997; Bierman and Caffee, 2002), and used to make global compilations (Portenga and Bierman, 2011; Covault et al., 2013).

Using equation (7.22), if the decay constant of production rate with depth y^* is 0.7 m, and the surface production rate P_0 is 6 atoms per gram of quartz per year, a measured concentration of ^{10}Be of 0.6×10^6 atoms per gram gives an erosion rate, assumed to be steady, of 0.00525 mm yr^{-1} (Figure 7.14). Allowing for nuclide decay during the exhumation process, and using the inverse half-life λ given earlier, equation (7.23) gives an

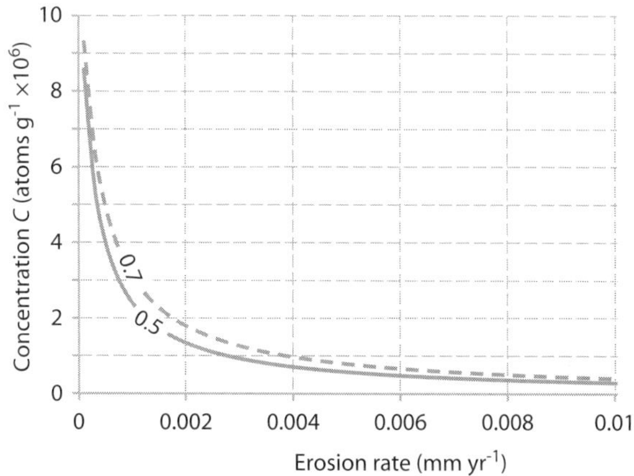

Figure 7.14 Concentration of the ^{10}Be nuclide in atoms per gram of quartz by the time the grain reaches the surface as a function of erosion rate ϵ in mm yr^{-1}, for two values of the depth decay constant y^*, where the surface production rate P_0 is 6 atoms g^{-1} yr^{-1} and the decay rate of the ^{10}Be nuclide λ is 5.0×10^{-7} yr^{-1}. Note that at very small erosion rates, the concentration varies strongly, but at higher erosion rates, concentration is no longer sensitive to erosion rate. This insensitivity is caused by the small time spent in the zone of in situ ^{10}Be production during movement of the quartz crystal or grain towards the Earth's surface, which limits the technique to landscapes undergoing rather low erosion rates.

almost identical value of 0.0049 mm yr^{-1}. Rapid erosion causes the quartz grain or crystal to ascend to the Earth's surface quickly, limiting the dosage of cosmic radiation received. Consequently, at erosion rates of greater than 0.005 mm yr^{-1} the concentration of ^{10}Be is no longer sensitive to erosion rate. Although this does not represent a problem in cases of low bedrock weathering rates, and where mean concentrations are calculated from stream samples at the outlet of large drainage basins with low relief, it limits the usefulness of cosmogenic nuclide dating to constrain erosion rates in many regions of active tectonics.

Since there are two unknowns, the cosmogenic age and the erosion rate, estimates of erosion rate can be improved by the use of multiple isotopes. The most common are ^{10}Be in combination with ^{26}Al (Lal, 1991). The ratio of the concentrations of two isotopes, such as ^{10}Be and ^{26}Al, changes sensitively with the erosion rate. The ratio of ^{26}Al/^{10}Be is conventionally plotted against the logarithm of ^{10}Be concentration (Lal, 1991; Ivy-Ochs and Kober, 2008). Since ^{26}Al has a shorter half life than ^{10}Be, the Al/Be ratio decreases as the erosion rate ϵ decreases. The ratio for a continuously exposed, non-eroding surface evolves along the thick black line in Figure 7.15, decreasing with increasing exposure age. Continuously exposed surfaces undergoing steady erosion, however, follow trajectories that splay downwards from the no-erosion trend, reaching a final value of the ^{26}Al/^{10}Be ratio depending on the erosion rate. These final positions follow a trend shown in the dashed grey line in Figure 7.15. The zone between the no-erosion line and the final positions of erosional trajectories is known as the 'steady-state erosion island' (Klein et al., 1986; Lal, 1991). Data from surfaces that have undergone continuous, simple exposure plot within the erosion

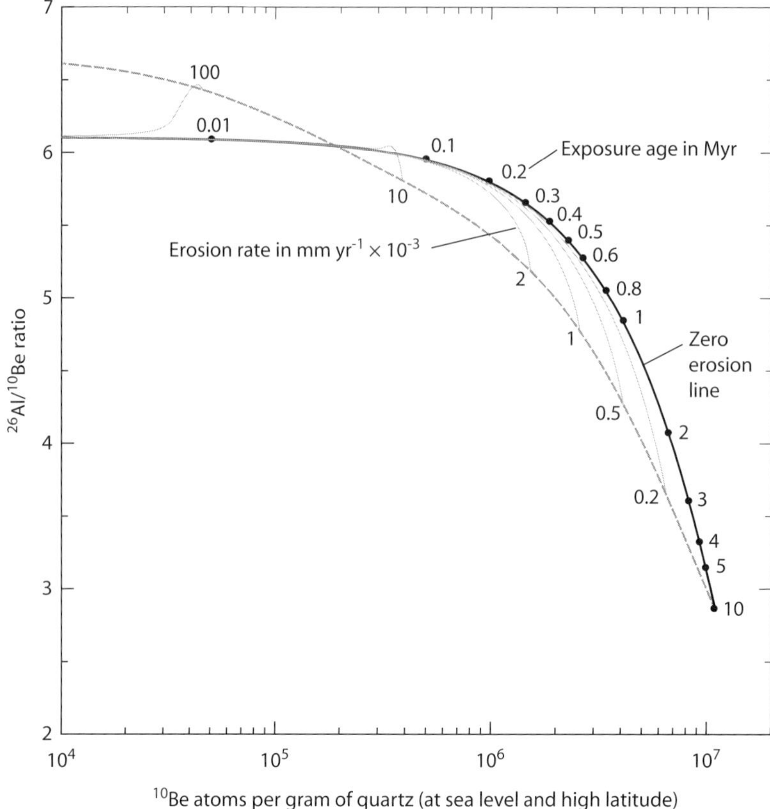

Figure 7.15 Erosion island plot of ^{26}Al/^{10}Be ratio versus ^{10}Be concentration in quartz, showing evolution of the ratio for continuously exposed (no burial) surfaces experiencing different rates of erosion (thin grey lines). Modified from Ivy-Ochs and Kober (2008) (fig.8) with permission of Quaternary Science Journal.

island, whereas data that reflect complex exposure histories involving burial plot below the Al/Be erosion island. The erosion island (or 'banana') plot allows a solution for both exposure time and erosion rate, particularly where erosion rates are low and exposure ages long (greater than 100 kyr), as in dry, tectonically stable settings such as the deserts of cratons. Erosion rates of rock surfaces in ancient landscapes typically range down to 1 mm kyr^{-1} (Cockburn and Summerfield, 2004).

7.6.1 Catchment-Averaged Erosion Rate from Cosmogenic Nuclide Analysis

In applications to stratigraphy, it is more important to estimate catchmentwide mean erosion rate than to have pointwise estimates of erosion. Such catchment-averaged rates can be obtained by analysis of many detrital grains sampled at the outlet of the catchment (Brown et al., 1995; Granger et al., 1996; von Blanckenburg, 2005). This is commonly carried out using ^{10}Be, sometimes in association with ^{26}Al. Catchment-averaged erosion rates can then be used to estimate sediment discharges into adjoining basins. We are effectively 'letting nature do the averaging', both spatially and temporally. The spatial averaging results from sediment being derived from throughout the catchment, mixed and deposited at the sampling point. The temporal averaging is over the time taken for grains to pass through the depth range y^* at the average velocity ϵ. This timescale is typically 10^4 to 10^5 years.

The analysis of catchment-averaged erosion rate from detrital quartz samples is a very useful technique, but suffers from a number of potential drawbacks and may be more complicated than interpreting nuclide concentrations in bedrock. Sediment grains may be mixed within regolith by processes such as freeze-thaw and soil creep, and especially by bioturbation, but this does not affect the average cosmogenic nuclide concentration (Granger et al., 1996). Detrital deposits may also contain mixtures of grains from areas in the catchment subject to different erosion rates. However, if these different areas contribute sediment volumes in proportion to their erosion rate, then the average nuclide concentration continues to represent the average erosion rate of the catchment. Alluvial sediment may also accumulate cosmogenic nuclides during periods of storage and intermittent transport, or be shielded by burial in alluvial stratigraphy. The net effect of storage and remobilisation is likely to be small if the mean residence time in storage is small compared to the erosional timescale y^*/ϵ, or alternatively, if the total volume of stored sediment is small compared to the sediment produced by erosion by an amount y^*, that is $A_c y^*$, where A_c is the catchment area. This condition may be satisfied in relatively small, steep upland catchments where storage is minimal, but is unlikely to apply to large alluvial systems with extensive floodplains. In the case of the latter, the cosmogenic nuclide concentrations in alluvial sediment reflect the inherited concentrations at the time the sediment was deposited, which will be affected by past erosion rates, and the cosmogenic decay during storage, which depends on the residence time, or age, of the stored alluvium. However, samples collected from the active floodplains of rivers, such as the Amazon, have cosmogenic nuclide concentrations that are essentially unchanged during alluvial transit and storage, making it possible to use them to assess catchment-averaged erosion rates in the area supplying sediment (Wittmann and von Blanckenburg, 2009; Wittmann et al., 2009).

Since altitude and topographic shielding affect the cosmic flux received by the target, the best results come from catchments with little relief of less than a few hundred metres. The distribution of quartz-bearing lithologies should also be relatively uniform in the catchment. Different erosional processes (from rapid, instantaneous landslides and debris flows to slow, quasi-continuous soil creep) also affect the timescale of cosmic irradiation.

The technique of using detrital samples has been used in a variety of settings. A pioneering study used in situ-produced ^{10}Be to calculate the average denudation rate using seven grain-size classes (between 0.063 and 8 mm representing very fine to very coarse sand) of river sediment in a small catchment in Puerto Rico (Brown et al., 1995). The upper catchment above the Icacos River gauging station is small (3.26 km^2) and the relief relatively low (150 m), which indicates that production rates in the drainage basin do not vary significantly, with few latitudinal, altitudinal or topographic shielding effects. An average production rate of 8 atoms g^{-1} yr^{-1} was therefore assumed by Brown et al. (1995). The catchment is underlain almost entirely by a quartz-diorite pluton, which is assumed to uniformly generate quartz during weathering. Storage of sediment on the valley floor is minor. Bioturbation of the surficial zone, despite changing the concentration profile and the surface exposure ages of soils, is unlikely to affect the calculation of erosion rates. River sediments were divided into grain-size classes, since concentrations of chemical constituents are known to be non-uniform with grain size. Be concentrations in sediment collected from the main stem of the Icacos River are delivered from stable ridge crests, and by mass wasting of hillslopes. Fine river sediments (less than 0.25 mm) have ^{10}Be concentrations indistinguishable from those of ridge crest soils, whereas coarse grain sizes (greater than 2 mm) have lower concentrations similar to those of landslide material moved in high energy mass transport events. The average ^{10}Be concentration was calculated by weighting the fluxes of quartz in each grain-size class using hydrological and suspended sediment data at gauging stations and direct sediment sampling. Suspended load (silt and fine sand) contains only 5–20% quartz, whereas bedload is dominated by quartz (90%) in the sand and gravel grade. The weighted average ^{10}Be concentration corresponds to a mean denudation rate of 0.043 mm yr^{-1}, which compares reasonably well with a rate of 0.075 mm yr^{-1} based on mass balance studies.

The values of catchment-averaged erosion rate derived from concentrations of cosmogenic nuclides in river sediment can be compared with estimates from other methods such as the volume of sediment deposited in dated basin-margin fans (Granger et al., 1996). Two small catchments in the Fort Sage Mountains in northeastern California are underlain by granodiorite and discharge sediment into a basin formerly occupied by Lake Lahontan. The sediments can be dated because they overlie well-sorted beach sands deposited as the lake retreated from 16.4 ± 0.2 ka. Fan volumes were calculated using thicknesses from 19 core locations combined with surveying of the depositional surface of the fans. The catchments were also surveyed to give the areas supplying sediment to the fans. Granger, Kirchner, and Finkel (1996) found that the two catchments had average denudation rates of 0.058 ± 0.014 mm yr^{-1} and 0.030 ± 0.005 mm yr^{-1} over the last 16 kyr. Concentrations of ^{26}Al and ^{10}Be were measured from 0.25–2 mm quartz grains recovered from stream bed samples. Erosion rates were calculated using equation (7.22) as 0.060 ± 0.014 mm yr^{-1} averaged over the timescale y^*/ϵ and 0.036 ± 0.009 mm yr^{-1} averaged over the last 17 kyr for the two catchments. These estimates are very close to those calculated from fan sediment volumes and represent averages over very similar time periods (Table 7.3). Cosmogenic nuclide concentrations in river sediment therefore provide a reliable estimate of the erosion rate of the contributing catchment area, at least at the erosional timescale y^*/ϵ (approximately 10^4 years).

Table 7.3 *Comparison of estimates of catchment-averaged denudation rates derived from cosmogenic nuclide concentrations with those from fan sediment volumes, Fort Sage Mountains, California. After Granger et al. (1996) (tab.1, p.253) with permission of University of Chicago Press.*

	Units	Catchment A	Catchment B
Fan volume	10^3 m^3	188 ± 34	301 ± 52
Contributing area of catchment	10^4 m^2	132 ± 20	408 ± 12
Erosion rate from nuclides	mm yr^{-1}	0.06 ± 0.014	0.036 ± 0.009
Timescale for averaging	kyr	10	17
Erosion rate from fan volumes[a]	mm yr^{-1}	0.058 ± 0.014	0.030 ± 0.005
Timescale for averaging	kyr	16	16

[a] Fan volumes are calculated since the retreat of Lake Lahontan with corrections of bulk density for fan (1,700 kg m^{-3}) and catchment bedrock (2,600 kg m^{-3}).

Schaller et al. (2001) and Schaller et al. (2004) used ^{10}Be concentrations in quartz grains in old river terraces and in the modern bedload of a number of European rivers. Concentrations of ^{10}Be due to inheritance, plus a component accumulated since abandonment of the active river terrace, were accounted for. Schaller et al. (2004) calculated catchmentwide average erosion rates of $0.02 - 0.1$ mm yr^{-1} over the last 1.3 Myr, with a twofold acceleration in erosion rate during the early to middle Pleistocene.

The estimates of erosion rate from cosmogenic nuclide concentrations provide information on the time-integrated production history of quartz grains, whereas the suspended sediment loads measured at gauging stations are prone to being affected by short term variations, including those caused by man. A comparison of denudation rates obtained from the two methods allows the role of man, the buffering capacity of floodplains and the probability density function of sediment transporting events to be speculated upon (Covault et al., 2013). Covault et al. (2013) compiled historic gauging station data on sediment loads using the global database of Milliman and Farnsworth (2011) involving 774 data sources from rivers with catchments exceeding 100 km^2 in area, and added a further 219 examples with smaller catchments using data from the U.S. Geological Survey National Water Information System and an additional 248 sources based on sundry publications. These data were combined with calculations of catchment-averaged denudation rate derived from cosmogenic nuclide analysis from 1252 examples cited in scientific literature, chiefly from Portenga and Bierman (2011), most of which were from measurements of ^{10}Be.

When sediment discharges derived from short-term river gauging data are plotted against sediment discharges derived from cosmogenic nuclide data (Figure 7.16), the data cloud is centred on a 1:1 relationship, with the notable exception that highly anthropogenised catchments have much higher short-term stream gauging station loads.

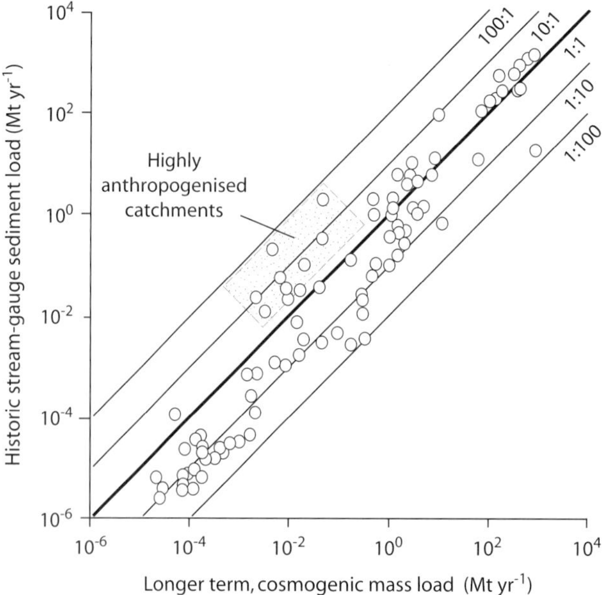

Figure 7.16 Short term sediment loads derived from stream gauge records versus longer term mass loads derived from cosmogenic nuclide analysis at approximately the same locations ($n = 103$). Highly anthropogenised catchments have sediment loads that far exceed the sediment mass derived from cosmogenic nuclides. After Covault et al. (2013) (fig.11) with permission of University of Chicago Press.

7.7 Sedimentation: Patterns, Rates and Hierarchies

7.7.1 Life-Span of Sedimentary Basins

The topography of the Earth's terrestrial surface, the bathymetry of the sea bed and broad patterns of sediment thickness are fundamentally controlled by processes acting in the lithosphere and underlying mantle (Allen and Allen, 2013). Maps of the global or plate-scale distribution of sediment thickness reveal strong variations (Laske and Masters, 1997; Divins, 2003). At both the global scale and the plate scale much of the area of the continental interiors is devoid of any sedimentary cover, with Precambrian crystalline rocks exposed at the surface. Elsewhere, the greatest sedimentary thicknesses are found in particular geological settings such as at extensional continental margins and fringing the world's great collisional mountain belts. These regions of large sedimentary thickness have undergone extensive and prolonged subsidence (Bally and Snelson, 1980). The complexities of geological history have resulted in a patchwork of currently subsiding active basins and their ancient counterparts. Sedimentary basins, ancient and modern, are the primary archive of information on the evolution of the Earth over billions of years. Further details of sedimentary basins and their relationship to plate motion and mantle circulation are found in Allen and Allen (2013).

Different basin types have a typical time span of existence (Ingersoll and Busby, 1995), after which they may be uplifted and eroded, locally inverted, or modified into another basin type to become polyhistory in nature. Each phase of existence is recognised by a megasequence, that is, a package of stratigraphy deposited during a phase controlled by a particular plate tectonic process or set of processes. Time spans of megasequences may range over 4 orders of magnitude (Woodcock, 2004) (Figure 7.17). Oceanic trench basins have the shortest life span (0.1–0.8 Myr) due to rapid subduction velocities and early incorporation into an accretionary wedge. Trench-slope, intra-arc, extensional back-arc and strike-slip basins also have relatively short life spans in the range 2–25 Myr, reflecting their tectonically active settings. Rift, forearc, back-arc and foreland basins have longer time spans in the range 4–125 Myr. The longest life spans (60–440 Myr) are found in intracratonic, passive margin and ocean basins, since they are situated in regions of low strain rate, experiencing prolonged thermal subsidence.

Different sedimentary basin types also have characteristic subsidence histories (Xie and Heller, 2009) and characteristic planform areas caused by their formative mechanisms. In general, at one extreme strike-slip basins are short-lived and localised but involve very rapid subsidence. At the other extreme, cratonic basins are very long-lived, extensive and involve slow subsidence (Figure 7.18).

7.7.2 Sedimentation Rates and Hierarchies

Sedimentation rates vary over more than 11 orders of magnitude, from 10^{-4} to 10^{7} mm yr^{-1}, dependent on the timescale over which the rate is calculated (Figure 7.19). This inverse relationship between sedimentation rate and timescale is due to the presence of gaps, large and small, in the stratigraphic record. Preservation of sediment in the geological record has been termed 'frozen accidents' (Bailey and Smith, 2010; Miall, 2014), while Allen and Allen (2013) referred to 'gating' as the mechanism by which sediment is preserved, like an on/off switch in electronics.

Preserved stratigraphic units and gaps in the record are both proposed to be fractal in nature, so that hierarchies of cycles are illusory (Section 10.4) (Plotnick, 1986; Schlager, 2004; Bailey and Smith, 2010). However, Miall (1991) suggested that a natural hierarchy of processes and preserved deposits does exist, but only if considered at the correct hierarchical level. The natural hierarchy is termed the 'Sedimentation Rate Scale' (Miall, 2010, 2014). Nevertheless, there is evidently more 'gap' than record, and since this is generally the case, some care needs to be taken in attributing 'structure' in time series of events preserved in the depositional sink to driving mechanisms in the erosional source region of sediment routing systems.

Nature exhibits complicated behaviour, but displays striking simplicity in its forms, whether in the shape of a wind-blown dune, a river meander or a snowflake. This combination of complicated behaviour arising from a simple form is a classic symptom of a complex system (Werner, 1999). Such complexities arise from the non-linearities and

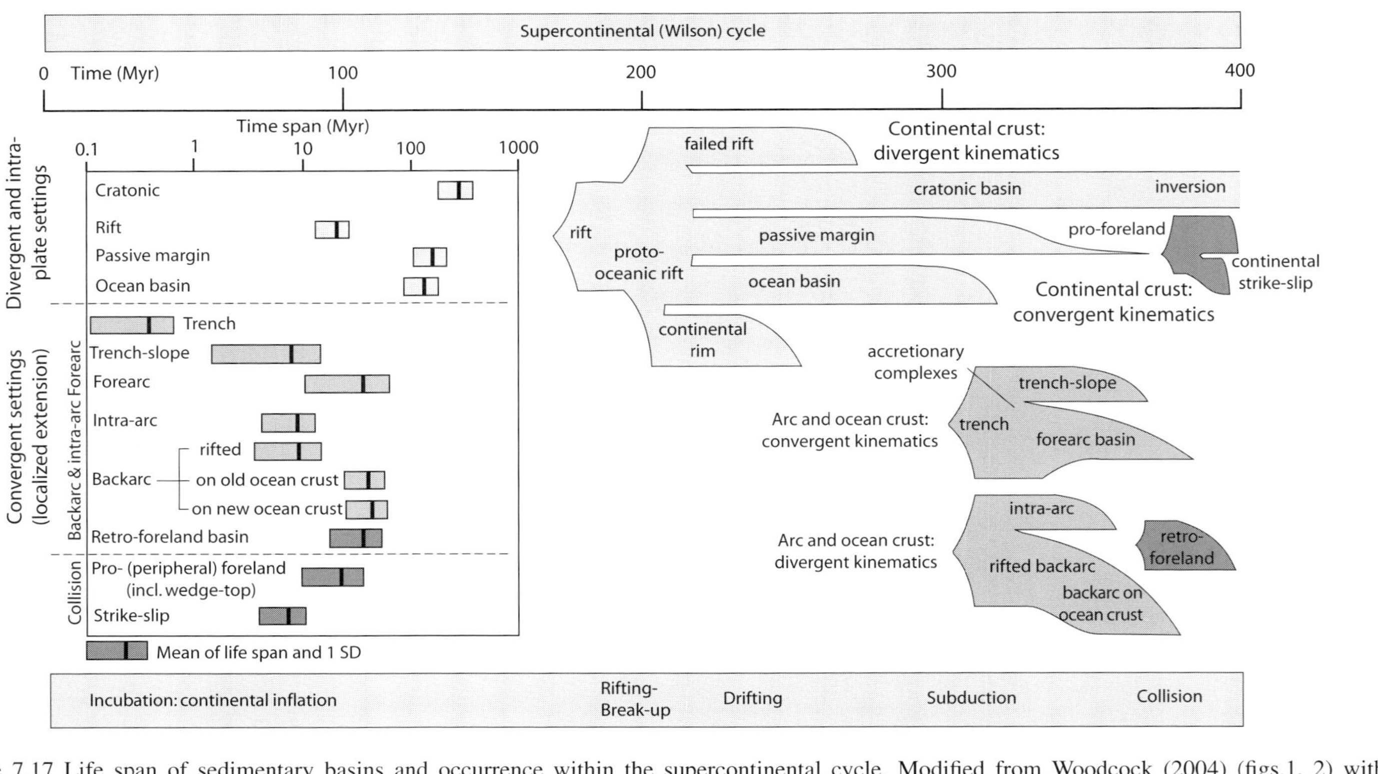

Figure 7.17 Life span of sedimentary basins and occurrence within the supercontinental cycle. Modified from Woodcock (2004) (figs.1, 2) with permission of Geological Society of America.

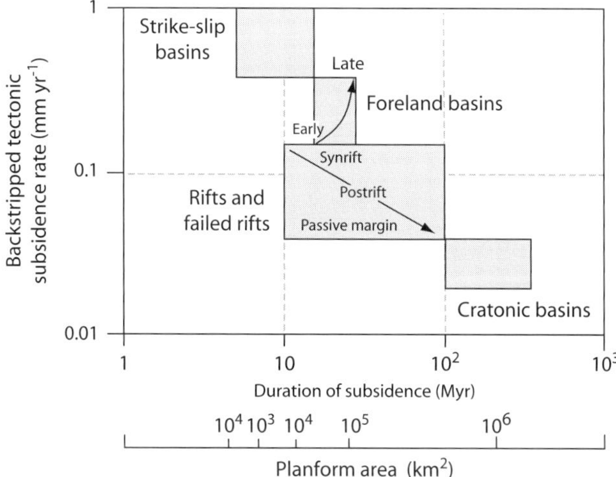

Figure 7.18 Plot of duration of subsidence *versus* typical tectonic subsidence rate, backstripped to remove the effects of sediment loading, allowing strike-slip basins, foreland basins, rifts and failed rifts, and cratonic basins to be discriminated. Modified from Allen and Allen (2013) (fig.9.17b) with permission of John Wiley & Sons Inc.

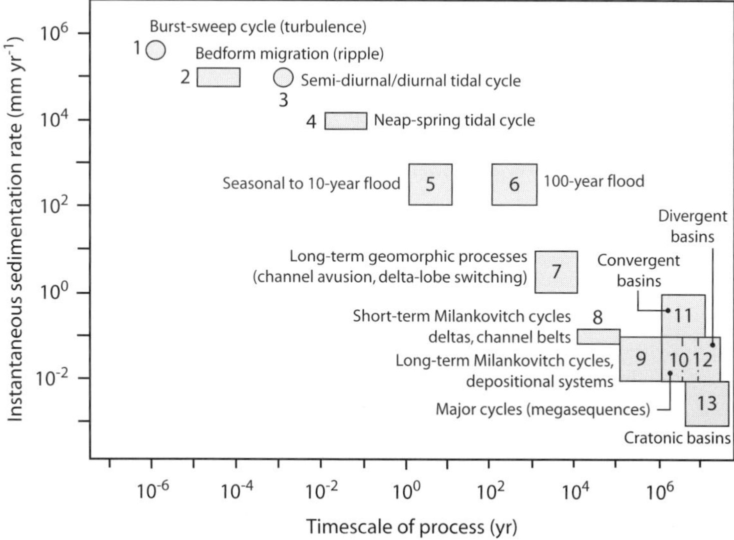

Figure 7.19 Rates and durations of sedimentary processes and products. Numbers refer to position in the hierarchy of the 'Sedimentation Rate Scale'. Sedimentation rates range over more than 11 orders of magnitude. Modified from Miall (2014) (fig.3) with permission of the Geological Society.

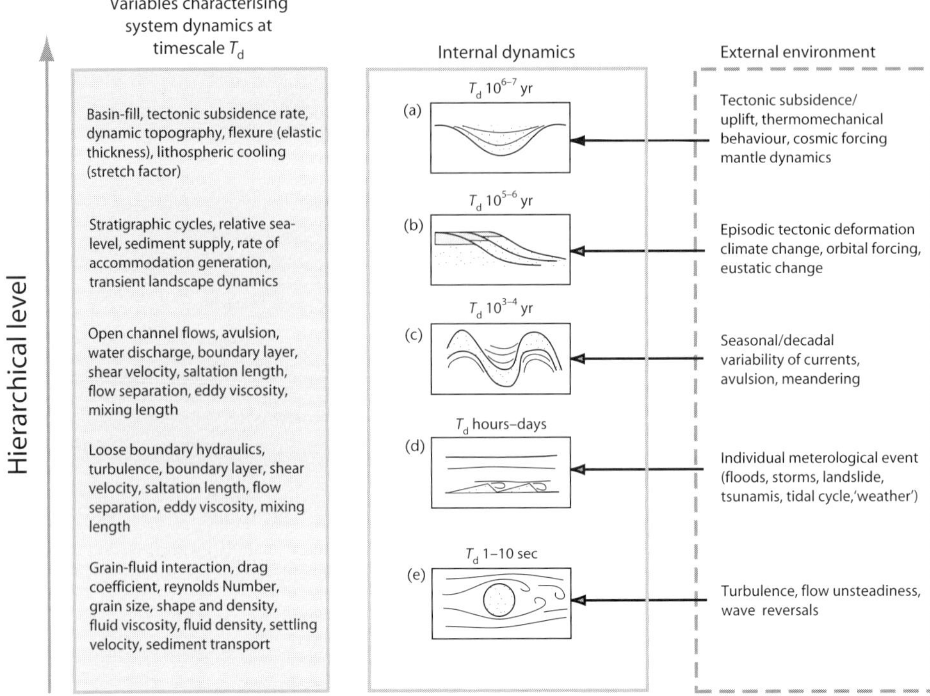

Figure 7.20 Hierarchical levels: the external environmental forcings and internal dynamical variables at different timescales T_d. From Allen and Allen (2013) (fig.8.42), inspired by Werner (1999). Published with permission of John Wiley & Sons Inc.

feedbacks in natural systems that are open to exchanges of energy across their boundaries, so that they are sensitive to external forcings. Complex systems become more organised as they evolve, and by the interactions within them, new behaviours emerge that were not previously evident. These self-organised new behaviours are described by new dynamical variables. Variables with short (fast) timescales may become slaved to longer timescale (slow) variables, losing their status as independent variables. The best example of this is the slaving of small-scale and fast variables determining the physics of motion of sand grains over the floor of a desert to the larger-scale and slower variables determining the motion of large barchan dunes, or the even larger-scale and slower-still motion of large draas with 200 m-high slip-faces.

Complex natural systems can be investigated by identifying the different dynamical levels from small to large, thereby forming a hierarchical system of interacting levels (Figure 7.20). Moving from the short-term scale of individual sediment-transporting meteorological or geophysical events to the long-term timescale of the stratigraphy of the entire basin-fill is challenging. Hierarchical modelling suggests that this upscaling from short-term to long-term cannot be achieved by simply using the same dynamical parameters for longer time periods, but instead that we should be searching for new sets of dynamical

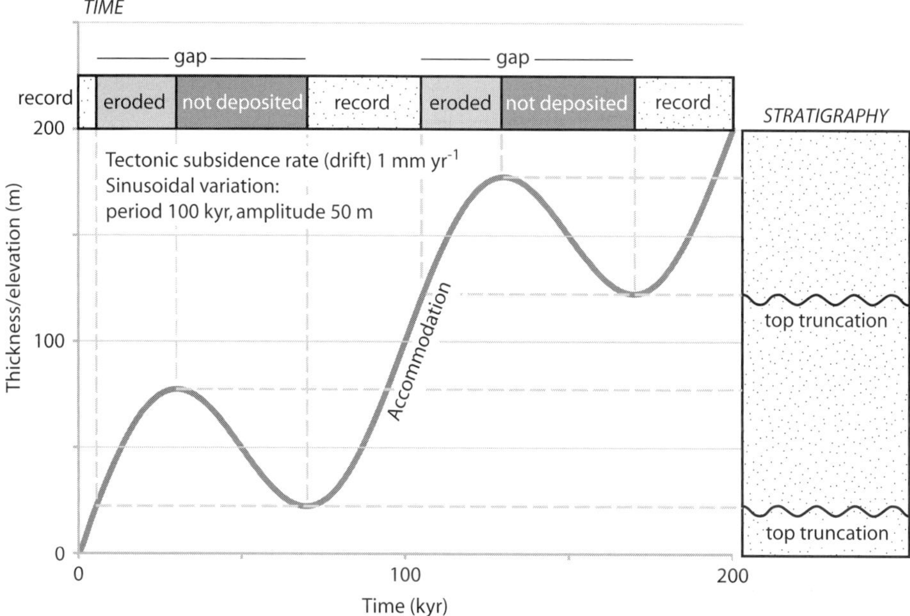

Figure 7.21 Relative sea level, equivalent to accommodation, for a sinusoidal sea level change superimposed on a background tectonic subsidence or 'drift', based on the pioneering work of Barrell (1917). It is assumed that sedimentation continually keeps pace with accommodation generation. Erosional and non-depositional gaps are present due to loss of accommodation, represented in the stratigraphic column by unconformities. From Allen and Allen (2013) (fig.8.39, p.320) with permission of John Wiley & Sons Inc.

variables. These new variables will operate at a hierarchical level with different external forcings than at the short-term level. Figure 7.20 gives some examples of different hierarchical levels and their controlling variables. A more detailed discussion is found in Allen and Allen (2013), section 8.4.3.

7.7.3 Stratigraphic Completeness

Barrell's experiments at the turn of the twentieth century involved a sinusoidal sea level change superimposed on tectonic subsidence (Barrell, 1917) (Figure 7.21). Depending on the relative magnitude of the tectonic 'drift' and the eustatic component, part of the time represented by the stratigraphic succession is taken up by erosional gaps when accommodation is decreasing (Allen and Allen, 2013) (fig.8.39, p.320). This produces a stack of top-truncated cycles. When plotted in a Wheeler diagram of time versus horizontal distance coordinate (Wheeler, 1958, 1959), stratigraphic gaps would recur with the frequency of the eustatic change, which means that the average sediment accumulation rate is smaller than the instantaneous rate actually experienced locally during a period of adequate accommodation and plentiful sediment supply. Stratigraphy is therefore incomplete (Sadler, 1981).

Andrew Miall (Miall, 2014) used this concept as the basis for his 'sedimentation rate scales', which range from the very short timescales of localised sediment transport under the macro-turbulent structures of a fluid flow through a wide range of longer timescales, to the km-scale stratigraphic units of entire basin-fills. These sedimentation rate scales are effectively the hierarchical levels introduced in Section 7.7.2.

Deposition in the sediment routing system is therefore far from continuous. It is self-evident that if we extrapolate rates of deposition measured in modern environments to the long timescales of the geological past, stratigraphic sections are less thick than predicted, so must contains gaps or hiatuses (Reineck, 1960; Newell, 1962). The rate of sedimentation derived from stratigraphic successions is inversely related to the time span over which the rate is measured (Schindel, 1980; Sadler, 1981; Sadler and Strauss, 1990). This has critical importance to the assessment of evolutionary rates and patterns of extinction (Dingus and Sadler, 1982; Dingus, 1984), but also to the practice of age assignment by interpolation from biostratigraphic, magnetostratigraphic and geochronologic boundaries. However, the duration of gaps in the stratigraphic record is difficult to assess, particularly as the timescale of observation becomes small. It is therefore easier to estimate the time missing across a 'megasequence' or 'supersequence' boundary than at a sequence boundary, and progressively more difficult still as the timescale reduces to that of the deposition of a single sedimentary lamina. So how complete is the stratigraphic record? Peter Sadler (1981) (p.569) wrote:

Continuity and steadiness of sedimentation are troublesome notions.

Stratigraphy is a selective and partial record of events that took place in the past, rather than a faithful recorder of everything that happened. Philip Allen stated (2008a) (p.275):

Time transforms sediment routing systems into geology, and like history, selectively samples from the events that actually happened to create a narrative of what is recorded.

Stratigraphic incompleteness clearly influences the reliability of the sedimentary archive to say meaningful things about environmental changes in the past, but also brings into question how representative modern depositional environments are of the fragmentary mosaics constituting the stratigraphic record.

Peter Sadler formalised this problem of stratigraphic completeness (Sadler, 1981). Imagine sampling a stratigraphic succession deposited over a time period T, which is sampled at a time interval t^*. A 'complete' stratigraphic section is one where there is a stratigraphic record in each time interval t^*, so that any hiatuses present have a duration of less than t^*.

Consider a stratigraphic section where the sampling interval is the same as the duration of the stratigraphic succession, that is, $t^* = T$. The completeness must be equal to 1, since the section contains a preserved sedimentary record. But, as the time intervals become finer, the likelihood arises that some intervals t^* will fall within gaps, so completeness reduces, and becomes a minimum when there is only one time interval left in the whole stratigraphic section that contains a sedimentary record. The fraction of time intervals that have left a record is then t^*/T. The stratigraphic completeness therefore depends on the time interval

t^*, so sections might be regarded as 'highly complete' or 'poorly complete' dependent on the length of the time interval t^* alone. The history of sediment accumulation derived from dated sections of rock of known thickness becomes an increasingly distorted record of the true rate of accumulation with lower values of completeness.

Completeness at a given timescale t^* can be modelled from:

- the duration of the stratigraphic section, which is generally known from biostratigraphic, palaeomagnetic and geochronological data
- the long-term net accumulation rate or drift, which is derived from sediment thickness
- the unsteadiness of the sedimentation rate, which is best modelled statistically in terms of wavelength and amplitude (Schwarzacher, 1987; Sadler and Strauss, 1990)

Completeness is then a positive function of timescale, drift and wavelength of fluctuations, and an inverse function of standard deviation and amplitude of sedimentation rate fluctuations. In other words, thick stratigraphic sections of long duration with slow fluctuations in sedimentation rates tend towards being complete, whereas thin stratigraphic sections of short duration with rapid, high-amplitude fluctuations tend towards being incomplete.

Different basin types, with different histories of subsidence and sediment supply, should be characterised by different levels of stratigraphic completeness. If the rate of sediment accumulation is plotted in logarithmic space against the time span for which it was determined, there is a clear inverse relationship with considerable scatter, but with a number of clusters or modes (Sadler, 1981) (Figure 7.22) that reflect the common ways of estimating accumulation rates. The variability is caused by differential compaction, non-uniformity at a given location, and unsteadiness in time, which reflect in some complex way the sediment hydrodynamics of the depositional environment and its uplift/subsidence history. Consequently, accumulation rate versus time-span relationships should vary according to depositional environment and basin type (for example, fluvial, lacustrine, carbonate platforms and reefs, siliciclastic shelves, deep-marine minibasins and abyssal plains). Environments above wave base characterised by limited accommodation have similar trends in accumulation rate *versus* time span, especially at time spans greater than 1 kyr, suggesting that they have a common influence, which is tectonic processes of stretching, cooling and flexure. In contrast, sediments deposited below wave base, where accommodation is not limited, have a different relationship between accumulation rate and time span (Figure 7.23).

Sadler (1981) (p.580) proposed that completeness should be calculated as the long-term rate of sediment accumulation of the whole section S compared to the average rate of accumulation at the given timescale t^*, denoted S^*. The measured deposition rate decreases as a power-law function of the interval of time over which it is measured. The expected completeness is then

$$(S/S^*) = \left(\frac{t^*}{T}\right)^{-m} \tag{7.24}$$

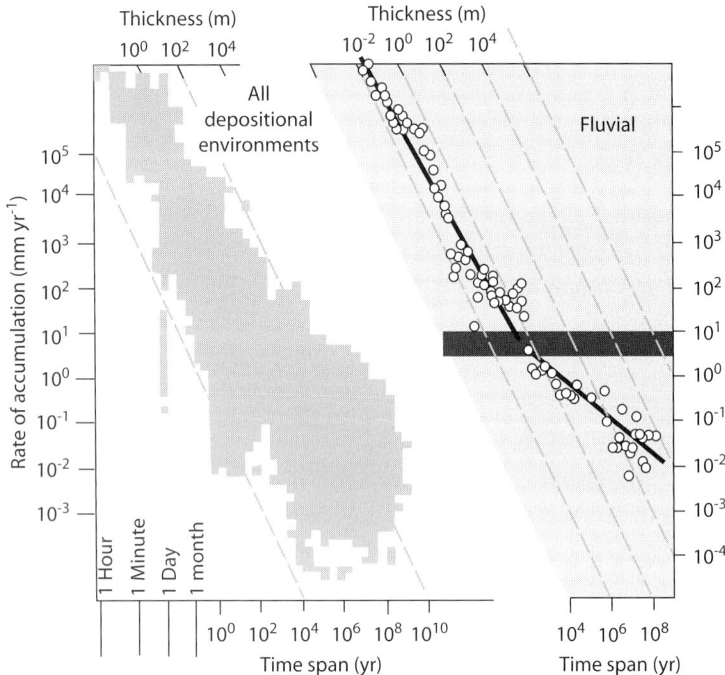

Figure 7.22 'Sadler' plot for all depositional environments, showing (left) relationships between bed thickness, time span of observation, and rate of accumulation, and (right) a subset for fluvial environments, showing a decrease in net rate of accumulation for longer time spans. Note the different gradients of the best-fit lines through the fluvial data. Modified from Sadler (1981) (fig.1, p.571), with permission of University of Chicago Press.

where the exponent m is the gradient of the trends in Figure 7.22. The value of m differs in different depositional environments and basin types (Figure 7.23) and is in the range $-1 < m < 0$. The peak in relative frequency of m values according to depositional setting is given in Table 7.4.

Consider a carbonate stratigraphic succession 1,000 m thick, deposited in a time period of 12 Myr, with time intervals of 1 Myr provided by ammonite stratigraphy. The expected completeness is therefore $(1/12)^{0.45} = 0.33$, meaning that only one third of the chosen time intervals will be represented by a sedimentary record. In comparison, let us take a marine basin 1000 m thick of duration 12 Myr, with time intervals of 2 Myr; the expected completeness is $(2/12)^{0.2} = 0.7$, meaning that the stratigraphic succession is 70% complete at the time resolution of 2 Myr.

With this view of stratigraphic completeness, the stratigraphic record is seen to be strongly filtered by the timescale and statistical properties of sediment dynamics over the Earths surface relative to the slower rates and lower statistical variability of accommodation generation. Only a small fraction of the sediment transport events taking place on the surface of the Earth survive this stratigraphic filter (Schumer and Jerolmack, 2009; Schumer, Jerolmack, McElroy, 2011).

Table 7.4 *Peak value of the frequency distribution of completeness exponent m for different depositional settings*

Fluvial	-0.7
Siliciclastic shelf	-0.4
Carbonate platform	-0.45
Lacustrine	-0.35
Marine minibasins	-0.2
Abyssal plain carbonate ooze	-0.2

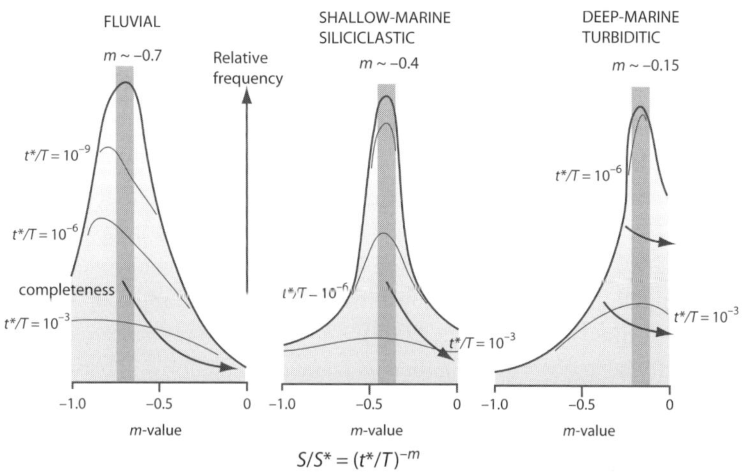

Figure 7.23 Value of the completeness exponent m for fluvial, shallow marine siliciclastic and deep marine turbiditic depositional environments and processes. Arrows indicate trend with increasing completeness. Redrawn after Sadler (1981) (fig.13) with permission of University of Chicago Press.

The duration of gaps in the stratigraphic record and the durations of depositional units may be fractal (Mandelbrot, 1983; Plotnick, 1986; Schlager, 2004, 2010). That is, the gaps are 'scale-invariant' or 'self-similar'. Plotnick (1986) (p.885) describes a model in which a pile of sediments of a certain thickness and time duration is broken down into its depositional components by intervening gaps. As stratigraphic gaps are inserted iteratively, so the sediment deposition rate increases in the successively smaller intervals of time represented by the depositional units. If the sedimentation rate and time span are both plotted logarithmically, the trend is a straight line with a slope that indicates the way in which gaps are inserted within the stratigraphic pile. If, in a stratigraphic succession, this intercalation of gaps changes either temporally in a secular trend, or coevally between different depositional environments, the differing styles of intercalation might potentially be recognised by different slopes in the power law of sedimentation rate versus time span. The style of intercalation of gaps demonstrates the completeness of the succession, which also varies with the time span of sampling (Sadler, 1981).

7.7.4 Bed Thickness Statistics

Statistical analyses of turbidite successions indicate that the distribution of bed thickness is related to both basin geometry (such as the degree of confinement) and sediment input (including the volume, type and dynamics of sediment-transporting flows), and so are useful in understanding the distal end of sediment routing systems. The scaling relationships of turbidite sandstones, particularly their bed thickness distributions, have been a particular focus of attention, since they comprise important deep-water targets for petroleum exploration (Flint and Bryant, 1993; Drinkwater and Pickering, 2001; Sinclair and Tomasso, 2002). Turbidite bed thickness distributions are commonly regarded as power laws (Malinverno, 1997), whereas an alternative model is of a log-normal mixture (Talling, 2001; Sylvester, 2007). The techniques used to study turbidite sandstones can potentially be used to investigate the scaling relationships of beds in different depositional settings (Malinverno, 1997; Sinclair and Cowie, 2003).

It has commonly been observed that there is an inverse relationship between the number of beds in a stratigraphic succession and bed thickness. That is, thin beds are much more frequent than thick beds. If the statistical distribution is a power law, a straight line will be obtained on a log-log plot of $N(h)$ versus h, where $N(h)$ is the number of beds whose measured thickness is greater than h, which is proportional to bed thickness h raised to an exponent $-\beta$, and a is a constant representing the total number of beds measured in the data set. That is

$$N(h) = ah^{-\beta} \tag{7.25}$$

Significant departures from a straight line on log-log plots are commonly interpreted in terms of segmented power laws (Rothman, Grotzinger, and Flemings, 1994; Rothman and Grotzinger, 1995; Sinclair and Cowie, 2003), made of two or more different populations of beds, or of truncated power laws. Breaks in slope between segments with different β on the log-log graph occur at threshold bed thicknesses η^1, η^2 etc. (Figure 7.24). The values of these threshold thicknesses may reflect the different mechanisms of emplacement of turbidites such as flow rheology, different source areas or different basin floor geometry.

Log-normal distributions for turbidite bed thickness, or log-normal mixtures of, for example, thin and thick bedded populations, appears to satisfy some bed thickness field data, such as from the extremely well documented Miocene *Marnoso arenacea* of Italy (Talling, 2001). The bimodality of turbidite bed thicknesses may reflect two modes of transport and depositional mechanics, such as dilute and highly concentrated variants,

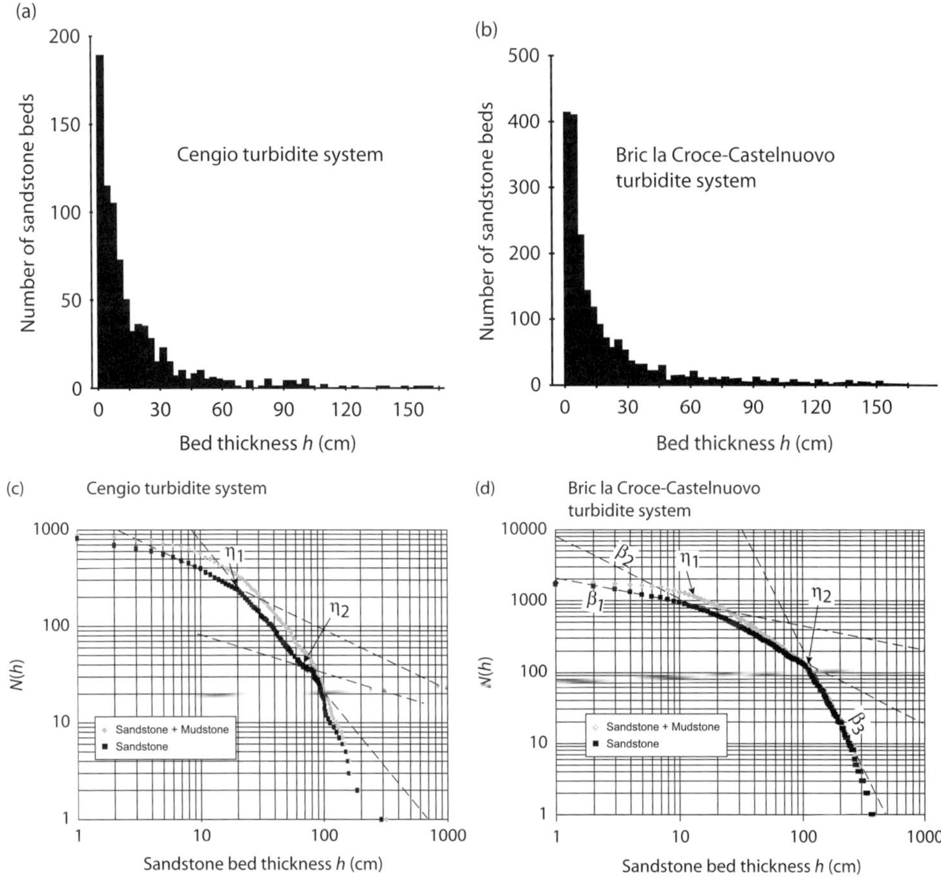

Figure 7.24 Frequency distribution of thickness of sandstone divisions in turbidite beds of the Cengio (a) and Bric la Croce-Castelnuovo (b) turbidite systems. Log-log plots are shown in (c) and (d), showing segmented curves at thickness thresholds η_1 and η_2, separating straight line fits with slopes β_1, β_2 and β_3. Modified from Felletti and Bersezio (2010) (fig.6), with permission of Elsevier.

or slow and rapid deposition rates. The bed thickness distribution of turbidites may also be used to assess the degree of basin confinement. Confined turbidite flows are forced to aggrade vertically, thereby changing the bed thickness distribution in favour of thick beds. Unconfined basins allow the turbidite bed to spread laterally, producing extensive but thin beds.

In the context of the sediment routing system, the distribution of bed volume is more informative than the distribution of bed thickness, since the variation of bed volume is an indicator of the magnitude-frequency characteristics of the sediment supply from up-system. Malinverno (1997) gives the scaling relationships between bed thickness and bed volume for disclike sandstone bodies in a sedimentary basin-fill. The power law distribution of turbidite bed volume has the form $N(v) \propto v^{-c}$, where $N(v)$ is the number of beds whose

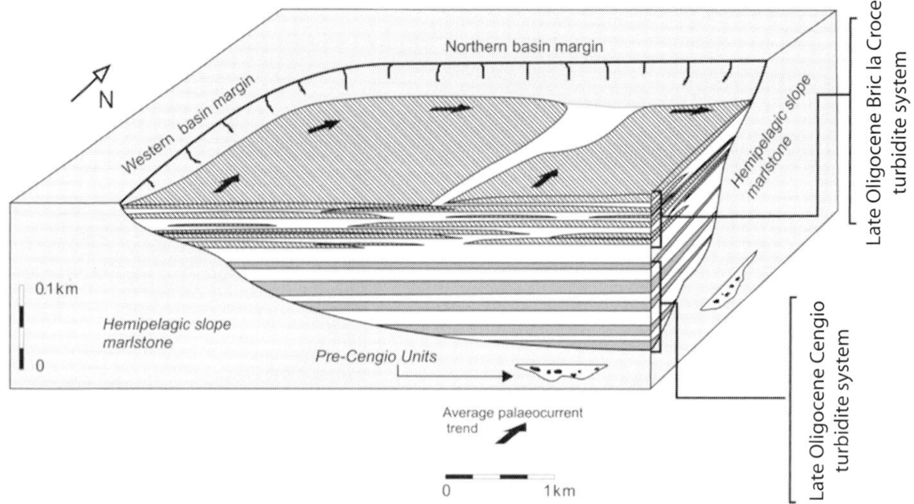

Figure 7.25 Contrasting architectural styles for the Cengio (ponded, thick turbidites) and Bric la Croce-Castelnuovo (unconfined) turbidite systems with the bed thickness statistics shown in Figure 7.24 and Table 7.5. After Felletti and Bersezio (2010) (fig.10) with permission of Elsevier.

volume is greater than or equal to v, and c is the exponent on bed volume. The exponent on bed volume c is related to the exponent on bed thickness β by

$$\beta = c + \gamma(2c - d) \tag{7.26}$$

where:

- γ is the scaling parameter between bed length and bed thickness given by $l = ah^\gamma$, where h is bed thickness and l is bed length, so when $\gamma = 0$ beds have the same diameter irrespective of thickness, and when $\gamma = 1$, bed length scales linearly with h. When $\gamma > 1$ bed length scales non-linearly with bed thickness.
- d is a parameter expressing the spatial distribution of depocentres relative to a sample line, which varies from 0 to 2. When $d = 0$, all the beds have centres located at the same point, which coincides with the sampling line. When $d = 1$ the beds have centres stretched out along a line. If $d = 2$, the bed centres are distributed randomly in the horizontal plane.

The volume of the largest turbidite bed v_{max} compared to the the total volume of a sequence of N_{tot} beds, v_{tot}, gives an index of the relative importance in terms of volume of the largest bed v_{max} compared to the smaller events. Megabeds of turbidite sand may comprise a large percentage of the total sand volume v_{tot} in a small basin.

Thick, ponded turbidites have higher values of the ratio v_{max}/v_{tot} than thin unconfined beds, as seen in the differences between the Cengio and Bric la Croce-Castelnuovo turbidite systems in the Italian Piedmont (Felletti and Bersezio, 2010) (Table 7.5). Both these turbidite systems were deposited in the same basin fed by the same source area, during a

Table 7.5 *Exponents c and γ and ratio v_{max}/v_{tot} computed for the Cengio turbidite system and the Bric la Croce-Castelnuovo turbidite system in the Italian Piedmont. From Felletti and Bersezio (2010) (tab.3, p.530) with permission of Elsevier.*

Turbidite system	Sandstone body	c	v_{max}/c_{tot}	γ
Cengio	I	0.65	0.51	0.31
	III	0.89	0.25	0.89
	IV	0.86	0.26	0.92
	V	0.88	0.23	0.57
	VI	0.73	0.27	0.53
Bric la Croce-	I	1.02	0.16	1.16
Castellnuovo	II	1.00	0.15	1.54
	III	0.91	0.18	1.42
	IV	-	-	0.92

period of quiescent intrabasinal tectonics. The Cengio turbidites are flat and aggradational, suggesting ponding. During deposition of the Bric la Croce-Castelnuovo system, the depositional area widened as the basin progressively filled (Figure 7.25), allowing the turbidity currents to spread laterally. These two scenarios showing different degrees of ponding are discriminated in the bed thickness and bed volume statistics:

- In the ponded Cengio turbidite system, η_1 corresponds to the separation between dilute turbidity flows and highly concentrated turbidity flows, and therefore reflects differences in rheology of the gravity flows. The step at η_2 however, is initially associated with a reduction in slope β_3, suggesting ponding of beds thicker than η_2.
- In the non-ponded Bric la Croce-Castelnuovo system, η_1 again reflects the change from dilute to highly concentrated flows, but there is no reduction in slope at η_2, suggesting that flows above this thickness were able to spread laterally.

These interpretations of the effects of ponding are supported by the scaling statistics. In the Cengio turbidite system the volume exponent c lies between 0.5 and 1, but increases to greater than 1 in sandbodies I and II of the Bric la Croce-Castelnuovo system (Table 7.5). The increase in c illustrates a reduction in the importance of thick beds in total sand volumes. This is mirrored in the progressive decrease in the ratio of v_{max}/v_{tot} from the base of the Cengio system to the top of the Bric la Croce-Castelnuovo system, indicating a reduction in the volume of the largest bed compared to the sum of all the bed volumes, consistent with a change from ponding to non-ponding. In the Cengio system, $\gamma < 1$, indicating that gravity flows were unable to freely spread laterally. In the Bric la Croce-Castelnuovo system $\gamma > 1$, indicative of a still confined but less ponded basin. The effect of ponding is therefore seen in the thicker rather than thinner beds, depending on the relationship between flow volume and basin width. As a result, small, structurally confined basins are more likely to exhibit ponded bed geometries than wide basin plains.

8

Dynamics of Sediment Routing Systems

8.1 Moving Boundaries

The segments of sediment routing systems are characterised by a particular set of physical processes and depositional environments (Sømme et al., 2009; Carvajal and Steel, 2012). Segments are commonly separated from other segments by boundaries or transition zones involving complex processes of energy and sediment transfer, which are examples of 'moving boundaries' (Swenson et al., 2000, 2005; Kim et al., 2006). The equivalent concept in sequence stratigraphy is the analysis of the shoreline or shelf edge trajectory (Helland-Hansen and Hampson, 2009) (Chapter 10). The position of moving boundaries is a sensitive indicator of the functioning of the sediment routing system. They are commonly filters or barriers to the propagation of sediment flux signals from upland catchments. There are therefore 'teleconnections' between different parts of the sediment routing system, which enable signals originating in one segment to be transferred to a different, adjoining segment (Hoth, Kukowski, and Oncken, 2008).

Moving boundaries such as the coast-shoreface and shelf edge migrate in response to external forcings and internal dynamics. Migration results in characteristic stratigraphic architectures, especially patterns of progradation or flooding of the clinoforms that are conspicuous in seismic reflection profiles. Clinoforms representing the shelf edge and smaller clinoforms representing the shoreface migrate independently and sometimes out of phase with each other, principally under the effects of sediment dynamics set up by river floods, wave energy and storm-driven currents (Swenson et al., 2005) (Section 5.3, Figure 5.10). If one clinoform set can grow while another of the same age shrinks, then oceanographic effects may clearly dominate over eustatic change in generating stratigraphic architectures.

8.1.1 The Gravel Front and Gravel Cline

It is well known that gravel-sand rivers exhibit an abrupt grain-size transition, termed the 'gravel front' or the 'gravel-sand transition' (Dubille and Lavé, 2015), generally accompanied by a rapid change of slope (Smith and Ferguson, 1995; Cui and Parker, 1998; Parker and Cui, 1998; Knighton, 1999; Ferguson, 2010). The gravel front is a moving boundary that changes in its downstream position in response to changes in boundary conditions (Paola, 2000; Marr et al., 2000) or as a result of the unperturbed evolution of a drainage network (Gasparini, Tucker, and Bras, 1999). In present-day rivers, the gravel front coincides with a gap (at a median size of $1-10$ mm) in the grain size of sediment comprising the river bed. The gravel-bed section and the sand-bed section of the river are associated with distinctly different values of dimensionless bankfull shear (Shields) stress (Church and Rood, 1983). Whereas gravel-bed rivers have Shields stress values at slightly higher than critical for sediment motion, which indicates that the gravel moves almost entirely as bedload, sand-bed rivers have Shields stresses considerably higher than the threshold, implying that the dominant transport mode is by suspension (Dade and Friend, 1998).

The gravel front has been variously attributed to the response to a break in slope, an excess supply of sand, physical breakdown of clasts and hydraulic sorting (Ferguson, 2010; Jerolmack and Brzinski, 2010). It can be identified in ancient fluvial stratigraphy by an abrupt down-system decrease in the gravel fraction and clast size, as in the mid-upper Eocene Escanilla Formation of the south-central Pyrenees (Michael et al., 2013, 2014a) and the Carboniferous Pocono Formation of the Central Appalachian Basin (Robinson and Slingerland, 1998). In fluvial stratigraphy, two moving boundaries may be recognisable: a proximal 'gravel cline', which is a relatively narrow zone over which gravel percentage reduces rapidly in the downstream direction, and a distal 'gravel front' marking the exhaustion of gravel in the sediment supply (Michael et al., 2013, 2014a).

The steady progradation of a wedge of sediment shed from a mountain belt can result in an abrupt upward coarsening in stratigraphy as a result of the existence of a gravel front. Such an abrupt upward coarsening is found in the Neogene Xiju Formation adjacent to the Tian Shan of northwest China (Charreau et al., 2009) and the Miocene-Pleistocene Siwalik sediments deposited in the Indo-Gangetic foreland basin of the Himalayas (Dubille and Lavé, 2015). In the Siwalik stratigraphy, median grain sizes (D_{50}) change from less than 0.2 mm to ~ 20 mm, a 100-fold increase, over a sharp, $80-100$ m thick transition zone (Figure 8.1). A similar sharp gravel-sand transition is found in the longitudinal profiles of modern rivers, such as the Churre River, in the proximal part of the Gangetic foreland basin in central Nepal (Figure 8.2). Grain size (D_{50}) remains constant at ~ 20 mm in the reach from the topographic front of the source region to 12 km downstream, then drops by two orders of magnitude in a 4 km-long transitional reach. The gravel fronts of present-day Nepalese rivers are therefore a close analogue for the coarsening-up trend in Middle to Upper Siwalik stratigraphy.

In the mid-upper Eocene Escanilla palaeo-sediment routing system of the south-central Pyrenees, a gravel cline is situated at about 10 km from the depositional apex of stratigraphic interval 2 (Michael et al., 2013), and is characterised by a sharp reduction of grain size and gravel fraction over a ~ 5 km-long transitional reach. In a mass balance framework, the gravel cline represents a sharp drop in the gravel fraction from between 0.7 and 0.8 to ~ 0.3 at a mass balance (χ) value of 2%. That is, the gravel cline occurs when about 2% of the total sediment budget has been deposited. When the mass balance coordinate χ is transformed to a mass balance coordinate for the gravel fraction only χ_g (Figure 8.3a), the

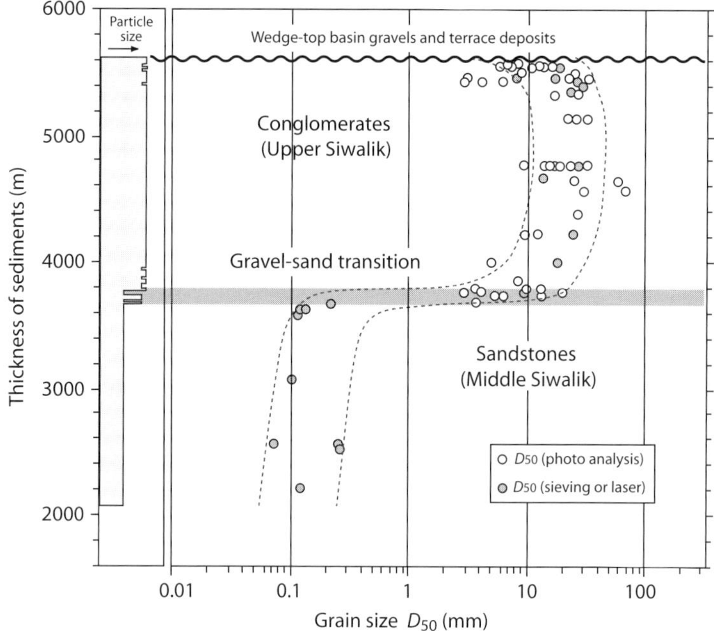

Figure 8.1 Grain-size evolution of the Middle-Upper Siwalik stratigraphy along the Churre section, central Nepal. The transition zone, representing the gravel-sand transition in rivers draining the Himalaya, is 80−100 m thick. Redrawn from Dubille and Lavé (2015) (fig.4). Published with permission of John Wiley & Sons Inc.

gravel cline occurs at 30% of the gravel fraction. That is, roughly a third of the total gravel content is deposited upstream of the gravel cline.

The gravel cline is displaced downstream to a distance of 20−40 km from the depositional apex in stratigraphic interval 3, corresponding to 5% of the total mass balance. Since stratigraphic interval 3 has a higher gravel fraction in the sediment supply than stratigraphic interval 2 (13%), the gravel cline once again represents the point at which roughly one-third of the total gravel content has been deposited (Figure 8.3).

The scale of the grain-size decrease and abruptness of the transition zone suggest that the proximal moving boundary of the gravel cline in the Escanilla system is equivalent to the sharp gravel-sand transition in Siwalik stratigraphy and in present-day Nepalese rivers. The distal moving boundary of the gravel front recognised in the Escanilla stratigraphy therefore represents the down-system exhaustion of gravel in the sediment supply, occurring at mass balance χ values of 0.4, 0.15 and 0.4 in stratigraphic intervals 1, 2 and 3 respectively. There is therefore a potentially important distinction between a distal gravel-sand transition that is linked to the gravel supply running out, as proposed in the model of Cui and Parker (1998), and a proximal, abrupt gravel-sand transition that exists despite an abundance of gravel, and which therefore requires an alternative explanatory mechanism. Quantitative models

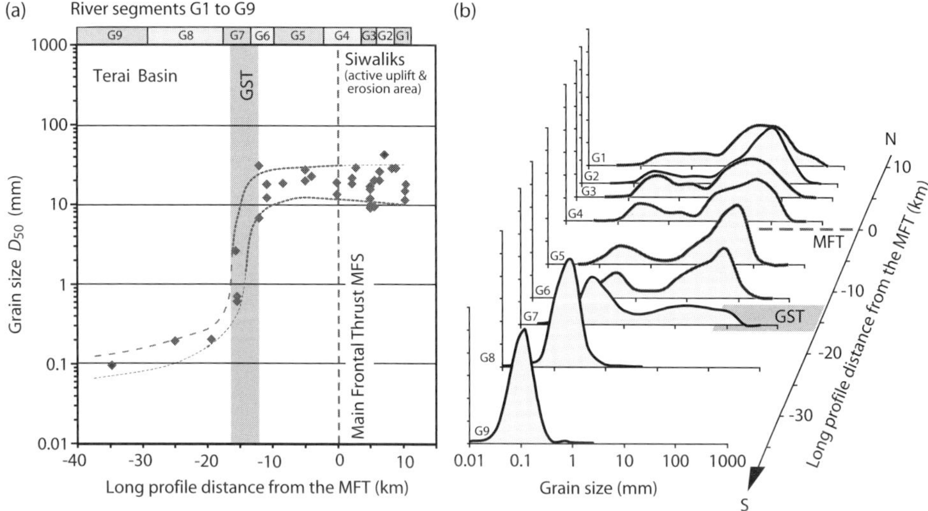

Figure 8.2 (a) Grain-size measurements from gravel and sand bar deposits in the Churre River, at different downstream distances from the Main Frontal Thrust, showing the position of the gravel-sand transition. (b) Downstream evolution of the grain-size distribution from nine river segments, showing the bimodal nature of the gravel in active transport, which becomes unimodal downstream of the gravel-sand transition. Redrawn from Dubille and Lavé (2015) (fig.7). Published with permission of John Wiley & Sons Inc.

of basin-filling of a gravel-sand mixture (Marr et al., 2000) conventionally place the gravel front at the point of exhaustion of gravel and invoke perfect sorting of sand from gravel, so that stratigraphy deposited upstream of the gravel front is 100% gravel and stratigraphy downstream of the gravel front is 100% sand.

Quantitative models suggest that the pattern of migration of the gravel front is strongly influenced by the timescale of the external driver compared to the 'intrinsic' or 'equilibrium' timescale of the basin (Paola et al., 1992; Marr et al., 2000) (Figure 8.4). The intrinsic or equilibrium response time for a diffusive system is $\tau_{eq} = L^2/\kappa$, where κ is the diffusivity and L is the basin length. In a mixed sand-gravel system, sand and gravel may have their own diffusivities. Using realistic values of diffusivity for gravel κ_g of 0.01 km^2 yr^{-1} and for sand κ_s of 0.1 km^2 yr^{-1} and a basin length of 100 km, the basin response times become 10^6 and 10^5 yr respectively (Marr et al., 2000; Allen and Allen, 2013). When the timescale of the forcing mechanism T is short (10^5 yr) compared to the equilibrium response time, it can be thought of as rapid forcing, whereas if the timescale of the forcing mechanism is long (10^7 yr), it can be thought of as slow forcing. The relative speed of the forcing determines how moving boundaries shift in position (Marr et al., 2000) (Figure 8.4).

In simulations of slow forcing, the positions of moving boundaries such as the gravel front are in phase with variations in the sediment supply (Figure 8.4a). However, rapid forcing causes the gravel front to migrate out of phase with the variations in the sediment supply

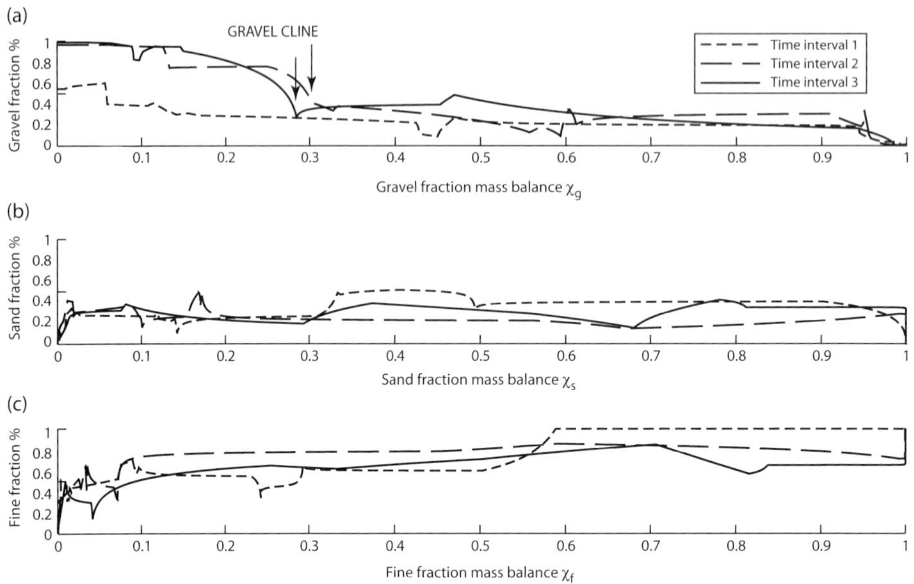

Figure 8.3 Relative abundance of gravel (a), sand (b) and fines (c) for the three stratigraphic intervals of the Escanilla sediment routing system, with a transformed χ_g coordinate for the mass balance of the gravel fraction, χ_s coordinate for the mass balance of the sand fraction, and χ_f coordinate for the mass balance of sediment finer than sand. The gravel cline is at 0.3 of χ_g for both stratigraphic intervals 2 and 3, despite significant differences in the gravel content of the sediment supply. The gravel front, by definition, is at $\chi_g = 1$. The gravel cline is less well defined in stratigraphic interval 1. From Michael et al. (2013) (fig.15) with permission of University of Chicago Press.

(Figure 8.4b). With fast forcing, the driver changes more rapidly than the basin can respond, so the gravel front fails to closely follow variations in the sediment supply or gravel fraction. If applied to the fluvial toe, or the coastline, progradational or retrogradational patterns in stratigraphic architecture would be a function of the relative timescale T/τ_{eq} (where T is the period of the perturbation). In addition, since river basins vary greatly in size, and therefore in basin response time, each river entry point into the ocean would experience a different phase of sediment supply in response to a regional or even global driver.

8.1.2 The Shoreline

The shoreline is the most dynamic of all moving boundaries. It can be modelled similarly to the movement of a phase front in a Stefan problem (Stefan, 1891) such as the solidification of magma or a lava flow, and the freezing of ice in a lake (Swenson et al., 2000). The set-up is of a subaerial alluvial plain coupled to a subaqueous delta with a shoreline in between (Figure 8.5). The stratigraphic evolution of the fluvio-deltaic basin, and the trajectory of the shoreline, are driven by the boundary conditions of sediment supply to the basin

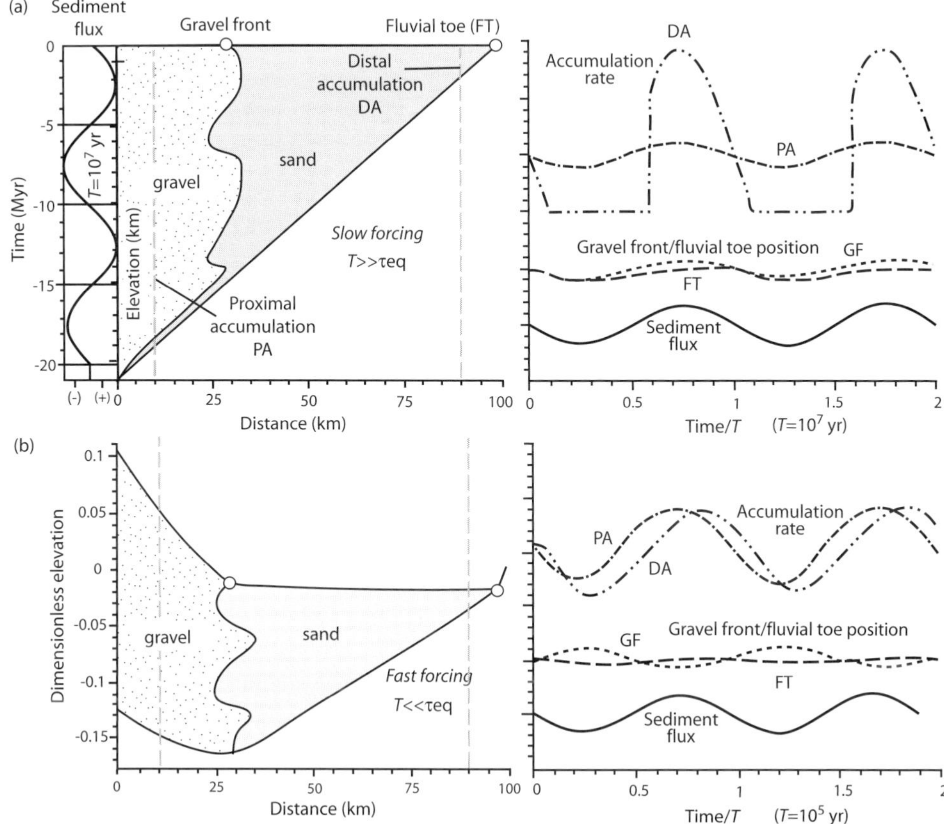

Figure 8.4 Cross sections of alluvial stratigraphy under slow (a) and fast (b) sinusoidal forcing of sediment flux. Diagrams on right show time variation of the position of the gravel front. In the case of slow forcing, the gravel front varies in phase with the sediment flux. In the case of fast forcing, the gravel front is out of phase with the sediment flux. Apart from the timescale of the forcing, all other variables (table 2 in Marr et al. (2000)) are the same in the simulations in (a) and (b). Redrawn from Marr et al. (2000) (figs.8, 9) with permission of John Wiley & Sons Inc.

q_{s0}, the spatial distribution and rate of tectonic subsidence $\sigma(x)$ and the position of base level $y_{bl}(t)$.

In the mathematical model, a steady discharge of sediment is delivered to the basin (q_{so} at $x = 0$). The basement of the basin, $b(x,t)$, subsides under regional steady, spatially variable tectonic subsidence $\sigma(x)$ where σ is measured positive downward and has a mean value $\bar{\sigma}$. Base level is allowed to vary over time with an amplitude and period of A and T. The position of the shoreline and the intersection of the delta toe with basement are denoted by $s(t)$ and $u(t)$ respectively, while the elevation of the sediment-air or sediment-water interface is $\eta(x,t)$. The subaqueous delta has a slope α.

Details of the mathematical model and the numerical solution are found in Swenson et al. (2000) and are not repeated here. The sediment transport in the submarine section

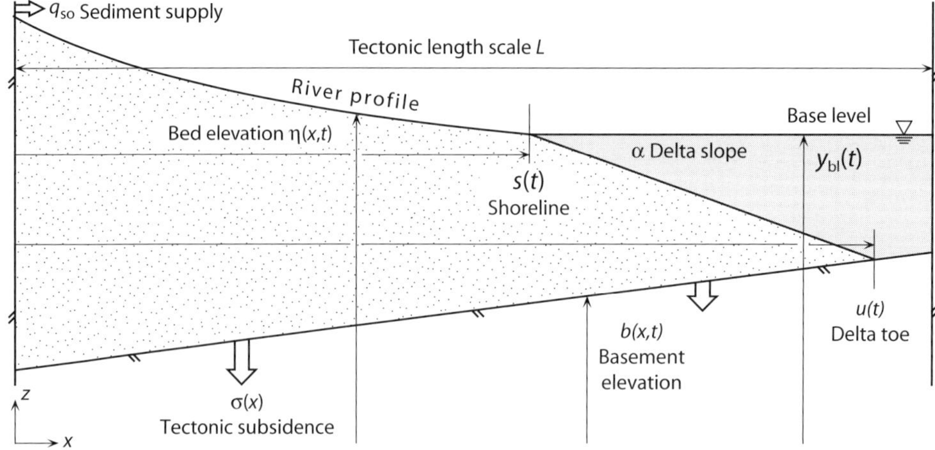

Figure 8.5 Set-up and notation for a cross-section of fluviodeltaic sedimentation with the shoreline treated as a phase front in a generalised Stefan problem. After Swenson et al. (2000) (fig.2) with permission of Cambridge University Press.

of the delta is dominated by avalanche processes and maintains a linear profile of slope α. The offshore advection of sediment by waves and tidal currents is ignored, so the fluviodeltaic model is most relevant to effluents entering low energy marine basins (Section 5.1). Sediment transport in the fluvial regime is assumed to be diffusional so that the sediment flux $q_s(x, t)$ varies linearly with the slope.

The stratigraphic response with steady base level is a brief period of regression followed by an extended period of transgression, giving an autoretreat trajectory (Muto and Steel, 1997) (Figure 8.6). This trajectory is driven by the long-term volume imbalance between the sediment supply and the rate of generation of accommodation ($\bar{\sigma}L$), rather than being driven by changes in boundary conditions. In particular, note that there was no change in base level (sea level). The effect on the model of changes in boundary conditions is investigated using the following groups (Figure 8.7):

- marine efficiency: the relative efficiency of marine sediment transport relative to fluvial transport, $e_m = q_{s0}/\alpha\kappa$, where κ is the sediment diffusivity, which depends on the water discharge in the river system. Order of magnitude changes in marine efficiency strongly affect the life span of the fluvial part of the model. The fluvial life span increases with decreases in e_m. This initially counterintuitive result is geometric. For a given water depth, as e_m increases, so does the delta length. The shoreline flux necessary to feed the subaqueous delta and cause progradation therefore also increases, which for a constant sediment flux causes the fluvial life span to be short.
- capture ratio: the global balance between sediment supplied to the basin and the rate at which space is created by tectonic subsidence, $\Lambda = q_{s0}/\bar{\sigma}L$, where $0 < \Lambda < 1$, applicable to starved rather than overfilled basins. Changes in capture ratio affect the shoreline trajectory strongly. An emaciated basin ($\Lambda << 1$) has a limited fluvial life

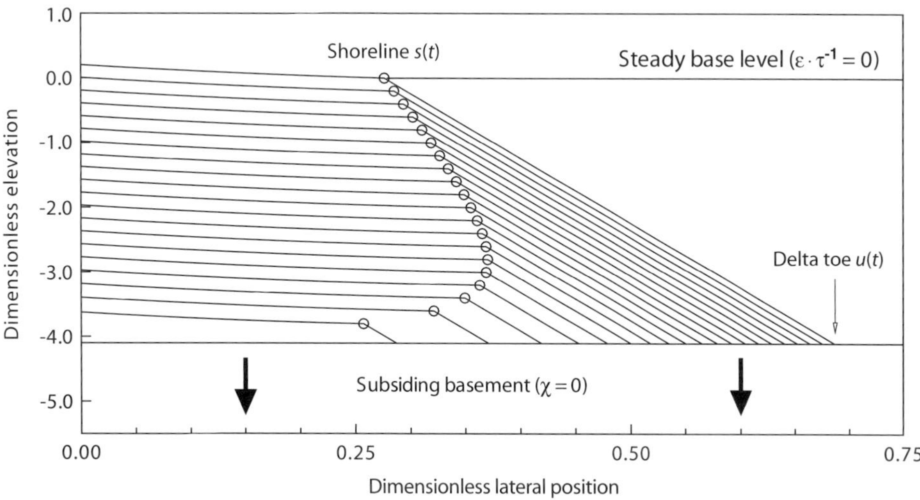

Figure 8.6 Cross-section of a fluvio-deltaic basin showing the shoreline trajectory. As the delta lengthens, foreset thickness reduces and the shoreline becomes transgressive. After Swenson et al. (2000) (fig.3) with permission of Cambridge University Press.

span and an early onset of autoretreat, whereas a more nourished basin has a longer fluvial life span and a delayed onset of autoretreat.

- tectonic style. the spatial distribution of subsidence, χ, where $\chi = 0$ for a basin with no spatial variation in tectonic subsidence. If subsidence is focussed in the proximal part of the model, $\chi < 0$, whereas if it is focussed distally, $\chi > 0$. Progradation is limited and the onset of autoretreat is early when $\chi \to -1$ relative to the case where $\chi \to 1$.
- initial water depth b_0: variation in the initial water depth primarily affects the initial rate of shoreline advance, but has little effect on autoretreat parameters.

8.2 Mass Balance

Sediment delivered to a basin is extracted from the surface flow by net deposition, thereby building the stratigraphic record. The extraction of sediment can be viewed relative to the total mass budget of the sediment routing system (Figure 1.4). A mass balance framework enables systems of different size and shape to be quantitatively compared, but also facilitates an explanation of generic stratigraphic trends, such as downstream fining of grain size and fluvial style, to be understood in terms of mass extraction effects (Strong et al., 2005; Paola and Martin, 2012; Michael, Whittaker, and Allen, 2013).

The overall mass balance for the sedimentary basin can be expressed as a capture ratio Λ, which represents the balance between accommodation and sediment supply (Section 8.1), given by

$$\Lambda = \frac{1}{q_{s,0}} \int_0^L \sigma(x)dx \tag{8.1}$$

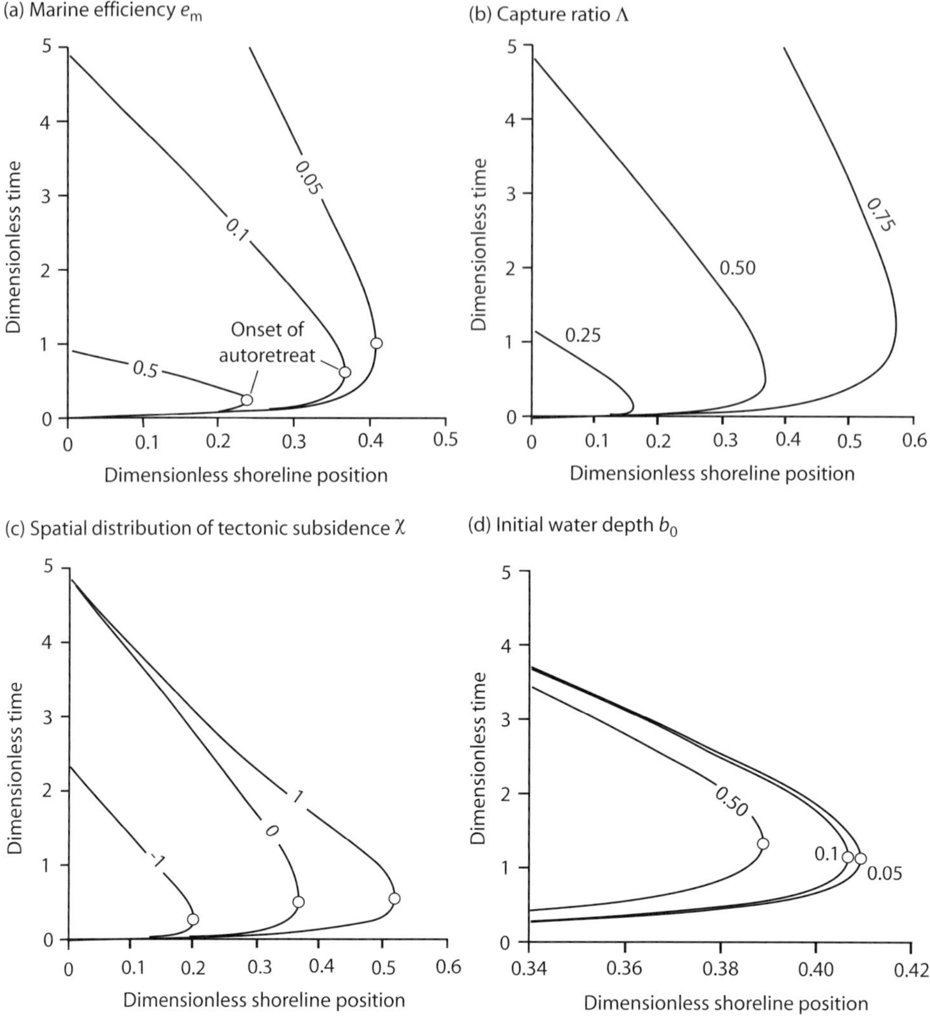

Figure 8.7 Sensitivity of the autoretreat shoreline trajectory to variations in (a) marine efficiency, (b) capture ratio, (c) spatial distribution of tectonic subsidence and (d) initial water depth. After Swenson et al. (2000) (fig.5) with permission of Cambridge University Press.

where $q_{s,0}$ is the unit width sediment supply at $x = 0$, $\sigma(x)$ is the tectonic subsidence rate, x is the down-system distance, and L is the total length of the basin at which point the sediment flux becomes exhausted. If the basin is completely filled to the brim with sediment, then the capture ratio $\Lambda = 1$. When the basin is starved of sediment, $0 < \Lambda < 1$.

To track the sediment routing system in a mass balance framework, the downstream distance x is replaced as a coordinate by a dimensionless fraction of the sediment flux lost to deposition up to position x. This dimensionless fraction is denoted by χ following Strong et al. (2005) and Paola and Martin (2012). The flux extracted from the surface flow to deposition is the integral of the net rate of deposition r over the distance x. The value of χ for a given initial sediment flux at any location is then given by the total sediment flux lost to deposition normalised by the total sediment budget per unit width q_{total}, which over a time increment is equivalent to the initial sediment supply $q_{s,0}$:

$$\chi(x) = \frac{1}{q_{total}} \int_0^x r(x)dx \tag{8.2}$$

Equation (8.2) can be adapted to transport systems of variable width B to give the total initial sediment supply Q_{total}:

$$\chi(x) = \frac{1}{Q_{total}} \int_0^x B(x)r(x)dx \tag{8.3}$$

The spatial pattern of cumulative deposition is strongly affected by the cross-sectional shape of the basin (Figure 8.8). A basin with a linear subsidence profile, such as a hanging-wall basin associated with tilted extensional fault blocks, produces a mass extraction profile that is concave-up. An exponential subsidence profile, as found in flexural foreland basins, produces a mass extraction profile that closely tracks the trend of tectonic subsidence.

If compaction is ignored so that any elevation change of the sediment-water interface $\partial\eta/\partial t$ is due only to the combined effects of erosion/deposition and tectonic subsidence, we can write

$$r = \sigma + \frac{\partial\eta}{\partial t} \tag{8.4}$$

for incorporation in either equation (8.2) or (8.3).

The profile of mass loss $\chi(x)$ is called the mass extraction profile. In an anologous way, a mass balance framework can be constructed for the different grain-size fractions in the sediment supply (for example, gravel, sand and fines) by dividing the cumulative deposited volume of that fraction at a specific downstream point x by the total sediment volume of that fraction Qf_{total} (Michael, Whittaker, and Allen, 2013):

$$\chi f(x) = \frac{1}{Qf_{total}} \int_0^x Qf_{dep}(x)dx \tag{8.5}$$

where $Qf_{dep}(x)$ is the depositional flux of a grain-size fraction at any downstream point x.

Mass extraction from the total sediment supply reduces the surface sediment discharge in the downstream direction $Q_s(x)$. The ratio of the surface sediment discharge to the depositonal flux at any point in the downstream direction is an indication of the amount of bypass. It is denoted by a dimensionless parameter β (Paola and Martin, 2012) and is given by

$$\beta(x) = \frac{q_s(x)}{Lr(x)} = \frac{Q_s(x)}{Q_{dep}(x)} \tag{8.6}$$

A low bypass ratio implies a locally high rate of sediment extraction and vice versa, and the value of $\beta \to \infty$ at very high bypass rates.

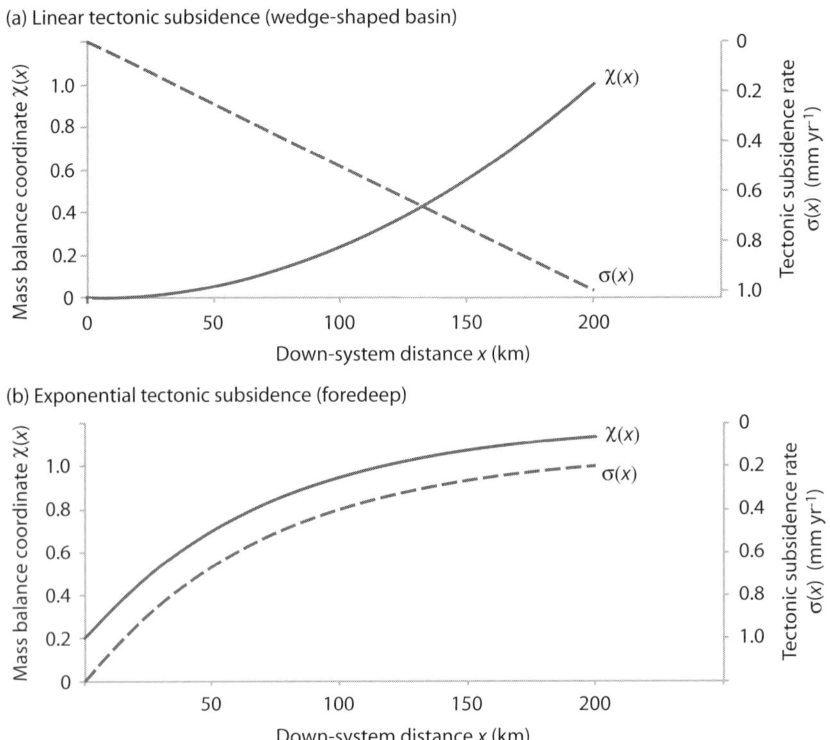

Figure 8.8 Transformation from Cartesian coordinates into a mass balance framework for different basin cross-sectional shapes: (a) linear distribution of tectonic subsidence rate producing a wedge-shaped basin; (b) exponential distribution of tectonic subsdience in a foredeep-type basin. The down-system coordinate x is transformed by normalising the depositional volume (or mass) by the total volume (or mass) deposited in the entire source-to-sink system. The mass balance coordinate χ therefore varies between 0 (at the origin, or apex, of the depositional system) and 1 at the depositional length L_d where sediment is exhausted. This allows sediment routing systems of different sediment budgets, accommodation, and size to be directly compared. See also Figure 1.4.

Transforming the mid to upper Eocene Escanilla Formation of the south-central Pyrenees into mass balance coordinates (Figure 8.9), Michael et al. (2013) were able to track the movement of the gravel cline, gravel front, shoreline and sand front through three time intervals of the palaeo-sediment routing system.

8.3 Stochastic Theory of Particle Trajectories

Sediment en route to a long term depositional sink commonly experiences temporary storage in transient states. Temporary storage modulates and slows down the long-term transport of sediment through the routing system. In the alluvial environment, sediment may spend time stored in alluvial fans, river channel beds and bars, floodplains and deltas. While

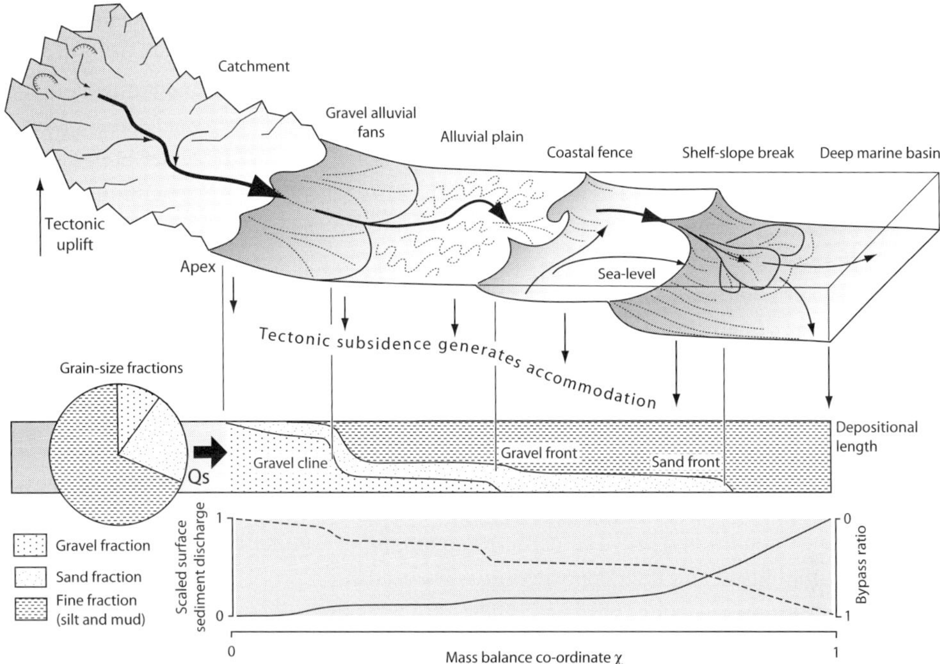

Figure 8.9 Schematic illustration of a sediment routing system supplied with a discharge Q_S and a grain-size mix of gravel, sand, and fine fractions, derived by erosion of a mountain catchment area, derived from the Eocene Escanilla palaeo-sediment routing system of the south-central Pyrenees. Tectonic and climate forcings cause the various moving boundaries of the system to migrate, including the gravel cline, gravel front, sand front, coast, and shelf-slope break. Stratigraphy is built up by mass extraction from the surface flux. Consequently, the surface flux reduces downstream. The depositional flux depends on the spatial distribution of accommodation and in this terrestrial to marine example increases beyond the coast. The bypass ratio therefore decreases markedly in the ocean. The coastal zone acts as a fence for the seaward transport of gravel and sand. From Michael, Whittaker, and Allen (2013) (fig.17) with permission of University of Chicago Press.

in storage constituent grains suffer degradation during weathering. The annual rates of exchange between channels and floodplains may exceed the annual downstream sediment flux, as in the well-studied Amazon basin (Dunne et al., 1998) (Section 4.5).

Over the appropriate time and space scales, the trajectory of a particle through an alluvial valley is a random process consisting of transport events separated by periods of storage of varying duration (Malmon et al., 2003). Storage zones may vary in their mobility from highly active to highly stable. Exchanges of sediment between temporary stores with differing mobility can therefore be expressed as transition probabilities. A Markov chain describes such a process (Kelsey, Lamberson, and Madej, 1987). The methodology described here is used in the context of an alluvial valley, but can be applied to other segments of the sediment routing system.

The most downstream boundary is treated as an absorbing state, which means that no particle that reaches it can return to any of the other states in the alluvial valley. Within each transient state, all particles are equally susceptible to future erosion, sediment transport and deposition. A particle in a transient state i has a fixed probability of moving to state j after a unit of time. This probability is denoted $p_{ij} \geq 0$. Transition probability values are assigned based on the rates of sediment transfer within the alluvial tract. Each transition consists of an erosion event from i, E_i, and a deposition event into state j, D_j (Figure 8.10). If the alluvial floor has two transient states i and j, there are three possible outcomes: a particle may be eroded from i and deposited in j; a particle may eroded from i and deposited in the same state i; finally, a particle may be eroded within the alluvial tract but reaches the absorbing state.

The transition probability for erosion from i and deposition in j for a unit time step p_{ij} is given by

$$p_{i,j} = P(E_i)P(D_j|E_i) \tag{8.7}$$

where | means 'conditional on'. The particle can remain in i, either by not being mobilised or by being mobilised and then deposited in i. The transition probability is the sum of the two probabilities representing these outcomes:

$$p_{ii} = P(E_i^c) + P(E_i)P(D_i|E_i) \tag{8.8}$$

where E_{ij}^c is the complement of E_{ij}. Equation (8.8) can therefore be expressed as

$$p_{ii} = [1 - P(E_i)] + P(E_i)P(D_i|E_i) \tag{8.9}$$

The transition probability for a particle that is not deposited in any of the transient states but reaches the absorbing state x is the fraction of area inside E_i not occupied by either D_i or D_j (Figure 8.10). Generalising to the situation where there is an arbitrary number b of transient states and a single absorbing state at the downstream boundary, the transition probability per time unit is

$$p_{ix} = P(E_i) \left[1 - \sum_{j=1}^{b} P(D_j|E_i) \right] \tag{8.10}$$

The probability of a particle being mobilised per unit time is the inverse of the mean residence time of the sediment. If all particles within a storage site are equally mobile, the erosion probability per unit time is the mass rate of erosion divided by the total mass of that deposit, that is

$$P(E_i) = \frac{Q_{E_i}}{m_i} \tag{8.11}$$

where Q_{E_i} is the erosion rate of deposit i in units of mass/time, and m_i is the mass of sediment in unit i. The assumption of equal mobility may not be valid. For example, a floodplain close to an active channel may have a greater chance of being eroded than

(a) Venn diagram

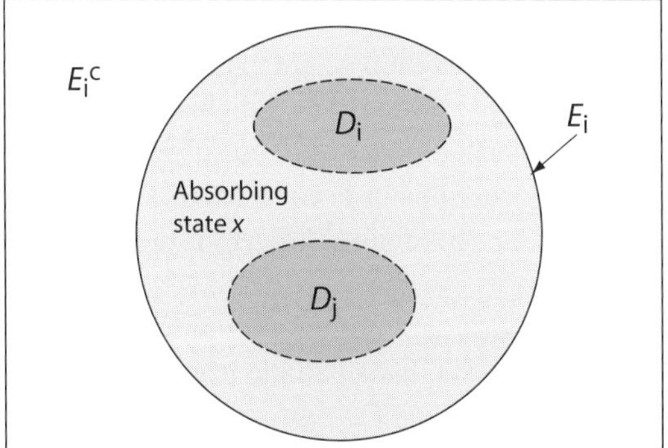

(b) Transition probabilities

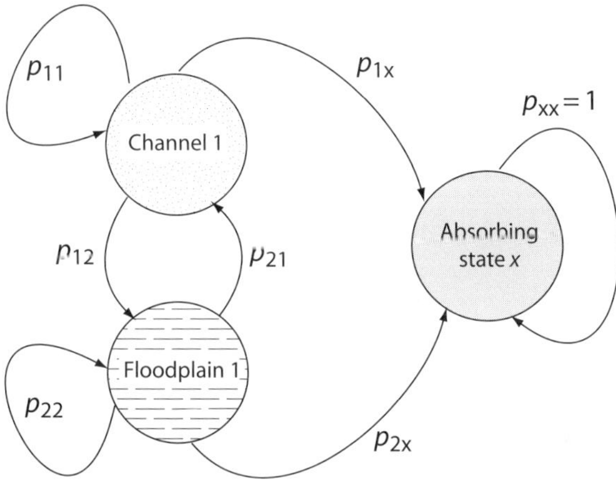

Figure 8.10 (a) Venn diagram illustrating the derivation of transition probabilities for a model with 2 transient states i and j and a single absorbing state x. The rectangle is the universe of all possible outcomes after an increment of time for a particle initially stored in transient state i. Circle E_i is the event that the particle is eroded from i, and ovals D_i and D_j are the events that the particle is deposited in i and j respectively conditional on first being removed from i. E_i^c is the complement of E_i. (b) Schematic diagram of a very simple Markov model of an hypothetical alluvial valley floor composed of a single reach with a channel, a floodplain and an absorbing state at the down-system limit of the reach. Modified from Malmon, Dunne, and Reneau (2003) (fig.2) with permission of University of Chicago Press.

a floodplain far from the channel. To incorporate this possibility, additional floodplain transient states would need to be included at different distances from the main channel.

For the probabilities of deposition, in a reach r, the conditional probability that a particle will be deposited in j given that it was eroded from i is

$$P(D_j|E_i) = \frac{Q_{D_i}}{Q_{O_r} + \sum_{k \in B_r} Q_{D_k}} \tag{8.12}$$

where Q_{D_j} is the mass rate of sediment deposition into deposit j, B_r is the portion of the transient state space that is located in reach r, and Q_{O_r} is the sediment flux out of reach r at its downstream end. The summation $\sum_{k \in B_r} Q_{Dk}$ represents the total rate of sediment deposition into all the units located within reach r, including state j. A particle in transport within reach r will either be deposited in one of the storages in B_r or will leave the reach at its downstream boundary. The denominator in equation (8.12) is the total mass of sediment in transport within reach r, and the probability $P(D_j|E_i)$ is the mass fraction of that sediment that is deposited in j.

If the particle is not deposited in any of the units in reach r, it enters the downstream reach $r + 1$. The probability $P(O_r|E_i)$ that a particle leaves r given that it was eroded from deposit i in reach r is

$$P(O_r|E_i) = \frac{Q_{O_r}}{Q_{O_r} + \sum_{k \in B_r} Q_{Dk}} \tag{8.13}$$

The conditional probability that a particle in state i in an upstream reach is eroded and transported into state j immediately downstream of reach r and is deposited in reach $r + 1$ is

$$P(D_j|E_i) = P(O_r|E_i) \frac{Q_{D_j}}{Q_{O_{r+1}} + \sum_{k \in B_{r+1}} Q_{Dk}} \tag{8.14}$$

where i is in B_r and j is in B_{r+1} and the conditional probability $P(O_r|E_i)$ is found from equation (8.13).

To calculate the transition probabilities, information is required about the sediment budget. This information includes (1) the erosion and deposition rates Q_{E_i} and Q_{D_j} of each geomorphic unit, (2) the sediment flux Q_{O_r} at the downstream boundary of each reach r, and (3) the mass m_i of each of the storages. As an example, we can take the hypothetical case given by Malmon et al. (2003) where in each of three reaches A, B and C, the mass of sediment in the channel $m_{1,3,5}$ is 1×10^6 tons, the mass in the floodplain $m_{2,4,6}$ is 20×10^6 tons, the downstream sediment flux Q_O is 2×10^6 tons yr^{-1}, the channel erosion/deposition rate $E_{1,3,5}$ and $D_{1,3,5}$ is 0.5×10^6 tons yr^{-1} and the floodplain erosion/deposition rate $E_{2,4,6}$ and $D_{2,4,6}$ is 1×10^6 tons yr^{-1}. The probability of a particle moving from the floodplain in reach A to the channel in reach C (that is, p_{25}) in any one year is

$$p_{25} = P(E_2)P(O_A|E_2)P(O_B|O_1)P(D_5|O_2) \tag{8.15}$$

which gives a solution of $p_{25} = 0.0023$ (0.23%).

The transition probability for a particle being eroded from a floodplain in state i and deposited in a channel in state j decreases linearly with an increase in the floodplain mass. Using the parameter values given previously, the floodplain erosional and depositional turnover is 5% of the floodplain mass, and the channel erosional and depositional turnover is 50% of the channel mass. If these percentages change to lower turnover percentages, the probability of a particle advecting to a down-system reach is reduced, increasing residence times. A reduction in percentage of turnover might result from an increase in the area (and mass) of floodplain in the alluvial valley, or a reduction in the erosion/deposition rate caused by a change in river hydrology driven by climate change. In other words, alluvial systems with large floodplains relative to their erosion/deposition rate exhibit low probabilities of downstream advection.

The residence time of a deposit can be defined as the inverse of the erosion probability in equation (8.11). For the previous example, the residence time of the floodplain in reach A becomes 20 years. If the floodplain turnover rate per year is decreased from 5% to 1%, the residence time for particles in floodplain storages increases to 100 years. The transit time, on the other hand, is the time it takes for a particle to reach an absorbing state, starting from some initial deposit i. Some particles will travel quickly through the alluvial valley while others will spend protracted periods of time in storage. The mean particle transit time for each deposit/storage can be calculated (Kelsey et al., 1987). However, the statistical distribution of transit times is strongly right-skewed, with some transit times extending to infinity.

Over many time steps, the entire alluvium initially present in the valley will be transported to the absorbing state, its place being taken by new sediment from upstream or from adjacent hillslopes. In a steady-state alluvial valley, the cumulative mass flux of valley-stored sediment delivered to the absorbing state remains constant over time. The evacuation time is scale-dependent (length of reach, width of floodplain) and also dependent on the geomorphic processes at work. Using the previous example, the time to achieve 90% evacuation (T_{90}) is 91 years, but doubling the channel and floodplain masses and halving the geomorphic process rates gives $T_{90} = 365$ yr. Evacuation times of large, lowland river floodplains with low rates of sediment exchange between channel and floodplain are high. Small, steep and narrow upland rivers with restricted floodplains should have short evacuation times.

The Markov model of Kelsey et al. (1987) and Malmon et al. (2003) assumes a single grain size. Sediment storages could, however, be divided into grain-size fractions, such as gravel, sand and fines (silt and clay), each with their own probabilities for erosion and deposition. Field data for such an approach requires grain-size fractions to be calculated along the sediment routing system, as in the geological example of the Escanilla palaeo-sediment routing system from the Eocene of the south-central Pyrenees (Michael et al., 2013, 2014a). Alternatively, separate budgets could be made for bedload and suspended load grain-size classes.

Stochastic models of particle trajectories are currently restricted to relatively small, modern systems where long-term tectonic subsidence, base-level change and transient driving mechanisms for sediment supply can be neglected. Nevertheless, these stochastic approaches highlight the impact of storage on particle trajectories and on particle-population transit and valley evacuation times. They point to avenues of fruitful future research.

8.4 Towards a Vocabulary for Tectonic Landscapes

Our vocabulary for landscapes should take account of the relationship between the response timescales of landscapes compared with the nature of its perturbations. Beaumont, Kooi, and Willett (2000) and Marr et al. (2000), for example, discriminated slow forcing, intermediate forcing and rapid forcing (Section 8.1.1). The response is commonly damped and phase-shifted, so that a lag time is apparent. This lag time, for example between tectonic uplift and sediment discharge, may be very long (10^6-10^7 years) for large orogenic systems (Kooi and Beaumont, 1996), but shorter for trains of folds in fold-thrust belts (order 10^5 years; Tucker and Slingerland (1996)), which immediately suggests that some caution should be used in inferring the timing of tectonic events on the basis of the age of the stratigraphic response. Beaumont et al. (2000) also identify an 'impulsive forcing' where the response follows an approximately exponential decay curve, allowing a 'relaxation time' to be calculated. Densmore et al. (2007a) used the term 'analytical response time' for the response calculated from a broad theory (diffusive or wave-like) of system behaviour for the time to attain full equilibrium.

The forcing function, or perturbation, may be of different forms (Figure 8.11):

- It may take the form of a function with a characteristic timescale for its duration t and amplitude A and it may repeat with a period σ.
- It may also be a discrete, spike-like function with negligible duration but amplitude A and period σ.
- The forcing may be a step change, in which case we let the time since a change in boundary conditions be T.

In a first scenario, the response time of the landscape or sediment routing system is very long compared with the repeat time (period) of the perturbation, which is equivalent to the rapid forcing case of Beaumont et al. (2000). In this case, the landscape never fully adjusts to a single tectonic perturbation, so the tectonic events are strongly buffered by the sluggish reaction of the landscape. An example is the buffering of individual slip events on a range-bounding fault, which might involve a slip of $1-2$ m repeated every 10^3-10^4 years, whereas the relaxation time for the longitudinal profile of transverse rivers is expected to be in the region of 5×10^4 years or more, and the timescale for attainment of uniform relief approximately 10^6 years.

In a second scenario, a tectonically driven sediment efflux signal of a catchment, with amplitude A and duration t, is modified by transfer through a large channel-floodplain system before entering the ocean. This is equivalent to the intermediate forcing of Beaumont et al. (2000). If the alluvial channel-floodplain system has a characteristic timescale τ, the signal will reduce to an amplitude At/τ at the entry point into the ocean. If τ is large ($\approx 10^6$ years), as suggested for the large alluvial systems of southeast Asia (Castelltort and Van Den Driessche, 2003) and $t/\tau \ll 1$, the sediment routing system is highly buffered. Buffered systems therefore occur where $\tau \gg \sigma$ and where $t \ll \tau$ (Figure 8.12).

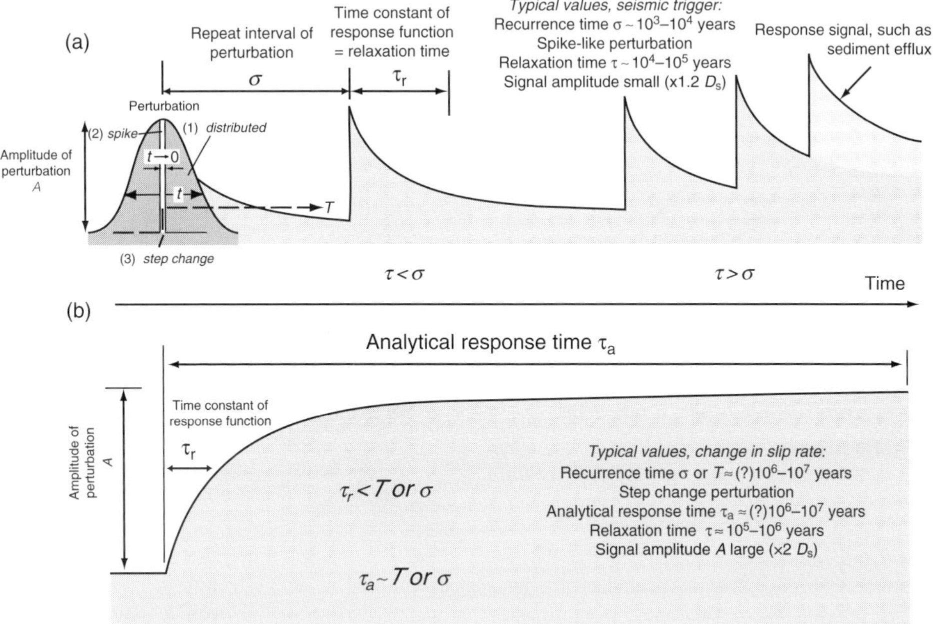

Figure 8.11 The concept of a perturbation and its response function. (a) Perturbation may be (1) distributed, in this case in the form of a sinusoid with timescale t, (2) a spike of negligible time duration or (3) a step change in boundary conditions. The perturbation has a characteristic repeat interval σ, and the response function (such as sediment efflux) has a response time, here shown as a relaxation time τ_a. Two contrasting situations are shown, where $\tau < \sigma$ and $\tau > \sigma$. (b) A step change in boundary conditions provokes a long-term response, with an analytical response time for the full attainment of equilibrium at the asymptote τ_a, and a relaxation time attained when the response function has achieved a fraction $(1 - e^{-t/\tau})$ of the total change in signal τ_r. From Allen (2008b) (fig.8) with permission of the Geological Society.

Third, the response time of the landscape may be fast compared with the long repeat time of the perturbation, which is partly equivalent of the 'slow forcing' case of Beaumont et al. (2000). In this case the landscape responds relatively rapidly to the perturbation, and can be termed 'reactive'. An example of a reactive landscape is a humid mountainous landscape with steep, narrowly spaced rivers, as on the western flank of the Southern Alps of New Zealand. Such a landscape responds quickly to a change of base level by rapid landsliding of hillslopes and bedrock river incision. Similarly, if the alluvial timescale τ is relatively small, as may occur on fan delta systems, sediment efflux signals of mountain catchments will be transferred with little modification to the ocean, and the shallow marine (and perhaps deep marine) stratigraphic record may closely record the tectonic driver in the source region both in terms of magnitude and timing. Reactive systems therefore occur where $\tau < \sigma$ and where $t > \tau$ (Figure 8.12).

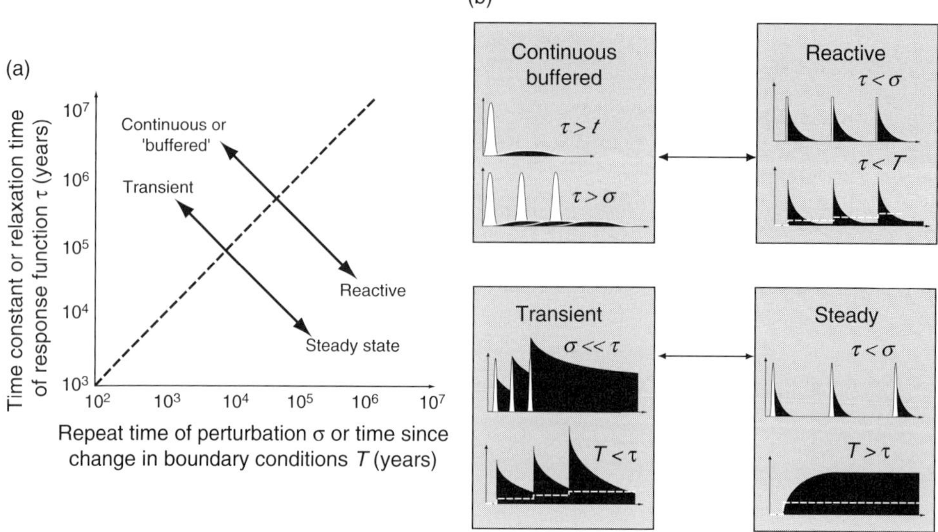

Figure 8.12 Concept of a vocabulary for landscapes coupled to active tectonics. Landscapes are shown varying along two axes, from transient to steady, and from buffered to reactive. These theoretical types depend on a number of parameters: the timescale of the perturbation t; the time since a change in boundary conditions T; the response time and relaxation time of the response function τ; and the repeat time of the perturbation σ. From Allen (2008b) (fig.9) with permission of the Geological Society.

The idea of buffered versus reactive landscapes and sediment routing systems is related but not identical to the concept of steady state or equilibrium. A landscape can be said to be steady or in steady state if, through the operation of all of its linked geomorphic subsystems, it is statistically in balance with the prevailing boundary conditions (Willett and Brandon, 2002). Examples of steady landscapes are the catchments of the central parts of extensional fault segments. Here, the combination of steep slopes and adequate precipitation causes bedrock incision rates to keep pace with the tectonic fluxes caused by cumulative slip on the range-bounding fault. If secular changes in fault slip rate are infrequent (T is large), the catchment response time τ is relatively rapid (5×10^4 yr) and the footwall topography is steady. Aggradation on adjoining fans can also be considered steady under these conditions. At a larger scale, landscapes in steep mountain belts in a wet climate grow to equilibrium relatively quickly with slowly changing plate velocities. Any changes in the rate of tectonic advection of rocks to the orogenic surface are quickly adjusted to in an aggressive erosional environment. The wet flanks of the Southern Alps (New Zealand), Taiwan and central-eastern Himalayas are examples.

If, however, the response time τ is long compared with the time since a change in boundary conditions T, the landscape is out of equilibrium with the prevailing conditions and can be termed 'transient' (Figure 8.12). An example is provided by the small catchments located within the tip-zone region of an extensional fault segment propagating at its tip. Despite their small size, these low gradient catchments respond slowly to their incorporation into the footwall block, and are in a transient stage with respect to their prevailing tectonic environment. Slip rate variations caused by glacial-interglacial changes in surface loads, such as ice and lake water masses, which operate at the 10^4 year timescale (Hetzel and Hanpel, 2005), are also likely to produce transient landscapes in uplifting footwalls. Alluvial basins and bedrock rivers subjected to eustatic changes in base level at the outlet are also commonly in a transient condition following high-frequency Quaternary glacial cycles. Downstream low gradients cause knickpoint celerities to be insufficient to establish steady conditions following high-frequency eustatic variation at the 10^4 year timescale. The result of partial adjustment to a series of high-frequency eustatic changes is a staircase of knickpoints each migrating at a different celerity (Bishop et al., 2004).

8.5 Transient Responses within Sediment Routing Systems

There is no single response time of a landscape to a change in prevailing conditions, nor a single index of landscape sensitivity. In order to understand how landscapes and sediment routing systems interact with a tectonic displacement field or a climatic regime, we need to investigate the varied transient responses of a linked set of geomorphic subsystems (Figure 8 13) The responses of landscapes and sediment routing systems are neither immediate nor linear, which led Philip Allen (2005) (p.961) to write:

Landscapes and their associated sediment routing systems respond polychromatically, like the striking of a chord. Coupled tectonic-surface process models therefore are geared to hearing the chords, discriminating the component frequencies and evaluating the strings and fret or finger positions used to create them.

Fundamental to the understanding of landscapes is the idea of steady state (Willett and Brandon, 2002), which implies that properties of the landscape, such as longitudinal river profiles, dynamically adjust towards equilibrium with prevailing boundary conditions. If these boundary conditions are perturbed, it follows that the landscape, or its component geomorphic subsystems, respond over a characteristic timescale. Attainment of or approach to adjustment to new ambient conditions does not require that the landscape becomes static, only that statistically (Ellis, Densmore, and Anderson, 1999) its dynamic subsystems (such as hillslopes) and pointwise events (such as landslides) operate to achieve a steady-state in terms of the mean values of chosen parameters, such as relief, long profile and sediment efflux, or in terms of a uniform scaling between, for example, catchment slope and drainage area. Consequently, the concept of steady state is complex, and a landscape may, for example, be in steady state with regard to one chosen parameter while it is in a transient state with regard to another.

The discussion of transient responses within sediment routing systems could potentially encompass a wide range of contexts. A range of situations is briefly considered: a diffusive timescale applied to first alluvial rivers (Section 8.5.1) and then hillslopes (Section 8.5.2);

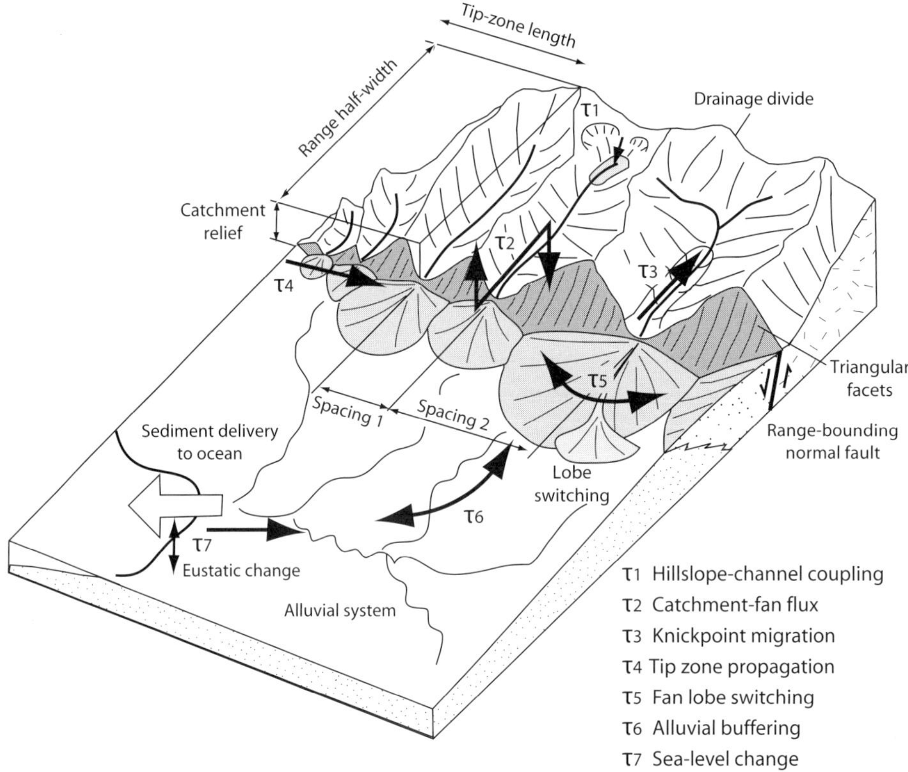

Figure 8.13 Different response times in a sediment routing system characterised by an extensional fault block in the source region, and an alluvial system in the transport-depositional region. τ_1, Coupling between hillslope erosion and channel incision; τ_2, time for the attainment of equilibrium conditions in catchment denudation rate and fan aggradation following a step change in the slip rate on the range-bounding fault; τ_3, time for the development of new equilibrium profiles in bedrock rivers by knickpoint migration following a base level fall at the outlet; τ_4, timescale for the attainment of uniform relief in the footwall estimated from the length of the tip-zone region in a laterally propagating fault segment; τ_5, timescale for fan resurfacing by lobe switching; τ_6, transformation in a buffered alluvial system of a sediment efflux signal from its source region; τ_7, response time for the upstream propagation of a knickpoint from the coast caused by a eustatic change. For a similar concept see Carrétier and Lucazeau (2005). Modified from Allen (2008b) (fig.10) with permission of the Geological Society.

a relaxation time represented by the response of sediment flux to tectonic and climatic perturbations of a catchment-fan system (Section 8.5.3); the time required for fault segment linkage and the transient landscape response (Section 8.5.4); an analytical response time for the lateral growth of a fault segment (Section 8.5.5); knickpoint migration in rivers incised into an uplifting fault block (Section 8.5.6); and a lag time for unroofing of tectonic folds as a function of rock erodibility (Section 8.5.7).

8.5.1 Alluvial Rivers

The response times of alluvial systems have been approximated from measurements of the sediment discharge leaving the catchment. River basins may have thick terraces in incised river valleys in their upstream portions and extensive floodplain areas in their downstream portions, which act as buffers to changes in any forcing variables, such as base level change at a river mouth, or changes in sediment flux near their headwaters, and therefore may have long response times. If the system is assumed to be overall diffusive in character, and the discharge of water varies systematically with floodplain width W, the mass effective diffusivity κ of the channel-floodplain system is

$$\kappa = \frac{Q_s}{W < \partial y/\partial x >} \tag{8.16}$$

where Q_s is the sediment discharge and $<>$ denotes the spatial average of the slope. If L is the downstream length of the river-floodplain system and H is the maximum relief between its upstream and downstream ends, the response time becomes

$$\tau = \frac{L^2}{\kappa} = \frac{L^2 W < \partial y/\partial x >}{Q_s} = \frac{LWH}{Q_s} \tag{8.17}$$

The large Asian river systems have typical values of $L \sim 10^6$ m, $W \sim 10^5$ m, $H \sim 1$ to 2×10^2 m (slopes of 10^{-3} to 10^{-4}) and sediment discharges Q_s of 10^{7-8} m^3 yr^{-1} (Métivier and Gaudemer, 1999). The characteristic response time is therefore in the region of 10^5 to 10^6 years. Castelltort and Van Den Driessche (2003) carried out a similar analysis on 93 of the world's major rivers, and found that response times varied between 10^4 yr to more than 10^6 yr (Figure 8.14). The response time depends on the scale of the channel-floodplain system. Large alluvial systems therefore potentially strongly buffer any variations in sediment supply with frequencies of less than 10^{5-6} years. This has strong implications for the detection of high frequency driving mechanisms in the stratigraphy of sedimentary basins with extensive alluvial floodplain tracts (Section 10.2).

Nevertheless, it should be noted that the long response times of buffered alluvial systems suggested by the diffusive approach described earlier are highly generalised, system-scale results that are unlikely to be closely matched by observational data on sediment transit times. For example, since the end of the LGM, it is estimated that $4050-5675$ km^3 of sediment has been deposited in the Indus sediment routing system (Clift and Giosan, 2014). Since 10 ka, most sediment has been derived from deep gorges in the Himalayan region and from incision of the upper alluvial plain. Only half of the total sediment reaching the delta since the LGM originated directly by weathering of bedrock. Clift and Giosan (2014) estimated that zircon grains took $7-14$ kyr to travel to the delta as bedload, introducing a lag of $\sim 10^4$ yr in the sediment delivery following the initial climatic trigger.

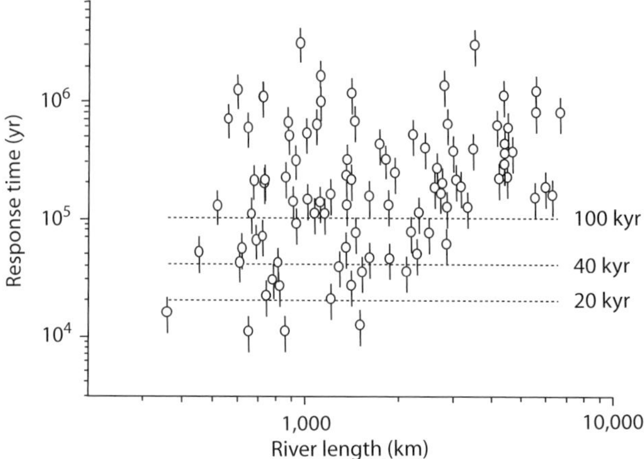

Figure 8.14 Response times of some modern drainage basins. Hydrological and geomorphological data from Hovius (1998). Vertical bars show uncertainties based on the unknown bedload sediment transport rates of the rivers shown. Note that the majority of response times are greater than 10^5 yr and some are greater than 10^6 yr. From a data set compiled by Castelltort and Van Den Driessche (2003) (fig.2A), with permission of Springer.

8.5.2 Hillslope Erosion

An assumption is commonly made that hillslopes respond instantaneously to a local base level fall, change in uplift rate or climate change. If this assumption holds, the hillslope subsystem can be ignored from the consideration of the response time of a catchment to a tectonic perturbation, since hillslope erosion is not the rate-limiting process. This assumption is worth briefly evaluating.

Hillslopes have traditionally been regarded as diffusive (Carson and Kirkby, 1972; Anderson, 1994; Fernandes and Dietrich, 1997), which means that the form of a hillslope profile is dependent on the diffusivity κ, but the amplitude of the profile is set by the rate of channel incision and the length of the hillslope from the ridge crest to the valley bottom L. The time constant for a diffusive system such as this is given by $\tau = L^2/\kappa$, from which it can be seen that, for the slow processes of rainsplash and creep, where κ most commonly falls in the range $0.05-0.01$ m^2 yr^{-1} (Anderson and Humphrey, 1990; Rosenbloom and Anderson, 1994; Burbank and Anderson, 2001), hillslope time constants on small hillslopes where $L = 50$ m range from 50 to 250 kyr. However, the hillslopes in the steep transverse catchments of the Basin and Range province are dominated by landsliding rather than soil creep, giving a non-linear dependence of the mass flux on the topographic gradient (Roering et al., 1999).

Effective diffusivities in humid, tectonically active regions dominated by landsliding, such as the western flank of the Southern Alps, New Zealand, range between 15 and 50 m^2 yr^{-1} (Koons, 1989; Hovius et al., 2007). Numerical modelling (Densmore et al., 1998) and field studies (Pearce and Watson, 1986) suggest that the response time of landslide-dominated hillslopes is rapid compared with the time required for bedrock rivers to adjust to base level change by knickpoint migration (Section 8.5.4). Consequently, it is justifiable to treat the rate of bedrock incision and velocity of knickpoint migration as the key variables determining the response time of a mountain catchment to a base-level change.

8.5.3 Catchment-Fan Systems

Densmore et al. (2007a) constructed a mass balance numerical model involving a transport-limited (where the rate of vertical lowering of the stream profile is limited by the rate at which sediment can be transported away) catchment, a spatially variable rock uplift pattern and an erosional law combining diffusive and concentrative terms to capture both hillslope diffusion and channel incision (Smith and Bretherton, 1972; Simpson and Schlunegger, 2003) (equation (10.2)). In the numerical model, sediment derived from the catchment fills in hangingwall accommodation up to a certain linear surface slope. Consequently, fan slope is determined by precipitation rate as a proxy for the climatic impact on erosivity, and by slip rate as a measure of the geometry of incremental accommodation in the hangingwall and of incremental tectonic uplift in the footwall.

The impacts of an increase and a decrease in precipitation on the catchment-fan system are illustrated in Figures 8.15 and 8.16 (Densmore et al., 2007a). A 50% instantaneous increase in precipitation from 1 to 1.5 m yr^{-1} causes an immediate increase in sediment flux from the catchment, accompanied by a strong reduction in fan slope. The relaxation time for the sediment flux is up to 0.5 Myr. A 50% decrease in precipitation rate from 1 to 0.5 m yr^{-1} also has an immediate and strong impact of the sediment efflux of the catchment, accompanied by an increase in fan slope. The relaxation time for sediment flux is, however, $\sim$1 Myr, considerably longer than for the case of increased precipitation. In a similar numerical model, Armitage et al. (2013) found a response time of 0.8 Myr for a step change in precipitation in a catchment-fan system.

An instantaneous doubling of the slip rate (from 1 to 2 mm yr^{-1}) results in an increase in fan slope, an increase in sediment flux at the catchment mouth, and initial backstepping of the fan toe, followed by a progradation of the fan toe (Figure 8.17). A halving of the slip rate (from 1 mm yr^{-1} to 0.5 mm yr^{-1}) results in the deposition of a sheet-like unit immediately after the perburbation, accompanied by a progradation of the fan toe (Figure 8.18).

A single perturbation, such as a step-change in slip rate on a range-bounding fault, therefore causes a number of effects on different aspects of the sediment routing system, each with its own response time. Whereas the fan slope and sediment efflux approach new equilibrium values with a time constant of approximately 600 kyr, the duration of the backstepping-prograding activity of the fan toe, from the time of the slip rate perturbation until the fan toe reaches its previous extent, is significantly longer (2.5 Myr). In other words, one should avoid talking of response times too generally, since different parameters and indices respond at different timescales to the same perturbation, like the striking of a chord

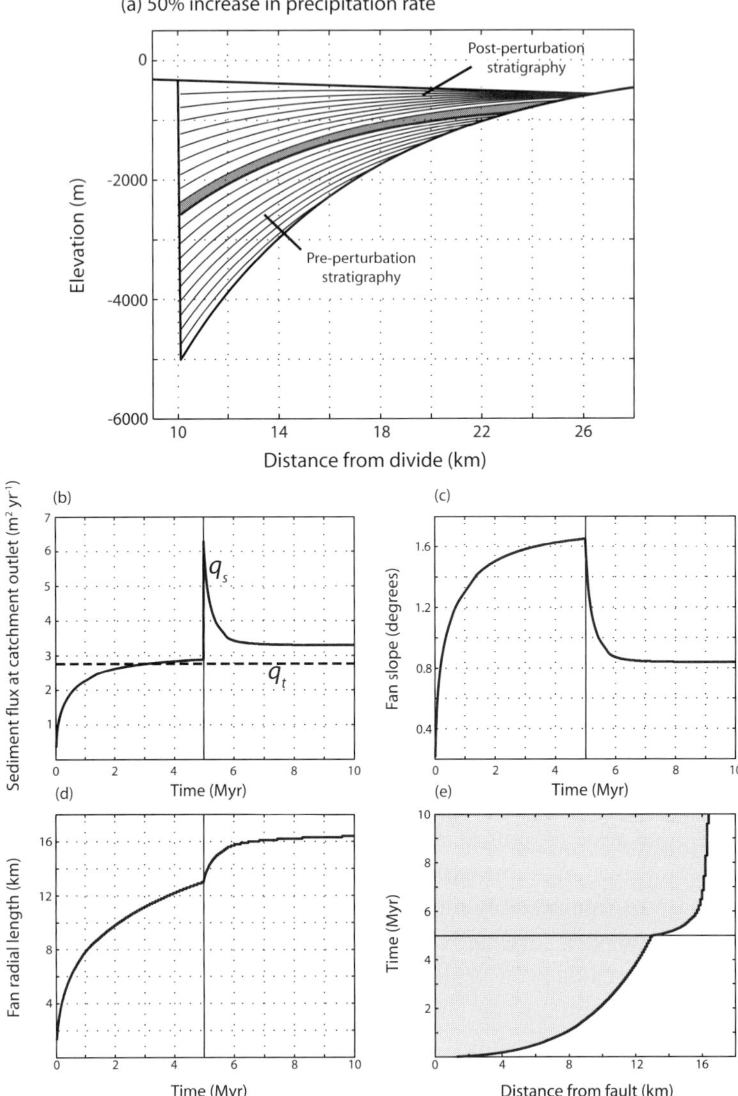

Figure 8.15 Experimental results following a 50% increase in the precipitation rate, to 1.5 m yr^{-1}, after 5 Myr of model run time. (a) Hangingwall stratigraphy before and after the increase in slip rate. Deformed stratigraphic time lines are recorded every 500 kyr. Fan deposition rate is unaffected by the perturbation. Increased precipitation rate leads to deposition of sheet-like unit (light grey) immediately after perturbation. Fan toe progrades, then stabilises at an equilibrium position. (b) Sediment flux at the catchment outlet. Vertical line here and in subsequent panels marks the increase in precipitation rate at 5 Myr. Dashed line shows the tectonic flux into the footwall q_t, which remains constant during the run. (c) Fan surface slope before and after the perturbation. (d) Fan radial length before and after the perturbation. Note minor progradation of the fan toe, followed by stabilisation. (e) Chronostratigraphic diagram of the basin. After Densmore, Allen, and Simpson (2007a) (fig.7) with permission of American Geophysical Union.

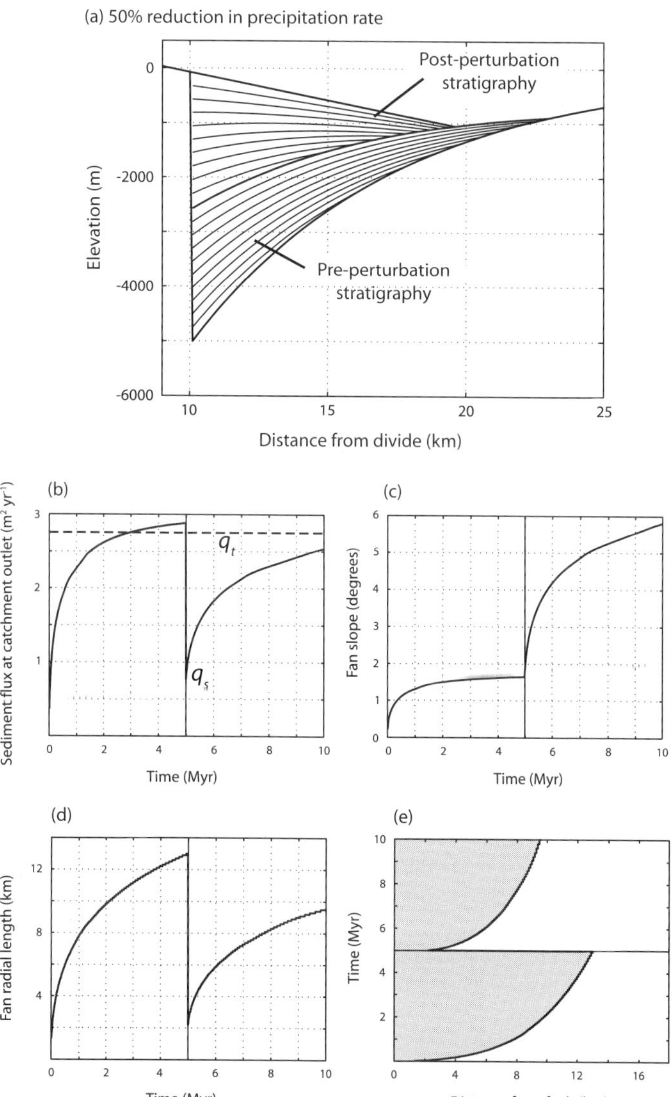

Figure 8.16 Experimental results following a 50% decrease in precipitation rate, to 0.5 m yr^{-1}, after 5 Myr of model run time. (a) Hangingwall stratigraphy before and after the decrease in precipitation rate. Deformed stratigraphic time lines are recorded every 500 kyr. Fan deposition rate is unaffected by the perturbation. Note back-stepping-prograding behaviour of fan toe and marked angular unconformity following the perturbation. (b) Sediment flux at the catchment outlet. Vertical line here and in subsequent panels marks the decrease in precipitation rate at 5 Myr. Dashed line shows the tectonic flux into the footwall q_t, which remains constant during the run. (c) Fan surface slope before and after the perturbation. (d) Fan radial length before and after the perturbation. Note backstepping of the fan toe by more than 10 km, followed by renewed progradation. (e) Chronostratigraphic diagram of the basin. After Densmore, Allen, and Simpson (2007a) (fig.8) with permission of American Geophysical Union.

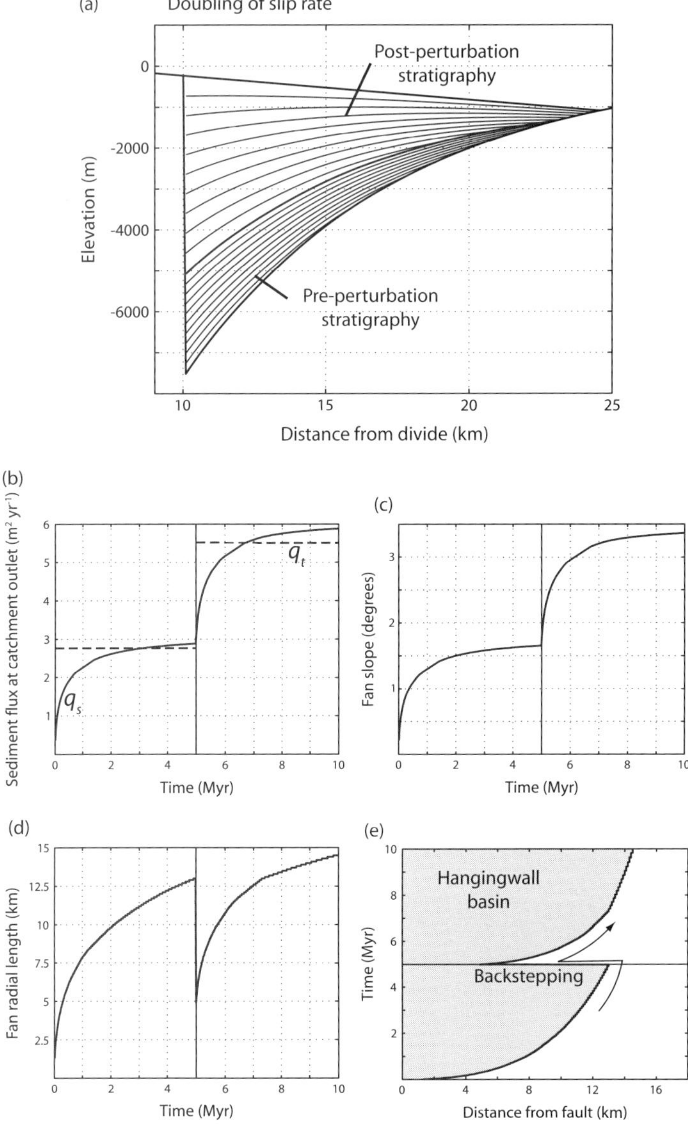

Figure 8.17 Experimental results following a doubled slip rate, to 2 mm yr^{-1}, after 5 Myr of model run time. (a) Hangingwall stratigraphy before and after the increase in slip rate. Deformed stratigraphic timelines are recorded every 500 kyr. Note increase in deposition rate and back-stepping-prograding behaviour of the fan toe following the perturbation. (b) Sediment flux at the catchment outlet. Vertical line here and in subsequent panels marks the increase in slip rate at 5 Myr. Dashed line shows the tectonic flux into the footwall q_t. (c) Fan surface slope before and after the perturbation. (d) Fan radial length before and after the perturbation. Note back-stepping of the fan toe by about 8 km, followed by renewed progradation. (e) Chronostratigraphic diagram of the basin. Grey area shows lateral extent of basin sedimentation as a function of time. After Densmore, Allen, and Simpson (2007a) (fig.5) with permission of American Geophysical Union.

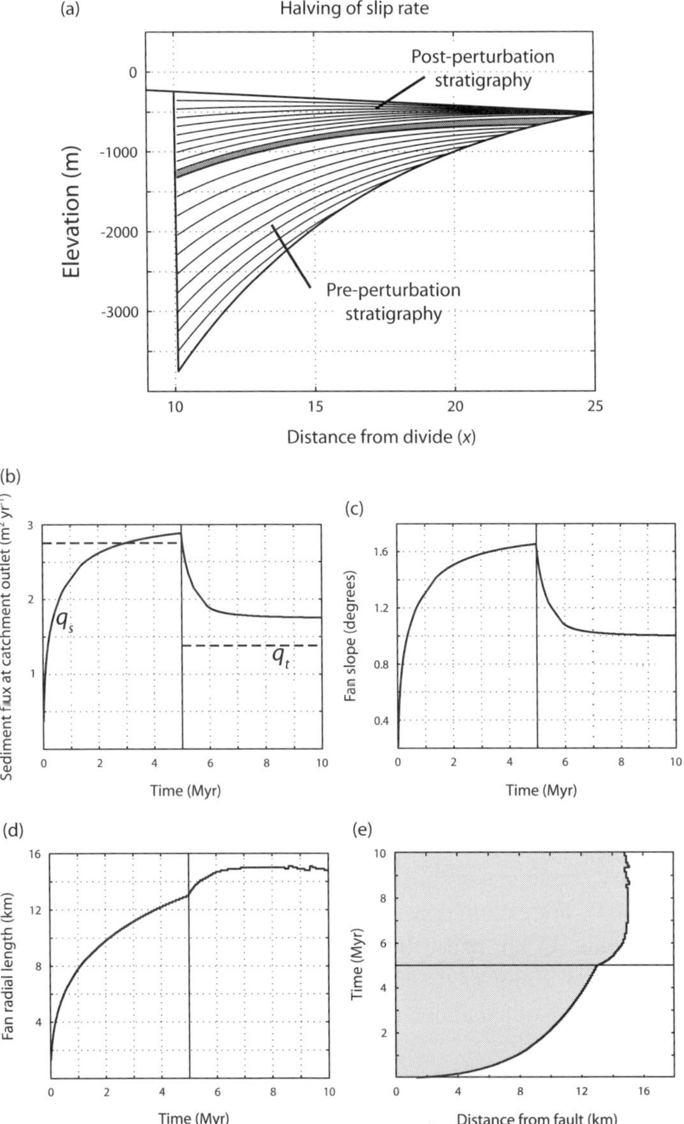

Figure 8.18 Experimental results following a halved slip rate, to 0.5 mm yr^{-1}, after 5 Myr of model run time. (a) Hangingwall stratigraphy before and after the decrease in slip rate. Deformed stratigraphic time lines are recorded every 500 kyr. Note decrease in deposition rate and deposition of sheet-like unit immediately after perturbation. Fan toe progrades, then stabilises at an equilibrium position. (b) Sediment flux at the catchment outlet. Vertical line here and in subsequent panels marks the decrease in slip rate at 5 Myr. Dashed line shows the tectonic flux into the footwall q_t. (c) Fan surface slope before and after the perturbation. (d) Fan radial length before and after the perturbation. Note minor progradation of the fan toe, followed by stabilisation. (e) Chronostratigraphic diagram of the basin. Grey area shows lateral extent of basin sedimentation as a function of time. After Densmore, Allen, and Simpson (2007a) (fig.6) with permission of American Geophysical Union.

(Figure 8.13). Catchment sediment efflux and fan slope decay to lower equilibrium values with a time constant of 500 kyr when the slip rate is halved, a value very similar to the case of doubling the slip rate.

The response time of a catchment-fan system must depend on the system diffusivity, which in the case of a diffusive-concentrative law is the square-bracketted term (units of m^2 yr^{-1}) in the expression for the rate of surface elevation change (equation (10.2)). The response time constant therefore decreases following an increase in precipitation rate, and since equation (10.2) is essentially a diffusion equation, the time constant varies approximately as the inverse square of the precipitation. However, somewhat surprisingly, the time constant is independent of variations in the slip rate. Despite strongly affecting surface slope, hangingwall accommodation, base level change and tectonic fluxes in the footwall, the slip rate (within the geologically reasonable range of 0.1 and 2 mm yr^{-1}) has an insignificant effect on the response time.

Densmore et al. (2007a) fitted an expression of the form $[1 - \exp(-t/\tau)]$ to the curves for the sediment flux q_s following a tectonic or climatic perturbation. The time constant varies with the inverse square of the precipitation rate (Figure 8.19a). However, the time constant is invariant at 0.5 Myr for slip rates in the range 0.1−2 mm yr^{-1}, with precipitation held constant at 1 m yr^{-1} (Figure 8.19b). The rock uplift rate therefore appears to have little impact on the response time of the catchment-fan system, despite that tectonic uplift/subsidence controls accommodation in the hangingwall basin and sets the topographic slope.

The occurrence of alluvial sedimentation (Carrétier and Lucazeau, 2005) and extensive piedmonts (Pelletier, 2004) at the range-front affects the response of mountainous catchments to changes in tectonic and climatic boundary conditions. The presence of an alluvial apron increases catchment response times to a tectonic perturbation relative to catchments with fixed boundary conditions at the outlet, a result supported by sand box experiments (Babault et al., 2005). For example, using a tectonically uplifting block 20 km wide with fringing alluvial plains 15 km wide, the mean denudation rate of the block following an initial tectonic uplift of 2 mm yr^{-1} (precipitation at 1 m yr^{-1}) shows a slower response for a range front with a flanking alluvial piedmont compared to one without (Carrétier and Lucazeau, 2005).

Once the drainage network is established, the mean denudation rate follows an exponential trend, from which relaxation times can be calculated. Whether the river incision model is detachment- or transport-limited, and whether the alluvial apron is characterised by channelised flow or sheet flow, relaxation times are 0.2−0.5 Myr for the range fronts without alluvial piedmonts and 0.7−1 Myr for those with range front basins. Carrétier and Lucazeau (2005) also found a relaxation time of approximately 1 Myr after an instantaneous increase in the tectonic uplift rate from 2 to 2.5 mm yr^{-1}, with precipitation held constant at 1 m yr^{-1}. Although the details of the modelling are different, these values are very similar to the time constants of Densmore et al. (2007a) and give confidence that for systems of this size, relaxation times are of order 1 Myr, with a significant amplification (1.5−10) caused by the presence of a dynamic sedimentary piedmont.

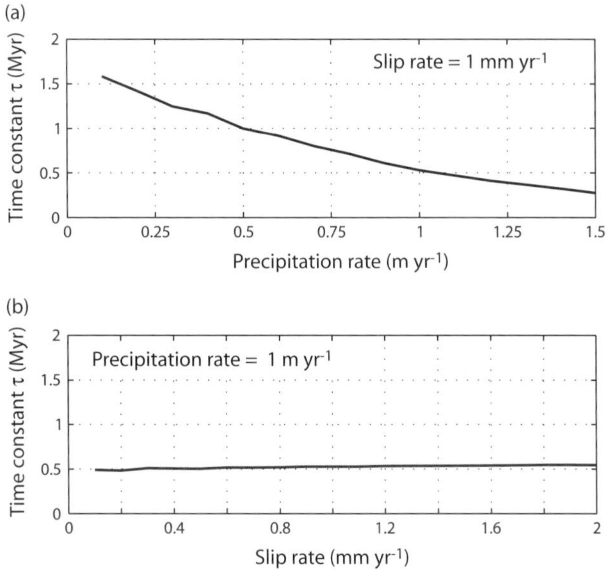

Figure 8.19 Response times as a function of precipitation rate (a) and slip rate (b). Model is run from a flat initial topography until a steady sediment flux at the catchment outlet is achieved. Response time for achievement of this equilibrium is expressed as exponential time constant τ, defined by fitting the sediment flux curve with an expression of the form $1 - \exp(-t/\tau)$. (a) shows time constant as a function of precipitation rate, which is varied from 0.1 to 1.5 m yr^{-1} in 0.1 m yr^{-1} increments, for a fixed slip rate. Note approximately quadratic decay of time constant with increasing precipitation rate, consistent with linear dependence of time constant on system diffusivity. (b) shows time constant as function of slip rate, which is varied from 0.1 to 2 mm yr^{-1} in 0.1 mm yr^{-1} increments, for a fixed precipitation rate. Response time is insensitive to slip rate over this geologically reasonable range. After Densmore, Allen, and Simpson (2007a) (fig.11) with permission of American Geophysical Union.

8.5.4 Growth of Extensional Fault Arrays

Faults in rifts are not infinite in extent: instead there is a displacement-length relationship, with most of the slip being taken up on a small number of interacting major fault segments. Fault displacement dies out towards the tips of fault segments (Figure 8.20). A single fault segment, whose length is commonly between 10 and 100 km, has a maximum cumulative displacement (d)-length (L) relationship (Schlische, 1991; Dawers and Anders, 1995) given by

$$d \approx kL \tag{8.18}$$

where k is a constant of proportionality that varies between 0.01 and 0.05 for many fault systems.

The growth of populations of interacting faults is affected by a stress feedback that promotes the growth of a small number of major through-going border faults at the expense

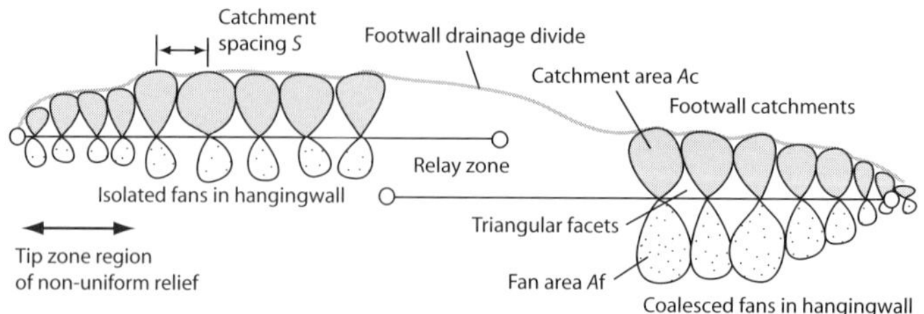

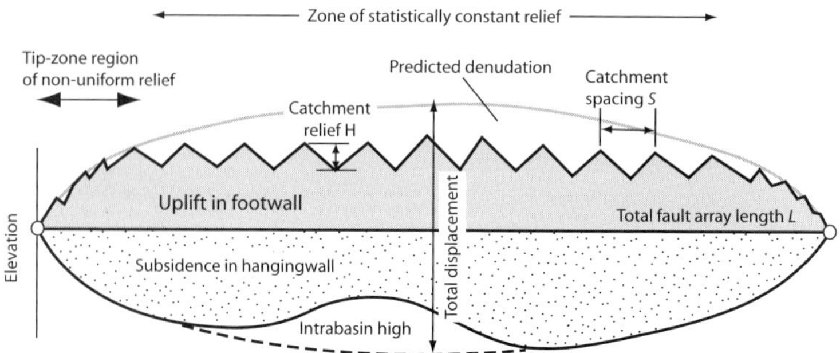

Figure 8.20 Linked extensional fault segments and their catchment-fan systems. (a) Footwall catchments and hangingwall fans along two fault segments connected by a relay zone. (b) Anticipated footwall topography and relief, and hangingwall subsidence along the same fault segments, showing a tip-zone region of non-uniform relief. Based on Densmore et al. (2004) (fig.11). Published with permission of American Geophysical Union.

of smaller displacement faults that become dormant (Schlische, 1991; Dawers and Anders, 1995; Cowie, Gupta, and Dawers, 2000). The evolution of the fault array has a profound effect on the topography and river drainage of rifted regions (Gupta et al., 1999; Cowie et al., 2000, 2006) (Figure 8.21), and therefore on the gross depositional environments and sediment thicknesses of the basin-fill (Gawthorpe and Leeder, 2000). Relay ramps between fault segments along the basin edge are commonly the entry points of river systems into the hangingwall depocentres (Gawthorpe and Hurst, 1993; Gupta et al., 1999). Footwall evolution, however, controls the establishment of the main drainage divide and therefore the size and sediment discharge of river catchments draining into the rift (Cowie et al., 2006). The Jurassic Brent-Strathspey-Statfjord fault array system in the North Sea basin (McLeod, Dawers, and Underhill, 2000), the Neogene fault array of the Gulf of Suez in eastern Sinai (Sharp et al., 2000) and the modern fault arrays of the Lake Tanganyika and Lake Malawi basins (Rosendahl et al., 1986; Specht and Rosendahl, 1989) are all excellent examples.

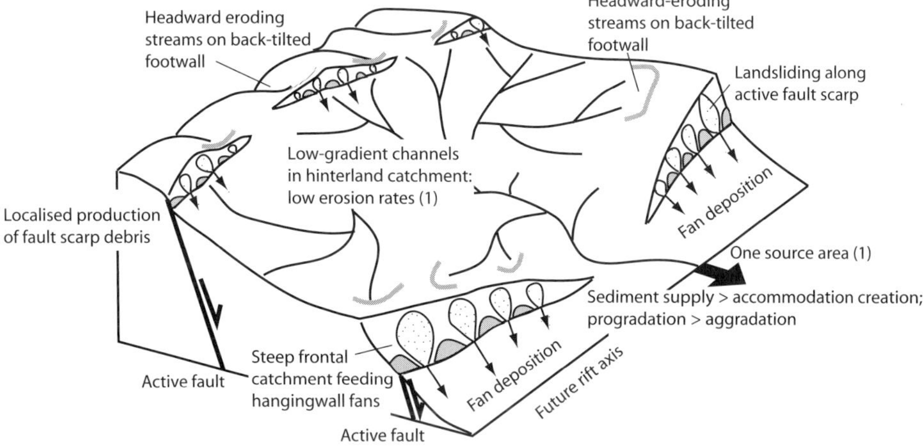

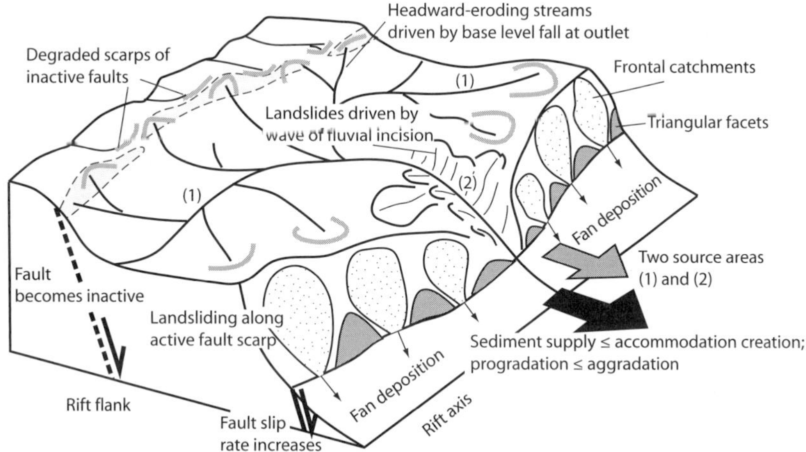

Figure 8.21 Schematic diagram showing the effect of fault linkage on surface topography and drainage in the fault propagation and interaction stage (a) and after fault linkage (b), which transfers extension to the main border fault. In (a), there is one source area for sediment, with a catchment outlet located between the tips of two fault segments. In (b), an additional source region of sediment is a convex, strongly incised reach caused by upstream knickpoint migration driven by fault segment linkage. From Cowie et al. (2006) (fig.20) with permission of John Wiley & Sons Inc.

The time taken for faults to initiate, grow, interact and link can be estimated from the transient landscapes sculpted by rivers incising into actively deforming extensional fault systems. In the Sperchios Basin and northern Gulf of Evia, Greece, the throw of middle Pliocene (3.6 Ma) to Recent active faults is 2 to 3 times the footwall relief (Whittaker

Table 8.1. *Response times to climatic and tectonic forcing.*

Perturbation	Response time	References
Orogenic topographic decay	$\sim 10^8$ yr	Pelletier (2004)
Knickpoint migration following long-wavelength tectonic uplift	$> 10^7$ yr	Roberts et al. (2012)
Change of tectonic fluxes into collisional orogen	$10^6 - 10^7$ yr	Willett (1999)
Sediment flux from unroofing thrust-related anticlines	0.1–4 Myr	Tucker and Slingerland (1996)
Hillslope diffusive timescale	50–250 kyr	Burbank and Anderson (2001)
Response of sediment flux to slip-rate change on fault	500–600 kyr	Densmore, Allen, and Simpson (2007a)
Precipitation change on catchment-basin system	0.8 Myr >700 ky	Armitage et al. (2013) Forzoni et al. (2014)
Achievement of steady state in footwall topography	10^6 yr	Densmore et al. (2007b)
Change of uplift rate in range without fringing piedmont	0.2–0.5 Myr	Carrétier and Lucazeau (2005)
Change of uplift rate in range front with alluvial apron	1 Myr	Carrétier and Lucazeau (2005)
Analytical response time due to uplift of a block:		Whipple (2001)
(i) Steep bedrock rivers	10^5 yr	
(ii) Badlands & alluvial rivers	$>10^6$ yr	

and Walker, 2015). Two sets of knickpoints in river long profiles indicate that the fault system has experienced two changes in slip rates: one due to the initiation and growth of fault-generated topography, and another due to fault linkage (Goldsworthy, Jackson, and Haines, 2002), when slip rates increased by a factor of ≥ 3. Using a fault throw enhancement factor of 3, it is estimated that fault linkage took place at ~ 1.6 Ma in the Sperchios Basin and at ~ 1.4 Ma along the Coastal Fault of the Gulf of Evia. Since the faults initiated in the middle Miocene, the timescale for linkage is approximately 2 Myr, which is similar for the time taken for knickpoints to propagate though the landscape in other Mediterranean regions (Whittaker et al., 2007; Whittaker and Boulton, 2012).

In the Celano catchment of Abruzzi, central Italy, simulations using the spatially lumped numerical model PaCMod show that the peak in the magnitude of sediment export from the catchment has not yet been attained after 700 kyr since the base level perturbation (Forzoni et al., 2014) (see Table 8.1). Such response times become longer with progressively smaller fault slip rates. Analytical response times calculated for hillslopes in three catchments in Abruzzi ranged from 4.6 Myr for the tectonically active Celano

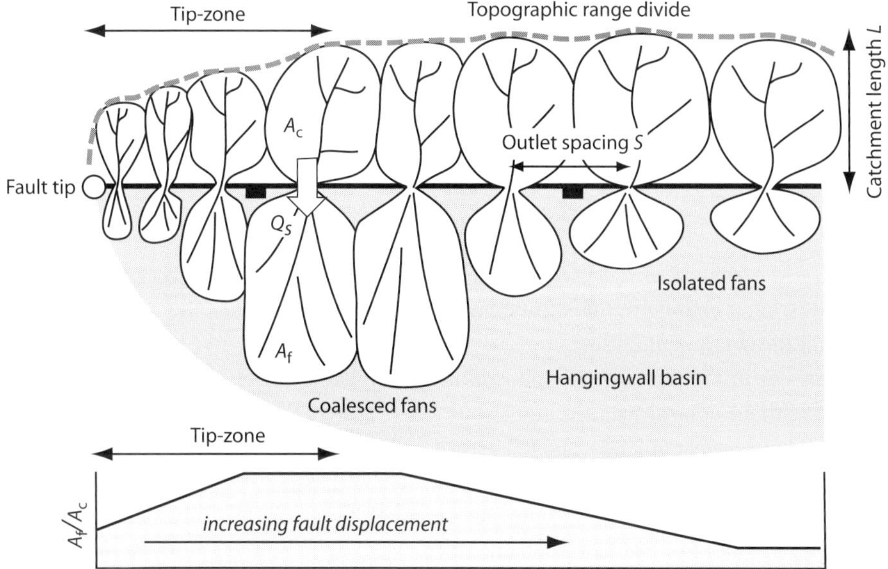

Figure 8.22 Concept of variation in the ratio of fan area to catchment area along a growing fault segment. Catchment areas provide variable discharges of sediment along the fault strike Q_s, which is distributed over the fan surface. A_f/A_c can be expected to vary depending on fault slip rate, geometry of fan surface and equilibrium of catchment-fan systems to ambient slip rates.

catchment ($U = 1.6$ mm yr^{-1}) to 17 Myr for the tectonically inactive L'Apa catchment ($U = 0.3$ mm yr^{-1}).

Tectonics play a major role in the creation of uplifting footwalls as source regions for sediment, but also in determining the spatial pattern and rate of hangingwall subsidence, and thereby in controlling fan thickness and fan progradation distance. Not surprisingly, plots of fan area versus catchment area from examples in the arid southwestern United States (Figure 4.10) (Allen and Hovius, 1998) show that areas with high fault displacement rates and areas with low fault displacement rates are discriminated by their value of A_f/A_c (Section 4.2, Figure 4.10). This implies that coarse-grained fan bodies are stacked against active range-bounding faults, whereas fans coalesce and prograde basinwards where fault displacement rates are smaller.

8.5.5 Lateral Growth of a Fault Tip

Since the ratio between fan area and catchment area A_f/A_c reflects the balance between sediment supply and hangingwall accommodation (Section 4.2) (Figure 4.10), it is anticipated that there will be a lateral variation in A_f/A_c along a fault segment propagating at its tip (Figure 8.22). There is a contrast between the analytical response times of catchments located in the tip-zone of growing faults and in the central zone of uniform relief (Figure 8.20). In order to test whether catchments along Basin and Range fault segments are in equilibrium with their tectonic displacement field, Densmore et al. (2007b) introduced a normalised response time τ^* given by

$$\tau^* = \frac{\tau_a}{t_{onset}} \tag{8.19}$$

where τ_a is the analytical response time for the propagation of knickpoints (see Section 8.5.6) following a change in the slip rate and t_{onset} is the time since the initiation of rapid fault slip at the catchment outlet.

Catchments in the tip-zone of the Lemhi Fault, Idaho, have very high values of τ^* compared with the central sector of uniform relief (Figure 8.23). This suggests that tip-zone catchments have transient landscapes, whereas those in the central sector are in equilibrium with the present-day tectonic displacement field. The time since the onset of faulting ranges from 0 Myr at the fault tip to 7.5 Myr at the centre of the segment. Assuming a steady rate of lateral tip propagation, the length of the tip zone gives an indication of the time required to achieve steady state, which is of order 10^6 yr.

8.5.6 *Knickpoint Migration*

When bedrock rivers are perturbed tectonically by either a block-type uplift of rocks, or an instantaneous basel level fall at the catchment outlet, they work to attain equilibrium with the new boundary conditions by the upstream wave-like migration of erosional knickpoints (Whipple and Tucker, 1999; Whipple, 2001). The response time is defined as the time required for an erosional knickpoint to reach the upstream end of the channel network. Using values of concavity (m/n) typical of steep bedrock rivers, these analytical response times are of order 10^5 yr irrespective of system size, but where concavities are low, as found in alluvial and badlands type rivers, analytical response times rise to significantly greater than 10^6 years for moderate to large system sizes (Figure 8.24). The implication is that statistical steady state is unlikely to be achieved where catchments are perturbed rapidly by changing tectonic or climatic boundary conditions.

Phases of accelerated rock uplift may be recognised by the presence of knickzones in the longitudinal profile of a river. Compilation of long profile information from many streams in a region experiencing uplift may reveal otherwise unrecognised long-wavelength doming of the continental surface.

Consider an elevation profile of a river $y(x)$ affected by the migration of knickpoints generated by spatially uniform tectonic activity $U(t)$ (Pritchard et al., 2009; Roberts and White, 2010). Erosion is assumed to be made of two components, one a diffusive lowering, and the other a result of knickpoint migration. Knickpoints migrate upstream with a celerity C, which causes a change in the elevation of the stream channel. The knickpoint migrates upstream at a rate controlled by the water discharge and the erodibility of rocks comprising

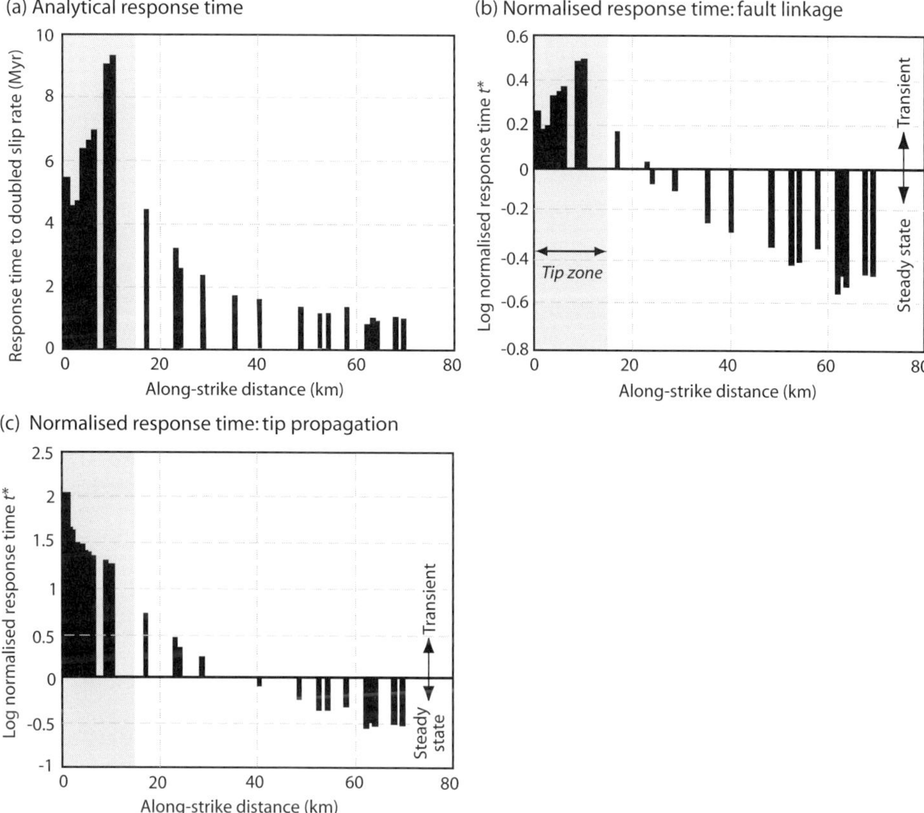

Figure 8.23 Response times of catchments along the southern half of the footwall of the Lemhi Fault, Idaho, following a doubling of the slip rate. The pre-perturbation slip rate is assumed to vary from 0 at the fault tip to 0.5 mm yr^{-1} at the centre. The grey area is the tip-zone, characterised by increasing footwall relief. (a) Analytical response time calculated using the solution for knickpoint migration given in Whipple (2001). (b) Normalised response time τ^* (logarithmic scale) for fault growth by linkage, with the onset of fault slip uniform at 6 Myr. (c) Normalised response time τ^* (logarithmic scale) for fault growth by tip propagation at 10 mm yr^{-1}, so that the time since the onset of faulting varies from 0 at the fault tip to 7.5 Myr at the centre. Note that catchments in the tip zone are transient, whereas catchments in the central part of the footwall are likely to be in steady-state. Modified from Densmore et al. (2007b) (fig.10) with permission of American Geophysical Union.

the stream bed. The change of elevation of the bed resulting from the migration of the knickpoint is

$$\frac{\partial y}{\partial t} = C\frac{\partial y}{\partial x} = (k_e \rho g)q_w \frac{\partial y}{\partial x} \tag{8.20}$$

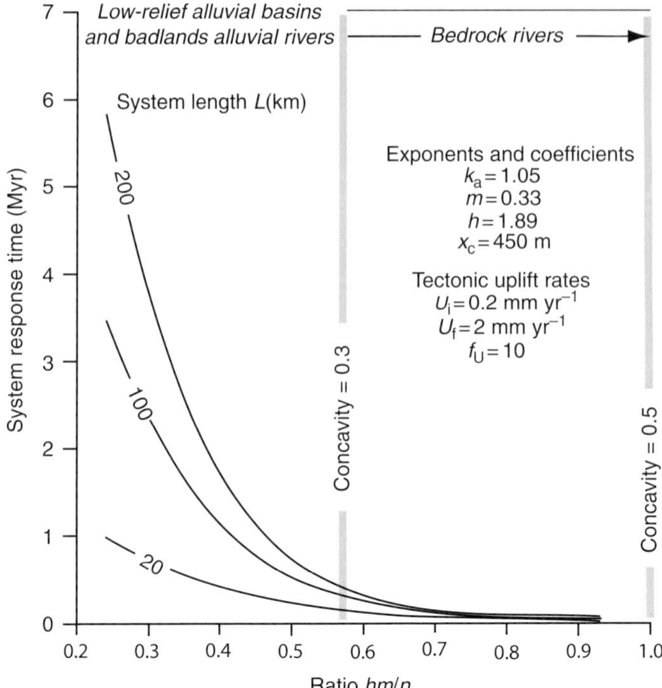

Figure 8.24 The analytical response time calculated using the stream power rule (Whipple, 2001) versus hm/n for different system lengths. System response times are less than 1 Myr for system lengths less than 200 km for the high concavities typical of bedrock rivers. From Allen (2008a) (fig.3) with permission of the Geological Society.

where the celerity is proportional to the water unit discharge q_w and k_e is the erodibility coefficient, dependent on rock strength. Since water discharge varies with the downstream distance x, equation (8.20) can be expressed as

$$\frac{\partial y}{\partial t} = v_0 x^m \left(\frac{\partial y}{\partial x} \right)^n \tag{8.21}$$

where v_0 is a knickpoint migration parameter, equal to $k_e \rho g$, which equals the celerity C when $m = 0$ and $n = 1$. Assuming diffusion to be slow compared to advection (knickpoint migration), the tectonic uplift is given by

$$U(t) = \frac{\partial y}{\partial t} + v_0 x^m \left(-\frac{\partial y}{\partial x} \right)^n \tag{8.22}$$

where the first term is the change of bed elevation over a time increment, and the second term is a wave-like advection describing the upstream migration of a knickpoint.

The inversion of uplift rate history from the variation of the stream gradient is given in full in Pritchard et al. (2009). Following rescaling of the horizontal coordinate and the stream gradient, a time variable τ can be introduced to describe the time since the origination of the wave-like knickpoint at $x = L$. The uplift rate as a function of time can be inverted from topographic information on river longitudinal profiles (such as provided by a digital elevation model). The tectonic uplift history as a function of time is then given by

$$U(\tau) = v_0 x^m \left(-\frac{\partial y}{\partial x} \right)^n \tag{8.23}$$

where the time variable is

$$\tau = -\frac{L^{1-m}}{(n-m)v_0} \left[\frac{1 - (x/L)^{1-m/n}}{(x/L)^{m(n-1)/n}} \right] \left(-\frac{\partial y}{\partial x} \right)^{1-n} \tag{8.24}$$

There is clearly a need for m, n and v_0 to be calibrated. In their study of the uplift of the Bié Dome, Namibia, Pritchard et al. (2009) used $m = 0.5$ and $n = 1$, in which case equation (8.24) simplifies to

$$\tau = -\frac{\sqrt{L} - \sqrt{x}}{0.5 v_0} \tag{8.25}$$

Consequently, if the advective erosion coefficient v_0 can be approximated, the timescales for the propagation of a knickzone up the river profile starting at $x = L$ can be found, and the uplift history can be calculated from stream gradients extracted from a digital elevation model. Pritchard et al. (2009) used $v_0 = 50 \text{ m}^{1-2m} \text{ Myr}^{-1}$ from their study of the Bié Dome, whereas Roberts et al. (2012) used $v_0 = 200 \text{ m}^{1-2m} \text{ Myr}^{-1}$ for the Colorado River.

As an example, the Colorado River has a total length of $L = 2,300$ km. Two knickzones are found at $x = 200$ km and 1,000 km. The characteristic timescales for the migration of these knickzones to their present position can be found. Using $m = 0.25$, $n = 1$, and $v_0 = 84 \text{ m}^{0.5} \text{ Myr}^{-1}$ derived from an area-slope analysis, and using the knickpoint positions at 200 km and 1,000 km, the characteristic timescales are 25 and 12 Myr respectively. In other words, the time taken for a long river to adjust to a phase of tectonic uplift, as shown by the migration of a wave of incision, is of order greater than 10^7 yr.

8.5.7 Unroofing of Tectonic Folds

In areas of thrust-related folding, with high rates of tectonic uplift of rocks U, the topographic relief is essentially controlled by the ratio U/c_v, where c_v is the bedrock erodibility. Streams are known to change their gradients markedly over substrates of different erodibility (Hack, 1973). Tucker and Slingerland (1996) estimated bedrock erodibilities c_v to vary over an order of magnitude in the arid landscapes of the Zagros. Using a numerical landscape evolution model, they successfully simulated the progressive unroofing of stratigraphy with strongly varying erodibility related to the growth of a series of tectonic folds (Figure 8.25). Sediment flux from the fold-thrust belt reflects both tectonic growth of the fold structures and also the extent of exposure of resistant versus weak lithologies. For

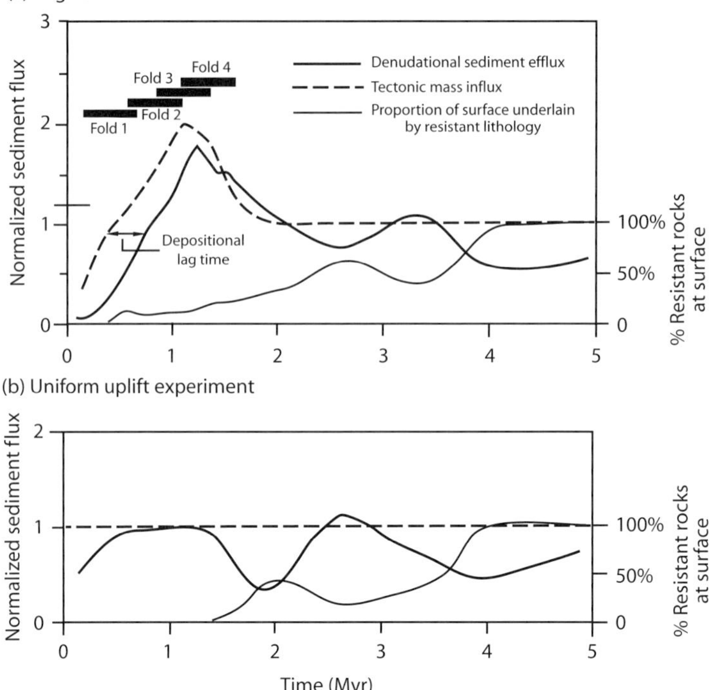

Figure 8.25 Landscape evolution model results of the unroofing of folds in the Zagros fold-thrust belt. (a) Tectonic influx from uplift of rock (dashed line) and sediment efflux out of the model grid (solid line) versus time. In this simulation, four anticlinal folds grow progressively in time. Note the lag time between tectonic influx and depositional response of less than 0.5 Myr during the growth phase of the folds. The thin solid line shows the percentage of resistant rocks exposed at the land surface as a function of time. Note that the decrease then increase of sediment efflux is due to this variation in erodibility of outcropping rocks. (b) Sediment efflux in which the variations are solely due to lithological variations in the stratigraphy being unroofed. Tectonic influx is spatially and temporally uniform. After Tucker and Slingerland (1996) (figs. 11 and 12) with permission of John Wiley & Sons Inc.

a landscape dominated by bedrock channels, the time required to reach equilibrium (90% of equilibrium sediment flux) with a given rate of tectonic uplift is proportional to the rock erodibility c_v, the uplift rate U and the spatial scale (width) of the uplifting region L,

$$\tau = k \frac{U^{\frac{1}{n}-1}L}{c_v^{1/n}}$$

(8.26)

where n is the exponent for slope, commonly assumed to be 2/3. Calibrated against model output from a landscape evolution model used to simulate the topography of the Zagros (using $c_v = 0.02$ to 0.2 m yr^{-1}, $L = 40$ km, and $U = 1$ mm yr^{-1}), the coefficient of proportionality k is $7-10$. The geomorphic response time τ for sediment flux is between

0.1 Myr and 4 Myr for the weak and resistant lithologies respectively. For the resistant lithologies, the geomorphic response time is very long (greater than 1 Myr). In this case, it is most likely that periods of different thrust displacement rate in the fold-thrust belt would not be discernible in the stratigraphic record of neighbouring foreland and thrust-sheet-top basins. On the other hand, small thrust-related anticlines composed of weak lithologies should respond quickly to changes in tectonic boundary conditions, and produce a recognisable pulse of sediment in the basin.

The list of response times to climatic and tectonic perturbations listed in Table 8.1 is a summary of the results of the analysis given in Section 8.5 and is far from exhaustive.

8.6 Coupling of Tectonics and Surface Processes

Paul Hoffman and John Grotzinger wrote in Hoffman and Grotzinger (1993) (p.198):

Savor the irony should those orogens most alluring to hard-rock geologists owe their metamorphic muscles to the drumbeat of tiny raindrops.

In this visionary sentence the authors draw attention to the surprising connection, or two-way coupling, between the mighty forces of tectonics and the seemingly insignificant effect of erosion by drops of rain. Yet that irony has become a reality, as the coupling between tectonics and surface processes becomes ever more evident at a range of temporal and spatial scales.

Sediment routing systems are superimposed on a mobile crustal template subject to horizontal and vertical stresses and deformations, caused by the relative motion of plates and the underlying mantle circulation. These deformations are manifested in patterns of net uplift of source regions and net subsidence of sedimentary basins, as well as in horizontal translations of orogenic loads and their associated flexural basins and bulges (Allen and Allen, 2013). Tectonics therefore plays an important role in providing forcing mechanisms and boundary conditions for the operation of sediment routing systems (Castelltort, Whittaker, and Verges, 2015).

The interaction of sediment routing systems and tectonics takes place at a wide range of scales. The formation and dispersal of supercontinental assemblies takes place at the very long timescale of $\sim 10^8$ years. At an intermediate spatial and temporal scale (10^0-10^1 Myr), below that of the formation of supercontinental assemblies and the evolution of plate margins, tectonic processes are highly influential in causing uplift of sediment source areas, in guiding sediment transport pathways and in driving subsidence in sedimentary basins. This interaction between active tectonic structures, topography and sediment routing systems is particularly well illustrated by extensional fault arrays, as in the Basin and Range province of the United States and the Aegean domain of Greece and Turkey, and by contractional fold-thrust belts such as the Zagros province of Iran and the Tian Shan of central Asia. Gravity-driven submarine fold-thrust arrays also deflect turbidite pathways travelling down the continental slope and cause a ponding of their deposits. Sediment routing is also affected by discrete movement on individual faults and the ground shaking of individual earthquakes, which may cause landslide damming of mountain valleys (Li et al., 2014), the avulsion of rivers, and the triggering of massive submarine gravity-driven slides, slumps, debris flows and turbidites (Talling et al., 2014).

An exhaustive bibliography exists on the role of tectonics in constructing submarine and subaerial topography, the formation and evolution of sedimentary basins, the interaction of tectonically active structures with sediment transport and the chemical composition of sediments dispersed from eroding landscapes. Summaries and reviews are found in Dickinson and Suczek (1979), Bull (1984), Ingersoll (1988), Weltje and von Eynatten (2004), Najman (2005), Garzanti et al. (2007), Burbank and Anderson (2001), Busby and Pérez (2011), Leeder (2011), Weltje (2012), Whittaker (2012), Allen and Allen (2013), Allen et al. (2015a) and Castelltort et al. (2015), in which details of more specialist literature can be found.

8.6.1 *Orogenic Wedges and Foreland Basin Systems*

Many of the world's most vigorous sediment routing systems are found in convergent mountain belts and their adjacent foreland basins. Relatively young convergent mountain belts comprise the most dramatic topographic features on the surface of the Earth, but older mountain belts are commonly still recognised topographically. The fact that old Palaeozoic orogenic belts, such as the Appalachians and Urals, still exist as elevated topography today suggests that mountain belts may persist over very long timescales. The rate of lowering of a mountain belt at mean elevation H following the cessation of active uplift (Ahnert, 1970; Pelletier, 2004) can be expressed

$$\frac{\partial H}{\partial t} = -\frac{1}{\tau_d} H \left(\frac{\rho_c}{\rho_m - \rho_c} \right) \tag{8.27}$$

where τ_d is the characteristic timescale for the decay of mountain topography, and the density terms in brackets (for crust and mantle) incorporate the effects of Airy isostasy. If the rate of erosion (m yr^{-1}) scales linearly on the mean elevation (m) by a coefficient equal to 0.61×10^{-7} (Pinet and Souriau, 1988), τ_d becomes 70 Myr, which is substantially smaller than the age of some ancient orogens. Baldwin, Whipple, and Tucker (2003) and Pelletier (2004) suggested that the presence of extensive piedmonts, whether composed of bedrock or alluvium, could extend the timescale of decay of mountain belt topography.

Of greater interest in the present context is the coupling of tectonics and surface processes in currently active orogens such as the Himalayas and Southern Alps of New Zealand. At the large scale, mountain belt landscape is perturbed by underlying tectonic fluxes of rock driven by continental convergence. There is therefore a balance between the rate of erosional unroofing and the flux towards the surface through tectonic advection. Willett (1999) expressed this balance by a dimensionless 'erosion number' N_e

$$N_e = \frac{kL}{U} \tag{8.28}$$

where U is the tectonic uplift rate of rocks, L is the half-width of the uplifting piece of crust, and k is a coefficient proportional to the bedrock incision efficiency and precipitation, with units of $[T]^{-1}$. If N_e tends to infinity, erosion ruthlessly planes off the topography despite high rates of tectonic uplift. If N_e is small (less than 1), erosion weakly offsets tectonic uplift rates, causing lower rates of exhumation. The erosion number therefore strongly controls the time taken for a mountain belt to achieve equilibrium with prevailing tectonic fluxes. If the timescale for the attainment of steady-state topography is t then its dimensionless form can be written

$$t^* = tV_p/h \tag{8.29}$$

where V_p is the plate convergence velocity and h is crustal thickness. Numerical experiments (Willett, 1999) indicate that $t^* < 2$ to ≈ 12 for erosion numbers of 10 to 2 respectively. For typical values of the convergence velocity V_p ($0.005-0.01$ m yr^{-1}) and crustal thickness h ($30-35$ km), this timescale is of order $10^6 - 10^7$ years. Such a timescale is important when considering the response of mountain belts to changes in tectonic uplift rates in the form of changes in sediment and solute effluxes.

Foreland basin systems typically take the form of an asymmetric basin (foredeep) close to the load and a broad, low-amplitude, uplifted forebulge far from the load. In the simplest terms, foreland basins develop at the front of active thrust belts where the thrust transport direction is towards the evolving basin. Foreland basins located on the subducting, lower plate are known as pro-foreland basins, whereas those located on the upper overthrusting plate are known as retro-foreland basins (Sinclair, 2012). Since the thrust load is inherently mobile, the foreland basin itself becomes involved in the deformation. To what extent the basin becomes deformed at its orogenic margin, thoroughly dissected or completely detached depends on a number of variables including the propagation rate of the thrust tips, availability of subsurface easy-slip horizons underlying the basin and the angle of convergence. The foreland basin system (DeCelles and Giles, 1996) contains four depositional zones: (1) basins that rest on moving thrust sheets as a thrust-sheet-top (Ori and Friend, 1984), wedge-top or piggyback basin, which receive sediment principally from the eroding orogenic wedge with a minor amount of backshedding from frontal anticlines; (2) basins ahead of the active thrust system in a broad foredeep, which is supplied with sediment from both the continental foreland and the orogenic wedge; (3) sediment may also accumulate on the flexural forebulge if accommodation is available, for example because the foreland lithosphere is submerged below sea level as a result of negative dynamic topography (Allen and Allen, 2013), and (4) a shallow, broad back-bulge basin filled with shallow marine and continental sediments; ongoing convergence should cause the back-bulge depozone to be uplifted and eroded in the flexural forebulge as the orogen advances over the foreland plate.

Individual foreland basins may contain examples of these four depozones, but the importance of particular depozones varies according to the geodynamics of each case. In the Tertiary North Alpine Foreland Basin (NAFB) of Switzerland and eastern France, the foreland basin system was dominated by a well-developed foredeep and minor wedge-top basins (Allen, Crampton, and Sinclair, 1991; Sinclair, 1997). Post-depositional tectonics (late Miocene-Pliocene) detached the entire basin in western Switzerland, using Triassic salt as an easy slip horizon, as deformation progressed into the Jura province (Homewood, Allen, and Williams, 1986; Burkhard and Sommaruga, 1998). In other settings, wedge-top deposition was particularly important in basin evolution, as in the linked system of inner wedge-top basins and outer foredeeps in the Apenninic chain of Italy (Ricci Lucchi, 1986; Artoni, 2007) (Figure 8.26). The Apenninic foreland basin depocentre has now extended into the Adriatic Sea where penecontemporaneous thrust deformations have produced submarine structural culminations (Ori, Roveri, and Valloni, 1986).

Tectonics have a primary influence on sediment dispersal patterns. Uplifting thrust fronts may not act as major sediment suppliers but may instead form barriers or guides to basinward sediment transport (Ramos, Busquets, and Vergés, 2002; Clevis et al., 2004a; Vergés, 2007; Barrier et al., 2010) (Figures 8.27, 8.28). The preservation of wedge-top basins depends on the balance between tectonic uplift of the thrust wedge V_{vert} and the flexural subsidence due to orogenic loading $w(x)$ (Clevis et al., 2004a). The balance is expressed by

$$\sigma(x) = V_{vert} - w(x) \tag{8.30}$$

where the vertical uplift velocity is a function of the horizontal velocity and thrust ramp angle θ

$$V_{vert} - V_{horiz}\tan\theta \tag{8.31}$$

If the horizontal velocity is 10 mm yr^{-1}, the angle of the frontal ramp is 20° and the angle of the thrust flat under the wedge-top basin is 2°, the vertical uplift rates are 3.64 and 0.35 mm yr^{-1} for the frontal hangingwall anticline and basin respectively. Assuming the flexural subsidence varies laterally, with $w(x) = 0.5$ mm yr^{-1} under the frontal ramp and $w(x) = 0.2$ mm yr^{-1} under the thrust flat, there is a net uplift of the frontal anticline ($\sigma = 3.44$ mm yr^{-1}) and net subsidence of the wedge-top basin ($\sigma = 0.15$ mm yr^{-1}). This spatial variation of uplift rate of rocks results in topography upon which surface processes act.

Structural topography is commonly provided by lateral or oblique ramps, as in the southern Pyrenees. The apices of major fluviatile systems in the southern Pyrenees are located at structural lows or re-entrants in the thrust front, whereas small, locally developed fans with highly restricted drainage basins typify the structural salients in the thrust front (Hirst and Nichols, 1986; Barrier et al., 2010). This same bimodal picture of small, locally sourced fans coexisting with large megafans sourced from deep within the orogen is also found in the Andes (Horton and DeCelles, 1997) and Himalayas (Gupta, 1997).

Foreland basins undergo a gross stratigraphic evolution related to the geodynamical controls on subsidence and sediment supply (Sinclair et al., 1991). The oldest deposits of foreland basins, referred to as 'Flysch', are commonly predominantly fine-grained, often turbiditic sediments that accumulated in sub-shelf water depths, which pass distally into shallow water carbonates deposited close to the flexural forebulge (Dorobek, 1995; Sinclair, 1997; Allen et al., 2001). The younger deposits of foreland basins are, in contrast, predominantly shallow-water or continental and typify the term 'Molasse'. In the case of

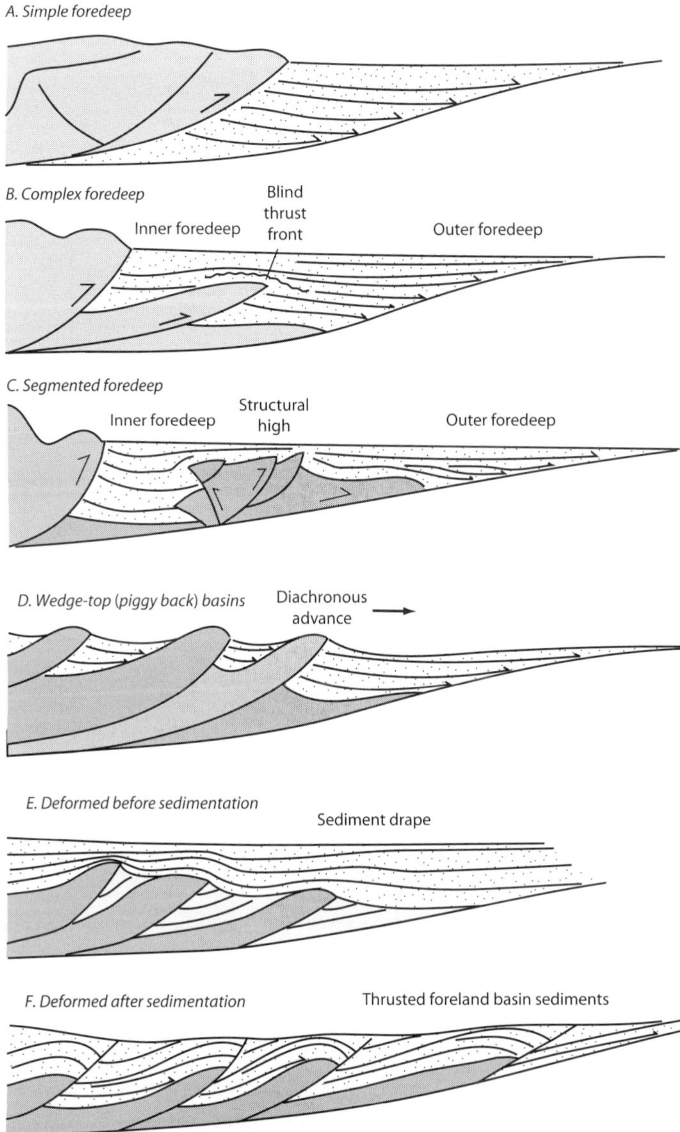

Figure 8.26 Foreland basin/thrust belt interactions, based mainly on seismic records from the Apennines and foreland of Italy, modified from Ricci Lucchi (1986). Basins may be simple, asymmetrical wedges with stratigraphic onlap onto the foreland plate, as in the North Alpine Foreland Basin of Switzerland (a). Basins may be complex as a result of tectonic advance into the basin, causing an inner foredeep separated from an outer foredeep by buried (blind) thrusts (b) or by more complex zones of faulting (c). The wedge-top may be disconnected from the foredeep, consisting of an array of wedge-top basins separated by thrust culminations (d). Some basins are draped over a previously deformed substrate (e), whereas others are dissected and shortened after the period of foreland basin deposition (f). From Allen and Allen (2013) (fig.4.36) with permission of John Wiley & Sons Inc.

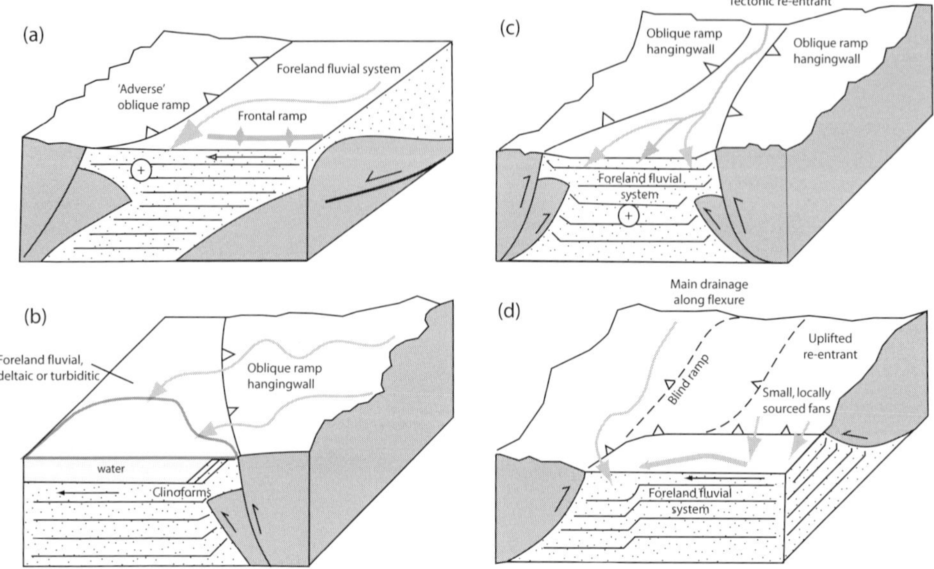

Figure 8.27 Sediment dispersal patterns by rivers at mountain fronts, based on Pyrenean examples. (a) River guided by frontal ramp flows axially, and breaks through into outer foreland basin when faced with adverse topographic slope of oblique ramp hangingwall. (b) River originating on oblique ramp hangingwall flows down topography into fluvial system or marine embayment. (c) River system constrained by tectonic re-entrant between two oppositely verging oblique ramps, flows transversely into foreland fluvial system. (d) River flows down flexure associated with major oblique ramp, whereas smaller systems drain small catchments on thrust-related hangingwalls and turn to flow longitudinally. Adapted from Vergés (2007) (fig.1) with permission of John Wiley & Sons Inc.

the Western Alps, during the underfilled stage the topography of the orogenic wedge was most likely relatively subdued, sediment delivery rates were low and the orogenic advance rate relatively high, which collectively caused deep-water conditions in the foreland basin (Sinclair and Allen, 1992). After the mountain belt had grown to a steady state, rapid erosion counterbalanced tectonic uplift, the advance rate was slow, and the basin was filled to the spill point with detritus. During this phase any excess sediment was removed from the foreland basin by fluvial and/or shallow marine processes. Present-day analogues are the overfilling and sediment export to the ocean of the Indo-Gangetic foreland basin and the Po basin of northern Italy.

At a smaller scale, tectonics and erosion are coupled in the case of an uplifting contractional fold or an array of such features. Simpson (2004a) and Simpson (2004b) suggested that the cutting of canyons by rivers through doubly plunging anticlines at their highest structural and topographic position, as is classically seen in the Zagros (Oberlander, 1985) and Apennine (Alvarez, 1999) belts, is a strong indicator of the local (10 − 20 km) feedback between erosional unloading and crustal deformation. The prerequisite is that the crust is already deforming plastically under regional compressive deviatoric stresses.

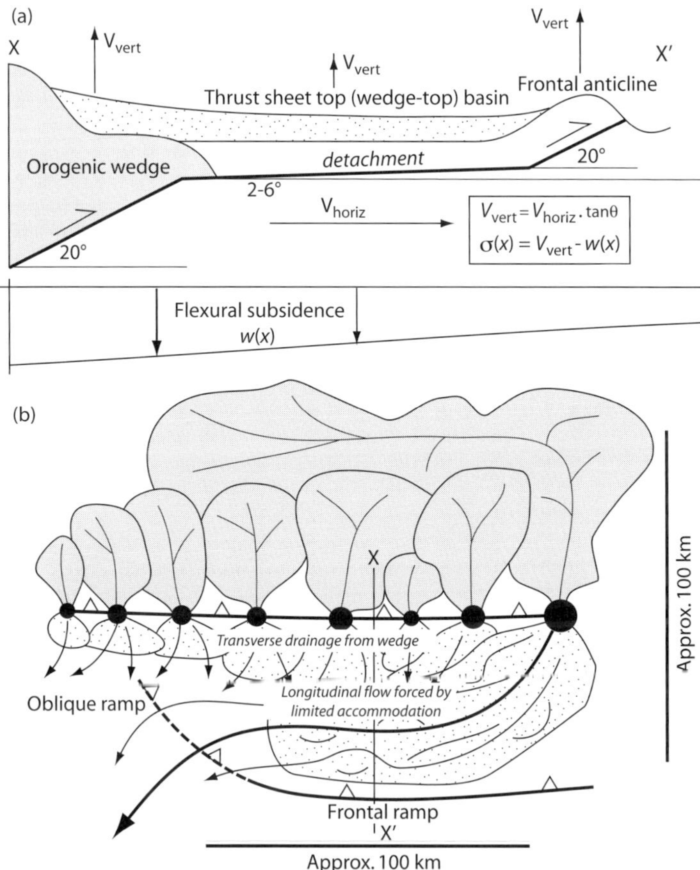

Figure 8.28 (a) Geometry of a wedge-top basin showing vertical and horizontal velocities. (b) Plan view, showing small fans supplied by transverse catchments, and a larger system deflected and guided by the growth of an anticline over the frontal ramp. Part (a) after Clevis et al. (2004a) (fig.3) with permission of John Wiley & Sons Inc.

8.6.2 Coupled Tectonic-Erosion Models of Foreland Basin Systems

Scaled analogue experiments using sandboxes indicate that erosion plays an important role in controlling structural style, localisation of deformation and exhumation trajectories in orogens and fold-thrust belts (Konstantinovskaya and Malavieille, 2005; Bonnet, Malavieille, and Mosar, 2008). Numerical models confirm that coupling may take place at a range of spatial and temporal scales (Willett, 1999; Simpson, 2006a).

Simpson (2006c) introduced a dimensionless parameter $\tilde{\kappa}$ reflecting the efficiency of surface processes in the face of tectonic deformation given by

$$\tilde{\kappa} = \frac{\kappa}{L^2 \dot{\epsilon}} \tag{8.32}$$

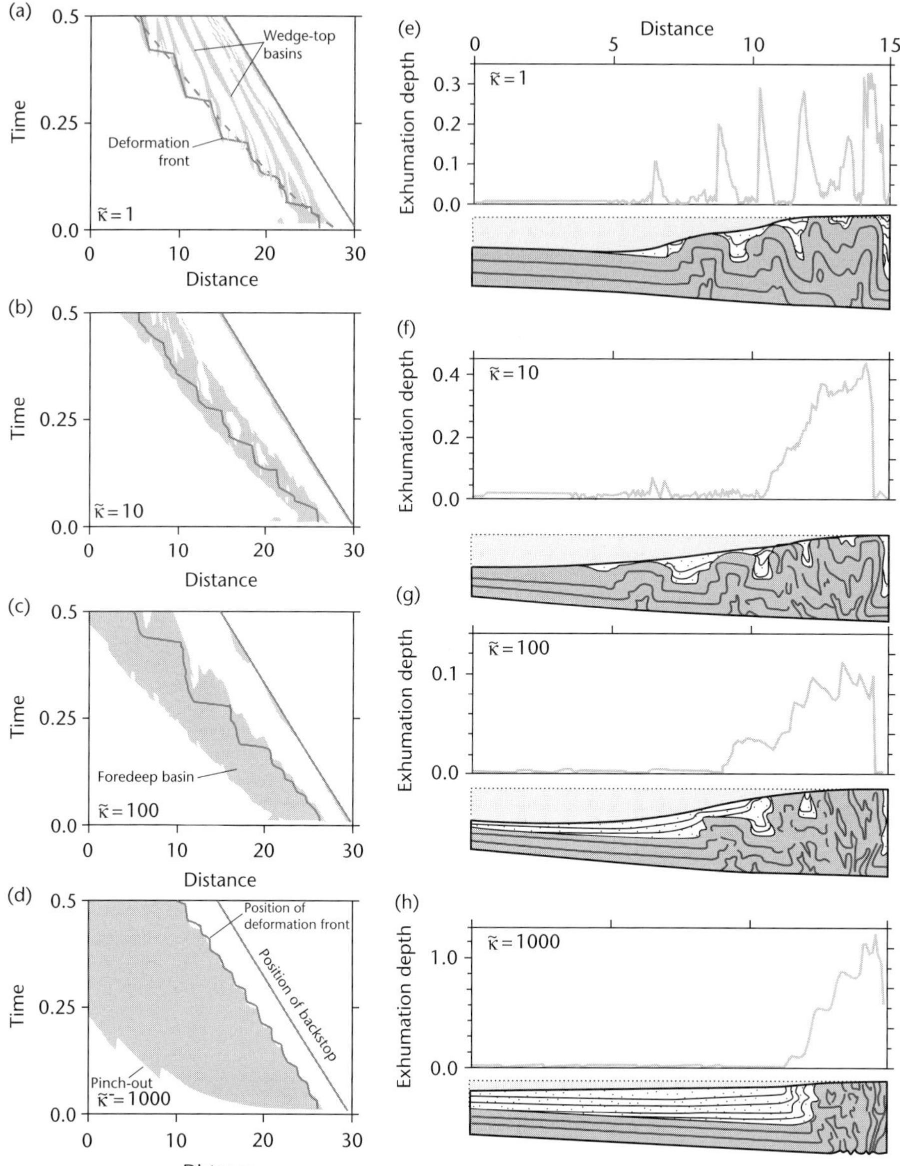

Figure 8.29 Effect of surface processes on orogenic wedge and foreland basin evolution. Left: Time-distance charts showing chronostratigraphy at different values of the dimensionless transport coefficient $\tilde{\kappa}$, from 1 to 1,000. Wedge-top deposition at low values of $\tilde{\kappa}$ evolves into extensive foredeep deposition at high $\tilde{\kappa}$. The timescale is dimensionless, and the horizontal distance is scaled by the original thickness of the deforming layer. Right: shallow exhumation of arrays of folds at low values of $\tilde{\kappa}$ evolves to deep exhumation of wedge at high $\tilde{\kappa}$. The exhumation depth and the horizontal distance are both scaled by the original thickness of the deforming layer. After Simpson (2006c) (fig.5) with permission of John Wiley & Sons Inc.

where κ is the surface process diffusivity, L is the length scale of the initial thickness of the elasto-visco-plastic layer in the model lithosphere, $\dot{\epsilon}$ is the initial imposed strain rate. The non-dimensional diffusivity $\tilde{\kappa}$ is the ratio of the deformation timescale ($1/\dot{\epsilon}$) to the timescale for surface mass transport (L^2/κ). Low values of $\tilde{\kappa}$, where tectonic deformation is essentially unaffected by surface processes, promote the formation of an array of thrust-related folds defining regularly spaced wedge-top basins, with exhumation peaking at each fold culmination. At very high values of $\tilde{\kappa}$, an extensive foredeep is formed, with no wedge-top deposition. Exhumation is concentrated in an intense region of shortening at the rear of the wedge (Figure 8.29). Transitional states exist between these two extremes. Values of $\tilde{\kappa}$ are strongly affected by changes in erosion rate at the surface of the orogenic wedge, or by the emergence of the orogen above sea level (Simpson, 2006b).

Wedge rheology and the strength of the basal detachment may play a primary role in determining the geometry of thrust belt deformation and basin development (Ford, 2004). The proportion of wedge-top deposition to foredeep deposition has been linked to the coupling between a deforming elasto-visco-plastic fold-thrust belt, flexural subsidence and diffusional surface processes (Simpson, 2010) (Figure 8.30). Depending on the strength of the basal detachment, two end members of foreland basin system are apparent:

1. When the basal detachment is strong, the foreland basin system is associated with a highly asymmetrical orogen with thrusts concentrated in the pro-wedge. Sedimentation on the retro-side occurs in one major, simple, foredeep basin that grows throughout orogenic evolution, whereas deposition on the pro-side begins in a foredeep but transfers to wedge-top basins that eventually are advected backwards, uplifted and exhumed.
2. When the basal detachment is weak, as with the presence of salt, the foreland basin system is associated with a broad, low-relief orogen showing little preferential vergence and a predominance of folding relative to faulting. Deposition is complex, mostly taking place in wedge-top basins, often displaying growth strata, and is discontinuous as the locus of active deformation jumps from structure to structure.

Natural examples of foreland basin systems fall within the range between the two end-members described previously (Figure 8.31). The Central Himalayan basin and the eastern sector of the North Alpine Foreland Basin formed adjacent to thrust belts with strong detachments. More complex foreland basins such as those in the northern Apennines and Zagros have a high proportion of wedge-top basins and are associated with thrust belts with weak basal detachments composed of salt. Wedges formed above weak detachments have low surface slopes, such as the Jura of the western sector of the North Alpine Foreland Basin, the northern Apennines and the Salt Range of Pakistan. In contrast, strong detachments are associated with high surface slopes, such as the Central Himalaya, Taiwan and the eastern sector of the North Alpine Foreland Basin. Wedge rheology therefore has a profound effect on basin development and topography, and thereby influences the sediment routing systems of mountain belts. Allen et al. (2013) illustrated the sediment routing systems in the North Alpine, south Pyrenean and northern Apennine systems as examples within this range of possible wedge rheologies.

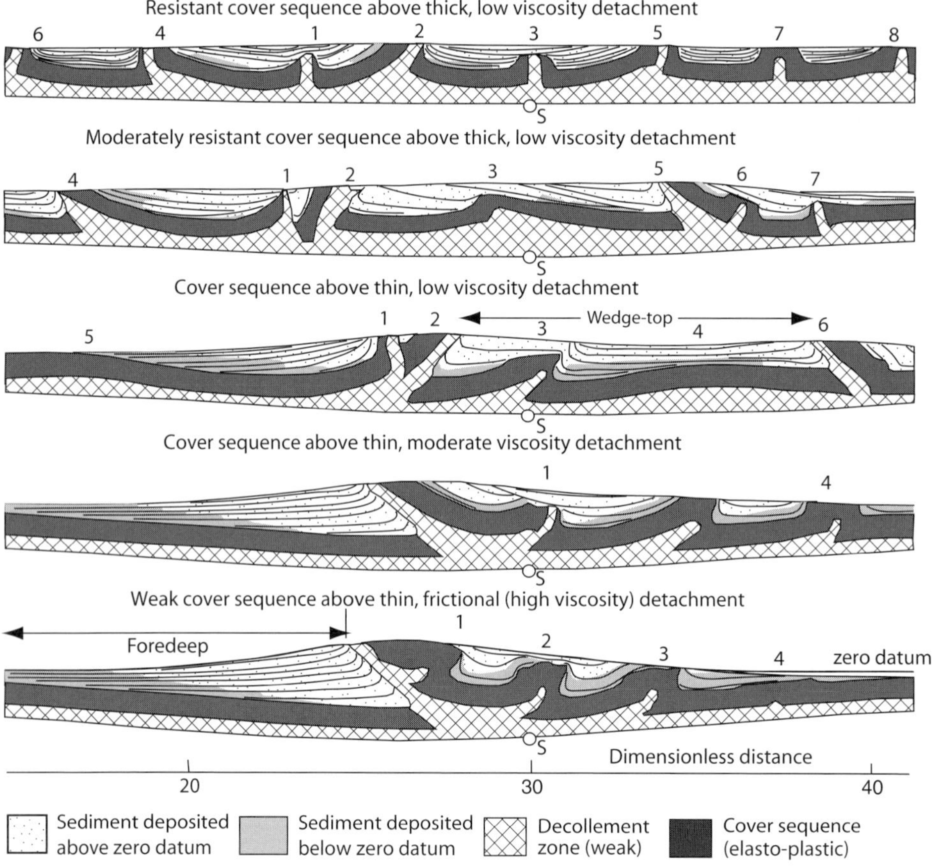

Figure 8.30 Comparison of results of numerical experiments with different mechanical behaviour of the wedge after 30% shortening. Stratigraphy is shown in terms of altitude of deposition below and above the zero datum (initial surface). Sediments are commonly deposited above the zero datum in wedge-top basins and in the proximal parts of foredeeps. Horizontal scale is a normalised distance based on the initial total thickness above the base of the detachment. Numerals show sequence in which structures developed. From Simpson (2010) (fig.9) with permission of John Wiley & Sons Inc.

8.6.3 Arrays of Contractional Folds

Erosion and deposition influence the deforming plate by modifying the distribution of vertical surface loads and by changing the plate thickness and therefore the strength distribution. The interaction between tectonic deformation and surface processes is particularly evident at the scale of contractional folds (wavelength $\sim$10 km, height 1–3 km). The strength of tectonic deformation relative to fluvial erosion can be indexed by the dimensionless ratio denoted R (Simpson, 2004a) (Figure 8.32)

$$R = \frac{c\alpha^{n}L^{n-1}}{\dot{v}} = \frac{\tau_d}{\tau_e} \tag{8.33}$$

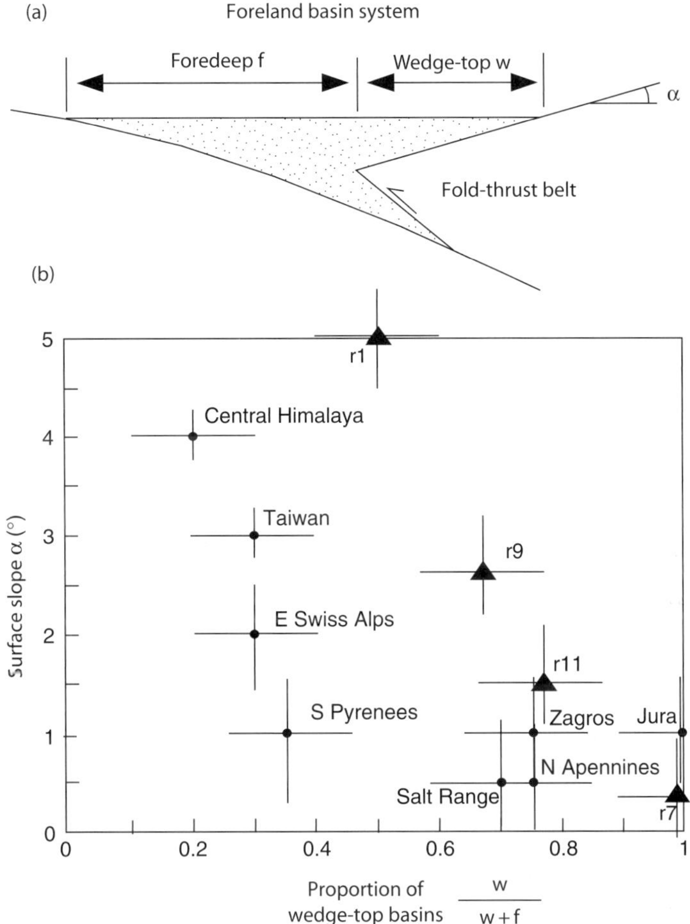

Figure 8.31 Estimates of the wedge surface slope α and the proportion of wedge-top sediments in the adjacent foreland basin system. Natural examples of foreland basins from various sources cited in Simpson (2010). In addition, results are shown from four numerical simulations (r1, r7, r9, r11) of Simpson (2010). From Simpson (2010) (fig.11) with permission of John Wiley & Sons Inc.

where L is the characteristic length scale, $\dot{v}$ is the imposed boundary displacement rate, the deformation timescale is $\tau_d = L/\dot{v}$, and the fluvial erosion timescale $\tau_e = L^{2-n}/(c\alpha^n)$, where α is the rainfall rate in excess of infiltration. Consequently, when R is large, fluvial processes are fast relative to the rate of deformation, and the reverse is true where R is small.

Simpson (2004b) also varied the initial regional topographic slope (denoted by β) between 0% and 2% (Figure 8.32). His numerical model indicates that when β and/or R are low (that is, the initial regional topography is flat and/or relative rates of deformation compared to fluvial processes are low), folding is two-dimensional with small-scale

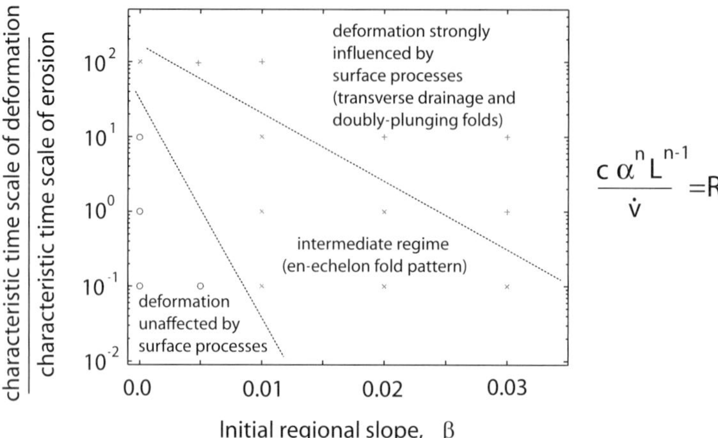

$$\frac{c\,\alpha^n\,L^{n-1}}{\dot{v}} = R$$

Figure 8.32 Phase diagram showing how the ability of surface processes to influence tectonic deformation depends on two critical parameters: the initial regional slope β and the ratio of the characteristic timescale for the imposed deformation to the characteristic timescale of fluvial erosion R. Surface processes exert no influence on folding when β and R are relatively low, whereas they strongly amplify and localise deformation when β and R are relatively large. For intermediate values, complex *en echelon* fold patterns are observed. After Simpson (2004b) (fig.8) with permission of American Geophysical Union.

drainage networks transporting sediment locally from anticlines to adjacent basins. This may be characteristic of the Zagros, for example, When β and/or R are relatively high, a large-scale transverse river network cuts across growing structures and strongly influences tectonic deformation, causing amplification and localisation of folding. The largest transverse rivers are therefore associated with the greatest structural and topographic relief of doubly plunging folds.

The interaction of tectonic deformation and surface processes is ideally studied where deformation propagates into the margins of foreland basins, as in the Junggar Basin of north-central China (Fu et al., 2003). The thrust-related anticlines are cut through by rivers, forming transverse canyons, as is commonly observed in fold-thrust belts from as far afield as the Apennines, Zagros, Pyrenees, Swiss and French Alps, Himalayas and central Andes (Oberlander, 1985; Alvarez, 1999; Simpson, 2004b). Some streams cut transversely through folds, whereas others are deflected around their tips. For example, in the Marche region of the Italian Apennines, a series of rivers drain to the Adriatic coast by cutting straight through the points of maximum tectonic uplift in NW-SE oriented folds (Alvarez, 1999). The impact of growing tectonic structures on drainage patterns has been discussed by a number of authors (Jolley et al., 1990; Burbank and Vergés, 1994; Talling et al., 1995; Burbank, Beck, and Mulder, 1996; Gupta, 1997; Vergés, 2007; Barrier et al., 2010). One of the major impacts for the analysis of sediment routing systems is that sediment entry points into basins may be shifted by growing tectonic structures, causing heterogeneity in basin stratigraphic architecture.

The simplest approach to understanding the interaction between growing folds and surface processes is to consider a river that is in equilibrium between channel incision and tectonic uplift rates. For bedrock streams, a stream power rule for the case of equilibrium is

$$\frac{U}{c_v} = \left[\frac{Q_w}{Q_*} \right]^m S^n \tag{8.34}$$

where Q_* is a characteristic channel discharge (equal to the total area of the catchment times the average precipitation rate), Q_w is the water discharge, S is the slope, c_v is the efficiency of bedrock incision expressed as a velocity and U is the tectonic uplift velocity (Tucker and Slingerland, 1996). If the uplift rate of rocks in the fold crest region is high ($U = 0.01$ m yr^{-1}) and the stratigraphy is strongly resistant ($c_v = 0.025$ m yr^{-1}), the dimensionless ratio U/c_v is high (0.4), indicating that a stream is likely to be deflected. If, however, the uplift rate of rocks in the fold crest region is low ($U = 0.001$ m yr^{-1}) and the rocks are weakly resistant to erosion ($c_v = 0.25$ m yr^{-1}), U/c_v is low (0.004), indicating that the discharge-slope product of the stream may be sufficient to cut through the growing anticline.

8.7 Transformation of Signals in Sediment Routing Systems

The morphology of landscapes and the stratigraphy of sedimentary basins preserve the record of the action of sediment routing systems. But how reliable are these geomorphic and stratigraphic products for inversion of their environmental drivers? Environmental signals are propagated through sediment routing systems at a range of time and space scales and are changed in the process (Figure 8.33) (Table 8.2). The nature of the transformation of environmental signals depends critically on their timescale (Romans et al., 2015) (Figure 8.34). Events recorded at the historical timescale include gravity waves, mass flows (turbidites, debris flows, landslides), earthquakes, river floods and tides. They are recorded by direct measurement, from historical accounts and by the use of very short half-life isotopic systems. Fundamental climatic (rather than meteorological) forcings of the sub-Milankovitch and Milankovitch wave band occur at the intermediate timescale ($10^2 - 10^6$ yr). These forcings are dated by cosmogenic nuclides, some low-temperature thermochronometry such as (U-Th)/He and some radioisotopes such as ^{14}C. Most tectonic drivers occur at the deep time temporal scale, including the linkage of faults, the growth of mountain belts, subsidence in sedimentary basins and plate tectonics. Processes are measured using long half-life isotopic systems such as U-Pb and by apatite and zircon fission track analysis.

Modelling studies suggest that some catchments may be reactive to single step changes of precipitation but act to buffer long-term cyclic forcings in the Milankovitch band (Armitage et al., 2013). Other models indicate that landscapes may possess a resonance frequency that causes sediment discharges to be amplified when subject to cyclic climatic forcing (Godard et al., 2013). Rivers may respond faster to changes in water discharge

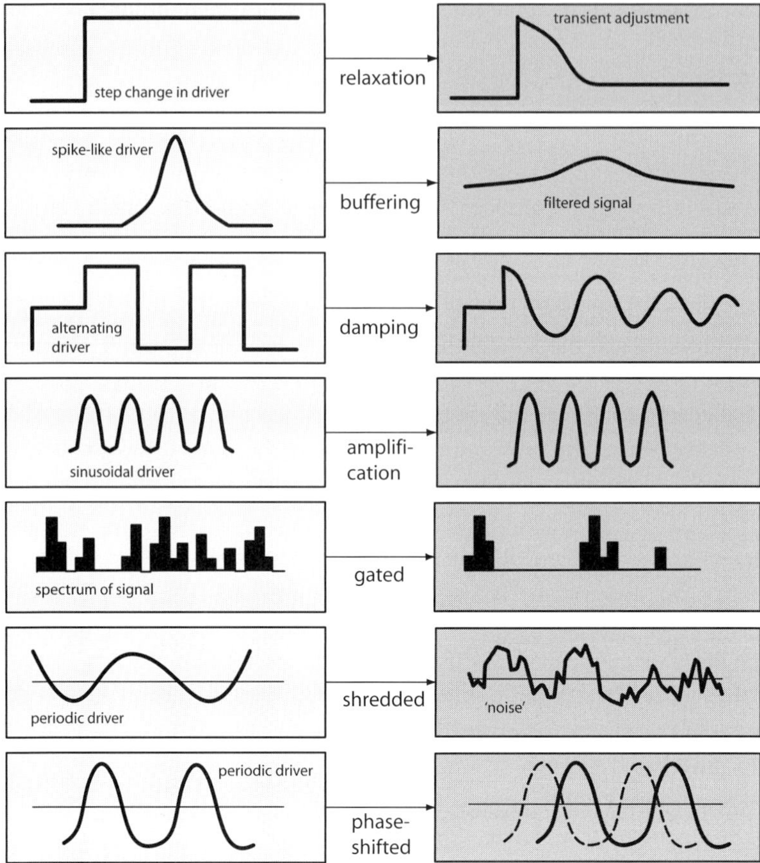

Figure 8.33 Schematic diagram illustrating various forms of signal transformation applicable to sediment routing systems. Modified from Allen and Allen (2013) (fig.8.37) with permission of John Wiley & Sons Inc.

than to changes in upstream sediment supply (Van den Berg van Saparoea and Postma, 2008; Simpson and Castelltort, 2012). Consequently, high-frequency cyclicity discernible in delta-shelf stratigraphy may be driven by climate variations, whereas low-frequency patterns may result from changes in sediment supply driven by tectonics.

Climate or tectonic-induced signals originating in a catchment may be masked or 'shredded' by the autogenic dynamics of the river-floodplain system (Jerolmack and Paola, 2010). The separation of autogenic effects from external environmental forcings is problematical. Ganti, Lamb, and McElroy (2014) suggested that the scale over which autogenic processes operate is given by

$$l_a = uh_s/w_s \tag{8.35}$$

Table 8.2 *Signal transformations in sediment routing systems.*

Transformation	Example	References
Relaxation	Response of catchment-fan to step change in climate or slip rate	Allen and Densmore (2000) Densmore, Allen, and Simpson (2007a)
Buffering	Loss of periodic signal by propagation through alluvial valley	Métivier and Gaudemer (1999) Castelltort & v.d. Driessche (2003)
Damping	Reduction in amplitude of precipitation-driven oscillations with Milankovitch periodicities	Armitage et al. (2013)
Amplification	Increase in amplitude of sediment flux driven by climatic oscillations	Simpson and Castelltort (2012) Godard et al. (2013)
Gating	Selection of record by preservation potential	Allen and Allen (2013)
Shredding	Masking or loss of signal by autogenic dynamics	Jerolmack and Paola (2010) Want, Straub, and Hajek (2011) Ganti, Lamb, and McElroy (2014)
Phase-shifting	Moving boundaries out of phase with forcing	Marr et al. (2000)

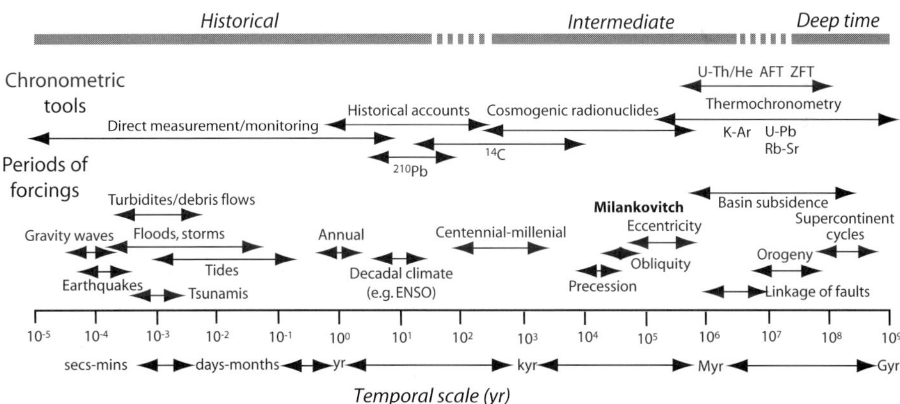

Figure 8.34 Summary of typical timescales of events and cycles relevant to sediment routing systems, together with the tools used to date events and cycles. Modified from Romans et al. (2015) (fig.2) with permission of Elsevier.

where l_a is the advection length of sediment transport before settling out takes place, u is the flow velocity, h_s is the average settling height and w_s is the settling velocity. Autogenic feedbacks operate when the appropriate length scale, for example, system size, is greater than l_a.

Autogenic dynamics are also dependent on the roughness scale l_r of the sediment routing system (Want, Straub, and Hajek, 2011). The timescale of stacking of sedimentary deposits caused by the infilling of negative roughness T_c is given by

$$T_c = l_r/S \qquad (8.36)$$

where T_c is termed the compensation timescale, and S is the basin-wide long-term sediment accumulation rate. If the roughness length is the mean channel depth of a river such as the Lower Mississippi, and the sediment accumulation rate is 0.26 mm yr^{-1} (Straub et al., 2009), the compensation timescale is 115 kyr. Since the rate of recurrence of avulsions is much smaller than T_c, it can be argued that they are part of the autogenic compensational stacking of stratigraphy in the Lower Mississippi valley.

Part IV

The Stratigraphic Record of Sediment Routing Systems

9

Sediment Production, Evolution and Provenance

9.1 The Formation of Sediment

There is a complex web of processes affecting the generation of sediment from bedrock weathering, through transport and intermittent storage, to eventual long-term deposition in a sedimentary sink. This web of processes imparts a certain textural and compositional imprint on the final deposit, yet joined-up approaches to the pathway from source to sink are rare (Ibbeken and Schleyer, 1991). Gert Jan Weltje wrote in Weltje (2012) (p.4):

Current approaches to modelling of surface processes treat the coupled evolution of source areas and sedimentary basins in terms of bulk mass transfer only, and do not take into account compositional and textural sediment properties.

In such mass balance approaches, the generation of sediment in source regions is commonly treated as a boundary condition for sediment supply, rather than as a dynamic system in its own right. Consequently, the properties of weathering products, both texturally and compositionally, entering the sediment cascade remain poorly constrained. Rare examples of where a mass balance approach is integrated with a provenance study are taken from modern and Quaternary environments (for example, Weltje and Brommer (2011)) rather than from ancient sediment routing systems. The advantage of integrating sediment generation with Earth surface process models is that the properties of sediment entering the transport network of sediment routing systems can be predicted and parameterised under different conditions of climate and tectonics, instead of being simply treated as a volumetric or solid mass boundary condition.

The evolution of sediment involves important modifications to the original parent rock (Johnsson, 1993). Such modifications might be chemical and mechanical weathering of bedrock and regolith, or the fractionation of grains during transport. Compositional variation within a sediment routing system is due to (1) mixing of grains from different sources and (2) chemical and mechanical weathering of the grain population, which may take place during regolith formation at source, or during transport and storage en route to a depositional sink (Weltje, 2012).

Figure 9.1 illustrates a simple mixing model of a fault-bounded sedimentary basin with two catchments feeding sediment to an adjacent basin. There is a lateral variation in

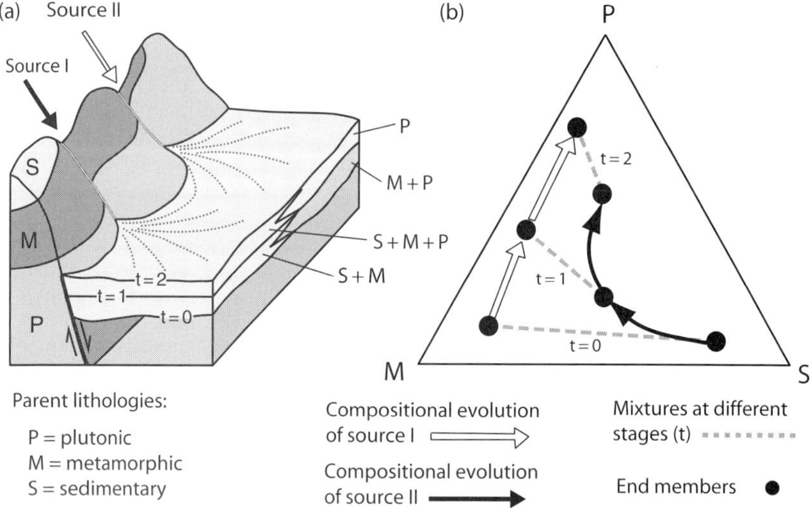

Figure 9.1 Lateral and vertical variation of sediment composition in a fault-bounded basin fed by two catchments. (a) Uplift of lithologically heterogeneous source areas results in progressive unroofing of bedrocks; (b) Compositional evolution of the sources and the range of mixtures in the basin-fill at three points in time (*t*1 to *t*3) are shown in the ternary diagram. Modified from Weltje (2012)(fig.3) with permission from Elsevier.

composition caused by the differences in the lithologies being eroded in the two source catchments and dispersed into the basin, and a vertical variation caused by progressive unroofing over time of rocks in the source catchments. Conglomerate clast compositions can be used to investigate the progressive erosion of lithologically distinct units in the source catchments (Graham et al., 1986; Steidtmann and Schmitt, 1988; Pivnik, 1990). Sediment compositions can therefore potentially be used to estimate the proportional contributions of different sediment sources to rivers or other agents of sediment transport such as the wind. The proportional contributions can be converted to absolute rates of sediment flux by using independent estimates of the total sediment discharge, for example, from measured sediment loads of rivers (Vezzoli et al., 2004; Vezzoli, 2004).

Different source area lithologies have a range of capacities for generating sediment by weathering and erosion, and different bedrock types have different capacities for breaking down into particular grain-size ranges. The Sand Generation Index (SGI), for example, quantifies the relative efficiency of sand production from different lithologies (Palomares and Arribas, 1993; Arribas and Tortosa, 2003). The SGI is a variable that reflects the relative sand-generating capability of a certain bedrock when compared to another bedrock type that is generating sand-grade sediment under the same topographic and climatic conditions. The SGI of a given bedrock type A of a dual source A+B, denoted $SGI_{A(A+B)}$, is expressed in terms of the outcrop area of bedrock type A, denoted S_A, needed to produce a sand composed of equal amounts of both A and B bedrocks:

$$\mathrm{SGI}_{A(A+B)} = [S_A + S_B]/S_A \qquad (9.1)$$

where $[S_A + S_B]$ is the total surface area of the source region (100%), and S_A and S_B are the outcrop areas of bedrocks A and B within the source region required to produce sand with the average modal composition observed. Likewise, the SGI for bedrock type B is

$$\mathrm{SGI}_{B(A+B)} = [S_A + S_B]/S_B \qquad (9.2)$$

Each rock type has a different potential to generate sand, depending on properties such as its mineralogy, average crystal size and microfabric (cleavage, fractures) (Palomares and Arribas, 1993). Studies of metamorphic and granitic terrains show that the SGI of granitoid rocks is 14–20 times the SGI of slates and schists, and the SGI of gneisses is five times that of slates and schists. In sedimentary terrains, siliciclastic sources are the most productive sedimentary source of sand in the fine, medium and coarse fractions of river sands analysed by Arribas and Tortosa (2003) from catchments in the Iberian Range, Spain (Table 9.1) (Figure 9.2). The SGI of siliciclastic sources varies from 20–4, whereas the SGI of carbonates varies from 1.0–1.3. Siliciclastic sources are therefore 3–20 times more efficient at producing medium sand-grade sediment than carbonates.

Vezzoli et al. (2004) modified the original concept of the Sand Generation Index in order to assess the relative contributions of seven end-member sources of detritus in the Dora Baltea Basin, Western Alps (Table 9.2). The seven bedrock types are shown in ternary diagrams of light fraction and heavy fraction (Figure 9.3). The Monte Bianco Massif has the highest SGI index in the Dora Baltea Basin (SGI of 16.4), followed by the Austroalpine

Table 9.1 *Sand Generation Index (SGI) values from compound sedimentary source areas calculated from the medium sand fraction of river deposits in the Iberian Range. From Arribas and Tortosa (2003)(tab.3, p.289), with permission of Elsevier.*

Compound sedimentary source	Surface area corresponding to average composition[a] (%)	SGI
Siliciclastics+carbonates		
Siliciclastics	5–5	20–4
Carbonates	95–75	1.0–1.3
Dolostones+limestones		
Limestones	18–30	5.5–3.3
Dolostones	82–70	1.2–1.4
Carbonate lithofacies		
Micritic limestones	35–45	2.8–2.2
Micritic dolostones	(100)[b]	(1)
Sparitic limestones	5–35	20–2.8
Sparitic dolostones	40–75	2.5–1.3

[a] Percentages estimated visually from diagrams such as Figure 9.2.

[b] In parentheses: unequivocal assignment of grain type to source lithology in the case of dolomicrite.

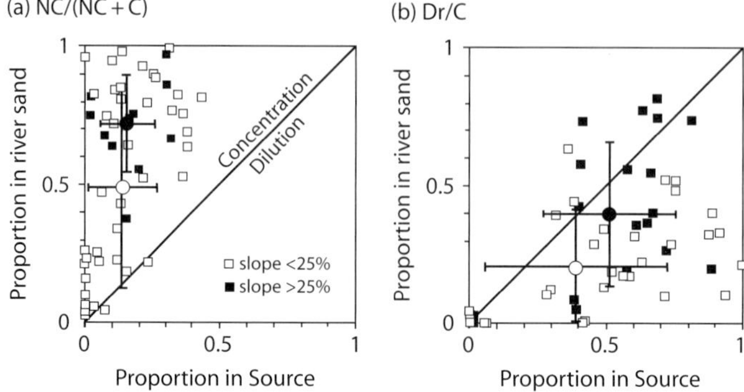

Figure 9.2 Plots of two petrographic indices versus indices of the surface exposure of lithologies in the source region for the medium sand fraction of low-order stream deposits from the Iberian Range, Spain. (a) NC, Non-carbonate (siliciclastic) rocks; C, Carbonate rocks. (b) Dr, sparitic dolostones. Means with standard deviation of each group of samples is shown. Samples are grouped according to slope. Dilution is the case where the petrographic ratio is lower in river sands than in surface area of exposure of lithologies, as in (b), showing that sparitic dolostones are under-represented in medium sand fractions. Concentration occurs where the petrographic ratio is higher in river sands, as in (a), showing that siliciclastic rocks are a highly productive source of sand. After Arribas and Tortosa (2003) (fig.9) with permission of Elsevier.

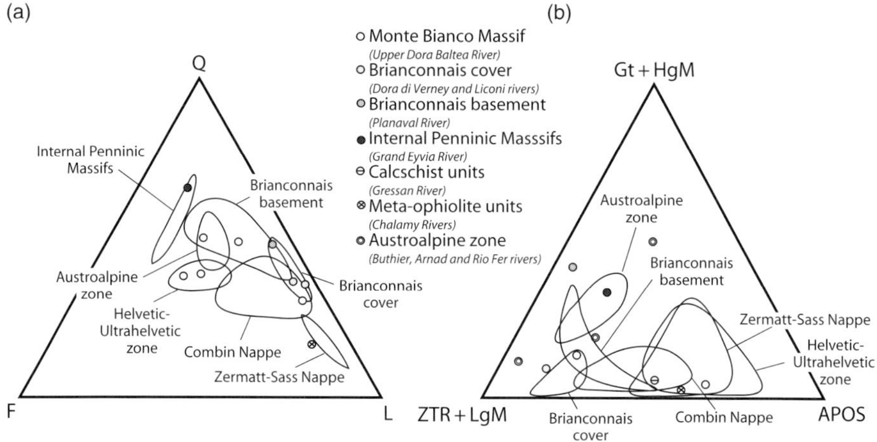

Figure 9.3 (a) Framework composition (Quartz-Feldspar-Lithics) of detritus from the end-member sources of the Dora Baltea basin compared to statistically rigorous confidence regions about the mean compositions of (mainly) Alpine domains. (b) Mineralogical signatures, with vertices representing Gt+HgM (garnet and high-grade metamorphic minerals), ZTR+LgM (ultrastables and low-grade metamorphic minerals), and APOS (amphiboles, pyroxenes, olivine and spinel). Modified from Vezzoli et al. (2004) (figs.3, 4) with permission of Elsevier.

domain (SGI of 1.4). The high capacity for the production of sand-grade sediment of the Monte Bianco Massif, despite occupying just 3% of the whole catchment, is promoted by its granitoid lithology, heavy glaciation and extreme relief - it is the highest peak in Europe at 4,810 m.a.s.l.

Table 9.2 *Contribution from various tectonic units to the bed load of rivers in the Dora Baltea Basin. From Vezzoli et al. (2004) (tab.6, p.243) with permission of Elsevier.*

Tectonic unit	Contribution based on bulk composition (%)	Contribution based on Sand Generation Index
Monte Bianco Massif	50 (2)	16.4
Briançonnais cover	1 (1)	0.1
Briançonnais basement	2 (1)	0.1
Penninic massifs	7 (2)	1
Calcschist units	1 (1)	0.1
Meta-ophiolite units	4 (2)	0.2
Austroalpine zone	35 (3)	1.4

Percentages based on bulk composition were averaged from several independent trials. Numbers in parentheses are 1σ standard deviation. Meta-ophiolites refers to Zermatt-Sass Nappe.

The total sediment flux of the Dora Baltea Basin is 1 Mt yr^{-1}, comprising 0.59 Mt yr^{-1} of bedload and 0.41 Mt yr^{-1} of suspended load. Analyses of sediment composition indicate that greater than 75% of the Dora Baltea bedload is derived from tributaries with sources in the highest granitoid peaks above 4,000 m.a.s.l., though these tributaries drain only 36% of the total catchment area (Vezzoli, 2004). The Monte Bianco Massif alone produces 0.294 ± 0.006 Mt of arkosic sands per year, representing 50% of the total bedload flux of the Dora Baltea Basin. Based on the total sediment flux and drainage basin area, the sediment yield for the entire Dora Baltea catchment is 306 t km^2 yr^{-1} and the equivalent denudation rate is 0.12 mm yr^{-1}.

Di Giulio et al. (2003) also analysed the composition of river bed sands in relation to the surface exposure of sedimentary (and minor metamorphic) lithologies in small catchments in the northern Apennines. Siliciclastic rocks in source areas are under-represented (diluted) in river sands compared to their outcropping surface area. Carbonate rocks are also under-represented. Some rock types, such as metamorphic rocks, occur in quantities of up to 20% of sand samples despite the fact that they are absent in the source area catchments, indicating that grains are *recycled* from older siliciclastic rocks (Zuffa, 1987). When the expected amount of siliciclastic rocks in the drainage area is recalculated to account for recycling of grains from older coarse-grained siliciclastic rocks, siliciclastic rocks re-plot as slightly over-represented in river sands, as anticipated from other studies (Arribas and Tortosa, 2003).

There is therefore a difference between the bulk composition of a parent rock and the composition of the grain assemblage in regolith or bedload of upland rivers. This difference increases over time as the sediment is dispersed through the sediment routing system (Weltje, Meijer, and de Boer, 1998).

The primary controls on the type of sediment released into the sediment cascade are the mineralogical composition, texture and crystal or particle size of the parent rocks.

The parent rocks are transformed into sediment by weathering, the extent of which depends on weathering rate and the residence time of material in the weathering environment. The rate of silicate weathering is fundamentally controlled by climate, specifically temperature and precipitation (White and Brantley, 1995). Tectonics provides relief, which controls the residence time for weathering by the onset of creep or mass flow of hillslope material downslope. The extent of chemical weathering can be monitored through the ratio of the abundance of quartz (Q) to feldspars (F). Preferential weathering of feldspars by acid hydrolysis causes an increase in $\log(Q/F)$ over time.

Weltje (1994) generalised the cumulative chemical weathering index of Grantham and Velbel (1988) to obtain a semi-quantitative index of the extent of weathering under different climatic and physiographic conditions. The distance in compositional space relative to the composition of parent rocks (w) is the integration of the rate of weathering V over the amount of time elapsed since weathering began t:

$$w = \int V \mathrm{d}t = \bar{V}t \tag{9.3}$$

where $\bar{V}$ is the time-averaged weathering rate. The weathering index of Weltje (1994) (I_w) relates w to present-day climate and physiography of sediment sources:

$$I_w = CR \simeq w \tag{9.4}$$

where C is a semi-quantitative indicator of climate (essentially rainfall) and R is a semi-quantitative indicator of relief. C is used as a proxy for time-averaged weathering rate and R is assumed to be inversely proportional to residence time in the source area. The weathering index I_w takes integer values between 0 and 4 in Weltje (1994) and Weltje et al. (1998), since S and R are both integers (with values of 0, 1 and 2) as shown in Table 9.3.

The weathering rate can also be expressed in the form of an exponential relationship

$$V = RSI\left(\frac{c_1 Q_i \exp(c_2/(c_3 T_a))}{H_{reg}}\right) \tag{9.5}$$

where RSI is a regolith supply index, with values of $0 - 1$, where a value of 1 indicates the availability of unlimited regolith for the ambient erosion rate, T_a is the year-average catchment temperature in $°C$, Q_i is the soil infiltration rate in mm yr^{-1}, H_{reg} is the regolith thickness in mm, and c_1 to c_3 are constants (Forzoni et al., 2014).

The extent of weathering of parent rock to produce sediment can be assessed by using the log ratios of principal framework elements. Examples are $\log(Q/F)$ and $\log(Q/R)$, where Q is quartz content, F is feldspar content and R is the rock fragment content of the sand. Simultaneous use of both logratios discriminates the parentage and weathering history of a sand. The values of both logratios correlate with I_w because quartz is more resistant to weathering than feldspar and rock fragments. The general trend of compositional evolution with increasing values of the weathering index is therefore towards high values of both $\log(Q/F)$ and $\log(Q/R)$ (Figure 9.4).

Table 9.3 *Integer values of climate (rainfall) parameter C, relief R and weathering index $I_w = CR$. After Weltje et al. (1998)(tab.1, p.134) with permission of John Wiley & Sons Inc.*

C	Climate
0	(Semi) arid and Mediterranean
1	Subhumid
2	Humid
R	Relief
0	High (mountains)
1	Moderate (hills)
2	Low (plains)
I_w	Weathering index
0	Unweathered
1	Slightly weathered
2	Moderately weathered
4	Intensely weathered

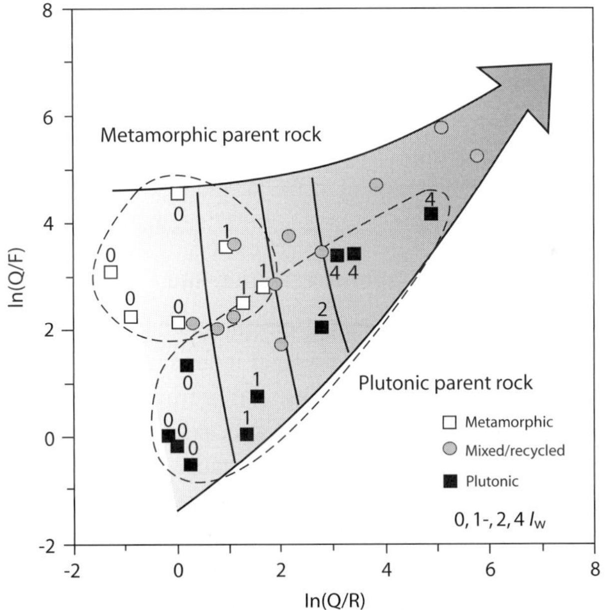

Figure 9.4 Plot of logratios for medium grained river sands, showing relationship between petrographic composition, parent rock type and weathering index I_w (proxy for climatic and physiographic conditions in catchment source areas *CR*) for two clusters of first-cycle sands derived from monolithologic catchments. Mixed/recycled samples are from Orinoco drainage basin. Modified from Weltje (1994) and Weltje et al. (1998) (fig.4) with permission of John Wiley & Sons Inc.

Changes in the rate of sediment supply can be detected in shifts in sediment grain size and composition, both of which are related to weathering history. Since weathering rates in humid-tropical conditions are fast ($\sim 10^4$ yr timescale) (Johnsson and Meade, 1990), and similar rates are found in other climatic zones (Nesbit, Fedo, and Young, 1997), high frequency variations in sediment supply are potentially attributable to changes in the extent of weathering, driven, for example, by Milankovitch-band climate change (Section 10.2). Variations in $\log(SiO_2/Al_2O_3)$ in Pliocene turbidites in Corfu (Greece) show that the extent of weathering decreases from the base to the top of each depositional lobe, indicating that weathering of sands in the subaerial part of the sediment routing system took place prior to lobe reactivation (Weltje and de Boer, 1993).

A combination of grain size and compositional characteristics in the non-biogenic portion of deep sea sediments makes it potentially possible to infer palaeoclimate in the source areas for sediment (Weltje and Prins, 2003). The grain-size distribution of a sediment sample represents a mixture of sediment populations corresponding to the different mechanisms of sediment production and sediment transport. The grain-size characteristics of a finite number of so-called dynamic populations represent a certain set of processes of sediment production and transport. Each dynamic population can be tied to a specific source and dispersal mechanism. The bulk composition or mineralogy of a sample is usually a function of the grain-size distribution, since each size fraction has a characteristic fingerprint in terms of composition. Dynamic populations are expected to be markedly different, for example, in turbidites, where transport and deposition are highly size-selective, and ice-rafted detritus, where processes of sediment production may dominate over processes of transport. Examples of palaeoclimate interpretations based on the recognition of dynamic populations in marine sediment are found in Weltje and Prins (2003).

9.2 Precipitation, Vegetation and Erosion

Climate, expressed in terms of mean annual precipitation, precipitation variability, vegetation cover, temperature or other climate-related indices, is a major control on the supply of sediment to sedimentary basins (Hooke, 2000). Climatic influences are commonly expressed in terms of precipitation or effective precipitation, as in the seminal study of Langbein and Schumm (1958). The effective precipitation is the annual rainfall that would be required, on average, to produce the observed run-off. The Langbein–Schumm curve of effective precipitation versus sediment yield shows a maximum in yield at ~ 300 mm yr^{-1} after which sediment yield decreases with increasing precipitation (Figure 9.5), which is thought to be due to the binding effect of vegetation at high values of precipitation. The curve presented by Dendy and Bolton (1976) has a similar shape, but with the yield maximum at ~ 500 mm yr^{-1}, whereas the curve of Douglas (1967) is more complex, with a peak in yield at ~ 450 mm yr^{-1} but with an increasing trend in yield at high levels of precipitation (Figure 9.5). Wilson (1973) reported similar results. The difficulty in interpreting the plots of effective precipitation and sediment yield is the possible impact of other factors such as relief, drainage basin size and bedrock lithology.

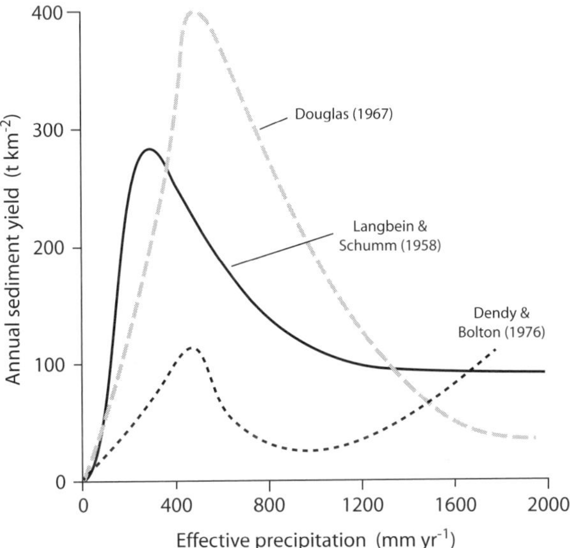

Figure 9.5 Relationships between denudation rate measured as sediment yield and effective precipitation, for three authors. After Hooke (2000)(fig.2) with permission of Geological Society of America.

Precipitation varies seasonally and on shorter timescales, which contributes to the intensity of precipitation and the erosivity of run-off. Rainfall variability can be modelled using the relation p^2/P, where p is the maximum mean monthly precipitation and P is the mean annual precipitation (Fournier, 1960). Alternatively, the storm run-off rate P_s can be related to the mean annual run-off rate $\bar{P}$ by a 'total flood duration ratio' ζ, defined as the total fraction of a given year that the drainage basin is subject to flooding (Tucker and Slingerland, 1997). For example, if rainfall fell continually at a uniform rate throughout the year, ζ would be equal to 1. If it fell on one day only, ζ would be equal to 1/365. The instantaneous storm run-off rate is then

$$P_s = \frac{\bar{P}}{\zeta} \tag{9.6}$$

Tucker and Slingerland (1997) varied $\bar{P}$, P_s and ζ in model runs using a landscape evolution model. Catchments responded strongly to run-off intensity, as opposed to mean precipitation, increasing in denudation rate and sediment efflux with increasing rainfall intensity. However, some authors have found no significant correlation between rainfall variability and sediment yield in global data sets (Summerfield and Hulton, 1994; Hovius, 1998). The effects of variability are most likely at their strongest in semi-arid and arid regimes. Precipitation variability may be more important than mean precipitation in long-term landscape evolution (Tucker and Bras, 2000), but the effect of short-term rainfall variability on long-term landscape development is likely to be smaller than the impact of human activities (Baartman et al., 2013).

The distribution of storm (or daily) rainfalls is commonly expressed as an exponential function, in which the number density of days with rainfall greater than r is estimated as

$$N(r) = N_0 \exp(-r/r_0) \tag{9.7}$$

where N_0 is the number of days on which rain fell, and r_0 is the mean rain per rain day, so $r_0 N_0$ is the total annual rainfall R. Erosion of soil and regolith requires overland flow, which can only be generated above a certain storm size with a run-off threshold, which is a storage capacity of the soil or regolith h. In a single storm the overland flow per unit area j is:

$$j = r - h \tag{9.8}$$

Summing equation (9.8) over all the rainfall events exceeding the run-off threshold gives

$$J = \int_h^\infty (r - h)N(r)\mathrm{d}r = N_0 r_0 \exp(-h/r_0) \tag{9.9}$$

Assuming sediment yield is proportional to the square of the water discharge, the yield from a single rainfall event y is

$$y \propto (r - h)^2 \tag{9.10}$$

which when summed over the frequency distribution gives the total sediment yield Y:

$$Y \propto \int_h^\infty (r - h)^2 N_0 \exp(-r/r_0)\mathrm{d}r = 2 N_0 r_0^2 \exp(-h/r_0) \tag{9.11}$$

which is known as the Cumulative Erosion Potential (CEP) (DePloey, Kirkby, and Ahnert, 1991). The CEP combines the effects of rainfall distribution through N_0 and r_0 and a soil hydraulic parameter h_0 to give a measure of the climatic component in soil erosion. As an example, if there are 100 rain-days per year (N_0), the mean rainfall per rain-day (r_0) is 10 mm, the annual rainfall (R) is 1000 mm. Letting the run-off threshold (h_0) be 30 mm, the CEP becomes $\sim$1,000. If we reduce the number of rain-days to 50, but increase the mean rainfall per rain-day to 20 mm, which reflects a more seasonal regime, CEP doubles to $\sim$2,000.

The CEP model has been extended to incorporate the effects of the growth of vegetation and the impact of biomass on soil strength. The CSEP (Cumulative Soil Erosion Potential) of Kirkby and Cox (1995) gives a global picture that shows highest erosion potential in monsoonal climates and lowest erosion potential in deserts and cool temperate belts.

Plots of CSEP versus annual rainfall show significant differences between non-seasonal and seasonal patterns of rainfall, and as a function of mean annual temperature (Figure 9.6). The overall form of the curves is similar to that proposed by Langbein and Schumm (1958), with maximum erosion rate in semi-arid climates, a minimum in temperate climates and a renewed rise at very high rainfalls. In seasonal Mediterranean-type climates, rainfall seasonality has opposite effects at high and low temperatures. At low temperatures, seasonality increases erosion potential, whereas at high temperatures, seasonality reduces erosion potential. Clearly, there is a limit to the applicability of the full CSEP model to ancient sediment routing systems where parameters describing the rainfall regime and the

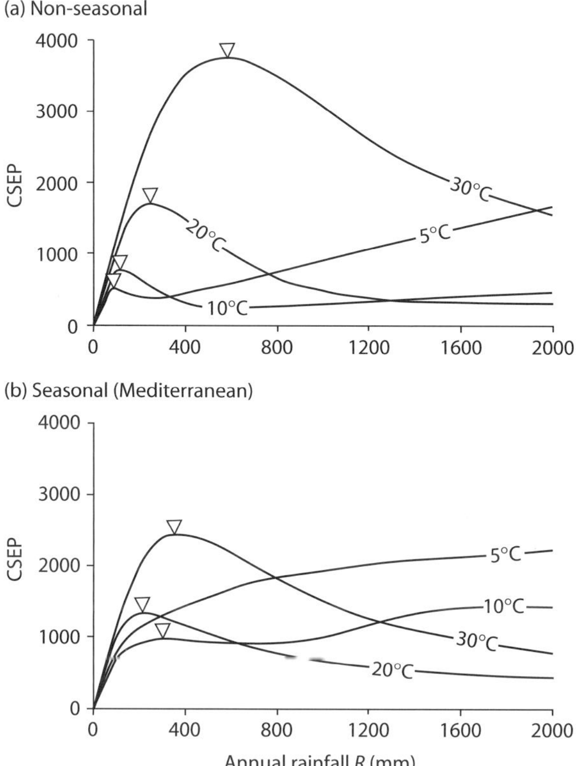

Figure 9.6 Estimated CSEP using equation (9.11). Triangles show erosion maxima. (a) CSEP for rainfall in non-seasonal climates (each month has the same amount of rainfall), with a strong semi-arid maximum and temperate zone minimum, particularly at high temperatures. (b) CSEP for rainfall with strong seasonality (wet winter, Mediterranean climate). The temperate zone minimum seen in (a) is reduced, except at high temperatures. After Kirkby and Cox (1995)(fig.6) with permission of Elsevier.

vegetation-organic growth dynamics are unknown. However, the curves in Figure 9.6 are a starting point for making educated guesses of the likely impact of climate change on erosion and sediment flux in the geological past, especially where the palaeoclimate can be well constrained (Van der Zwan, 2002). The CSEP model has been used to understand variations of sediment production linked to Quaternary glacial-interglacial cycles in the Great Basin, United States (Leeder et al., 1998).

A global scale view can be gained by implementing a soil erosion model developed from the results of plot-scale ($\sim$1 m^2 to 30 m^2) erosion experiments and has the form

$$E = kJ^2 S^{1.67} \exp(-0.07v) \qquad (9.12)$$

where E is erosion in mm day^{-1}, k is a soil erodibility constant with values betweeen 0.02 and 0.69, J is the overland flow in mm day^{-1}, S is the slope and v is the vegetation cover expressed as a percentage (Zhang, Drake, and Wainwright, 2002). Rates of overland flow can be expressed by a modification of the Carson and Kirkby (1972) model, assuming that daily rainfall amounts approximate an exponential frequency distribution within a time period and that the real soil water storage capacity is affected by the initial soil moisture:

$$J_i = P_i \exp(-(rc - Q_{Ti})/(P_i/N_i)) \tag{9.13}$$

where J_i is the overland flow (run-off) in the given time period (mm), N_i is the number of rain-days in the given time period, Q_{Ti} is the total initial soil moisture (mm), rc is the potential water storage capacity (mm) equal to 300 mm, representing a rooting zone, and i is the ith time period from 1 to 12 for monthly intervals (Zhang et al., 2002). If $k = 0.4$, $S = 0.05$, $v = 0.6$ and $J = 730$ mm yr^{-1}, the calculated erosion rate E is 0.06 mm yr^{-1}.

A soil erosion model of the form in equation (9.12) has been used at timescales appropriate for long-term landscape dynamics, where the erosion rate E is termed the maximum achievable erosion rate, applying to the case of unlimited regolith availability (Forzoni et al., 2014). Regolith production rates may be significantly lower than the potential erosion rate (Tucker and Slingerland, 1997). Forzoni et al. (2014) therefore introduced a regolith supply index *RSI* with values of 0 to 1 in order to simulate both detachment-limited and transport-limited conditions.

Regolith production rates can be approximated by the rate of conversion of bedrock to regolith by physical and chemical weathering. This conversion rate is commonly taken as inversely proportional to the thickness of regolith cover C (Tucker and Slingerland, 1997):

$$\omega = k_w \exp(-C/C^*) \tag{9.14}$$

where ω is the vertical rate of descent of a weathering front, k_w is the weathering descent rate for bare bedrock (that is, at $C = 0$), and C^* describes the rate at which the weathering rate decays with increasing regolith thickness. The bare bedrock regolith production rate k_w and the weathering decay constant C^* were given values of 0.0005 m yr^{-1} and 0.5 m respectively by Tucker and Slingerland (1997).

Annual precipitation varies with latitude (Figure 9.7), so the trends of sediment yield versus effective precipitation in Figure 9.5 can be viewed as a latitudinal zonation. Precipitation values of greater than 1,000 mm yr^{-1} are mostly restricted to tropical latitudes, with the exception of wet temperate locations in the southern hemisphere such as New Zealand and southern Chile. Peaks in sediment yield, which occur at effective precipitation values of about 300–600 mm yr^{-1} are found principally in subtropical latitudes. Since rate of denudation is clearly mediated strongly by vegetation, changes in land vegetation over the Phanerozoic should modify the picture evident in Figure 9.5 (Schumm, 1968). Key developments were the advent of land vegetation (Late Silurian), the appearance of horsetails and ferns (Devonian), conifers (Carboniferous), flowering plants (end Jurassic) and the appearance of grasses near the end of the Cretaceous. Before the colonisation of the land surface by plants it is expected that there would be a simple linear trend of increasing sediment yield with increasing precipitation. If so, sediment routing systems in the Precambrian and the early Palaeozoic would have functioned considerably differently to those of today.

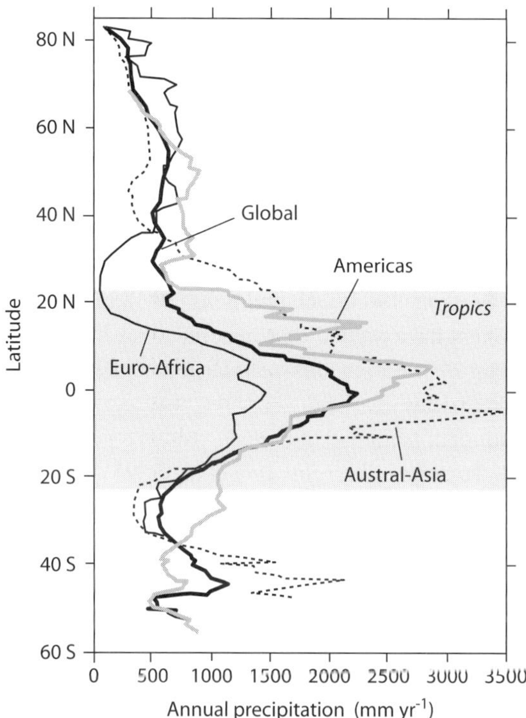

Figure 9.7 Latitudinal distribution of global precipitation and in the Americas, Euro-Africa and Austral-Asia. After Milliman and Farnsworth (2011)(fig.2.3) with permission of Cambridge University Press.

9.3 Grain-Size Mix of Sediment Supplied to Basins

A key factor in the downstream dispersal and fractionation of sediment is the grain-size distribution of sediment supplied by upstream catchments. Different distributions of grain size in the sediment supply to basins affect the pattern of downstream fractionation (Strong et al., 2005; Fedele and Paola, 2007; Duller et al., 2010; Whittaker et al., 2010; Armitage et al., 2011; Michael et al., 2013; Michael et al., 2014a; Schlunegger and Norton, 2015). Building an understanding of the impact of the changing distribution of grain size is vital for prediction of subsurface sedimentary architectures, gross depositional environments and sedimentary facies (Strong et al., 2005; Carvajal and Steel, 2012; Michael et al., 2013; Michael et al., 2014a) and forms a key element in the fields of sequence stratigraphy (Chapter 10) and basin analysis. It is important, therefore, to know the characteristics of and the controls on the grain-size mix serving as an initial condition for down-system dispersal (Allen et al., 2015b).

It is generally recognised that there is a paucity of information on the grain-size distributions of the weathering products that define the initial condition for sediment dispersal (Weltje, 2012). The crystal size distribution in crystalline parent rocks, for example, has been described as log-normal (Eberl, Drits, and Srodon, 1998), but it is not known whether this is a general rule for other rock types and in regolith comprising rock fragments and broken particles rather than individual mineralogical crystals. The size distribution of weathered crystalline rocks such as granites, which have undergone brittle disintegration, has also been found to be approximated by Rosin's (exponential) law (Krumbein and Tisdel, 1940; McEwen, Fessenden, and Rogers, 1959). Crushed materials and those undergoing mechanical disintegration such as scree, regolith (including on moons and other planets), pyroclastic material and subglacial tills have also been described by the Rosin law (Kittleman, 1964; Ibbeken, 1983; Deb and Sen, 2013) and the closely related Weibull distribution. Other workers have argued that materials that have undergone comminution commonly have a fractal distribution of size (Hartmann, 1969; Turcotte, 1997; Smalley et al., 2005; Allen et al., 2015b).

The formation of sediment from parent rocks is part of a general process of 'fragmentation' (Hartmann, 1969; Turcotte, 1997; Smalley et al., 2005). The most important process of fragmentation in the present context is the breakdown of parent rocks by the processes of weathering (Wu, Borkovee, and Sticher, 1993; Bitelli, Campbell, and Flury, 1999; Wells, Willgoose, and Hancock, 2008). Weathering is an example of fragment size reduction, resulting in soils, sediment and regolith containing fine-grained particles (Jefferson et al., 1997). This process of fragment size reduction, or comminution, is well known to result in fractal distributions of grain size (Hyslip and Vallejo, 1997) (Figure 9.8).

If the grain-size distribution of the sediment released into the cascade from source to sink (Burt and Allison, 2010) is to be modelled in a geomorphic context, it is necessary to

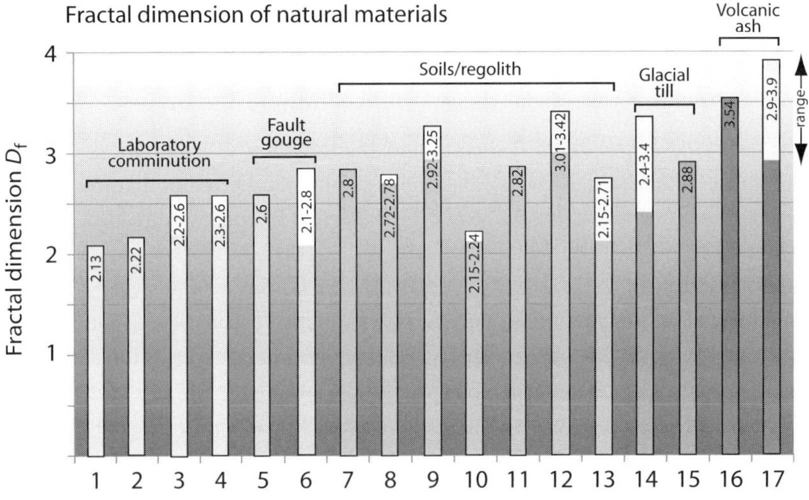

Figure 9.8 Average fractal dimension for a range of natural materials (rocks, soils, regolith, sediments). 1, disaggregated gneiss; 2, disaggregated granite; 3, 4, laboratory comminution of sand; 5, fault gouge; 6, cataclastic zone; 7, soils (general); 8, 52 gravelly and sandy soils; 9, semi-arid soils; 10, temperate soils; 11, loamy soils; 12, loess (quartz silt); 13, terrace sands and gravels; 14, 69 glacial tills; 15, glacial till; 16, volcanic ash and pumice; 17, 62 ash fall and ash flow deposits. From Allen et al. (2015b) (fig.1) with permission of University of Chicago Press.

account for the combined effects of parent rock lithology and microstructure, topography, climate and the passage of time on the fragmentation process. However, in the context of geological sediment routing systems, the range of controls on the particle size distribution of sediment exported to sedimentary basins needs to be reduced to as few parameters as possible so that the method can be used where little is known about the contributing catchment areas, including their location, size, morphometry, climate and lithological make-up (Sømme et al., 2009).

If a parent rock breaks into smaller pieces, which themselves break into smaller pieces, and so on, we should expect the number of fragments N with a size greater than a certain linear dimension x to decrease as the linear dimension increases. The number of fragments with a linear dimension greater than x can be expressed by the fractal number relation

$$N(X > x) = Cx^{-D_f} \tag{9.15}$$

where D_f is the fractal dimension associated with fragmentation and C is a coefficient. D_f is typically between 2 and 4 for a range of fragment types over 4 to 5 orders of magnitude of x (Figure 9.8). The number of fragments can be substituted by the volume or mass of sediment with particle sizes greater than a certain linear dimension to enable comparison with empirical plots of grain-size distribution derived from sieving (Tyler and Wheatcraft, 1992; Hyslip and Vallejo, 1997). In such a case, equation (9.15) is modified from its number-specific form to a weight-specific form:

$$P(X < x) = \frac{M(X > x)}{M_T} = \left(\frac{x}{x_{max}}\right)^m \tag{9.16}$$

where P is the proportion by weight of particles smaller than the sieve size x, M is the weight (or mass) smaller than the sieve size, M_T is the total weight of sediment sieved, x_{max} is the maximum sieve or screen size (through which all particles pass), and m is an index of the spread of the particle size distribution given by the slope of the power law represented by equation (9.15). Since particle volume or weight is related to the cube of particle number (Turcotte, 1997), the parameter m is related to the fractal dimension by

$$D_f = 3 - m \tag{9.17}$$

Using equation 9.15, it can be seen that when the grain-size distribution is heavily loaded with fine particles the fractal dimension is large, and when the grain-size distribution is loaded with coarse particles the fractal dimension is small.

It can reasonably be expected that the grain-size distributions of sediment released from catchments will be similar to those of crushed or mechanically disintegrated materials rather than to sediments that have undergone profound reworking and sorting during transport. Hydraulic sorting during transport is thought to generate grain-size distributions of exponential or log-normal types (Section 9.4.2) (McEwen et al., 1959; Middleton, 1976; Ibbeken, 1983; Allen et al., 2015b). A number of different regolith and sediment types have therefore been analysed that might approximate the time-averaged characteristics of the sediment efflux of upland catchments on geological timescales. These materials include (1)

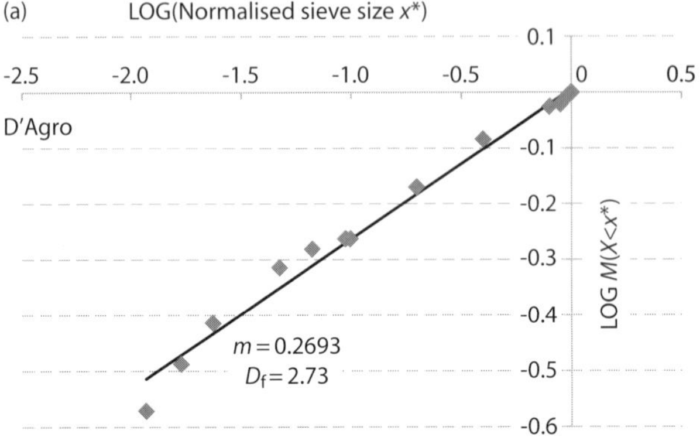

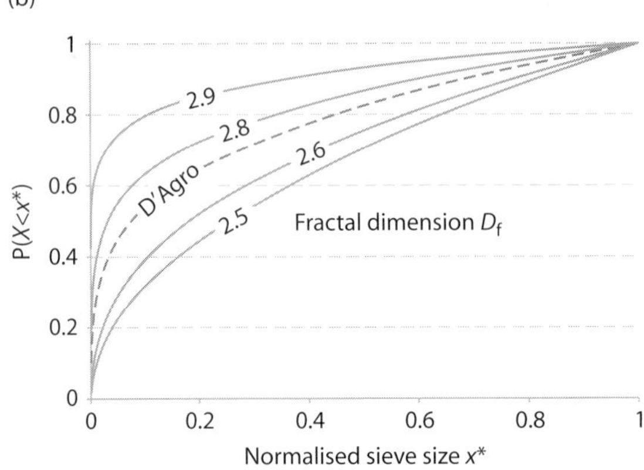

Figure 9.9 (a) Log-log plot of the normalised grain size $x^* = x/x_{max}$ versus the mass smaller than the sieve size with a straight line fit of gradient m. Data are for a sieved sample taken at a locality at the mouth of the D'Agro catchment, eastern Sicily. (b) Cumulative distribution function for grain size for a range of values of the fractal dimension. The cumulative distribution function that best fits the D'Agro data is shown ($m = 0.27$, $D_f = 2.73$). There is a greater contribution from fine grain sizes with higher values of the fractal dimension. From Allen et al. (2015b) (fig.3) with permission of University of Chicago Press.

bedload sediment in the main channels of the Pagliara, Fiumedinisi and Agro catchments etched into a region of active uplift at $1-1.5$ mm yr^{-1} in northeastern Sicily; (2) coarse sediment at the outlet points of catchments distributed along strike of topographic ranges in Sicily and Basilicata-Campania, southern Italy; and (3) weathering products (regolith) in catchments in southern Italy developed on a range of different rock types. Together, these data sets provide a good idea of the typical fractal dimensions of sediments and regolith stored in upland catchments prior to export to sedimentary basins.

Details of sample localities and the methods used to collect particle size data are given in Allen et al. (2015b). The fractal dimension can be calculated in two ways. First, taking the logarithm of both sides of equation (9.16) gives a straight line relationship between $M(X < x)/M_T$ and x/x_{max} if the distribution is fractal (Figure 9.9), where the slope is

$m = 3 - D_f$ (equation (9.17)). Equation (9.16) has been used to determine the fractal dimension from sieve data collected at 19 localities at the outlets of Sicilian catchments. Second, the value of the fractal dimension at each station along the main channel of 3 catchments in Sicily was interpolated by taking the proportion by weight at two characteristic grain sizes, D_{84} and D_{50}, using a modification of equation (9.16):

$$m = \frac{\log P(X < D_{84}) - \log P(X < D_{50})}{\log(D_{84}) - \log(D_{50})} \qquad (9.18)$$

where $P(D_{84})$ and $P(D_{50})$ are the proportions by weight smaller than the 84th and 50th percentiles respectively.

The estimated values of fractal dimension from weathering products can be compared with those for sediment at the outlets of catchments in the same region. Sediment at the outlets of 23 catchments along the Vallo di Diano and East Agri faults in Campania-Basilicata have an average fractal dimension of 2.42 using the interpolation method. This value reflects the mixed bedrock lithologies, and is very close to the average of the weathering products in the tectonically uplifting catchments, showing that weathering products are little modified during the process of sediment mobilisation before export to adjacent basins. In addition, the variation in fractal dimension at catchment outlets is small compared to the variation in the regolith caused by lithological controls. This suggests that although the average fractal dimension changes little from regolith to catchment outlet, the source materials undergo some homogenisation. It is expected that large catchments are more effective at homogenisation than small, but no relationship between fractal dimension and contributing drainage area is evident.

In summary, the fractal dimension of weathering products and sediment at catchment outlets in southern Italy are identical and are a good indicator of the grain-size mix of the sediment supply to sedimentary basins. Likewise, in Sicily, the fractal dimensions of sediment sampled at a number of stations along the main trunk channels of three catchments are indistinguishable from those calculated from sediment at catchment outlets (Allen et al., 2015b).

9.3.1 Simulations Using a Variable Grain-Size Mix in the Supply

Simulations are carried out using a physical sediment transport model in order to test the sensitivity of stratigraphic architectures to variations in the grain-size characteristics of the supply.

Although the fractal dimension has the advantage of describing the grain-size distribution of sediment in one number, fractal distributions do not have meaningful values of mean and standard deviation since they are scale invariant. The fractal power law, represented by equation (9.15), however, has the same form as the Pareto distribution, which is described by a shape parameter a representing the spread of the distribution and a scale parameter k representing the minimum grain size (Schroeder, 1991; Hastings and Sugihara, 1993; Vidondo et al., 1997). The unknown parameters of the Pareto distribution can be evaluated by interpolation between specified grain-size values. Here, interpolation is carried out using values of the 84th and 50th percentiles derived from the cumulative function distribution of grain size based on Wolman-type clast counts. If the proportion of a sample that has particles below a grain size D_b is denoted by P_b, and the proportion of a sample that has particles below a grain size D_c is denoted by P_c, the estimate of the shape parameter a is

$$a = \frac{\log(1 - P_b) - \log(1 - P_c)}{\log(D_c) - \log(D_b)} \qquad (9.19)$$

and the estimate of the scale parameter k_P is given by

$$k_P = \left\{ \frac{P_c - P_b}{(1/D_b^a) - (1/D_c^a)} \right\}^{1/a} \qquad (9.20)$$

Plots of fractal dimension D_f derived from sieved and clast count data versus Pareto shape parameter a derived from interpolation of D_{84} and D_{50} have a regression $D_f = 3 - 0.45a$. Making use of equation (9.17), it can be seen that $a \approx 2.22m$ (Figure 9.10).

The mean of the Pareto distribution μ_P is given by

$$\mu_P = \frac{a k_P}{a - 1} \qquad (9.21)$$

for $a > 1$. This mean value for a Pareto distribution with a shape parameter a and scale parameter k_P can be used to calibrate the input to a sediment dispersal model. For the sediment samples analysed, μ_P ranges between approximately 10 and 60 mm.

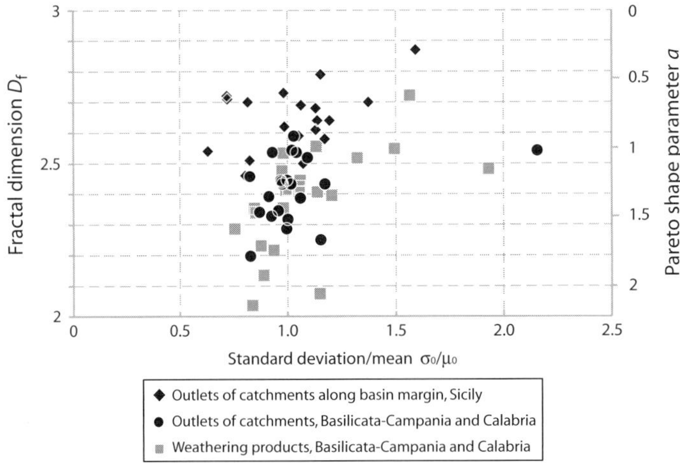

Figure 9.10 Plot of fractal dimension D_f and Pareto shape parameter a versus the ratio of σ_0/μ_0, for a range of sediments and regolith in the erosional engine. There are considerable uncertainties in associating fractal dimensions to particular values of σ_0/μ_0. The ratio σ_0/μ_0 is, however, centred on a value of about 1 and most values fall within the range 0.6–1.7. This range is incorporated in the simulations using a physical sediment transport model. After Allen et al. (2015b)(fig.8) with permission of University of Chicago Press.

Sediment is initially dispersed down-system using a simple Sternberg-type exponential function of the form

$$\mu(x^*) = \mu_0 \exp(-Cy^*) \tag{9.22}$$

where μ_0 is the mean grain size of the sediment supply, taken as the mean of clasts with a Pareto size distribution μ_P, $\mu(x^*)$ is the mean grain size as a function of the normalised down-system distance $x^* = x/L$, L is the total depositional length of interest in the sediment routing system, $y^*(x^*)$ is the cumulative down-system sediment deposition in the normalised (mass balance) coordinate system, and C is a coefficient that describes the rate of decrease in mean grain size in the down-system direction. If the initial mean grain size is 60 mm, and the normalised cumulative down-system deposition is 0.5 (that is, half of the total sediment has been deposited), the mean grain size of the deposit ranges from 36 to 47 mm for values of C between 1 and 0.5 respectively. C therefore reflects the fractionation of clasts from the surface flux, which depends on its variance of grain size. This in turn should be inversely dependent on the Pareto shape parameter of the sediment supply. C is taken to be 0.75 in the simulations. The value of μ_P is used as the mean of the sediment supply μ_0 in equation (9.22), which becomes

$$\mu(x^*) = \frac{ak_P}{a-1}\exp(-Cy^*) \tag{9.23}$$

Equation (9.22) is a simplified version of that proposed by Fedele and Paola (2007) and applied by Duller et al. (2010) and Armitage et al. (2011), described in Section 9.4.3, which includes terms for variance of the grain-size distribution.

A plot of σ_0/μ_0 (where σ_0 is the standard deviation in the sediment supply and μ_0 is the mean grain size in the supply) using clast count data from catchment outlets along the topographic range front in Sicily and Campania-Basilicata, and from weathering products in Campania-Basilicata and Calabria, versus fractal dimension and Pareto shape parameter is shown in Figure 9.10. As D_f increases and a decreases, there is an increasing contribution to the distribution from small fragments, which is reflected in an increase in the value of σ_0/μ_0. Most values of σ_0/μ_0 fall in the range $0.6 - 1.7$, which corresponds to the range of $2 < D_f < 3$ and $0 < a < 2.22$ found in the southern Italian and Sicilian data sets. This range of σ_0/μ_0 is employed in the numerical simulations.

Simulations of the downstream fining of mean grain size are carried out with background subsidence approximated by an exponential function, analogous to an extensional half graben bounded by a border fault or to a flexural foreland basin (Allen and Allen, 2013). The discharge of the sediment supply and the spatial distribution of tectonic subsidence are kept steady during model runs. The surface processes model (Armitage et al., 2013) calculates sediment transport using the formulation of Smith and Bretherton (1972) with values of coefficients derived from Simpson and Schlunegger (2003). The boundary conditions are of fixed elevation at the downstream limit of the system at 10 km and at the head of the system there is a constant unit width supply of sediment of 1 m^2 yr^{-1} required to fill the basin.

Two sets of simulations are carried out.

1. Set A: A simple exponential decline in mean grain size is combined with the physical sediment transport model with step-like (A1) and gradual (A2) changes in the starting mean grain size $\mu_P = \mu_0$ (Figure 9.11).
2. Set B: Sediment deposition is governed by the self-similar solution of equation (9.33) combined with the physical sediment transport model with step-like (B1) and gradual (B2) changes in the ratio of the standard deviation to the mean grain size in the sediment supply σ_0/μ_0 (Figure 9.12). C_v is held constant at 1.2 in the simulations and σ_0/μ_0 values are derived from field clast count data.

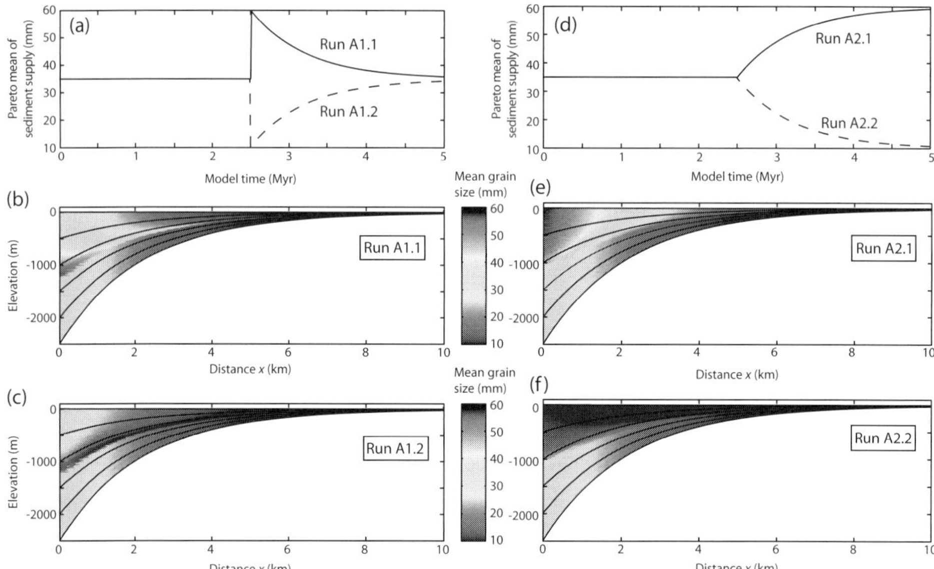

Figure 9.11 Simulation of grain-size variation over a 5 Myr time period in a half-graben or flexural-type basin with a maximum tectonic subsidence of 5 mm yr^{-1} using a Sternberg-type exponential decline in mean grain size. Four scenarios are modelled involving a step change increase (Run A1.1) and step change decrease (Run A1.2), and gradual increase (Run A2.1) and gradual decrease (Run A2.2) in the mean grain size of the sediment supply (μ_0), which has a distribution following the Pareto model. Step change and start of gradual change take place at 2.5 Myr. (b)-(c) and (e)-(f) show cross-sections of the basin with stratigraphy grey coded for mean grain size $\mu(x)$. (b) shows the effects of a step change increase in μ_0 from 35 to 60 mm, which causes an abrupt upward coarsening and an advance of the gravel front, followed by retrogradation. (c) shows the effects of a step change reduction in μ_0 from 35 to 10 mm, which causes an abrupt upward fining in grain size and upstream retraction of the gravel front, followed by gradual progradation. (e) shows the effects of a gradual increase in μ_0 from 35 to 60 mm, which causes an upward coarsening and prolonged progradation leading to a down-system extension of the gravel front. (f) shows the effects of a gradual reduction in μ_0 from 35 to 10 mm, which causes an abrupt upward fining in grain size and upstream retraction of the gravel front. From Allen et al. (2015b)(figs.9, 10) with permission of University of Chicago Press.

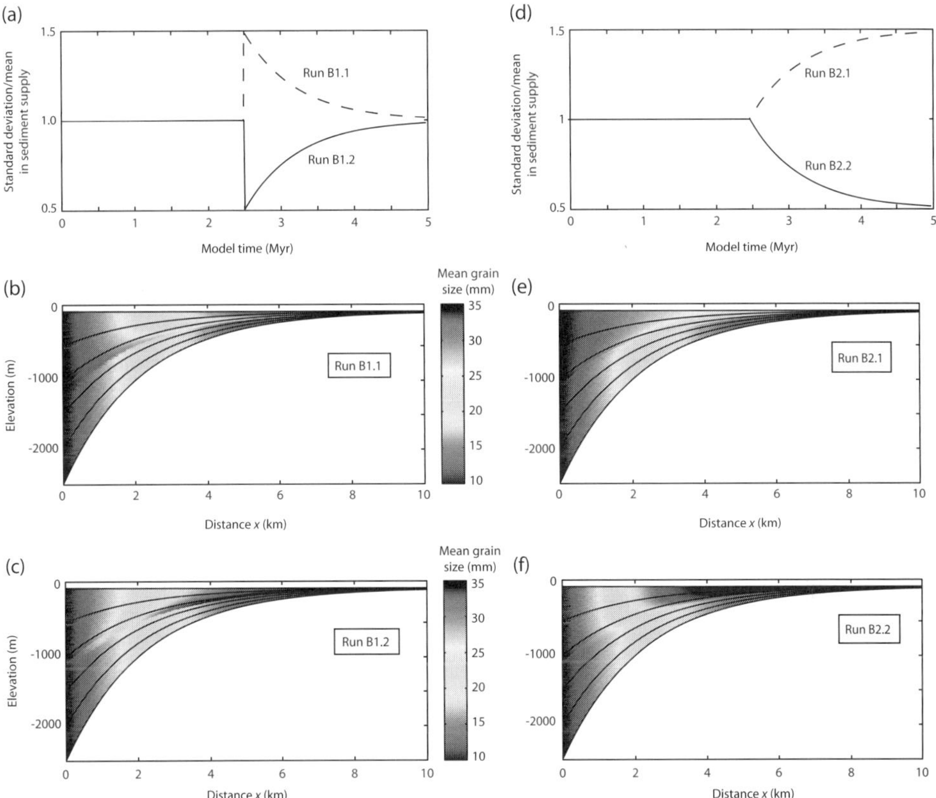

Figure 9.12 Simulation of grain-size variation in a half-graben or flexural type basin with a maximum tectonic subsidence of 5 mm yr^{-1} over a 5 Myr time period using the self-similar solution of Fedele and Paola (2007). Four scenarios are modelled involving a step change increase (Run B1.1), step change decrease (Run B1.2), gradual increase (Run B2.1) and gradual decrease (Run B2.2) in the ratio σ_0/μ_0 at 2.5 Myr. (b)-(c) and (e)-(f) show cross-sections of the basin with stratigraphy grey coded for mean grain size $\mu(x)$. (b) shows the effects of a step change increase in σ_0/μ_0 from 1 to 1.5, which causes an abrupt upward coarsening in grain size and a downstream extension of the gravel front. (c) shows the effects of a step change reduction in σ_0/μ_0 from 1.0 to 0.5, which causes an abrupt upward fining in grain size and a retreat of the gravel front, followed by slow progradation. (e) shows the effects of a gradual increase in σ_0/μ_0, which causes an upward coarsening in grain size and progressive down-system extension of the gravel front. (f) shows the effects of a gradual reduction in σ_0/μ_0, which causes a slight upward fining of mean grain size and minor retrogradation. From Allen et al. (2015b)(figs.11, 12) with permission of University of Chicago Press.

In the case of the simple exponential model, the Pareto mean grain size in the sediment supply is held constant at 35 mm before a step or gradual change at 2.5 Myr model time. The step change involves a doubling or halving of the Pareto mean followed by 2.5 Myr of relaxation. In the case of a gradual change, the Pareto mean increases or decreases to values of 60 mm and 10 mm after 5 Myr model time. In the simulations using the self-similar solution of Fedele and Paola (2007), the value of σ_0/μ_0 is set at 1.0 for the first 2.5 Myr, then increases to 1.5 or decreases to 0.5 in step or gradual changes. It is emphasised that the patterns of grain size in basin stratigraphy shown in the simulations are entirely driven by variations in the statistical properties of the grain-size mix of the sediment supply.

Simulations using the simple exponential model show marked shifts in grain size at the time of the step change. A step-like increase in the mean size of the sediment supply causes an abrupt upward coarsening and the down-system extension of the gravel front, followed by gradual retrogradation, producing a tongue or sheet of coarse gravel. A step-like decrease in mean grain size of the supply causes an abrupt upward fining and retraction of the gravel front. A gradual increase in the mean size of the supply causes an upward coarsening, but the trend is relatively slow and leads to long-term progradation. Likewise, the gradual reduction in the mean size of the supply results in a basin-wide fining in stratigraphy and long-term retrogradation.

Sedimentary architectures generated in simulations using the self-similar solution are consistent with those using the simple exponential model. A step-like increase in the ratio σ_0/μ_0 results in a pronounced gravel tongue, whereas the step-like descrease produces a surface marked by strong but short-lived retrogradation. A gradual increase in the ratio σ_0/μ_0 causes an increase in progradation, resulting in a long-term trend of upward coarsening. A gradual decrease causes long-term retrogradation and slow upward fining.

9.4 Grain-Size Fractionation in Sediment Routing Systems

9.4.1 Downstream Changes in Hydraulic Geometry

The downstream fining of grain size in river channels is accompanied by changes in the hydraulic geometry of the channels. Parker (1978b) suggested that gravel bed rivers configured themselves so that the shear stress on the bed at the centre of the channel is in excess of the critical stress required to entrain the median grain size. That is,

$$\frac{\tau_0}{\tau_{c\,50}} = \epsilon \tag{9.24}$$

This excess stress factor ϵ has been estimated at $1.2 - 1.8$ (meaning $20 - 80\%$ excess) in gravel reaches (Figure 9.13). Beyond the gravel front, however, the excess stress factor increases markedly to greater than 10, accompanied by an abrupt decrease in D_{50} and slope. By invoking the resistance equation for steady flow down an inclined plane and assuming that $\tau_{c\,50} \propto D_{50}$, equation (9.24) can be modified to

$$\epsilon = \frac{\rho g R S}{0.045(\rho_s - \rho)g D_{50}} \tag{9.25}$$

or

$$k = \frac{R S}{D_{50}} \tag{9.26}$$

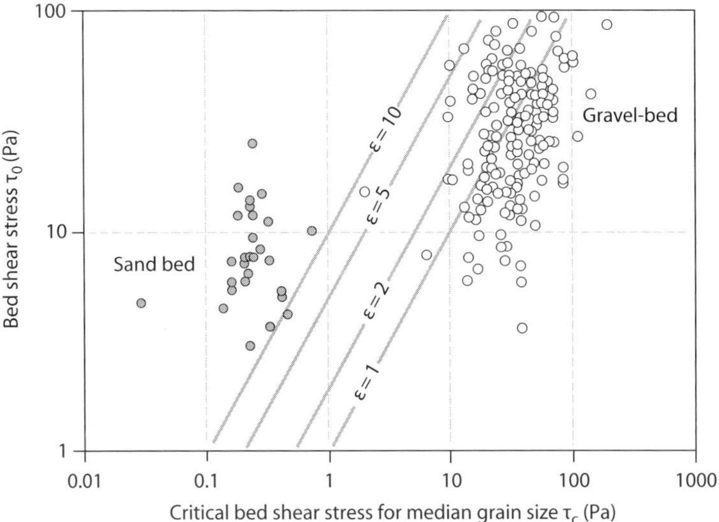

Figure 9.13 Excess shear stress relationship for 235 gravel bed and sand bed rivers for bankfull conditions, using data from Church and Rood (1983). The critical shear stress on the bed for the median grain size is assumed to be proportional to D_{50}. Modified from Robinson and Slingerland (1998)(fig.3) with permission of Society of Economic Paleontologists and Mineralogists.

where R is the hydraulic radius and k contains all the constants. That is, a downstream fining in the median grain size D_{50} only occurs where the product RS decreases downstream. Downstream variation in the hydraulic radius may be complex, particularly as a result of changes in channel width. Robinson and Slingerland (1998) compared a number of regime equations in terms of their success in predicting the variation in ϵ and channel width W across the gravel front. The best fit with observational data for 70 Canadian rivers was with

$$W = 3.83 Q_w^{0.528} D_{50}^{-0.07} \tag{9.27}$$

following Bray (1982), where the coefficient is likely to vary from approximately 1 to 4.5 depending on vegetation. A value of $\sim$2 is thought appropriate for rivers with cohesive, vegetated banks. In equation (9.27), a reduction of grain size causes flow width to increase, and an increase in discharge of water also causes the flow width to increase. For example, if the water discharge is 100 m^3 s^{-1} and the median grain size is 0.02 m in the gravel reach, channel width $W = 57$m. In the sand reach, if the median grain size decreases to 0.0005 m and the water discharge does not change, the river nearly doubles its width to $W = 107$ m.

9.4.2 Fractionation of Grain Size During Dispersal

It is not clear what changes take place in the grain-size distribution of sediment as it is mobilised from regolith, transported and sorted by hydraulic processes and eventually selectively extracted from the surface flux and transfered to the substrate. These various processes are likely to have a multiplicative effect (Dacey and Krumbein, 1979).

McEwen, Fessenden, and Rogers (1959) suggested that processes associated with transport such as abrasion and fracturing caused the grain-size distribution to be modified, stating that (p.492)

The cumulative effect of the various agents and processes acting on the sediment after its formation will, in general, impose a log-normal size distribution on the sediment.

If this were the case, it would be in agreement with the long-held view that the size distribution of transported sediments is commonly log-normal (Krumbein, 1938).

Allen et al. (2016) compared exponential (Rosin and Weibull 'laws'), Pareto and log-normal statistical models in fitting grain-size data collected from a range of sediment types involving different amounts of transport. Although these unimodal statistical models do not allow for mixing of end-member size distributions (Weltje and Prins, 2003), systematic variations in the goodness-of-fit of models to observed gravel-grade sediment size distributions allow the impact of mobilisation and transport to be assessed.

Sediment types were sampled from the regolith and bedload sediment of trunk channels in catchments serving as source regions (southern Italy and Sicily), proximal debris flow deposits that have suffered little granulometric change during emplacement (Owens Valley, California), and fluvial gravels that have been transported far from source (Eocene-lower Oligocene Escanilla and Antist systems of the southern Pyrenees of northern Spain and Miocene-Pliocene of the Great Plains, United States) (Figure 9.14).

The closeness of the cumulative distribution functions of the Pareto, Weibull and log-normal models to the observed cumulative function distribution of field data can be evaluated quantitatively using the Kolmogorov–Smirnov vertical difference parameter D_{KS}. This parameter is the greatest deviation of a cumulative distribution function of grain-size data from a reference or model cumulative distribution. Consequently, small values of D_{KS} indicate a close fit.

Ibbeken (1983) found that although jointed and weathered bedrock in southern Calabria, Italy, had a Rosin (Weibull) distribution, gravelly sediments sampled near the mouths of rivers where they enter the Ionian Sea, up to 20 km distant, were commonly bimodal with a sand fraction that was log-normally distributed. The log-normal property of sands is most likely imparted by transport processes causing hydraulic sorting (Middleton, 1976). A comparison of the goodness of fit of sediment in the erosional engine, at catchment outlets and as far-field fluvial gravel shows that the Rosin (Weibull) distribution is the poorest fit, particularly in far-field gravels (Figure 9.15). The log-normal distribution and the truncated Pareto distribution improve in the closeness of their fit with field data with increasing mobilisation and transport. This suggests that increasing fluvial transport enhances the log-normality of clast size distributions, as suggested by Krumbein (1938) and McEwen, Fessenden, and Rogers (1959).

Although fragment sizes in sedimentary materials from regolith to far-field gravel have a broad log-normality, increasing down-system, the Pareto model, base truncated to account

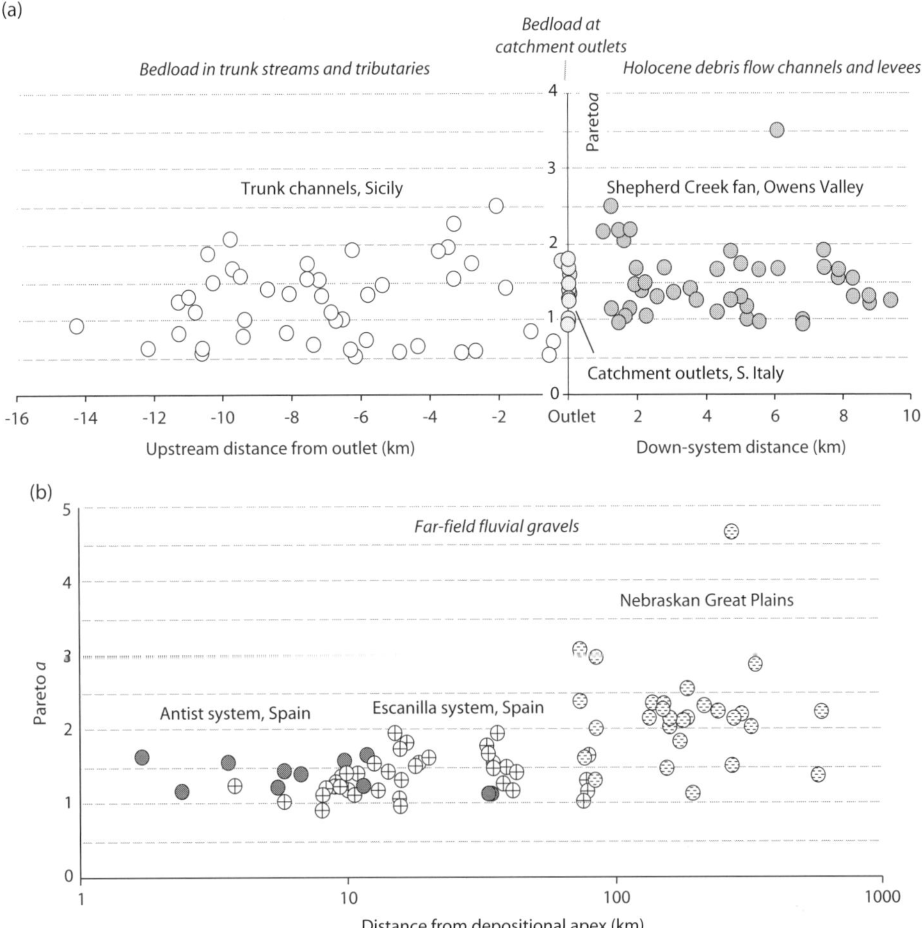

Figure 9.14 Compilation of estimates of the Pareto shape parameter *a* as a function of down-system distance. (a), Regolith, bedload in trunk channels and at catchment outlets, and proximal debris flow deposits. (b), Fluvial gravels, shown with a logarithmic scale for the down-system distance from the depositional apex. The average value of Pareto *a* increases with down-system distance, most likely driven by the loss of coarse particles by selective deposition. After Allen et al. (2016) (fig.26) with permission of John Wiley & Sons Inc.

for the finite minimum grain size (scale parameter k_P), also provides a good fit with the coarser half of the clast size distribution of field data from source to sink, and has the advantage of having a shape parameter that is potentially diagnostic of sediment transport and depositional processes. This shape parameter is well constrained in the erosional engine. On release into sediment transport systems, the clast size distribution may evolve predictably, moderated by the role of selective deposition. At short transport distances, shown by the Escanilla system, rapid deposition in proximal positions causes the extraction

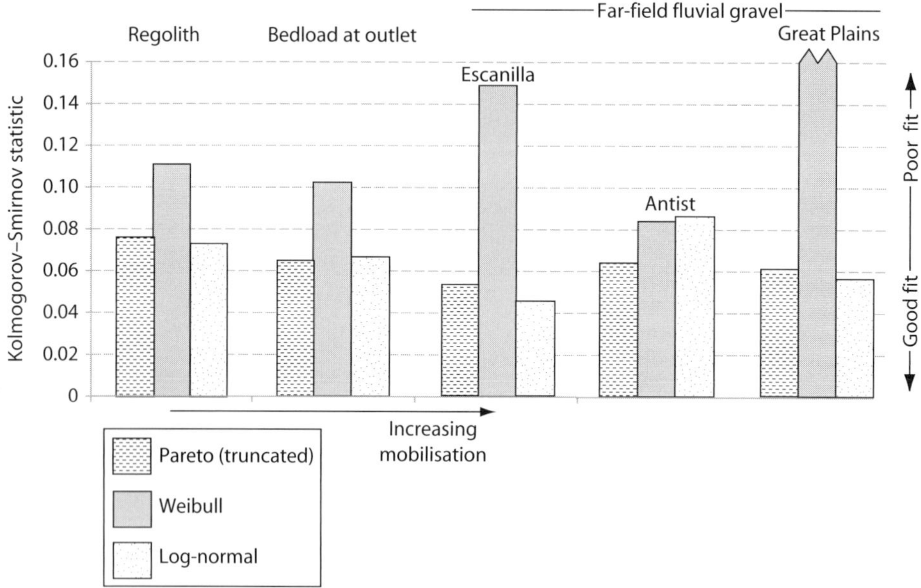

Figure 9.15 Kolmogorov–Smirnov goodness of fit statistic D_{KS} for weathering products (regolith in Basilicata, Campania and Calabria, southern Italy), coarse bedload in channels at catchment outlets (Pagiliara, Fiumedinisi and Agro catchments, Sicily) and far-field fluvial gravels (Escanilla palaeo-sediment routing system, south-central Pyrenees and Miocene-Pliocene of the Great Plains, United States). Whereas the Weibull model shows a worsening fit with degree of sediment mobilisation, the truncated Pareto and log-normal models show an improving fit with increasing mobilisation. Note that the Pareto statistics refer to the coarser half of the clast size distribution. After Allen et al. (2016)(fig.27) with permission of John Wiley & Sons Inc.

of the coarsest grains into stratigraphy, causing the shape parameter a to decrease from the value associated with the sediment supply from the erosional engine (Figure 9.16). At longer transport distances, or at positions further down-system in terms of the total mass balance (Martin et al., 2009; Michael et al., 2013), shown by the Great Plains system, the clast size distribution is enriched in relatively small particles compared to the sediment supply because of up-system selective deposition combined with abrasion during transport, increasing the shape parameter a (Figure 9.14). Consequently the Pareto shape parameter can potentially be used as an indication of proximality.

Despite the large down-system changes in mean grain size and gravel percentage from source region to depositional sink, particle size distributions appear to maintain log-normality over a wide range of transport distance. Consequently, full grain-size distributions can be predicted within the terrestrial segment of sediment routing systems given simple laws of downstream fining. These full grain-size distributions may help to inform the estimation of hydraulic properties of aquifers and oil and gas reservoirs, such as porosity and permeability. Use of statistical models therefore enables down-system fractionation of sediment released from source regions to be better understood and predicted and is a potentially valuable tool in source-to-sink approaches to basin analysis.

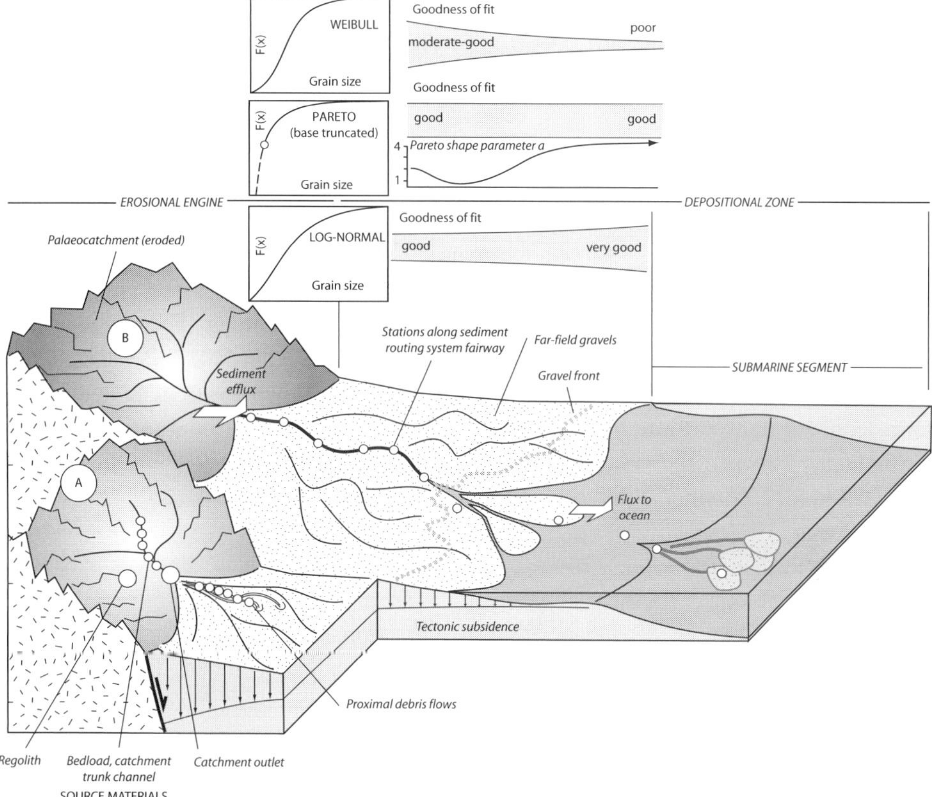

Figure 9.16 Schematic sediment routing systems showing the context of the samples analysed. A, Small coarse grained catchment-fan system with samples collected from regolith, channel bedload and sediment at the catchment outlet, which are indistinguishable from proximal debris flows in terms of Pareto shape parameter a. B, Large palaeo-sediment routing system with a catchment not preserved in the geological record. Stations are distributed along the sediment routing system fairway in the down-system direction. The truncated Pareto, Weibull and log-normal models are shown as cumulative distribution functions and in terms of goodness of fit, evaluated using the Kolmogorov–Smirnov statistic . The down-system variation of the Pareto shape parameter a is based on a compilation of all data sets and is partly speculative. After Allen et al. (2016)(fig.28), with permission of John Wiley & Sons Inc.

9.4.3 Down-System Fining of Gravel by Selective Deposition

In the absence of tributary inputs of sediment, grain size fines downstream in river systems. Basin-wide grain-size trends may be controlled by the relative mobility of mass, whereby the less mobile clasts are extracted first by a process of selective deposition. In many sedimentary systems, the observed rate of fining is too rapid to be explained by abrasion alone (Attal and Lavé, 2009) and consequently selective deposition must dominate. Field-based, numerical and laboratory studies (Toro-Escobar et al., 1996; Seal et al., 1997;

Sheets et al., 2002; Fedele and Paola, 2007; Paola et al., 2009; Duller et al., 2010; Armitage et al., 2011; Whittaker et al., 2011; Paola and Martin, 2012; Parsons et al., 2012; Rohais et al., 2012) have focused on understanding and decoding regional grain-size trends in this context.

Over many repeated sediment transport events, a portion of the mobile sediment flux is extracted to build stratigraphy. Stratigraphy is therefore a strongly filtered record of this succession of sediment transport events. The selective extraction of mass from the mobile sediment flux during downstream transport produces a characteristic trend in preserved grain size. This grain-size trend in turn affects fluvial style and depositional environments (Strong et al. 2005). Critical to the prediction of downstream fining (Fedele and Paola, 2007; Duller et al., 2010; Whittaker et al., 2011; Allen and Heller, 2012) is the influence of the spatial distribution of tectonic subsidence $\sigma(x)$ on the downstream trend in the long-term sediment transport rate $q_s(x)$ (Figure 9.17). This downstream trend in $q_s(x)$ reflects the interplay between the volumetric discharge of the sediment supply and the volume extracted to fill accommodation.

The impact of deposition is to reduce the sediment discharge q_s in the downstream direction from its initial value $q_s(0)$ at $x=0$. Eventually, the sediment supply is exhausted at a depositional length L_d. At each point in the downstream direction $x < L_d$ the solid sediment volume extracted by deposition as a fraction of the surface sediment flux is $(1 - \phi)\sigma(x)/q_s(x)$, where ϕ is the porosity of the deposited sediment. The mean grain size of the deposited stratigraphy therefore depends on the grain-size characteristics in the supply and the downstream extraction controlled by the spatial distribution of deposition. Consequently, granulometric trends are potentially valuable in constraining parameters describing the volumetric budget of sediment routing systems (Whittaker et al., 2011; Parsons et al., 2012).

Since both $q_s(x)$ and $\sigma(x)$ are determined by the tectonics and climate of the basin and its hinterlands, different basin types are expected to contain sediment routing systems with characteristic rates of downstream fining. For example, it is possible to discriminate between catchment-fan systems filling small wedge-top basins in thrust belts, which have rapid rates of downstream fining, from large, coalesced fluvial systems spreading sediment from the cores of mountains belts to distant foredeeps, which have low rates of downstream fining (Duller et al., 2010; Whittaker et al., 2011; Allen et al., 2013).

Allowing for the porosity (ϕ) of deposited sediment, and defining the depositional length L_d as the point at which the sediment discharge is exhausted, the downstream variation of the sediment discharge can be written

$$q_s(x) = q_s(0) - (1 - \phi) \int_0^{L_d} \sigma(x)\mathrm{d}x \qquad (9.28)$$

The sediment discharge can be calculated for different analytical expressions of the accommodation, or the model initialised with field observations. As an example, if the accommodation is given by an exponential model, decreasing away from the source region with a decay constant a, so that $\sigma(x) = \sigma_{max}\exp(-ax)$, equation (9.28) becomes

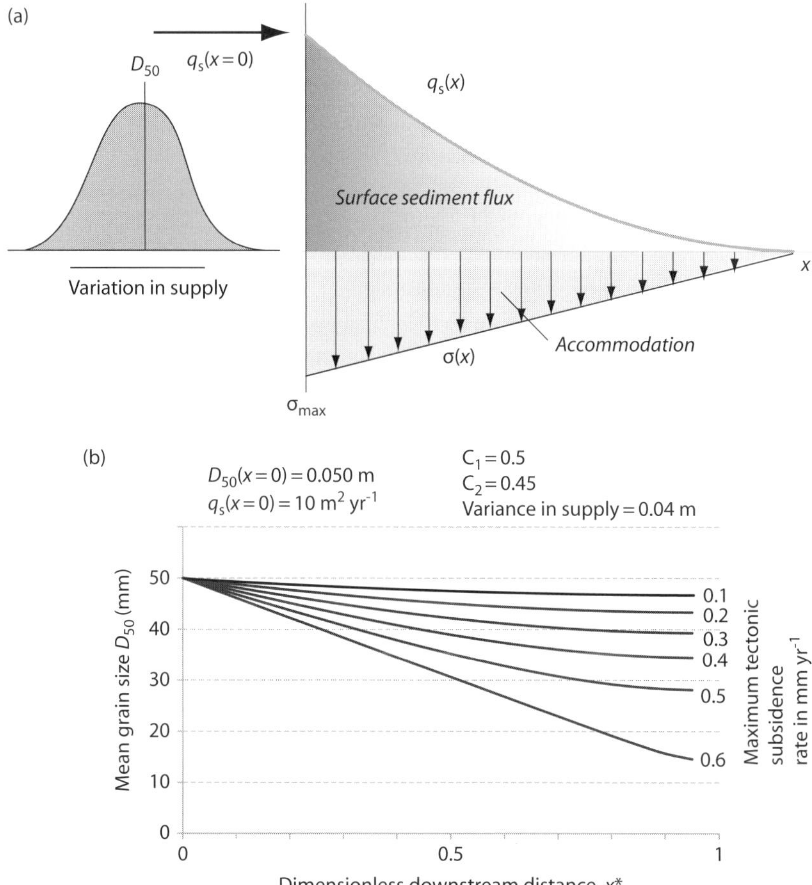

Figure 9.17 Downstream fining model with linear tilting for tectonic subsidence, showing downstream reduction of surface sediment flux q_s in (a). Mean grain size is shown for different values of maximum tectonic subsidence rate in (b).

$$q_s(x) = q_s(0) - (1 - \phi)\left[\frac{\sigma_{max}}{a}\exp(-ax) + 1\right] \tag{9.29}$$

Alternatively, for a sinusoidal distribution of accommodation, with a wavelength λ, increasing from zero at $x = 0$ to a maximum σ_{max} at $x = \lambda/4$, equation (9.28) becomes

$$q_s(x) = q_s(0) - (1 - \phi)\left[\sigma_{max}\frac{-\lambda}{2\pi}\cos\left(\frac{2\pi x}{\lambda}\right) - 1\right] \tag{9.30}$$

Equation (9.28) can easily be adapted to consider the sediment discharge for gravel only:

$$q_g(0) = f_g q_s(0) \tag{9.31}$$

for $x < L_g$, where f_g is the gravel fraction in the supply and L_g is the distance to the gravel front. Since at each point in the downstream direction $x < L_g$ the solid sediment volume extracted by deposition as a fraction of the surface sediment flux is $(1 - \phi)\sigma(x)/q_s(x)$, the non-dimensional parameter for the extraction from the surface flux by deposition R becomes

$$R(x) = (1 - \phi)L_g \frac{\sigma(x)}{q_s(x)} \tag{9.32}$$

by introducing the length term L_g in the numerator.

The x-axis can be transformed by setting $x^* = x/L_g$, in which case the integration of $R(x)$ over the normalised downstream distance x^* gives the cumulative downstream sediment deposition in relation to the available surface flux, $y^*(x^*)$. The dimensionless parameter $y^*(x^*)$ can be used to calculate the variation of mean grain size $D(x)$ from an initial mean grain size D_0 and variation ψ_0 in the sediment supply (at $x = 0$),

$$D(x) = D_0 + \psi_0 \frac{C_2}{C_1}\left[\exp(-C_1 y^*) - 1\right] \tag{9.33}$$

where C_1 and C_2 are constants related to the variation in the grain-size distribution (Fedele and Paola, 2007; Duller et al., 2010; Whittaker et al., 2011). C_1 is 0.5 and C_1/C_2 is 0.70 in the application in Duller et al. (2010).

The grain-size trend therefore reflects the grain-size distribution of the sediment supply, the sediment discharge from the upstream source, and the spatial distribution of accommodation (Duller et al., 2010; Parsons et al., 2012) (Figure 9.18). The rate of down-system fining decreases with increasing sediment supply, as indexed by increasing states of basin filling, F. F is a dimensionless parameter representing the ratio of q_{s0} to accommodation creation due to tectonic subsidence:

$$F = \frac{q_{s0}}{\int\limits_{0}^{L_d} \sigma(x)\mathrm{d}x} \tag{9.34}$$

so that the basin is underfilled when $F < 1$ and overfilled, leading to sediment bypassing the model basin at $F > 1$ (Figure 9.18).

9.4.4 Effects of Climate Change on Grain-Size Trends

In the rivers of the High Plains of the United States (Figure 9.19), which drain the Rocky Mountains, channel slopes have decreased since the Miocene (McMillan, Angevine, and Heller, 2002), suggesting that they have responded to climatic mechanisms rather than tectonic (see Wobus, Tucker, and Anderson (2010) for a discussion of the differentiation of climatic and tectonic forcing on river long profiles). There is a twofold increase in median grain size of sediment exported from the Rocky Mountains across the Miocene-Pliocene boundary (Duller et al., 2012) (Figure 9.20). This twofold increase is attributed to climatic

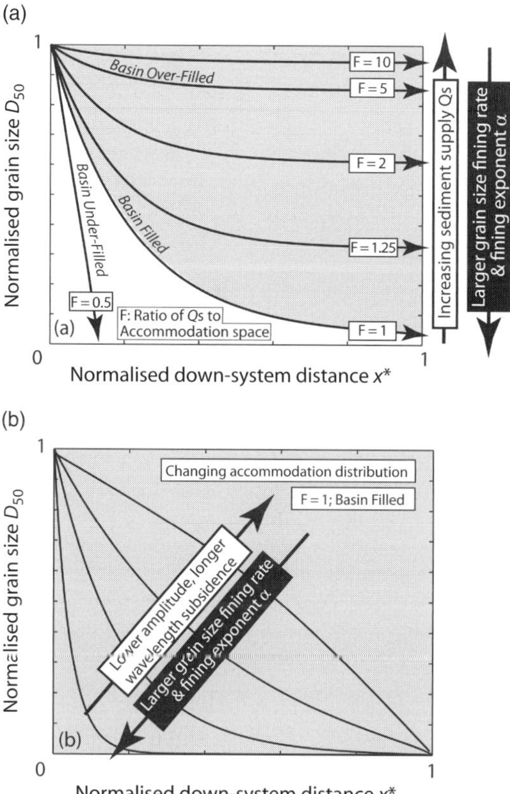

Figure 9.18 Down-system fining of grain size for different values of sediment supply (a) and tectonic subsidence (b). Perfect filling of the basin is at $F = 1$. Increased sediment supply leads to lower rates of downstream fining. For a filled basin, down-system fining is more rapid when the spatially exponential subsidence is higher amplitude and lower wavelength. After Duller et al. (2010) (fig.2) with permission of American Geophysical Union.

warming in the Pliocene, when there was higher precipitation and greater seasonality. River slopes $S(x)$ can be estimated from a Shields stress inversion of median grain size D_{50}

$$S(x) = \frac{1.4\tau_c^* \Delta\rho D_{50}}{H} \tag{9.35}$$

where τ_c^* is the critical Shields stress for entrainment of sediment on the river bed (0.05), H is channel depth and $\Delta\rho$ is the excess sediment density ($\rho_s - \rho_w$). Slope can be found by assuming that local shear stresses on the bed are 1.4 times the critical shear stress (Paola and Mohrig, 1996). Equation (9.35) can be combined with a resistance equation to derive estimates of unit width water dsicharge q_w (Figure 9.20b). D_{50} and H are constrained from field observations. The doubling of median grain size of the supply between the Miocene and Pliocene is associated with an increase in water discharge from $1-3$ m^2 s^{-1} to $3-6$ m^2 s^{-1}, and the thick succession of Pliocene Broadwater sediments suggest aggradation on the Great Plains after 3.7 Ma, consistent with high rates of sediment supply.

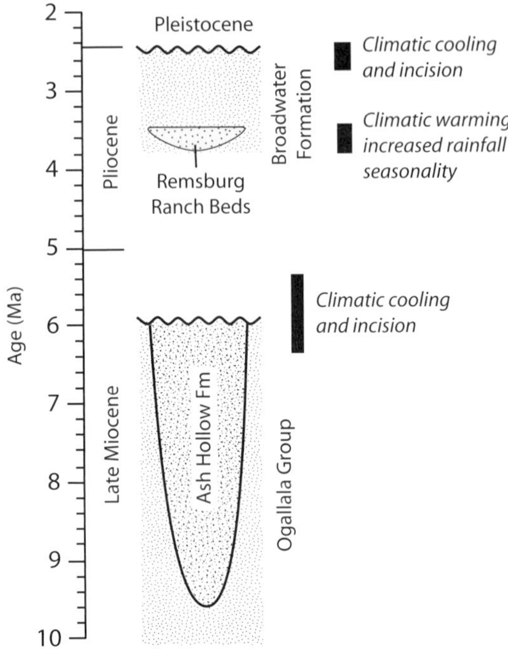

Figure 9.19 Fluvial stratigraphy of the Nebraskan High Plains from 10 Ma to the present day. Modified from Duller et al. (2012)(fig.1) with permission of Geological Society of America.

In numerical models where erosion has the form of a diffusive-concentrative equation (Smith and Bretherton, 1972; Simpson and Schlunegger, 2003) including a term for tectonic uplift, precipitation was doubled and halved (from a steady-state value of 1 m yr^{-1}) in order to investigate the effect of climate change on the grain-size distribution and stratal geometries in the basin (Whittaker et al., 2011). Doubling of the mean annual precipitation generated a sharp increase in sediment flux before returning to a steady-state value with a response time of about 0.5 Myr. The increased sediment flux causes a lengthening of the fan and an increase in median grain size. A halving of the precipitation rate causes a sharp reduction in sediment flux that returns to the steady-state value with a response time of approximately 1 Myr. The reduced sediment flux causes an abrupt backstepping of the fan toe and shrinkage of the fan. This backstepping is accompanied by an increase in the rate of down-system grain-size fining.

An increase in precipitation will increase the sediment grain size exported from the catchment because channel bed grain size scales on bed shear stress. The coupling between an increase in precipitation and an increase in sediment calibre leads to the marked progradation down-system of a gravel sheet with a duration of the order of the response time. The deposition of conglomeratic sheets down-system has previously been linked to increases in precipitation in the upstream catchment (Heller and Paola, 1992).

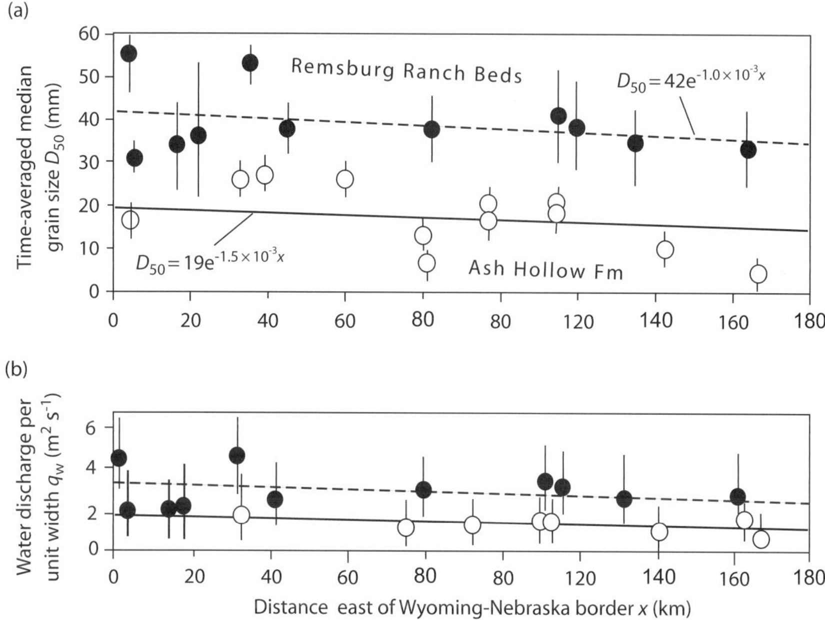

Figure 9.20 (a) Time-averaged median grain size of the upper Miocene Ash Hollow Formation and lower Pliocene Remsburg Ranch Beds. Grain sizes derived from Wolman point measurements ($n = 80$ for RRB and $n = 45$ for AHF). Vertical bars are the error associated with the combined median grain-size value. (b) Water discharge per unit width q_w predicted from measurements of median grain size and slope. Grain sizes and calculated water discharges are higher for the Pliocene Remsburg Ranch Beds than for the upper Miocene Ash Hollow Formation. From Duller et al. (2012) (figs.2, 4) with permission of Geological Society of America.

The Palaeocene-Eocene boundary (55 Ma) in the Tremp Basin of the Spanish Pyrenees coincides with the presence of the Claret conglomerate sheet (Figure 9.21). The Claret Conglomerate is about $1 - 4$ m-thick (locally up to 8 m) and bounded above and below by different paleosols that indicate an increase in precipitation rate over time (Schmitz and Pujalte, 2007). Climates and environments changed from semi-arid alluvial plains with small channels to a vast conglomeratic braidplain under a much wetter climate regime. The Claret Conglomerate is an extensive sheet with an erosive, low-relief base, covering an area of at least 500 km^2. However, the source of the conglomerate lay approximately 10 km to the northeast of the most proximal outcrops, so the braidplain may have covered an area of 2,000 km^2 at the Palaeocene-Eocene boundary. Maximum clast size is 65 mm, much coarser than pebbles in underlying channels, which rarely exceed 25 mm. The braidplain formed over a short time period of a few thousand years ($\sim$10 ky) directly after the Palaeocene-Eocene boundary. Such a short-lived pulse is consistent with the timescale of a transient response to a climatic perturbation in the model runs of Armitage et al. (2011).

The Palaeocene-Eocene thermal maximum (PETM) is registered in the marine time-equivalent stratigraphy of the Pyrenean Gulf in the form of a 2–4 m-thick (locally

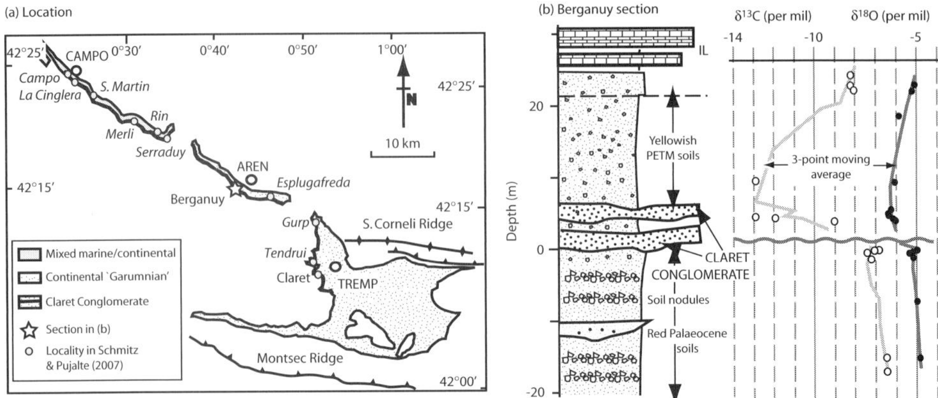

Figure 9.21 (a) Location of Claret Conglomerate at the Palaeocene-Eocene boundary in south-central Pyrenees. The Tremp Basin was an alluvial plain that passed westwards into a carbonate platform. (b) The stratigraphic section at Berganuy, showing isotopic data from carbonate nodules. Red paleosols with relatively positive carbon and oxygen isotopes are found in the Palaeocene, which pass up via the Claret Conglomerate into yellowish isotopically light paleosols of the Palaeocene-Eocene Thermal Maximum (PETM). IL refers to carbonates of the Ilerdian transgression. Modified from Schmitz and Pujalte (2007) (figs.1, 3) with permission of Geological Society of America.

up to 20 m) unit of clays intercalated within a carbonate-dominated succession (Pujalte, Baceta, and Schmitz, 2015). The clays represent a massive input of terrestrial siliciclastic sediment into the ocean attributable to the profound hydrological changes associated with the PETM. In addition, coarse-grained siliciclastics infilled incised valleys and a long-lived deep marine channel. The presence of coarse and fine-grained terrestrial siliciclastics in the Pyrenean Gulf coeval with the development of an extensive fluvial braidplain illustrates the down-system propagation of the climate signal through the sediment routing system.

9.5 Linking Source to Sink: Provenance Tools

The English poet William Blake wrote in 'Auguries of Innocence' (written 1803, published 1863):

To see a world in a grain of sand
And a heaven in a wild flower,
Hold infinity in the palm of your hand,
And eternity in an hour.

'Auguries of Innocence' contains a number of paradoxes where innocence stands alongside evil and corruption. An augury is a sign or omen. I doubt that Blake was thinking of provenance when he wrote those lines, but I do think he was marvelling at the big picture contained in small things, together with the contradictions and tensions. It seems improbable at first sight that global climate changes and tectonic movements might be reflected in the properties of a grain of sand. 'To see a world' of vanished source areas from a 'grain of sand' is an example of an inverse approach in the study of provenance. Provenance studies using the inversion of multiple proxies are a powerful way of linking sources to sinks, but do not capture the transformations that take place during weathering and transport. To develop models of such transformations, both compositional and textural, requires forward modelling calibrated by targeted field observations.

The chief inverse method of linking source to sink is to match the petrography, mineralogy, or chemistry of detrital grains to a source region using specific tracers. For many years a range of techniques has been applied to the study of the provenance of sand-grade sediment, including analysis of light fraction petrography (Zuffa, 1985) and heavy minerals (Mange and Maurer, 1992; Mange and Wright, 2007). Bulk and single-mineral geochemistry, isotopes and thermochronometers are increasingly used to constrain both the location and rock type of source areas and their exhumation history. U-Pb geochronology of detrital zircons allows the ages of eroding parent rocks to be identified and provides valuable insights into provenance.

The crystallisation age of zircons can be measured using the U-Pb decay series, usually with laser ablation ICP mass spectrometry. The distribution of U-Pb crystallisation ages can be compared with cooling ages obtained from fission track analysis of detrital and bedrock zircon samples. Double-dating allows zircon ages representing slow cooling during exhumation to be discriminated from ages representing rapid crystallisation due to volcanic or plutonic activity (Garver et al., 2000). Using U-Pb ages of detrital zircons to identify source regions and understand sediment trajectories is now a very well established and standard provenance tool. It is commonly used in conjunction with an analysis of heavy minerals, clast compositions and light fraction petrography (as an example, see Nicholson et al. (2014)).

Data sets containing U-Pb ages, heavy mineral point counts and petrographic point counts need to be extensive in order to be statistically meaningful. Consequently, statistical techniques are employed to quantify the pairwise distance between sample attributes that are fed into a multidimensional scaling (MDS) algorithm (Vermeesch, 2013). MDS visualises data as a map that preserves the difference between samples as relative distances between data points. An example is the analysis of sand samples from the Taklamakan desert to discriminate between a northerly provenance by wind and a southerly and westerly provenance by flowing water (Rittner et al., 2016). Consideration of heavy mineral, petrographic and U-Pb geochronological data collected from river bed and aeolian dune sands in the Taklamakan desert and from possible source areas surrounding the Tarim Basin (Figure 9.22) allowed the authors to conclude that the bulk of the sand originates from rivers draining the mountains in the southeast (Altun Shan), south (Kunlun Shan) and west (Pamir), but not by winds blowing predominantly from the Junggar Basin and Altai in the north.

Bulk petrography is analysed by point counting, using quartz, feldspar and lithic fragments as the main grain types. Lithic fragments are commonly subdivided into volcanic, sedimentary and metamorphic varieties. Mineralogy is conventionally illuminated from

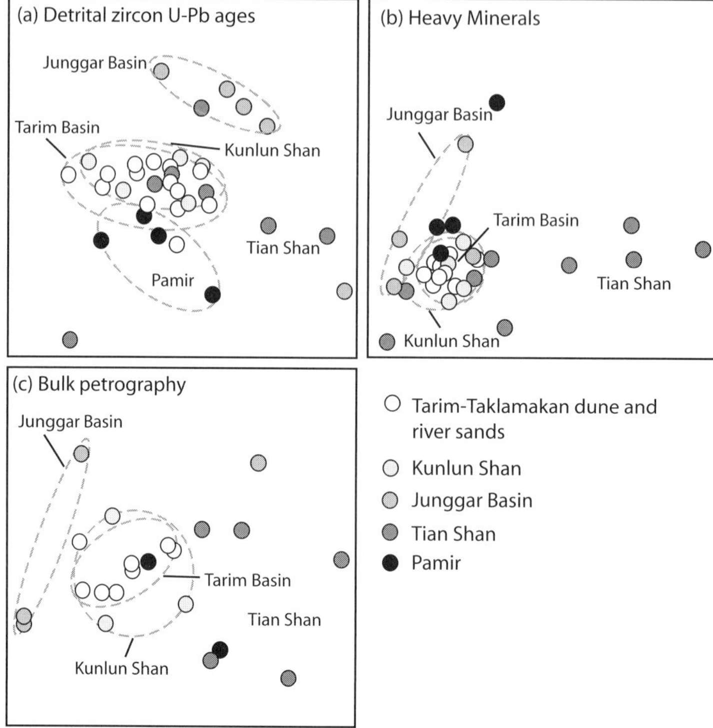

Figure 9.22 Multidimensional scaling maps of detrital zirocn U-Pb ages, heavy mineral assemblages and light-fraction petrography for modern river and aeolian dune sands. Tarim Basin sands form a distinct cluster that is overlapped by a field for the tributaries of the Kunlun range. Samples from rivers draining the Karakorum and Pamir form a cluster adjacent to the Kunlun and Tarim data. Samples from the Junggar Basin fall well outside the Tarim Basin cluster. Samples from the Tian Shan do not form a coherent cluster due to their heterogeneity and possibility of reworking. Modified from Rittner et al. (2016) (supplementary material fig.1) with permission of Elsevier.

ternary plots of heavy minerals: for example, pyroxene+olivine; the ultrastable group of zircon, tourmaline and rutile+garnet; and amphibole+the metamorphic trio andalusite, kyanite and sillimanite. Nicholson et al. (2014) used an apatite-tourmaline index ATi and a garnet-zircon index GZi to compare modern sands with older stratigraphy.

The sand in the Taklamakan desert can also be compared with the aeolian dust comprising the Chinese Loess Plateau and with other desert regions of NE China such as the Mu Us and Tengger deserts, and with the sediment load of the Yellow River (Figure 9.23). The Yellow River shows the closest affinity to loess sediments, as well as to the Mu Us desert, but there is no evidence to suggest that the Taklamakan desert is the source of dust for the Chinese Loess Plateau.

Drawing a linkage between sediment in the depositional sink and source regions can be carried out using other geochemical fingerprints. Weathering and sediment transport

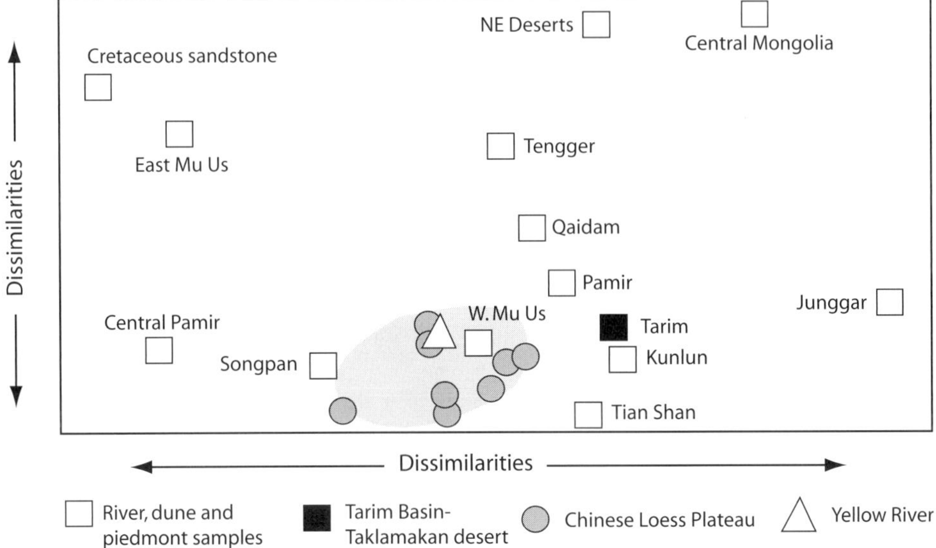

Figure 9.23 Multidimensional scale (MDS) map based on calculated Kolmogorov–Smirnov distances between U-Pb age spectra. Tarim Basin sands are compared with a range of possible source areas, modern sediment samples from the Chinese Loess Plateau and data from the upper reaches of the Yellow River. Modified from Rittner et al. (2016) (fig.5) with permission of Elsevier.

processes are not expected to result in isotopic fractionation, so the measured isotopic signature of any given sediment should reflect the bulk composition of the source. For example, ^{143}Nd/^{144}Nd isotopic values in clays recovered from the Bengal Fan are similar to those of modern sediments in the Ganga River (Goldstein, O'Nions, and Hamilton, 1984) and Bengal Fan sediments have been linked to denudation of source regions in the Himalaya (Bouquillon et al., 1990).

Eocene sedimentary rocks from the Gulf of Tonkin are uniformly less negative in ϵ_{Nd} than modern Red River sediment, indicating that Eocene sediments are more radiogenic and have younger crustal sources than the modern Red River. The Eocene samples are closer, however, to the ϵ_{Nd} of Tibetan sources for the Upper Yangtze River. Pb isotope analysis of single grains of feldspar allows further discrimination of the basement sources for the Eocene of the Gulf of Tonkin and supports the existence of Tibetan sources, but the differences with the modern Red River samples suggests that the Gulf of Tonkin received Yangtze Block detritus from the Middle Yangtze River before drainage capture by the Red River at some stage after the Eocene (Clift, Layne, and Blusztajn, 1994). A similar approach (Pb and Nd) was taken to the sediments of the River Indus (Clift et al., 2001b, 2002), supplemented by analysis of garnet geochemistry (Alizai, Clift, and Still, 2016). Provenance tools are therefore capable of highlighting major river drainage reorganisation events in contributing source areas.

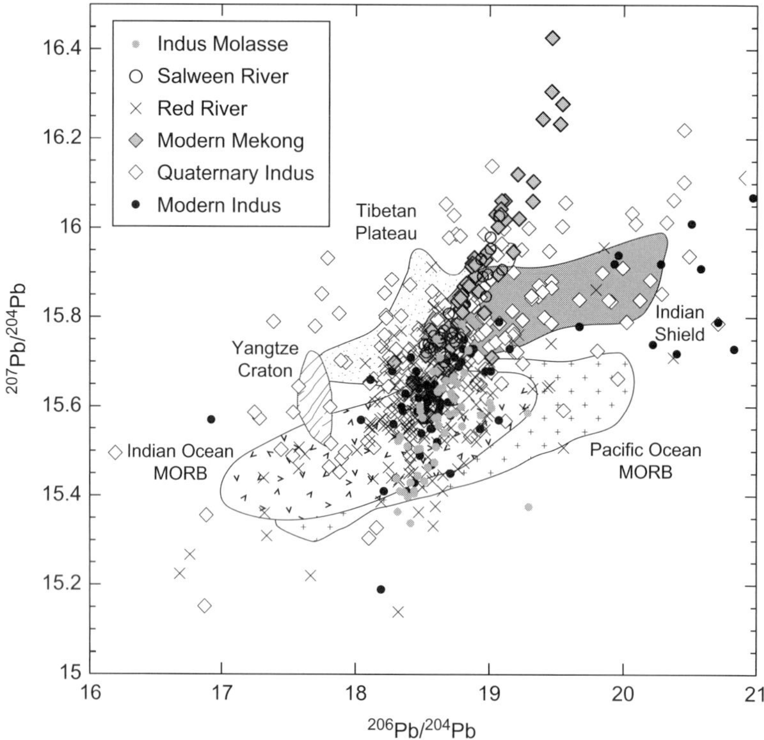

Figure 9.24 Lead isotope cross-plot of a number of detrital samples from modern rivers, Quaternary deposits and Neogene Molasse stratigraphy compared to possible source terranes. MORB; Mid-Ocean Ridge Basalt. Detrital grains plotting within the MORB fields are likely eroded from primitive arc-type sources within oceanic suture zones. Note that each set of sediments spans both primitive and more evolved continental sources, indicating a mixed provenance. Because the possible basement source rocks are not comprehensively surveyed many grains fall outside the defined fields (in this case the Yangtze Craton, Tibetan Plateau and Indian Shield). Grains with higher isotope ratios, especially ^{207}Pb/^{204}Pb tend to represent erosion of ancient continental crustal blocks. Data sources: Indus Molasse (Clift et al., 2001b); modern Indus River (Clift et al., 2002); Quaternary Indus River valley (Alizai et al., 2011); Red River (Clift et al., 2008b); Salween and Mekong Rivers (Bodet and Schärer, 2000); Indian and Pacific MORB (Stracke, Bizimis, and Salters, 2003). From Peter Clift (pers. comm.).

Lead isotopic ratios measured in detrital K-feldspars can be used to compare modern deposits from their ancient equivalents and correlatives (Figure 9.24). Modern sands and Quaternary and Neogene detrital equivalents are widely spread in a plot of ^{207}Pb/^{204} versus ^{206}Pb/^{204}, indicating mixing from more than one source. MORB, Indian continental crust, Yangtze craton and Tibetan Plateau occupy partly overlapping fields in terms of their Pb isotope characteristics.

10

Sediment Routing Systems and Sequence Stratigraphy

In 2008, Philip Allen wrote (2008b) (p.23):

The more we make steps in understanding landscapes and sediment routing systems, the more we discern their complex response to perturbations of all types. Had such information been available previously to guide stratigraphic models, it is arguable that the entire field of sequence stratigraphy … would have developed differently.

This is not to say that sequence stratigraphy and sediment routing system research are at war with each other, or that they occupy two distinct philosophical schools and 'social worlds' (Clarke, 1991). The two approaches to the understanding of stratigraphy are complementary and should ideally be used in tandem (Helland-Hansen et al., 2016). Nevertheless, the difference in approach of numerical modelling, which investigates through simulations the effect on stratigraphic architectures of variations in controlling parameters, and currently practised sequence stratigraphy, which, it is claimed, makes observations on stratigraphic architectures irespective of the controls, has been recently highlighted (Cantuneanu and Zecchin, 2016). With this purist methodology, sequence stratigraphy is rather like making a geological map in the field: boundaries are defined and mapped irrespective of their interpretation. Hence, Cantuneanu and Zecchin (2016) state (p.185) that

numerical modeling . . . has no bearing on the sequence stratigraphic workflow or methodology.

But what makes stratigraphy interesting and valuable is its use to unravel and understand the 'epic poem of the Earth' (see Preface) in an explanatory sense. Integrated system-type approaches to the fate of sediment from source to sink, including numerical modelling, have the power to fundamentally inform the sequence stratigraphy narrative, particularly in terms of the forcing mechanisms for observed architectural patterns and trends.

10.1 Insights from Quaternary Studies

Although the sequence stratigraphy paradigm was built principally on the interpretation of Mesozoic-Tertiary sedimentary rocks deposited at passive margins, the principles of sequence stratigraphy, and the dominant role of relative sea level in fashioning sequence architectures, are better displayed by the Quaternary record of continental margins (Lobo and Ridente, 2014; Ridente, 2016). High-frequency cycles are best preserved where high sediment supply rates combine with tectonic subsidence beneath continental shelves, and where the shallow depths of burial allow high-quality seismic reflection imagery.

The sea level history of the Quaternary under the influence of orbitally driven glacial-interglacial climate change is well understood. Sea level excursions in the Quaternary reflect Milankovitch climate cycles of period approximately 20, 40 and 100 kyr (Hays, Imbrie, and Shackleton, 1976; Schwarzacher, 2000). Such cycles are asymmetrical, involving abrupt warming during ice cap melting and gradual cooling during ice cap growth. They are also high in amplitude, of order 100 m sea level change in the case of 100 kyr cycles. These high-amplitude, asymmetrical cycles are recognised in the depositional architectures of continental margins (Yoo and Park, 2000; Lobo and Ridente, 2014). The most prominent feature is progradation of the shelf, which may occur during highstand, falling and lowstand phases. Rapid sea level rise is generally recorded by thinner, aggradational distal drapes and reworked deposits forming patchy clinoforms or channel-fill deposits. The stratigraphic architecture of Quaternary margins therefore differs from expectations from the prototype model in Vail et al. (1977) and its later derivatives (Wilgus et al., 1988) (Figure 10.1).

In Quaternary systems, such as the Korea Strait (Yoo and Park, 2000), where asymmetrical, high-amplitude sea level cycles are known to have taken place, sediment supply to the shelf does not decrease significantly during slow sea level fall, and sediment bypass of the shelf is restricted to the end of the falling stage. The lowstand stage is therefore marked by an elongate deltaic wedge of sandy muds and shelly sands deposited beyond the shelf edge, with no lowstand turbiditic fans. The transgressive deposits of sands and gravels are thin and retrogradational. The highstand is associated with the progradation of a muddy prodelta-shelf complex fed from rivers and is restricted to the inner shelf only. Fine sediment is unable to reach the outer shelf because of strong currents. Quaternary high-frequency sequences, such as those of the Korean Strait, therefore, are quite unlike the standard models based on Mesozoic-Cenozoic greenhouse cycles of 1–3 Myr period.

Despite the strong influence of relative sea level change on Quaternary margins, sequence architectures differ from place to place due to variations in tectonic uplift and subsidence and sediment supply driven by external forcings and/or internal dynamics (Figure 10.2). Such variability particularly affects lowstand deposition. Close to river input points, sediment discharges may be high enough to construct perched lowstand units composed of coarse-grained deltaic deposits, or to cause large-scale shelf margin and slope progradation (Porebski and Steel, 2003), as in the lowstand scheme of the conventional sequence stratigraphic model. Where sediment supply from deltaic sources is less important, slope progradation is impeded.

The sequence geometries found on Quaternary margins are strongly influenced by the amplitude and duration of sea level excursions. The predominance of 100 kyr cycles over the last 800 kyr has largely masked any stratigraphic expression of shorter-period cyclicity (20 and 40 kyr). During the Early Pleistocene and Holocene these shorter-period orbital

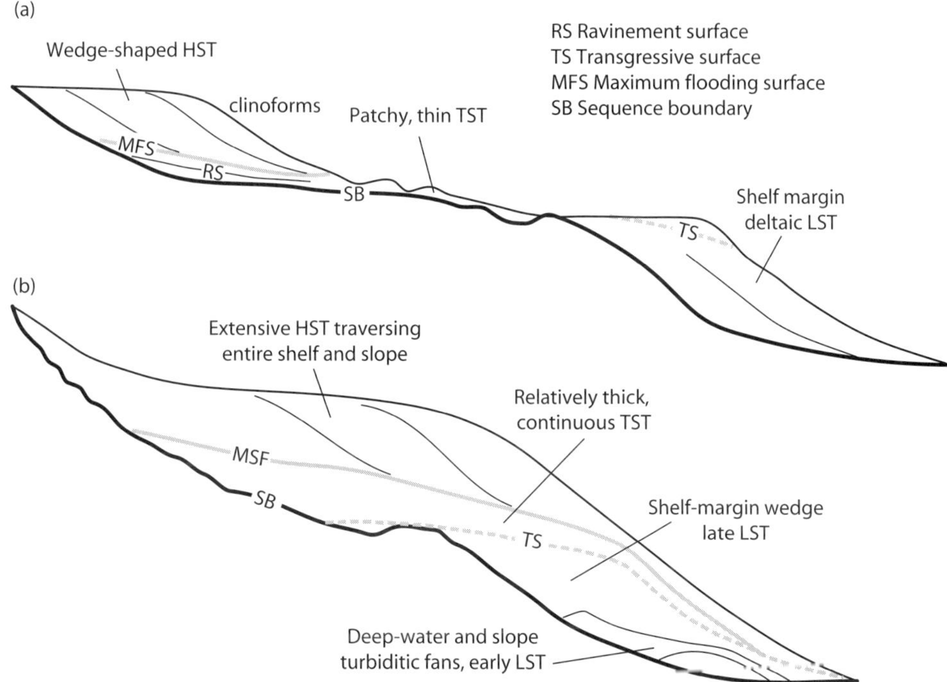

Figure 10.1 Schematic high-frequency depositional sequence architecture based on Quaternary continental margins, illustrated by the Korea Strait shelf (a) compared with the conventional sequence stratigraphy model (b). In (a), stratigraphy is dominated by the progradation of shelf-margin deltas during lowstand, a thin and patchy transgressive unit and progradation of a wedge-shaped highstand confined to the inner shelf. In (b), deep marine turbiditic basin-floor deposits and a shelf-margin wedge are deposited during the lowstand, with a relatively thick transgressive sheet and a progradational wedge extending across the shelf into deep water during sea level highstand. Modified from Yoo and Park (2000) (fig.13, p.308) and Lobo and Ridente (2014) (fig.2, p.2). Published with permission of Society of Economic Paleontologists and Mineralogists.

cycles are better expressed. A lengthening of cycle duration and a reduction of sea level amplitude would, according to some authors (Ridente, 2016), result in the sequence geometries proposed in the conventional sequence model, in which sediment dynamics and other allocyclic mechanisms take on an enhanced role compared to the Quaternary icehouse. The possibility exists, however, that greenhouse 'cycles' are not part of any ordered hierarchy (Schlager, 2004; Burgess, 2016) (Section 10.4).

The 100 kyr cycles that dominate Middle-Upper Pleistocene stratigraphy consist of sets of shelf sequences, each a few to tens of metres thick, enclosed by marked shelf-wide unconformities resulting from subaerial erosion and reworking during marine transgression (Figure 10.3). Individual sequences are dominated by regresssive deposits recording overall continuous progradation from highstand to lowstand of sea level. Since glacio-eustatic cycles involve a slow sea level fall and abrupt rise, the phase of progradation comprises

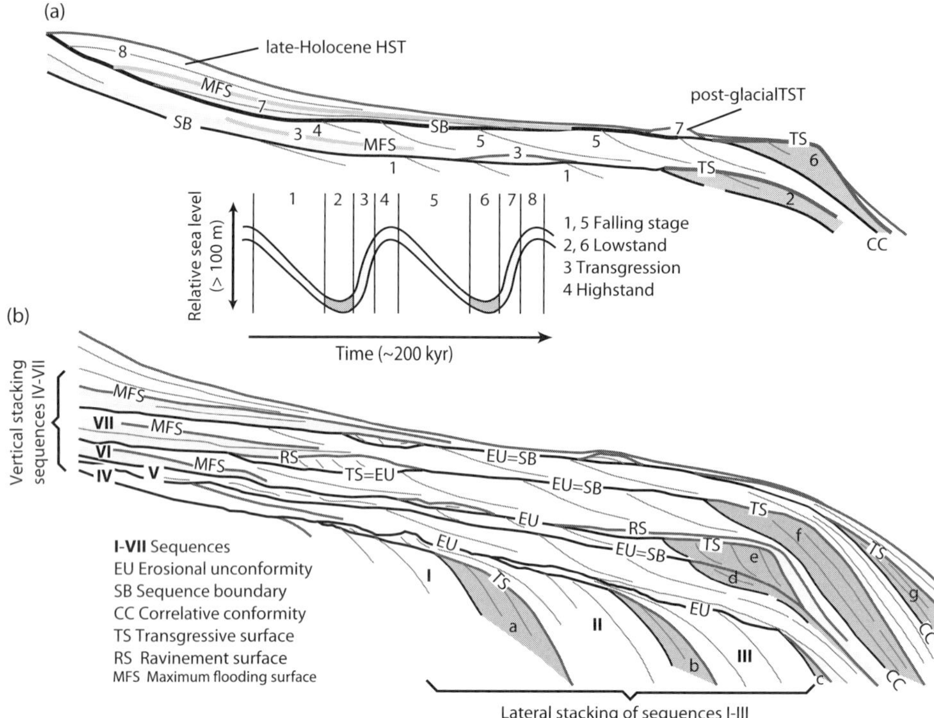

Figure 10.2 (a) Basic model representing 100 kyr sequences based on modern continental margins. Prograding clinoforms dominate. Highstand deposits (4) are limited to distal foresets and closely resemble falling stage deposits (5). (b) Ideal Quaternary margin recording multiple cycles, each exemplifying variants of the basic model. Sequences I to III stack laterally, typical of some pre-Middle Pleistocene narrow shelfal and tectonically uplifting settings, indicating a lack of accommodation generation on the shelf, forcing preservation of progradational units seaward of the shelf margin. Sequences IV to VII stack vertically, as is typical of subsiding margins, due to accommodation generation on the shelf during successive cycles. Lowstand deposits in darker grey: a-c, progradational slope; d, thin shelf margin; e, perched delta; f, progradational delta front; g, slope-confined 'draping' lowstand. Modified from Lobo and Ridente (2014) and Ridente (2016) (fig.4) with permission from the Geological Society.

greater than 70% of cycle duration. Deposition of transgressive deposits is short-lived, patchy and limited (Lobo and Ridente, 2014). Sediments deposited during falling sea level are relatively uniform, whereas those deposited during highstands and lowstands are more variable. Highstand units typically form at the base of or alternate with falling stage sediments, especially on muddy shelves. Lowstand sediments vary from shelf margin deltas to shorefaces, depending on the local supply of sediment and hydrodynamic setting of the shelf.

Quaternary marine sediment routing systems have the potential to yield quantitative data on the rates and periodicity of sediment supply to the deep sea, and on the external

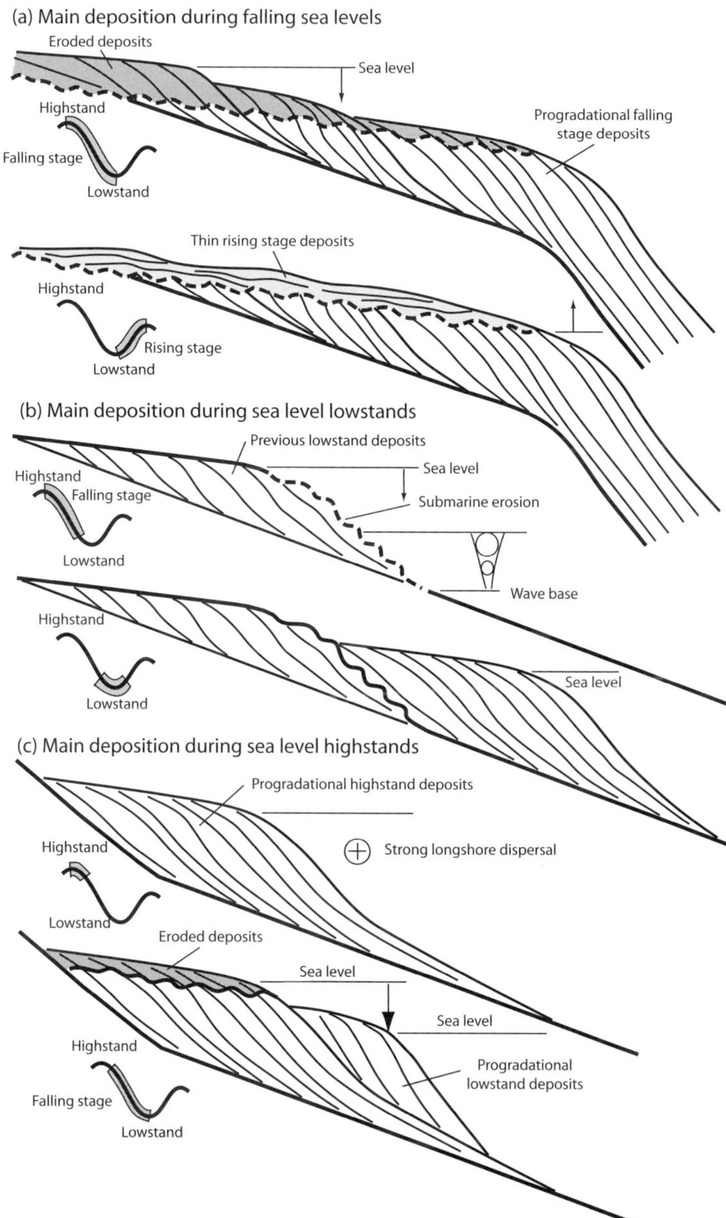

Figure 10.3 Evolutionary summary of high-frequency (100 kyr) sequences typical of the Middle-Upper Pleistocene stratigraphic record. (a) Main development of major sediment bodies during falling sea level, subsequent removal during subaerial erosion, followed by deposition of thin transgressive sheets. (b) Main development of major sediment bodies during relative lowstands, with front of lowstand clinoforms subject to submarine erosion. (c) Main development of major sediment bodies with strong alongshore sediment redistribution during relative highstands, followed by subaerial erosion of highstand wedge and deposition of falling stage prism. After Lobo and Ridente (2014) (fig.25, p.243) with permission of Elsevier.

or internal processes responsible. Geologically young marine records have the benefit of higher chronological resolution than the bulk of the Phanerozoic, provided by dating techniques including radiocarbon, and correlation with global oxygen isotope records. In addition, deep marine stratigraphy is more complete than terrestrial and shallow marine equivalents (Section 7.7.3). In the following paragraphs, Quaternary examples are given of the routing of sediment to the deep sea that illustrate the varied roles of sediment pathways, climate change and sea level change.

One of the best examples of deep marine successions with high chronological resolution is the Santa Monica Basin in the California Borderland (Romans et al., 2009) (Figure 10.4). The basin is fed with sediment from the Santa Clara River catchment (Warrick and Milliman, 2003) and the Santa Barbara littoral cell, which transports coastal sediment from west to east. The head of the Hueneme submarine canyon intersects the littoral cell. The Santa Monica Basin has an area of about 2,000 km^2 and a maximum water depth of 900 m. It is a closed basin with a ponded Quaternary fill (Normark, Piper, and Sliter, 2006), with no significant bypass to more distal settings. The Hueneme submarine fan occupies the western part of the Santa Monica Basin. Sea level has risen about 130 m since the LGM lowstand (18–25 ka) and reached its present level at ca. 7 ka. During lowstand, when the

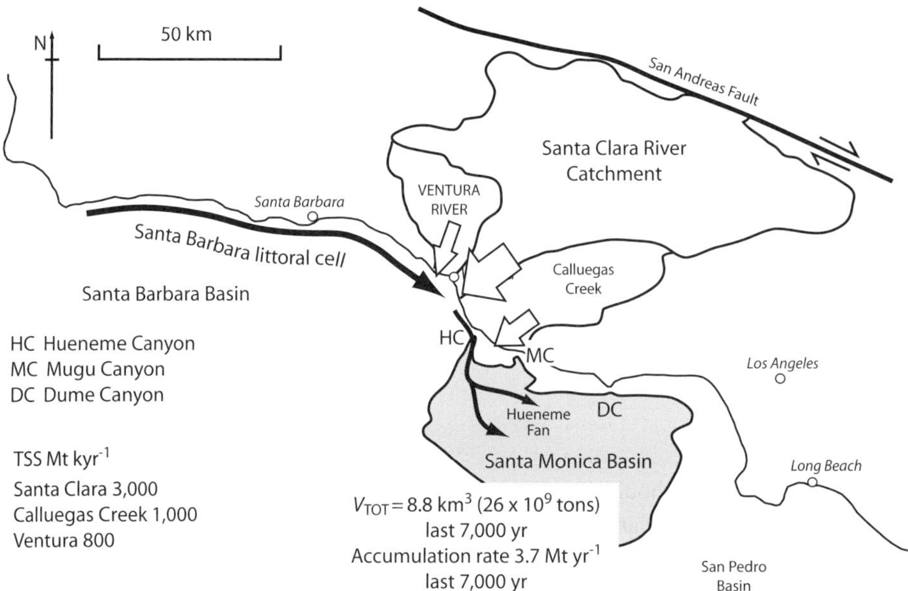

Figure 10.4 Sediment routing system of the Santa Clara River catchment and Hueneme submarine fan in the Santa Monica Basin, California. TSS, Total Suspended Solids for Ventura, Santa Clara and Calleugas rivers. V_{TOT} is total sediment volume of the Santa Monica Basin for the last 6.9 kyr. Modified from Romans et al. (2009) (fig.1, p.1396) with permission of the Geological Society of America.

shoreline was at the present shelf edge, the Santa Clara and other rivers delivered coarse grained sediment directly to deep water via submarine canyons. During post-glacial sea level rise, the Hueneme Canyon maintained a connection to the shallow marine sediment routing system, allowing Holocene sediment to continue to accumulate on the Hueneme Fan. New radiocarbon dates from core material, tied to seismic reflection data, were used to calculate a sediment budget over the last 7 kyr (Romans et al., 2009). The average accumulation rate of sediment for the total duration of five stratigraphic intervals identified from 6.9 ka to the present is 1.3 km^3 kyr^{-1} and the total mass of 3.7 Mt yr^{-1} compares well with the estimated sediment flux of the Santa Clara River over historical times (3.2 Mt yr^{-1}). The average accumulation rate, the average thickness of sand per turbidity current event, and the background hemipelagic deposition rate all increase from the older half to the younger half of the 7.9 kyr to present-day succession.

Historical records of the Santa Clara River show a correlation of increased sediment flux with increased frequency and magnitude of ENSO (El Niño Southern Oscillation) events. There was a marked increase in frequency at approximately 3 ka. In relation to sea level, sediment continued to be fed into deep water during the highstand period since 6 ka, since the Hueneme Canyon maintained its connection to sediment sources (Covault et al., 2007). The evolution through the last 7 kyr of the Holocene is thought to be a result of the change from direct river connection to the canyon head to littoral cell connection. The variability of sediment accumulation rate and sand percentage in the deep-sea Santa Monica Basin is therefore primarily due to the impact of climatic ENSO events on sediment flux in source area catchments combined with a switch in the connection of the canyon head with littoral sediment pathways in the staging area of the sediment routing system.

The role of climate change in the discharge of sediment into deep marine settings has also been tested in the Santa Ana River catchment, which feeds the Newport deep marine fan offshore California (Covault et al., 2000) (Figure 10.5). The Newport Fan continued to receive sediment during the Holocene marine transgression since it stayed connected to the Santa Ana River mouth. The Santa Ana River has historically delivered high sediment fluxes during ENSO-related flood events. It dominates the offshore fluxes to the head of the Newport canyon, with a small contribution from longshore currents of the San Pedro littoral cell. A variety of climate proxy information indicates the variability of precipitation in onshore sediment source areas over the last 9.5 kyr (Kirby et al., 2007), from a wet early Holocene followed by a long-term drying trend. It is inferred that the wet climate of the early Holocene should be reflected in high sediment discharges to the Newport canyon-fan system. Radiocarbon dates from piston cores taken across the Newport system allow sediment accumulation rates to be calculated. Average deposition rates calculated from individual radiocarbon cycles range widely in the latest Pleistocene-Holocene, but average rates for 1 kyr intervals since 10 ka range between 200–280 mm kyr^{-1}. The middle Holocene (7–3 ka) experienced the lowest deposition rates. Importantly, variation of offshore deposition rates is similar to onshore climate proxies, wetter climates being associated with enhanced deposition rates, and drier climates with reduced deposition rates. There is little evidence of a time lag between onshore climate change and deposition

               Sediment Routing Systems and Sequence Stratigraphy

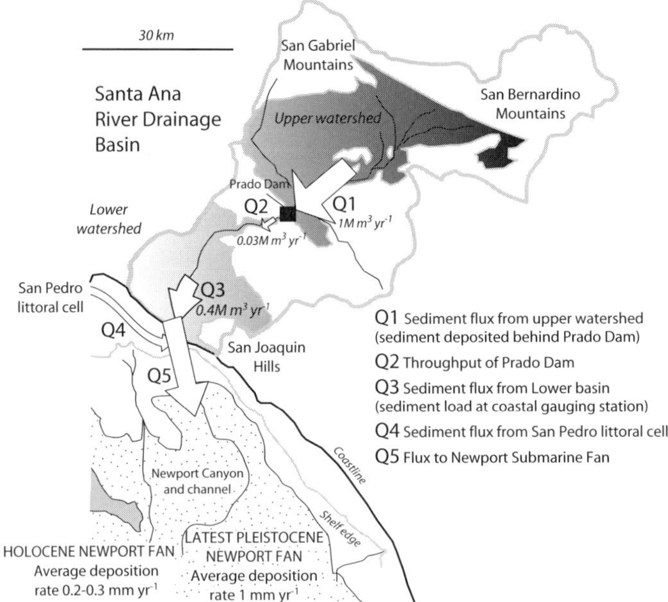

Figure 10.5 Sediment routing system of the Santa Ana River and Newport submarine fan, California, with sediment fluxes shown in terrestrial compartment and sediment accumulation rates in submarine sinks. From Allen and Allen (2013)(fig.3, p.229) with permission of John Wiley & Sons Inc., using information in Covault et al. (2000)(fig.1, p.249) and Warrick and Rubin (2007).

in the deep sea in the small scale, steep and tectonically active setting of the California Borderland.

The Golo sediment routing system is situated on the eastern margin of the island of Corsica in the western Mediterranean (Figure 10.6). The terrestrial sediment source area is the Golo River catchment, with an area of 1,005 km² and mainstem length of 90 km. The Golo River feeds 0.002 Mt yr⁻¹ of suspended sediment into the ocean. The continental shelf is narrow (10 km) and extends to a water depth of 110 m. It passes into a deep (700–900 m), narrow (45 km) confined basin known as the Corsican Trough, which acts as the absorbing state of the sediment routing system. The entire Golo deep sea fan system covers an area of approximately 500 km². A sediment budget (Sømme et al., 2011) shows that onshore storage of sediment has been relatively minor (about 13% of the overall budget) during the late Quaternary, and has taken place at all stages of relative sea level, including highstand, transgression and lowstand (Figure 10.7). The volume of sediment stored in the catchment (3.4 ± 1.8 km³) is roughly the same as that currently stored on the continental shelf, whereas the total sediment volume in deep sea fan lobes is 23.7 km³, of which 25–50% is sand. Deposition in the deep sea has ranged from being 50% higher than the river supply to 25% lower than the river supply during the late Quaternary, suggesting that the continental shelf must be acting as a sediment staging area (Sømme et al., 2011), alternately

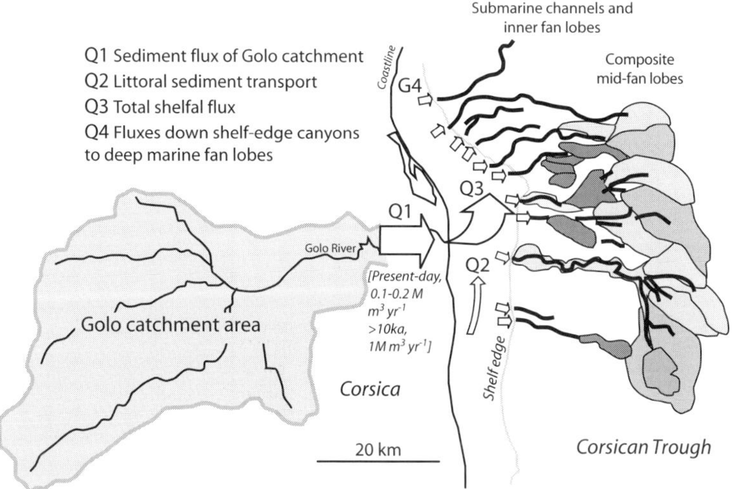

Figure 10.6 The Golo sediment routing system, which is fed from a catchment in eastern Corsica, delivers sediment to a narrow continental shelf before export to deep marine lobes, which are grey-coded according to age. Distribution of lobes and canyons from Deptuck et al. (2008). After Sømme et al. (2011)(fig.1) with permission of Society of Economic Paleontologists and Mineralogists.

storing sediment and releasing sediment to the deep sea. Delivery to the deep sea may or may not be in phase with variations in the river supply and in alluvial aggradation (marked by terrace formation), making correlation of depositional units within the sediment routing system and application of the 'standard' sequence stratigraphic model problematical.

The Golo system demonstrates that alluvial aggradation and the timing of sediment delivery to the offshore shelf, slope and deep sea is triggered when external forcing drives the system across internal thresholds. Consequently, the same external factors may result in different sequence architectures depending on these internal thresholds. There is no simple rule of thumb for predicting the timing of deposition of sediment in the deep sea and the timing of alluvial aggradation (Figure 10.7). The entire Golo system may aggrade along the entire onshore-offshore transect at any time, in contrast to sequence stratigraphic models that involve the seaward and landward migration of the main depocentre driven by relative sea level change.

10.2 Orbitally Driven Signals in Stratigraphy

Orbitally forced climate variations are widely accepted to be registered in stratigraphy, particularly in the pelagic and hemipelagic environments of lakes, seas and oceans. Cyclicity is commonly manifested in variations of the concentrations of the skeletons of microflora, detrital silts, clays and wind-blown dust, pollen records as well as in stable isotope ratios, trace element abundances and even in the gamma ray response of shales from well logs. In pelagic and hemipelagic environments, Milankovitch cyclicity reflects regular changes

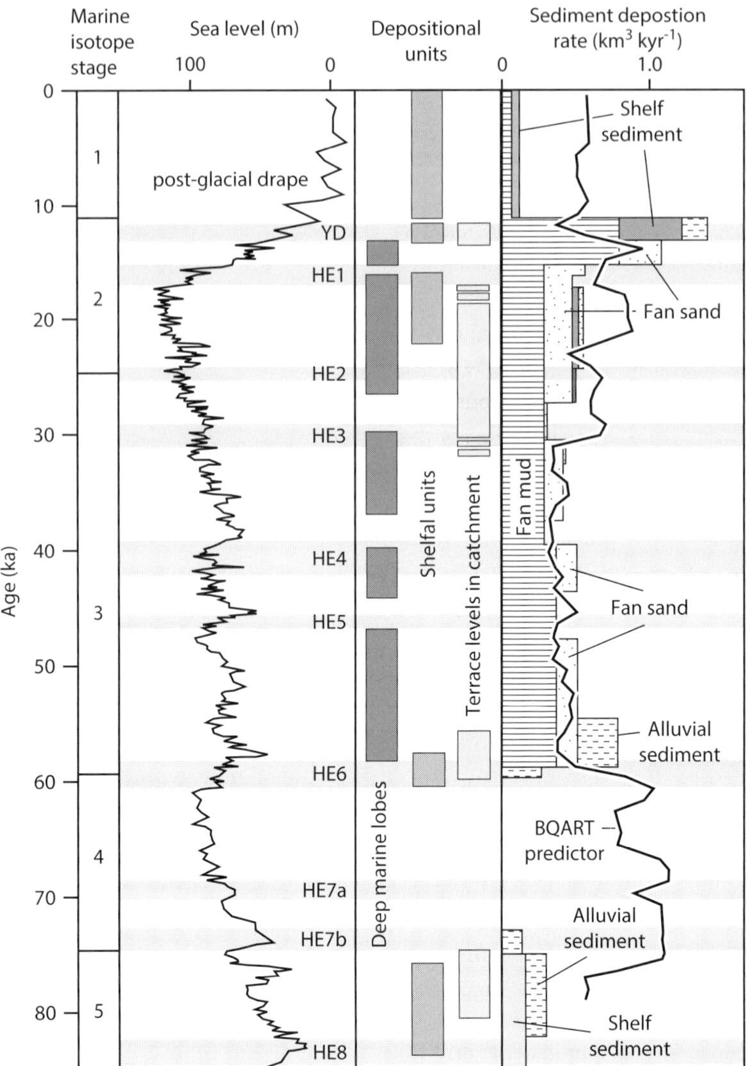

Figure 10.7 Correlation diagram showing eustatic and stratigraphic events in the Tyrrhenian Sea together with deposition rates of sedimentary units in the Golo sediment routing system. YD, Younger Dryas; HE1 to HE8, Heinrich events. Redrawn from Sømme et al. (2011) (fig.12, p.132) with permission of Society of Economic Paleontologists and Mineralogists.

in either organic productivity, dilution by detrital components, or chemical dissolution, oxidation and diagenetic overprinting (Fischer, 1986).

The sea level signal contained in $\delta^{18}O$ records from deep sea cores has an orbitally tuned chronology covering the last couple of millions of years (Pillans, Chappell, and Naish, 1998). The recognition of Milankovitch cyclicity (at 20, 40, 100, 400 kyr period) has been extended from the glacial-interglacial cycles of the Quaternary to older periods of Earth history, such as the Palaeocene-early Eocene (Hilgen, Kuiper, and Lourens, 2010), the Oligocene (Paike et al., 2006) and the Miocene-Pliocene boundary (Hilgen, 2007). Analysis of the gamma ray curves of well logs in the Permo-Carboniferous of the Jiyang depression, China suggests that orbital forcing can be detected in ancient icehouse periods of Earth history (Yu et al., 2008).

Some pelagic-hemipelagic and shelfal marine sedimentary sequences have been subjected to time series analysis of variations in bedding thickness to confirm a Milankovitch band cyclicity (Fischer and Schwarzacher, 1984; Weedon, 1989; Weedon and Jenkyns, 1990). In all these cases, the orbital control of sedimentary rhythms is overwhelmingly by a direct impact on the depositional environment of changing climate and seasonality, with little mediation by sediment transport systems. Cyclicity in shallow water marine environments, such as carbonate platforms, is more controversial, partly because biostratigraphic resolution is inferior to that in pelagic environments, partly that erosional gaps are more likely and partly that sedimentary rhythms may reflect sediment advection in complex, heterogeneous depositional mosaics. The classic succession of the Middle Triassic Latemar carbonate platform of the Alps, for instance, has been interpreted as an excellent example of Milankovitch-band orbital forcing (Hardie, Bosellini, and Goldhammer, 1986; Hinnov and Goldhammer, 1991; Zuhlke, Bechstadt, and Mundil, 2003), backed up by spectral analysis (Preto et al., 2001), whereas improved biostratigraphy and radiometric dating (Brack et al., 1996; Mundil et al., 1996) make an orbital forcing interpretation problematical.

Milankovitch cyclicity has also been interpreted in the growth of fan lobes in deep marine, turbiditic successions (Weltje and de Boer, 1993). The changes in the sediment supply to the delta-fed turbidite system in the Pliocene of Corfu (Greece) is directly attributed to precession-driven (23 kyr period) changes in precipitation and continental run-off, since the Mediterranean was cut off from the global ocean at this time and immune from glacial-interglacial eustatic fluctuations. The 500 m-thick turbiditic succession spans about 500 kyr and the average thickness of lobe units is 20–25 m, which means that the average time span for a lobe is 20–25 kyr. The average time span is therefore close to the precession period. Autocorrelation functions of bed thickness data, magnetic susceptibility and mineralogical maturity confirm a periodicity of 23–24 kyr. The Pliocene turbidites of Corfu show that cyclicity in deep marine successions may reflect palaeoclimatic changes in the terrestrial source region driving variations in sediment supply, rather than invoking glacio-eustatic sea level change (Section 10.2.1).

In terrestrial environments, Milankovitch-scale cyclicity is recorded in lake successions as cycles in 'non-glacial varves' comprising both organic-siliciclastic-carbonate sediments, as in the Eocene Green River Formation of Utah and Wyoming (Bradley, 1929), and evaporites, as in the Triassic Lockatong Formation of the Newark rift of New York-New Jersey (Van Houten, 1964; Olsen, 1984) and the Jurassic Todilto Formation of New Mexico (Anderson and Kirkland, 1960). Climatic influences attributed to Milankovitch band forcing are also recognised in trends in aridification (Dupont-Nivet et al., 2007), strengthening and weakening of monsoonal circulation (Xiao et al., 2010) and in the stacking of paleosols in floodplain successions (Aziz et al., 2008). Once again, in all these cases the influence of climate variability is to exert a direct control on the nature of the depositional environment through changing hydrology and chemistry.

A key question is whether external (allogenic) orbital forcing can be recognised in more complex sedimentary systems where time-averaged sediment discharges are affected by the transient response of erosional and depositional landscapes to climate perturbations (Allen, 2008a). Milankovitch forcing, for example, has been inferred to explain variations in water and sediment discharge on alluvial fans and in the supply and rate of accumulation of sediment in the deep sea (Van der Zwan, 2002; Waters, Jones, and Armstrong, 2010; Covault et al., 2011), and has been invoked to explain periodic development of anoxia is shallow Cretaceous seas controlled by river input of nutrients (Beckmann et al., 2005).

10.2.1 Effects of Sea Level Change

There is an abundant record of the effects of Pleistocene-Holocene climate change on fluvial systems (Bull, 1991), principally in the occurrence of aggradation-incision cycles. However, the timing of aggradation-incision cycles is not related to base level oscillations driven by eustatic change in any simple or linear way. The timing of aggradation events varies from place to place depending on topography and climate, and aggradation surfaces may be diachronous within a single drainage basin. The frequency of Pleistocene aggradation events is generally much lower than that of climatically driven sea level highstands and lowstands (Chappell and Shackleton, 1986). In the Mojave Desert and nearby San Gabriel Mountains, there are 3 aggradation events over the last 130 kyr, each coinciding with rising sea level, but there are 11 marine terraces formed at highstands of sea level over the same time period. These aggradation events coincide with the cutting of strath terraces in humid fluvial systems in New Zealand, such as the Charwell River (Bull, 1991). Aggradation events in the Charwell River catchment occurred at times of full glaciation in the Southern Alps, between the intervals of strath cutting. Aggradation appears to be a complex response to climate change involving the accumulation of sufficient stored hillslope detritus before stripping causes aggradation to take place.

Sediment delivery to the ocean reflects incision-aggradation cycles rather than simply conveyance of the products of erosion of upland catchments. There are therefore two contrasting situations for sediment delivery by river systems, which have been termed 'conveyor belt' and 'vacuum cleaner' modes (Blum and Hattier-Womack, 2009) (Figure 10.8). The effectiveness of the two different models can be investigated using the Late Quaternary history of sediment routing systems on the northern Gulf of Mexico continental margin (Blum and Törnquist, 2000; Blum and Garvin, 2010; Galloway et al., 2011).

The oscillation of sea level is accompanied by climate change in the erosional engines of sediment routing systems. Consequently, in principle, the sediment efflux of upland catchments should decrease during cold (glacial) climatic phases and increase during warm (interglacial) climatic phases. Use of the BQART predictor (Syvitski and Milliman, 2007)

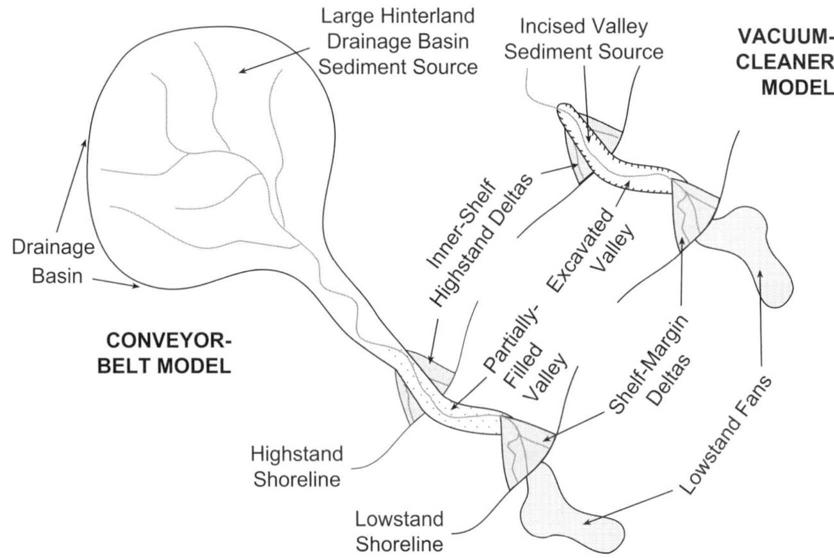

Figure 10.8 Conveyor belt versus vacuum cleaner models for sediment supply. The conveyor belt model involves a large inland drainage basin feeding large amounts of sediment directly to the shelf margin, whereas the vacuum cleaner model produces smaller quantities of sediment by excavating an incised valley fill. The vacuum cleaner model generates much smaller quantities of sediment than the conveyor belt model. From Blum and Hattier-Womack (2009) (fig.11, p.25) with permission of Society of Economic Paleontologists and Mineralogists.

(Section 7.3) suggests that sediment supply may be reduced by 10–40% during glacial periods corresponding to eustatic lowstands. Consequently, the Late Quaternary record of sediment discharge from upland catchments, assuming no transient behaviour, should follow an oscillating path in phase with temperature change during glacial-interglacial cycles. Periods of incision add little additional sediment to the sediment delivered to the ocean. Consequently, the sediment supply from the hinterland may be at a maximum when river mouths are in highstand positions, causing storage to be focussed on the coastal plain and inner shelf (Figure 10.9). Sediment supply from hinterlands may be at a minimum when river mouths are extended to the shelf margin during lowstands, when sediment is fed directly to the slope and deep basin floor.

Incised valleys form in a stepwise manner, with short periods of incision and extended periods of lateral channel migration and valley widening during which channel belt sands are deposited. The total volume of sediment exported during a period of incised valley formation is, however, small compared to the ongoing conveyor belt from upstream catchments. In contrast, periods of lateral channel migration and valley widening significantly increase sediment export (by 10–30%), so periods of fluvial aggradation correlate with increased sediment delivery to deep-water depocentres, the reverse of early sequence stratigraphic thinking (Posamentier and Vail, 1988).

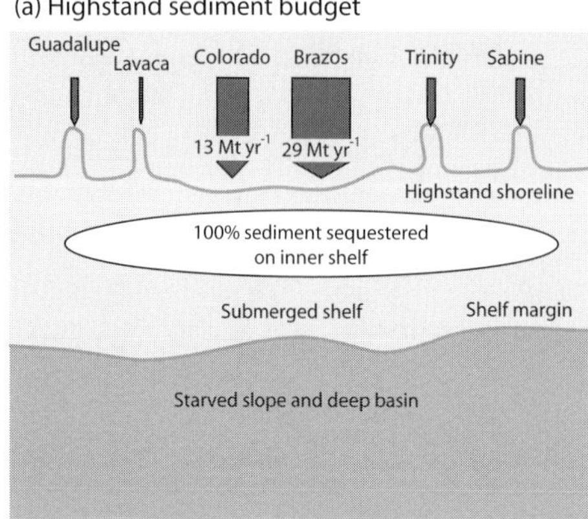

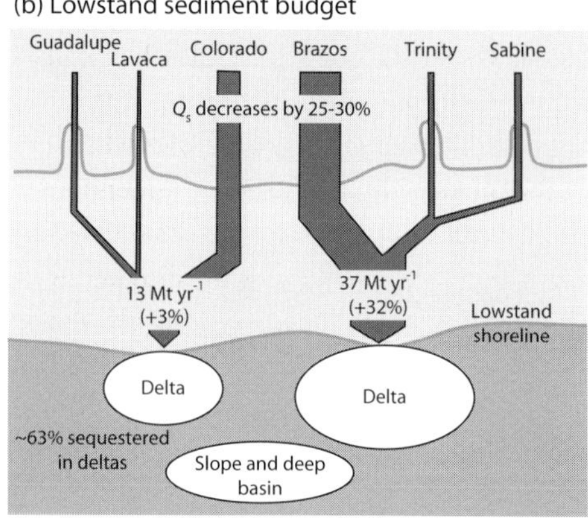

Figure 10.9 Schematic diagram illustrating sediment routing and budgets for the Gulf of Mexico fed by the Colorado and Brazos rivers for (a) the highstand typified by the present-day, and (b) the lowstand, typified by glacial maxima of the Quaternary. Redrawn from Blum and Garvin (2010) (p.32, 34 of 35) with permission of American Association of Petroleum Geologists.

During lowstands, when the shelf is subaerial, channel systems may merge, increasing drainage area and sediment flux, and therefore increasing delivery to the heads of canyons incised into the shelf margin (Figure 10.9). In the northern Gulf of Mexico during the Late Quaternary, the Guadalupe, Lavaca and Colorado systems merged, the combined system delivering 13 Mt yr^{-1} of sediment to the lowstand shoreline situated at the shelf edge. Likewise, the Brazos, Trinity and Sabine rivers merged into one combined system that delivered 37 Mt yr^{-1} to the ocean.

In summary, the vacuum cleaner mode of sediment delivery to the ocean is relatively unimportant (∼5%) compared to the conveyor belt when assessed in terms of their contribution of the total flux to the ocean. Processes such as the temperature-dependent efflux of weathering products from upland catchments and the merging of lowstand river systems on the exposed shelf are more important in determining the sediment fluxes and spatial pattern of sediment delivery to deep-water depocentres.

The relationship between Milankovitch cyclicity and sediment supply, mediated through climatic and vegetational changes in river catchments, has been discussed widely. In the upper Miocene EA Field in Nigeria, the main deltaic sand bodies are correlated with dry intervals with monsoonal rains, recognised from the prevalence of savanna pollen, recurring with a period of 100 kyr (Van der Zwan and Brugman, 1999). The impact of Milankovitch-scale climatic variations on sediment supply is more likely to be recorded in transition regions between two contrasting climatic zones. In transition regions, catchments may experience cyclical variations of climatic conditions.

Van der Zwan (2002) tested the Milankovitch forcing of the sediment supply by rivers in the Neogene (6.5 Ma to present) Niger sediment routing system using the forward modelling package STRATAGEM. Since the onset of northern hemisphere glaciation, climatic zones have shifted significantly, for example from rainforest to savanna, with an apparent Milankovitch frequency, which has caused a cyclicity in sediment supply to the delta region. Their simulations suggest that in an icehouse world subjected to oscillating eustatic base level, eustatic variations overwhelm the effect of climatic forcing of sediment release from upland catchments. In a greenhouse world, however, Milankovitch-scale sediment supply variations are expected to be more significant in the absence of significant sea level variations. The packaging of alluvial stratigraphy in coastal plain settings may therefore be at orbital frequencies as a result of climate-driven variations in sediment supply. For example, a cyclicity of ∼100 kyr was identified in the Eocene Lower Montanyana Group of the Tremp Basin, Spain (Marzo, Nijman, and Puigdefàbregas, 1988; Clevis, de Boer, and Nijman, 2004b). Such frequencies may, however, also be generated by autocyclic processes.

The 1-D transport-limited catchment-fan model (Armitage et al., 2011) can be expanded to focus on how the system responds to repeated, cyclical change in climatic forcing, manifested in variations in precipitation (Bar-Matthews et al., 2003; Ruddiman, 2006) and to test if it is reasonable to expect the time-integrated clastic stratigraphic archive of sedimentary systems to be a faithful recorder of high-frequency climate change. Simpson and Castelltort (2012) used a physically based numerical model to show that sinusoidal variations in water flux in an upland catchment were linked to alternate periods of deposition of aggradational sediment wedges and subsequent flushing out, thereby causing a markedly episodic history of sediment discharge from the catchment. In contrast, using a different set of boundary conditions, Armitage et al. (2013) found that over time, a sinusoidal variation in climate, expressed in terms of precipitation and run-off, is damped when measured in terms of sediment discharge, rather than recorded or amplified. In other words, the Milankovitch signal is 'lost in translation'. They concluded that high-frequency climate variations in the Milankovitch band are damped over time by the slow response time of the catchment relative to the timescale of the forcing.

10.3 Analogue and Numerical Experiments of Sequence Architectures

The sequence stratigraphy method originated from the analysis of distinctive geometries on seismic reflection profiles and was extended to outcrop and core data. It has, however, been relatively slow to adopt new results from the field of experimental stratigraphy using analogue experiments and numerical forward modelling. Such experiments have allowed the exploration of parameter space for external controls on stratigraphic architectures as well as revealing the possibilities of autogenic behaviour (Muto, Steel, and Swenson, 2007; Paola et al., 2009; Burgess and Prince, 2015).

Migration of moving boundaries, such as the shoreline, which generates a 'shoreline trajectory', is a common feature of numerical models of stratigraphy under external forcing from climate change and tectonics (Armitage et al., 2011; Hampson et al., 2014). Changes in precipitation as a proxy for climate and in tectonic uplift/subsidence rate generate grain-size trends and stratigraphic geometries that might be mistakenly attributed to changes in base (sea) level. Armitage et al. (2011), for example, used a simple sediment routing system of a 10 km-long frontal catchment and an up to 20 km-long alluvial fan, separated by a vertical normal fault. Tectonic uplift of the catchment, recorded as a slip rate on the range-bounding fault, causes erosion, which results in the transport of sediment into the hangingwall basin, as described in detail in Allen and Densmore (2000) and Densmore, Allen, and Simpson (2007a) and in Section 8.5.3. Sediment transport is in the form of a diffusive-concentrative equation (Smith and Bretherton, 1972)

$$\frac{\partial h}{\partial t} = -\frac{\partial q_s}{\partial x} + U(x,t) \tag{10.1}$$

where h is the elevation at time t, x is the horizontal distance, q_s is the unit width sediment discharge, and U is the tectonic uplift rate within the catchment. The sediment flux is split into hillslope diffusion and fluvial transport, assuming that fluvial transport is proportional to precipitation. The erosion therefore becomes

$$\frac{\partial h}{\partial t} = \frac{\partial}{\partial x}\left(\kappa + c(ax)^n\right)\frac{\partial h}{\partial x} + U(x,t) \tag{10.2}$$

where κ is the linear diffusivity, c is the non-linear sediment transport coefficient (taken as 1×10^{-6} m^2 yr^{-1} (Simpson and Schlunegger, 2003)), a is the catchment-average precipitation rate and $n = 2$ is the exponent that describes the dependency of sediment discharge on fluid transport. Equation (10.2) is made dimensionless by the length of the catchment L_c and the diffusive timescale L_c^2/κ, giving

$$h = L_c \bar{h} \tag{10.3}$$

$$x = L_c \bar{x} \tag{10.4}$$

$$t = \frac{L_c^2}{\kappa}\bar{t} \tag{10.5}$$

and equation (10.2) therefore becomes

$$\frac{\partial \bar{h}}{\partial \bar{t}} = \frac{\partial}{\partial \bar{x}}\left(1 + R\bar{x}^n\right)\frac{\partial \bar{h}}{\partial \bar{x}} + \bar{U} \tag{10.6}$$

where $\bar{U}$ is the dimensionless uplift and R expresses the relative importance of concentrative (fluvial) processes versus diffusive (hillslope) processes

$$R = \frac{c(aL_c)^n}{\kappa} \tag{10.7}$$

The unit-width sediment flux q_s is sensitive to the boundary condition at the catchment outlet. The absolute elevation of the depositional apex is free to move, but a continuity of gradient between catchment and fan is imposed, as observed in natural catchment-fan systems (Bull, 1964). The slope of the fan surface is assumed to be linear. Depositional architecture is calculated using a volume balance, assuming that no erosion takes place on the fan. The downstream fining of mean grain size is calculated using the self-similar solutions of Fedele and Paola (2007) and Duller et al. (2010).

The model can be applied to geological examples where the external forcing and grain-size characteristics of the sediment supply can be constrained. The Fucino Basin of central Italy is a recently drained lake containing fluvio-deltaic deposits that prograded into the basin from the faulted basin margin from the late Pliocene onwards (Cavinato et al., 2002). The Fucino Fault experienced a five-fold increase in slip rate from 0.3 to 1.5 mm yr^{-1} at 800 ka (Whittaker, Attal, and Tucker, 2007), accompanied by an increase in both the median ($D_{50} = 11-50$ mm) and coarse ($D_{84} = 20-110$ mm) grain-size percentiles exported from a catchment (Gole di Cetano) in the footwall (Figure 10.10). The increase in the median size of the sediment supply resulting from the five-fold increase in slip rate causes a wedge of gravel to extend into the basin in the model. However, at the same time, the larger D_{84}/D_{50} ratio promotes rapid downstream fining. As a result, the model output in Figure 10.10 shows different vertical grain-size trends depending on location in the basin. In proximal positions (marked A in Figure 10.10) the change in slip rate is marked by an abrupt progradation and coarsening of grain size. In medial positions (marked B in Figure 10.10) however, the same stratal surface is marked by a minor retrogradation and fining of grain size. In neither case does the sequence architecture reflect a change in base level.

The shoreline trajectory in the Fucino Basin can also be simulated with a heavily parameterised, spatially lumped numerical model (PaCMoD) (Forzoni et al., 2014) using a high-frequency sinusoidal climate (temperature and precipitation) signal and a low-frequency background sinusoidal tectonic subsidence. The tectonic history, climate and grain-size data of the adjacent Celano catchment were used to constrain model parameters. The results are very similar to those obtained by the forward model of Armitage et al. (2011), with a progradational lower unit, a rapid retrogradation of the shoreline at the base of an upper unit, followed by a second progradational phase.

10.3.1 Autostratigraphy Resulting from Internal Dynamics

The sequence stratigraphic method was initially built on the expectation that the lap-out relationships of stratal packages were diagnostic of external forcing mechanisms and that trends of shallowing and deepening were spatially consistent. Subsequently, it has been argued that sequence stratigraphic boundaries are not diagnostic of unique forcing mechanisms (Cantuneanu and Zecchin, 2016). The numerical model outlined previously shows that the landscape response, in terms of stratigraphic architectures and grain-size trends, to tectonic and climatic perturbations may be complex and non-linear. If natural systems behave similarly, it may be difficult to deconvolve the various effects of tectonics, climate, sea level change and unforced internal dynamics uniquely from the stratigraphic record (Burgess and Prince, 2015).

The movement of the shoreline in conventional sequence stratigraphic thinking is explainable with reference to the ratio of the accommodation generated A compared to the

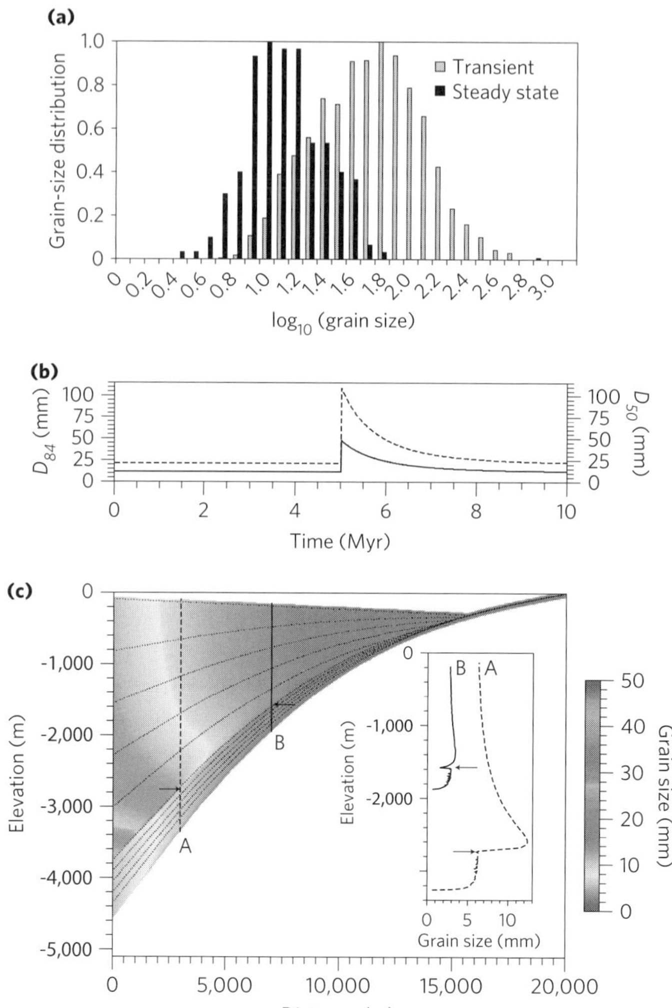

Figure 10.10 Model response to a five-fold increase in tectonic uplift rate, using data from the Gole di Celano fan, Fucino Basin, Italian Apennines. (a) Normalised grain-size measurements using Wolman clast count method, Gole di Celano, for regions that are responding to a slip rate increase from 0.3 to 1.5 mm yr^{-1} (transient), and those that are in steady state (Whittaker, Attal, and Allen, 2010). (b) Response of the input grain size (D_{50} and D_{84}) due to the change in slip rate. Dashed line is D_{84}. (c) Stratigraphy and grain size of the basin-fill for a five-fold increase in slip rate. Inset shows grain-size profiles. Arrows mark effects of slip rate perturbation in the stratigraphy. After Armitage et al. (2011) (fig.3, p.233) with permission of Nature Publishing Group.

sediment supply S. Using the *A/S ratio concept* (Shanley and McCabe, 1994; Muto and Steel, 1997; Kim et al., 2006), the turnaround would be attributed to an increase in A. The aggradational stage at the transition from regression to transgression would be at $A/S \sim 1$. Instead, the same geometry can be produced without any changes in external forcing in

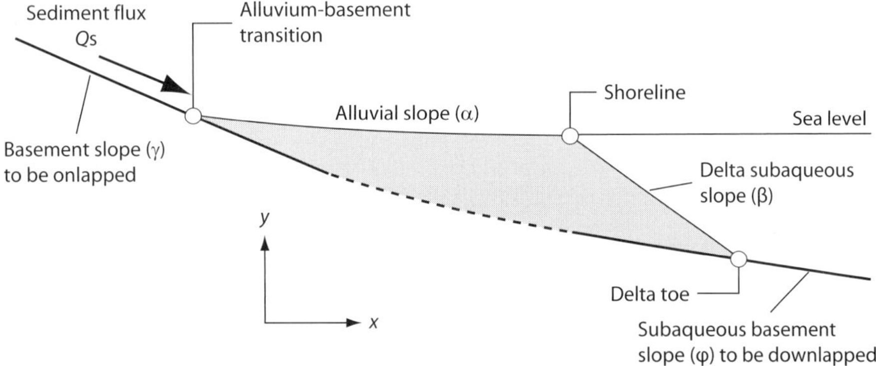

Figure 10.11 Geometrical parameters of the fluvio-deltaic system, showing the three moving boundaries. From Muto, Steel, and Swenson (2007)(fig.2, p.4) with permission of Society of Economic Paleontologists and Mineralogists.

laboratory experiments of the growth of a delta (Muto et al., 2007). The fundamental control on the migration of the shoreline is that the foreset of the delta must lengthen over time, since there is a finite basin slope, causing a change in the partitioning of sediment from topset to foreset.

In many sedimentary systems, there is a non-linear response to steady external forcing, which arises from internal feedbacks between sediment transport processes. In the autostratigraphy method, movement of the shoreline, defining trends in the position of coastal onlap (Vail et al., 1977) and the shoreline trajectory (Helland-Hansen and Hampson, 2009; Henriksen et al., 2011) (Section 8.1), reflects the non-equilibrium response of the fluvio-deltaic system to steady external forcing. The fluvio-deltaic system has three moving boundaries - the shoreline, the delta toe and the alluvium-basement transition (Figure 10.11), which are defined by the slopes β, γ and ϕ. The impact of autogenic processes compared to steady external forcing depends on the timescale of the fluctuations in external forcing T relative to the timescale τ of the the non-equilibrium response. Since sediment transport is in broad terms driven by slope, the autogenic timescale can be treated as determined by an effective diffusivity κ_e that encompasses the efficiency of various transport processes.

We firstly change the notation for the sediment supply to be consistent with elsewhere in this book, that is, $S \equiv Q_s$. If the sediment supply is steady (Q_s constant) and the rise or fall of relative sea level is also steady (A constant), the fluvio-deltaic system has a characteristic length scale given by

$$L = \frac{Q_s}{|A|} \tag{10.8}$$

and the characteristic timescale takes the familiar form

$$\tau = \frac{L^2}{\kappa_e} = \frac{Q_s}{|A|^2} \frac{Q_s}{\kappa_e} \tag{10.9}$$

If we further assume that the morphodynamics of the river system supplying sediment to the delta sets the overall timescale of the system, then κ_e becomes the fluvial diffusivity (Paola, Heller, and Angevine, 1992; Paola, 2000) and the term Q_s/κ_e is the characteristic slope α of the alluvial surface, so that

$$\tau = \frac{L^2}{\kappa_e} = \alpha \frac{Q_s}{|A|^2} \tag{10.10}$$

Low-gradient, small alluvial systems are therefore far more sensitive to autogenic responses than high-gradient, large systems (Muto, Steel, and Swenson, 2007).

Early sequence stratigraphy models held that the alluvial slope was in a state of 'grade' (Mackin, 1948), simply bypassing sediment to the shoreline along its entire length, and aggraded at times of relative sea level rise and degraded at times of relative sea level fall (Wilgus et al., 1988). In the non-equilibrium behaviour of autostratigraphy, however, alluvial rivers may attain grade and, in principle, be aggradational not only during rising sea level and at highstand, but also throughout falling sea level (Muto and Swenson, 2006). For the alluvial river to be in a sustained state of 'grade' depends primarily on the alluvial and basal slopes (α, ϕ).

With a steady rise in relative sea level, experiments show that there is an early stage of basinward migration of the shoreline, but this trend is reversed some time after the beginning of delta progradation, a phenomenon known as *autoretreat*. Autoretreat is caused by the progressive increase in storage capacity of the entire fluvio-deltaic system as relative sea level rises. Later still, there is a change in the fluvio-deltaic system, known as an *autobreak*, when the clear delta front configuration is lost (Figure 10.12). Autobreak occurs

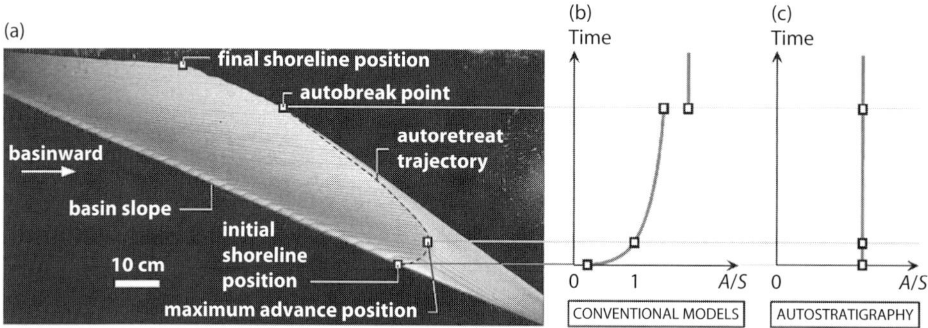

Figure 10.12 Autostratigraphic versus conventional sequence stratigraphic models for a regressive-transgressive turnaround. (a) Longitudinal section of an experimental run with a constant rate of sediment supply ($S = 1.029 \pm 0.020$ cm^2 s^{-1}), upstream water discharge ($q = 4.36 \pm 0.02$ cm^2 s^{-1}) and sea level rise ($A = 0.0251$ cm s^{-1}). (b) Variation of A/S over time with interpretation from conventional sequence stratigraphic models. (c) Interpretation from viewpoint of autostratigraphy, showing no variation in A/S over time. Note that without any change in dynamic parameters, shoreline advance (regression) is halted, changes to landward retreat (transgression) and eventually, after an 'autobreak point', the delta clinoform is no longer sustained. From Muto, Steel, and Swenson (2007)(fig.1, p.3) with permission of Society of Economic Paleontologists and Mineralogists.

when there is further growth of the surface area of the delta so that sediment supply is entirely consumed in building the subaerial coastal plain. If $\beta \geq \gamma$ and $T >> \tau$, both autoretreat and autobreak are inevitable. However, if $\beta < \gamma$, autobreak is not reached because any progressive increase in the surface area of the foreset is fully compensated by a progressive decrease in the surface area of the alluvial system.

Depending on the values of geometrical parameters for the receiving basin (γ, ϕ) and the river-fed delta (α, β), there are therefore potentially very different stratigraphic responses to the same external driver. Stratigraphic correlation between adjacent or distant fluvio-deltaic systems is hazardous without some understanding of the impact of internal autocyclic processes. Failure to deconvolve autogenic effects from external forcing would, for example, result in the construction of an erroneous sea level curve representing the external eustatic forcing.

Finally, any coastal depositional system evolving under the control of sea level has characteristic length and timescales for the manifestation of large-scale autogenic events (Muto, Steel, and Swenson, 2007). The characteristic autostratigraphic timescale τ_a and the characteristic autostratigraphic length scale Λ_a are given by

$$\Lambda_a = \sqrt{\frac{Q_s}{|\dot{E}|}} \tag{10.11}$$

and

$$\tau_a = \frac{\Lambda^2}{\kappa} = \frac{1}{\kappa} \frac{Q_s}{|\dot{E}|} \tag{10.12}$$

where $|\dot{E}|$ is the absolute speed of eustatic sea level change, Q_s is the sediment discharge $([L]^3[T]^{-1})$, and κ is a linear diffusivity parameter for alluvial sedimentation with dimensions $[L]^2[T]^{-1}$. Taking a value of the sediment discharge appropriate for a large river (Q_s = 1,000 Mt yr^{-1}), the characteristic autostratigraphic length scale ranges from 60 km for a sea level speed of 0.1 mm yr^{-1} to 6 km for a speed of sea level change of 10 mm yr^{-1}. The diffusivity for sand in alluvial basins is of order 10^5 m^2 yr^{-1} (Marr et al., 2000). Using this value of diffusivity, the characteristic autostratigraphic timescale ranges from 36 kyr to 364 yr for the large river with the same range of sea level change $|\dot{E}| = 0.1 - 10$ mm yr^{-1}. For a small river, let $Q_s = 10$ Mt yr^{-1}. Following the same procedure, τ_a ranges between $\sim$364$-$4 yr, showing that the response is almost instantaneous. Nevertheless, the characteristic response time varies by two orders of magnitude between the large- and the small-sediment discharge rivers. Two adjacent sediment routing systems whose coastal zones are fed by rivers of contrasting discharge may therefore show a nonequilibrium response to the same sea level rise differently, with different time lags, depending on the magnitudes of Q_s and κ.

Correlation of sequence architectures generated by diagnostic forcing mechanisms may therefore work only in restricted situations, for example where the timescale of the external fluctuations is very short relative to the timescale of the non-equilibrium response to steady forcing (that is, $T << \tau$). For the glacio-eustatic climate changes of the Quaternary,

$T \sim 100$ kyr, corresponding to the 'rapid forcing' of Marr et al. (2000) and the 'reactive' landscape dynamics of Allen (2008b). Under fast forcing, eustatic effects are likely to dominate the stratigraphic response, which explains why the Quaternary record contains strong evidence for glacio-eustatic cycles. On the other hand, if the fluctuation of the external control is slow compared to the timescale of the non-equilibrium response, that is, $T \geq \tau$, we have the case of 'slow forcing' of Marr et al. (2000) or the 'buffered' landscape dynamics of Allen (2008b). In this case, sequence architectures are unlikely to be attributable to particular external driving mechanisms and autogenic processes need to be considered. The sequence stratigraphy methodology and the process stratigraphy derived from theory, physical experiments and numerical models therefore have complementary relationships (Muto et al., 2007).

10.4 Cycles and the Global Sea Level Chart

Bearing in mind the central importance of sea level change in the original formulation of the sequence stratigraphy model, it is particularly instructive to understand the natural variation of sea level in the past. These sea level variations occur at a very wide variety of temporal scales. At short timescales below the range of interest for sequence architectures (Agnew, 1992), they can be attributed principally to waves and tides, and vary more or less continuously up to longer periods of greater geological interest (200 Myr) (Harrison, 1988). Long period variations in relative sea level come from studies of continental flooding (Bond, 1979; Harrison et al., 1981) where the driving mechanisms are variations in the volume of the ocean basins, losses of ocean water by subduction, and vertical movements of the continental surface, which may be caused by dynamic effects originating in the mantle (Lithgow-Bertelloni and Gurnis, 1997; Gurnis, 2001). Medium-scale period variations are revealed from oxygen isotope data as a proxy for sea level change over the last ca. 1 Myr, in which Milankovitch band peaks can be discriminated, especially that at 98 kyr (Imbrie et al., 1984; Muller and MacDonald, 1997). This timescale of sea level fluctuation is attributed to orbital variations driving global climate change.

10.4.1 An Ordered Hierarchy?

The global cycle chart of Haq, Hardenbol, and Vail (1987) gives a record of relative sea level extending back to nearly 600 Ma in which hierarchical orders overlap in their range of duration. Since publication in 1987, this overlap has not narrowed as more data have been collated (Williams, 1988; Van Wagoner et al., 1990; Carter et al., 1991; Vail et al., 1991; Reid and Dorobek, 1993; Duval, Cramez, and Vail, 1998; Lehrmann and Goldhammer, 1999; Schlager, 2004; Catuneanu, 2006), raising questions about whether the hierarchical levels are real. A sampling of the 'global' sea level curve at a frequency of 0.1 Myr gives a power spectrum that is indistinguishable from sources at higher and lower frequencies. That is, the global cycle chart itself suggests scale invariance for relative sea level. Indeed, Wolfgang Schlager stated (Schlager, 2010) (p.149) that the orders of sequence cycles are simply subdivisions of convenience in a stratigraphic continuum.

The vision of some authors is to establish a global cycle chart of sea level change that will serve as a chronometer of Earth history. Peter Vail and colleagues wrote in Vail, Mitchum, and Thompson (1977) (p.96):

One of the greatest potential applications of the global cycle chart is its use as an instrument of geochronology.

The global cycle chart derives estimates of global sea level history from the amount of coastal onlap on seismic reflection profiles (Vail, Mitchum, and Thompson, 1977; Haq, Hardenbol, and Vail, 1987; Haq, 1991; Haq and Schutter, 2008). Such global cycle charts have, however, attracted widespread criticism. They require high levels of precision in the dating of stratigraphic surfaces in order to act as a global standard, and the complexity of interacting processes has been under-appreciated in the zeal for recognising global controls.

The widely used global chart of Haq, Hardenbol, and Vail (1987) and its derivatives apparently reveal an ordered hierarchy of stratigraphic cycles, from first order cycles at timescales of 100 Myr, second order at 5–15 Myr, and third order at 1–10 Myr. As recently as 2008, Haq and Schutter (2008) (p.65) wrote that

The stratigraphic record is a composite of several orders of superimposed sedimentary cycles . . .

A hierarchical framework is based on the filling of accommodation at varying magnitudes and duration, giving in ascending order, parasequences, system tracts, depositional sequences, sequence sets, composite sequences, composite sequence sets and megasequences (Neal and Abreu, 2009). These hierarchical units are stacked into progradational, aggradational and retrogradation patterns, and are described as self-similar (Neal and Abreu, 2009) (p.782) and not associated with particular ranges of duration or amplitude.

The idea of an ordered hierarchy contradicts the view that sea level change is fractal (Harrison et al., 1981; Hsui, Rust, and Klein, 1993) and omits to recognise the possibility that some stratigraphic cycles may be due to unforced internal dynamics (Muto, Steel, and Swenson, 2007). The overlapping time ranges for 1st to 6th order cycles (Schlager, 2004)(p.185, fig.1) (Figure 10.13), casts doubt as to whether hierarchical order can be uniquely discriminated. In particular, some workers advocate an abandonment of the distinction between 2nd and 3rd orders (Hardenbol et al., 1998). In addition, the belief that depositional sequences (3rd order) are bounded by unconformities created by emergence, whereas parasequences (4th order and higher) are bounded by flooding surfaces (Van Wagoner et al., 1990) is not supported, especially in carbonate environments (Schlager, 2004). The ordered hierarchy of cycles may therefore be an illusion.

The critical aspect of the recognition of global cycles of sea level change throughout the Phanerozoic is the ability and accuracy of dating and correlation of stratigraphic events (Miall, 1994). In the Phanerozoic, biostratigraphic subdivisions require a calibration scale derived from radiometric dating, supplemented by chemostratigraphy and magnetostratigraphy. The various techniques and their uses in establishing an international chronostratigraphic chart (www.stratigraphy.org) are dealt with at length by Miall (2016).

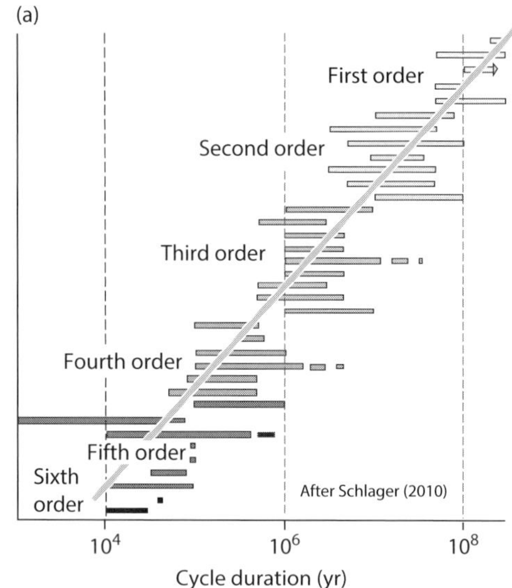

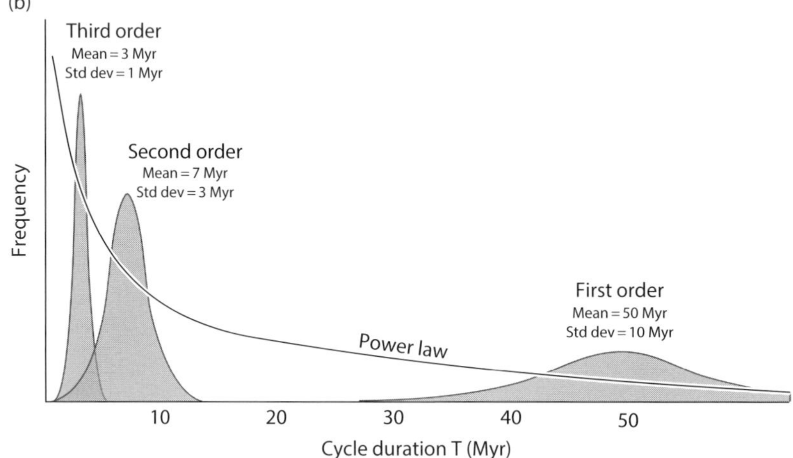

Figure 10.13 (a) Duration of first to sixth order cycles showing overlapping ranges and straight line indicating continuum of cycle duration. Compiled by Schlager (2010)(fig.9) with permission of Springer. (b) Frequency-duration plot of three normal distributions representing first, second and third order cycles compared to power law.

In the 2013 version, Palaeogene, Mesozoic and Palaeozoic dates are mostly given to the nearest 100,000 years, but the attainment of this level of accuracy depends on the availability of suitable datable material in the rock record and uncertainties may be up to 1 Myr.

So-called third-order cycles (depositional sequences), the fundamental unit for global correlation, are particularly problematical. It is clear that age assignments of sequence boundaries and maximum flooding surfaces are at or beyond the limit of chronological resolution. Age assignments serving as a basis for global correlation are to the nearest half million years in the chart given by Haq, Hardenbol, and Vail (1987), to the nearest 0.1 Myr in the Palaeozoic curve given by Haq and Schutter (2008) and in millions of years to two decimal places in some in-house industry schemes (e.g. British Petroleum), yet chronological resolution is commonly no better than ± 1 Myr (Hinnov and Ogg, 2007). As Ricken (1991) (p.773) has observed,

The range of error is often larger than the actual [cycle] time span considered.

To test whether the cycles of sea level illustrated in the global cycle chart form an ordered hierarchy, the duration of sequences (so-called 3rd order), supersequences (2nd order) and supersequence sets and megacycles (1st order) were read directly from the chart provided by Haq, Hardenbol, and Vail (1987), since this was the key publication from which alternative charts were inspired. The durations of sequence stratigraphic cycles are expressed in a log-log plot of the number of cycles of duration greater than T versus the duration T. This was performed *for each order of cycle* and then for the combination of all cycles. The null hypothesis is that the distribution of cycle durations for each order is normally distributed with a mean μ and standard deviation σ derived from inspection of the global cycle chart (Figure 10.14). In other words, if a certain cycle duration is known, it should be possible to say what the probability is that the cycle is first, second, third or higher in order. Failure of the null hypothesis would imply that the cycles drawn by Haq, Hardenbol, and Vail (1987) and its derivatives do not justify the idea of an ordered hierarchy.

The most straightforward expectation for an ordered hierarchy of cycles is that each order has cycle durations that are normally distributed about the mean. The probability density function for a normal distribution is

$$f(T, \mu, \sigma) = \frac{1}{\sigma \sqrt{2\pi}} e^{\frac{-(x-\mu)^2}{2\sigma^2}} \tag{10.13}$$

The number of cycles expected with a given duration is then the product of the total number of cycles and the probability function, which can be converted into a curve of number of cycles exceeding T in duration versus T. This procedure is carried out for each 'order', and then three normal distributions are combined for timescales of up to 80 Myr using the following data (Figure 10.13b):

Third order: $\mu = 3$ Myr; $\sigma = 1$ Myr
Second order: $\mu = 7$ Myr; $\sigma = 3$ Myr
First order: $\mu = 50$ Myr; $\sigma = 10$ Myr

The cumulative probability distribution for data forming a normal distribution is:

$$F(x) = \frac{1}{2}\text{erfc}\left(\frac{x - \mu}{2^{1/2}\sigma}\right) \tag{10.14}$$

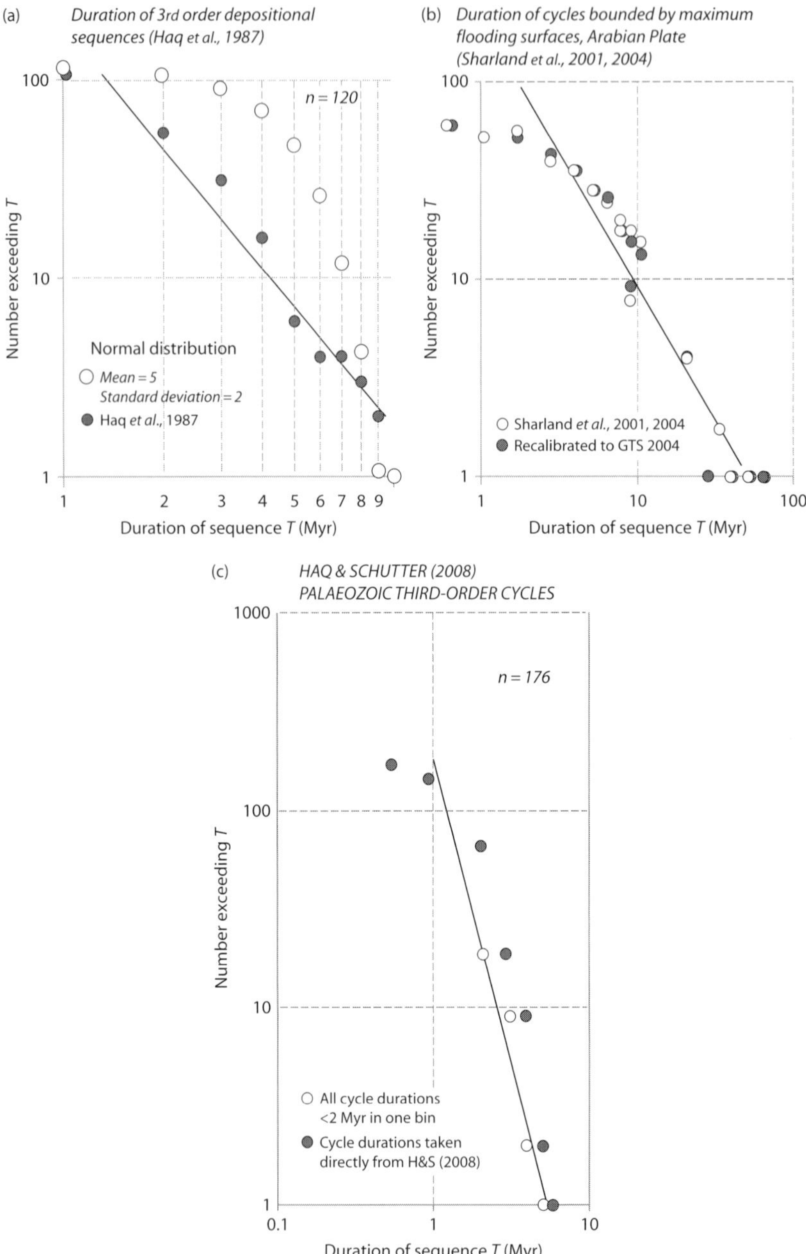

Figure 10.14 Number of cycles exceeding a chosen duration $N(t > T)$ versus sequence duration T for third-order depositional sequences. (a) Third-order sequences from Haq, Hardenbol, and Vail (1987) compared to a normal distribution. (b) Cycles bounded by flooding surfaces from the Arabian plate (Sharland et al., 2001, 2004). GTS is global timescale. (c) Palaeozoic third-order cycles from Haq and Schutter (2008). Straight lines are best power law fits of the form $N(t > T) = aT^b$ where t is the observed cycle duration and T is the chosen limiting duration.

where erfc is the complementary error function. The probability of falling within a particular bin of a normal distribution is then:

$$\Pr(x_1 < x \le x_2) = \frac{1}{2}\mathrm{erfc}\left[\frac{x_1 - \mu}{2^{1/2}\sigma}\right] - \frac{1}{2}\mathrm{erfc}\left[\frac{x_2 - \mu}{2^{1/2}\sigma}\right] \tag{10.15}$$

where x_1 and x_2 are the upper and lower limits of the bin. This can be compared with the probability of the observed distribution. For example, the probability of a cycle length being between 1.5 and 2.0 Myr if the cycle lengths are normally distributed with a mean of 3 Myr and a standard deviation of 1 Myr is 0.092. The probability of a cycle length being between 2.5 and 3.0 Myr is 0.191.

A compilation of 120 third-order cycles from the global cycle chart of Haq, Hardenbol, and Vail (1987) (Figure 10.14) shows that the best fit is a power law. The trend expected from a normal distribution is strongly different to the trend observed. By extending the sampling to second-order and first-order cycles in the global cycle chart, a similar picture emerges ($n = 158$). The data are better explained by a power law than by a model of three distinct normal distributions (Figure 10.13b). Substitution of the Haq, Hardenbol, and Vail (1987) data with cycle data derived from the Cenozoic and Mesozoic of the Arabian plate (Sharland et al., 2001, 2004) and with the same data recalibrated by the Global Time Scale (2004) by Haq and Al-Qahtani (2005) shows little difference.

The global curve for the Palaeozoic (542 − 251 Ma) of Haq and Schutter (2008) contains 172 third order cycles with an average duration of 1.7 Myr. Remarkably, cycle durations of as little as 0.4 Myr are proposed, and there are 35 cycles of less than 1 Myr duration. Since current estimates of the uncertainty of the geological timescale are ±1 Myr for the Pennsylvanian-Permian, and ±2−3 Myr for the remainder of the Palaeozoic (Hinnov and Ogg, 2007), the implied precision of the curve of Haq and Schutter (2008) is inexplicable. If we group all cycles of duration less than 2 Myr together, the log-log plot of number of cycles exceeding duration T versus T gives a power law fit rather than indicating a normal distribution centred on 1.7 Myr.

Cycle durations from less than 1 Myr to as much as 100 Myr are therefore best explained by a power law (Figures 10.14 and 10.15), suggesting that stratigraphic cycles, and the gaps that separate them, are essentially fractal (Schlager, 2004). Curves very similar to those put forward as representing the history of global sea level can be generated by a random walk (Harrison, 2002). This does not mean that stratigraphic prediction is impossible, but does suggest that the presence of cycles of duration greater than 1 Myr is fundamentally an outcome of a complexly interacting and self-organised system.

I conclude that there is no ordered hierarchy of cycle duration in the existing global sea level charts. This being the case, the search for driving mechanisms with a time-behaviour concordant with cycle repeat times is a futile exercise, since cycles occur at all durations. The subdivision of the stratigraphic record into a hierarchical order of cycles may be simply a result of perception (Drummond and Wilkinson, 1996). Sequence variability is more like statistical noise (Hergarten, 2002) than occurring at specified time and spatial scales in response to well-ordered, oscillating drivers.

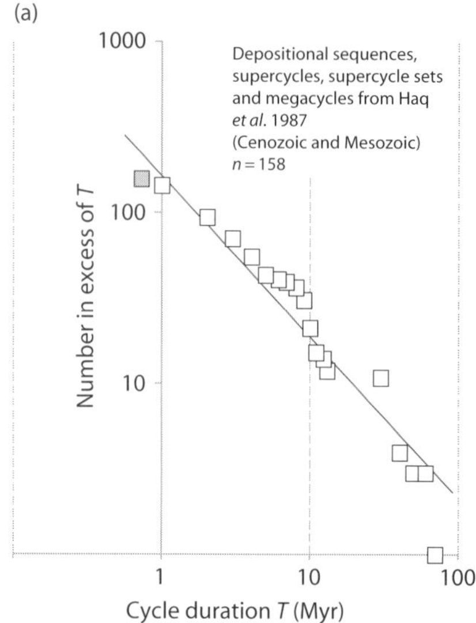

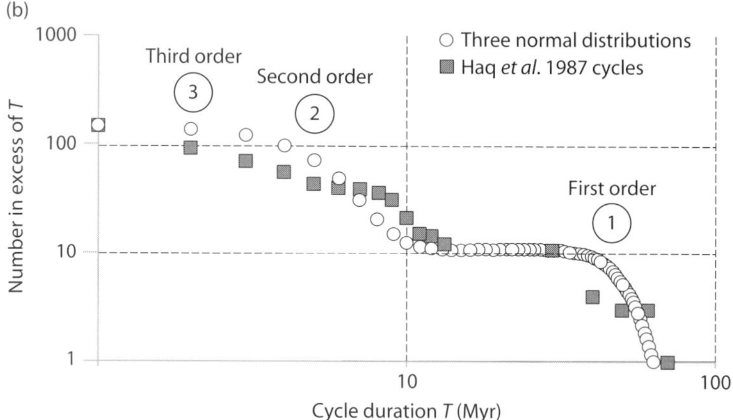

Figure 10.15 Number of cycles exceeding T in duration $N(t > T)$ versus cycle duration T. (a) For cycles of all orders in Haq, Hardenbol, and Vail (1987), showing good straight line fit indicating a power law distribution. (b) Cycles from Haq, Hardenbol, and Vail (1987) compared to three normal distributions representing first, second and third order cycles.

Stratal stacking patterns may not display clear cyclicity (Helland-Hansen and Hampson, 2009), and qualitative methods may not in any case be able to provide strong evidence of cyclicity (Wilkinson et al., 1997; Burgess, 2016). By comparing the expected upward transitions between facies, based only on their relative abundance, against the observed upward transitions, allows a Markov statistic to be calculated that describes the degree of order in a stratigraphic succession (Burgess, 2016). A 'runs order metric' also discriminates between ordered and disordered arrangements of facies thickness. None of the stratigraphic successions studied by Burgess (2016) showed evidence of order at the bed by bed scale, though there was some evidence for order at longer length and timescales. Ordered facies successions and thickness trends may be much less common than typically assumed.

There are therefore strong reservations about the chronological precision and global correlation of Phanerozoic sequence stratigraphic boundaries. Four aspects stand out. First, outside of the Quaternary, the 'global' cycle chart as presently formulated cannot be shown to be correct or otherwise, since many of the cycles proposed are shorter than the best available chronological resolution. The Phanerozoic 'global' cycle chart might therefore be regarded as plausible yet untestable. Second, the 'global' cycle chart, irrespective of chronological resolution, might be regarded as fundamentally flawed and internally inconsistent since its component cycles fail to represent an ordered hierarchy. Third, since sequence boundaries and flooding surfaces are likely to be variably diachronous with respect to a synchronous perturbation such as base level change, the search for global synchrony is fundamentally futile outside of those parts of the geological past dominated by high-amplitude, high-frequency glacio-eustatic change. Fourth, the order of stratigraphic successions, which forms the basis for recognising trends in facies and stratal thickness made use of in building sequence stratigraphic models, may be difficult to justify quantitatively.

References

Aalto, R., Dunne, T., and Guyot, J. L. 2006. Geomorphic controls on Andean denudation rates. *Journal of Geology*, **114**, 85–99.

Aalto, R., Maurice-Bourguin, L., Dunne, T., Montgomery, D. R., Nittrouer, C. A., and Guyot, J.-L. 2003. Episodic sediment accumulation on Amazonian flood plains influenced by El Niño/southern oscillation. *Nature*, **425**, 493–496.

Abrams, D. M., Lobkovsky, A. E., Petroff, A. B., Straub, K. M., McElroy, B., Mohrig, D. C., Kudrolli, A. and Rothman, D. H. 2009. Growth laws for channel networks incised by groundwater flow. *Nature Geoscience*, **2**, 193–196.

Adams, E. W., and Schlager, W. 2000. Basic types of submarine slope curvature. *Journal Sedimentary Research*, **70**, 814–828.

Adams, J. 1985. Large scale tectonic geomorphology of the Southern Alps, New Zealand: summary. Pages 105–128 of: Morisawa, M., and Hack, J. T. (eds), *Tectonic Geomorphology*. Boston: Allen and Unwin.

Agnew, D. C. 1992. The time-domain behavior of power-law noises. *Geophysical Research Letters*, **19**, 333–336.

Ahnert, F. 1970. Functional relationships between denudation, relief and uplift in large mid-latitude drainage basins. *American Journal of Science*, **270**, 243–263.

Alexander, C., Walsh, J., Sumner, B., Orpin, A., and Kuehl, S. 2006. Continental slope delivery and storage on an active margin: The Waipaoa margin example. *American Geophysical Union Fall Meeting, Abstract*, **OS12A-03**.

Alizai, A., Clift, P. D., Giosan, L., Van Laningham, S., Hinton, R., Tabrez, A. R., and Danish, M. 2011. Pb isotopic variability in the modern and Holocene Indus River system measured by ion microprobe in detrital K-feldspar grains. *Geochimica et Cosmochimica Acta*, **75**, 4771–4795.

Alizai, A., Clift, P. D., and Still, J. 2016. Indus Basin sediment provenance constrained using garnet geochemistry. *Journal Asian Earth Sciences*, **126**, 29–57.

Allen, P. A. 1997. *Earth Surface Processes*. Oxford, UK: Blackwell Science.

Allen, P. A. 2005. Striking a chord. *Concepts Essay, Nature*, **434**, 961.

Allen, P. A. 2008a. From landscapes into geological history. *Nature*, **451**, 274–276.

Allen, P. A. 2008b. Time scales of tectonic landscapes and their sediment routing systems. Pages 7–28 of: Gallagher, K., Jones S. J., and Wainwright, J. (eds), *Earth's Dynamic Surface: Catastrophe and Continuity in Landscape Evolution*. Special Publication Geological Society London 296.

Allen, P. A. 2014. *Earth Dramas: Ancient Mysteries and Modern Controversies*. Amazon Kindle Publications.

Allen, P. A., and Allen, J. R. 2013. *Basin Analysis: Principles and Application to Petroleum Play Assessment*. Oxford, UK: Blackwell-Wiley.

Allen, P. A., and Densmore, A. L. 2000. Sediment flux from an uplifting fault block. *Basin Research*, **12**, 367–380.

Allen, P. A., and Heller, P. L. 2012. The timing, distribution and significance of tectonically generated gravels in terrestrial sediment routing systems. Pages 111–130 of: Busby, C., and Azor, A. (eds), *Syntectonic Basin Development, Active to Ancient: Recent Advances*. Oxford: Wiley-Blackwell.

Allen, P. A., and Hovius, N. 1998. Sediment supply from landslide-dominated catchments: implications for basin-margin fans. *Basin Research*, **10**, 19–35.

Allen, P. A., Crampton, S., and Sinclair, H. D. 1991. Inception and early evolution of the North Alpine Foreland Basin, Switzerland. *Basin Research*, **3**, 143–163.

Allen, P. A., Burgess, P. M., Galewsky, J., and Sinclair, H. D. 2001. Flexural-eustatic numerical model for drowning of the Eocene perialpine carbonate ramp and implications for Alpine geodynamics. *Bulletin Geological Society America*, **113**, 1052–1066.

Allen, P. A., Bennett, S. D., Cunningham, M. J. M., Carter, A., Gallagher, K., Lazzaretti, E., Galewsky, J., Densmore, A. L., Phillips, A. W. E., Naylor, D., and Solla Hach, C. 2002. The post-Variscan thermal and denudational history of Ireland. Pages 371–399 of: Doré, A. G., Cartwright, J. A., Stoker, M. S., Turner, J. P., and White, N. (eds), *Exhumation of the North Atlantic Margin: Timing, Mechanisms and Implications for Petroleum Exploration*. Geological Society London Special Publication 196.

Allen, P. A., Armitage, J. J., Carter, A., Duller, R. A., Michael, N. A., Sinclair, H. D., Whitchurch, A. L., and Whittaker, A. C. 2013. The Qs problem: sediment volumetric balance of proximal foreland basin systems. *Sedimentology*, **60**, 102–130.

Allen, P. A., Eriksson, P. G., Alkmim, F. F., Betts, P. G., Catuneanu, O., Meng, Q., Mazumder, R., and Young, G. M. 2015a. Classification of basins, with special reference to Proterozoic examples. Pages 5–28 of: Mazumder, R., and Eriksson, P. G. (eds), *Precambrian Basins of India: Stratigraphic and Tectonic Context*. Memoir Geological Society London, 43.

Allen, P. A., Armitage, J. J., D'Arcy, M., Roda-Boluda, D., and Whittaker, A. C. 2015b. Fragmentation model for the grain size mix of sediment supplied to basins. *Journal of Geology*, **123**, 405–427.

Allen, P. A., Michael, N., D'Arcy, M., Boluda, D. C., Whittaker, A. C., Duller, R. A., and Armitage, J. J. 2016. Fractionation of grain size in terrestrial sediment routing systems. *Basin Research*, **27**, 1–23.

Aller, R. C., Blair, N. E., Xia, Q., and Rude, P. D. 1996. Remineralization rates, recycling, and storage of carbon in Amazon shelf sediments. *Continental Shelf Research*, **16**, 753–786.

Allison, M. A. 1998. Historical changes in the Ganges-Brahmaputra delta front. *Journal of Coastal Research*, **14**, 480–490.

Alvarez, W. 1999. Drainage on evolving fold-thrust belts: a study of transverse canyons in the Apennines. *Basin Research*, **11**, 267–284.

Amos, C. L., and Judge, J. T. 1991. Sediment transport on the eastern Canadian continental shelf. *Continental Shelf Research*, **11**, 1037–1068.

Anderson, R. S. 1994. Evolution of the Santa Cruz Mountains, California, through tectonic growth and geomorphic decay. *Journal Geophysical Research*, **99**, 20,161–20,179.

Anderson, R. S., and Anderson, S. P. 2010. *Geomorphology: The Mechanics and Chemistry of Landscapes*. Cambridge, UK: Cambridge University Press.

Anderson, R. S., and Humphrey, N. F. 1990. Interaction of weathering and transport processes in the evolution of arid landscapes. Pages 349–361 of: Cross, T. A. (ed), *Quantitative Dynamic Stratigraphy*. Englewood Cliffs, New Jersey: Prentice Hall.

Anderson, R. S., Repka, J. L., and Dick, G. S. 1996. Dating depositional surfaces using *in situ* produced cosmogenic radionuclides. *Geology*, **24**, 47–51.

Anderson, R. Y., and Kirkland, D. W. 1960. Origin of varves, and cycles of Jurassic Todilto Formation, New Mexico. *American Association of Petroleum Geologists Bulletin*, **44**, 27–52.

Anderson, S. P. 2005. Glaciers show direct linkage between erosion rate and chemical weathering fluxes. *Geomorphology*, **67**, 147–157.

Anderson, S. P., Drever, J. I., and Humphrey, N. F. 1997. Chemical weathering in glacial environments. *Geology*, **25**, 399–402.

Andrews, J. T. 2000. Icebergs and iceberg-rafted detritus (IRD) in the North Atlantic: facts and assumptions. *Oceanography*, **13**, 100–108.

Armitage, J. J., Duller, R. A., Whittaker, A. C., and Allen, P. A. 2011. Transformation of tectonic and climatic signals from source to stratigraphy. *Nature Geoscience*, **4**, 231–235.

Armitage, J. J., Duller, R. A., Dunkley Jones, T., Whittaker, A. C., and Allen, P. A. 2013. Temporal buffering of climate-driven sediment flux cycles by transient catchment response. *Earth and Planetary Science Letters*, **369**, 200–210.

Arribas, J., and Tortosa, A. 2003. Detrital models in sedimenticlastic sands from low-order streams in the Iberian Range, Spain: the potential for sand generation by different sedimentary rocks. *Sedimentary Geology*, **159**, 275–303.

Artoni, A. 2007. Growth rates and two-mode accretion in the outer orogenic wedge-foreland basin system of the Central Apennines (Italy). *Italian Journal Geosciences*, **126**, 531–556.

Attal, M., and Lavé, J. 2009. Pebble abrasion during fluvial transport: Experimental results and implications for the evolution of sediment load along rivers. *Journal of Geophysical Research*, **114**, F04023, 1–22.

Avouac, J. P. 1993. Analysis of scarp profiles: evaluation of errors in morphologic dating. *Journal Geophysical Research*, **98**, 6745–6754.

Aziz, H. A., Hilgen, F. J., van Luijk, G. M., Sluijs, A., Kraus, M. J., Pares, J. M., and Gingerich, P. D. 2008. Astronomical climate control on paleosol stacking patterns in the upper Paleocene-lower Eocene Willwood Formation, Bighorn Basin, Wyoming. *Geology*, **36**, 531–534.

Baartman, J. E. M., Temme, A. J. A. M., Veldkamp, T., Jetten, V. G., and Schoorl, J. M. 2013. Exploring the role of rainfall variability and extreme events in long-term landscape development. *Catena*, **109**, 25–38.

Babault, J., Bonnet, S., Crave, A., and van den Dreissche, J. 2005. Influence of piedmont sedimentation on erosion dynamics of an uplifting landscape: an experimental approach. *Geology*, **33**, 301–304.

Bailey, R. J., and Smith, D. G. 2010. Scaling in stratigraphic data series: implications for practical stratigraphy. *First Break*, **28**, 57–66.

Baldwin, J. A., Whipple, K. X., and Tucker, G. E. 2003. Implications of the shear stress river incision model for the time scale of post-orogenic decay of topography. *Journal Geophysical Research*, **108**, 2158.

Bally, A. W., and Snelson, S. 1980. Realms of subsidence. Pages 9–75 of: Miall, A. D. (ed), *Facts and Principles of World Petroleum Occurrence*. Canadian Society Petroleum Geologists Memoir, 6.

Bar-Matthews, M., Ayalon, A., Gilmour, M., Matthews, A., and Hawkesworth, C. J. 2003. Sea-land oxygen isotopic relationships from planktonic foraminfera and spelothems in the Eastern Mediterranean region and their implication for paleorainfall during interglacial periods. *Geochimica Cosmochimica Acta*, **67**, 3181–3199.

Barnes, J. B., and Heins, W. A. 2009. Plio-Quaternary sediment budget between thrust belt erosion and foreland deposition in the central Andes, southern Bolivia. *Basin Research*, **21**, 91–109.

Barrell, J. 1917. Rhythms and the measurement of geologic time. *Geological Society of America Bulletin*, **28**, 745–904.

Barrier, L., Proust, J. N., Nalpas, T., Robin, C., and Guillocheau, F. 2010. Control of alluvial sedimentation at foreland basin active margins: a case-study from the northeastern Ebro Basin (southeastern Pyrenees, Spain). *Journal Sedimentary Research*, **80**, 728–749.

Beamud, E., Muñoz, J. A., Fitzgerald, P. G., Baldwin, S. L., Garces, M., Cabrera, L., and Metcalf, J. R. 2010. Magnetostratigraphy and detrital apatite fission track thermochronology in syntectonic conglomerates: constraints on the exhumation of the South-Central Pyrenees. *Basin Research*, **13**, 309–331.

Beaumont, C., Kooi, H., and Willett, S. 2000. Coupled tectonic-surface process models with applications to rifted margins and collisional orogens. Pages 29–55 of: Summerfield, M.A. (ed), *Geomorphology and Global Tectonics*. Chichester, UK: Wiley.

Beckmann, B., Floegel, S., Hoffmann, P., Schulz, M., and Wagner, T. 2005. Orbital forcing of Cretaceous river discharge in tropical Africa and ocean response. *Nature*, **437**, 241–244.

Bernard, C. Y., Laruelle, G. G., Slomp, C. P., and Heinze, C. 2010. Impact of changes of river fluxes of silica on the global marine silica cycle: a model comparison. *Biogeosciences*, **7**, 441–453.

Bernard, C. Y., Dürr, H. H., Heinze, C., Segschneider, J., and Maier-Reimer, E. 2011. Contribution of riverine nutrients to the silicon biogeochemistry of the global ocean – a model study. *Biogeosciences*, **8**, 551–564.

Berner, R. 1989. Biochemical cycles of carbon and sulfur and their effect on atmospheric oxygen over Phanerozoic time. *Palaeogeography, Palaeoclimatology, Palaeoecology*, **75**, 97–122.

Berner, R. A. 1982. Burial of organic carbon and pyrite sulfur in the modern ocean: Its geochemical and environmental significance. *American Journal of Science*, **282**, 451–473.

Berner, R. A. 1999. A new look at the long-term carbon cycle. *GSA Today*, **9**, 1–6.

Best, T. C., and Griggs, G. B. 1991. A sediment budget for the Santa Cruz littoral cell, California. Pages 35–50 of: Osbourne, R. H. (ed), *From Shoreline to Abyss: Contributions in Marine Geology in Honor of Francis Parker Shepard*. Society of Economic Paleontologists and Mineralogists Special Publication 46.

Beusen, A. H. W., Bouwman, A. F., Dürr, H. H., Dekkers, A. L. M., and Hartmann, J. 2009. Global patterns of dissolved silica export to the coastal zone: Results from a spatially explicit global model. *Global Biogeochemical Cycles*, **23**, GB0A02.

Bhattacharya, J. P., and Giosan, L. 2003. Wave-influenced deltas: geomorphological implications for facies reconstruction. *Sedimentology*, **50**, 187–210.

Bickle, M., Bunbury, J., and Chapman, H., Harris, N. B. W., Fairchild, I. J., and Ahmad, T. 2003. Fluxes of Sr into the headwaters of the Ganges. *Geochimica Cosmochimica Acta*, **67**, 2567–2584.

Bierman, P., and Steig, E. 1996. Estimating rates of denudation using cosmogenic isotope abundances in sediment. *Earth Surface Processes and Landforms*, **21**, 125–139.

Bierman, P. R. 1994. Using *in situ* produced cosmogenic isotopes to estimate rates of landscape evolution: A review from the geomorphic perspective. *Journal of Geophysical Research*, **99**, 13,885–13,896.

Bierman, P. R., and Caffee, M. W. 2002. Cosmogenic exposure and erosion history of ancient Australian landforms. *Geological Society of America Bulletin*, **114**, 787–803.

Bierman, P. R., and Nichols, K. K. 2004. Rock to sediment - slope to sea with ^{10}Be - rates of landscape change. *Annual Reviews Earth and Planetary Sciences*, **32**, 215–255.

Bishop, P., Hoey, T. B., Jansen, J. D., and Artza, I. L. 2004. Knickpoint recession rate and catchment area: the case of uplifted rivers in eastern Scotland. *Earth Surface Processes and Landforms*, **30**, 767–778.

Bitelli, M., Campbell, G.S., and Flury, M. 1999. Characterization of particle size distribution in soils with a fragmentation model. *Soil Science Society of America Journal*, **63**, 782–788.

Blair, N. E., and Aller, R. S. 2012. The fate of terrestrial organic carbon in the marine environment. *Annual Reviews Marine Science*, **4**, 401–423.

Blair, N. E., Leithold, E. L., and Aller, R. C. 2004. From bedrock to burial: the evolution of particulate organic carbon across coupled watershed-continental margin systems. *Marine Chemistry*, **92**, 141–152.

Blatt, H., and Jones, R. L. 1975. Proportions of exposed igneous, metamorphic, and sedimentary rocks. *Geological Society of America Bulletin*, **86**, 1085–1088.

Blum, J. D., and Erel, Y. 1995. A silicate weathering mechanism linking increases in marine S-87/Sr-86 with global glaciation. *Nature*, **373**, 415–418.

Blum, M., and Garvin, M. 2010. Role of incised valley systems in source-to-sink sediment routing and storage: Examples from the Late Quaternary northern Gulf of Mexico margin. *American Association Petroleum Geologists Search and Discovery Article*, **50248**.

Blum, M., Martin, J., Milliken, K., and Garvin, N. 2013. Paleovalley systems: insights from Quaternary analogs and experiments. *Earth Science Reviews*, **116**, 128–169.

Blum, M. D., and Aslan, A. 2006. Signatures of climate vs. sea-level change within incised valley-fill successions: Quaternary examples from the Texas Gulf Coast. *Sedimentary Geology*, **190**, 177–211.

Blum, M. D., and Hattier-Womack, J. 2009. Climate change, sea level change, and fluvial sediment supply to deepwater depositional systems. Pages 15–39 of: Kneller, B., Martinsen, O. J., and McCaffrey, B. (eds), *External Controls on Deepwater Depositional Systems*. Society Economic Paleontologists Mineralogists, 92.

Blum, M. D., and Törnquist, T. E. 2000. Fluvial responses to climate and sea-level change: a review and look forward. *Sedimentology*, **47**, 2–48.

Bodet, F., and Schärer, U. 2000. Evolution of the SE Asian continent from U-Pb and Hf isotopes in single grains of zircon and baddeleyite from large rivers. *Geochimica et Cosmochimica Acta*, **64**, 2067–2091.

Bond, G. 1979. Evidence for some uplifts of large magnitude in continental platforms. *Tectonophysics*, **61**, 285–305.

Bonnet, C., Malavieille, J., and Mosar, J. 2008. Surface processes versus kinematics of thrust belts: impact on rates of erosion, sedimentation and exhumation – insights from analogue models. *Bulletin Geological Society France*, **179**, 297–314.

Bornhold, B. D., Yang, Z. S., Keller, G. H., Prior, D. B., Wiseman, W. J., Wang, Q., Wright, L. D., Xu, W. D., and Zhuang, Z. Y. 1986. Sedimentary framework of the modern Huang He (Yellow River) delta. *Geo-Marine Letters*, **6**, 77–83.

Boswell, P. G. H. 1933. *On the Mineralogy of Sedimentary Rocks*. London: Thomas Murby and Company.

Bouquillon, A., France-Lanord, C., Michard, A., and Tiercelin, J. 1990. Sedimentology and isotopic chemistry of the Bengal Fan sediments: the denudation of the Himalaya. *Proceedings Ocean Drilling Program Scientific Research*, **116**, 43–58.

Brack, P., Mundil, R., Oberli, F., Meier, M., and Rieber, H. 1996. Biostratigraphic and radiometric age data question the Milankovitch characteristics of the Latemar cycles (southern Alps, Italy). *Geology*, **24**, 371–375.

Bradley, W. H. 1929. The varves and climate of the Green River epoch. *U.S. Geological Survey Professional Paper*, **158**, 87–110.

Braun, J. 2005. Quantitative constraints on the rate of landform evolution derived from low-temperature thermochronology. *Reviews in Mineral Geochemistry*, **58**, 351–374.

Braun, J., van der Beek, P., and Batt, G. 2006. *Quantitative Thermochronology: Numerical Methods for the Interpretation of Thermochronological Data*. Cambridge, UK: Cambridge University Press.

Bray, D. I. 1982. Regime equations for gravel-bed rivers. Pages 517–542 of: Hey, R. D., Bathurst, J. C., and Thorne, C. R. (eds), *Gravel-Bed Rivers; Fluvial Processes, Engineering and Management*. Chichester, UK: Wiley.

Brewer, I. D., Burbank, D. W., and Hodges, K. V. 2003. Modelling detrital cooling age populations: insights from two Himalayan catchments. *Basin Research*, **15**, 305–320.

Bridge, J. S. 2003. *Rivers and Floodplains: Forms, Processes and Sedimentary Record*. Oxford, UK: Blackwell Publishing.

Bridge, J. S., and Demicco, R. V. 2008. *Earth Surface Processes, Landforms and Sediment Deposits*. Cambridge University Press.

Bridge, J. S., and Mackey, S. D. 1993. A revised alluvial stratigraphy model. Pages 319–337 of: Marzo, M., and Puigdefàbregas, C. (eds), *Alluvial Sedimentation*. International Association of Sedimentologists Special Publication 15.

Brocklehurst, S. H., and Whipple, K. X. 2004. Hypsometry of glaciated landscapes. *Earth Surface Processes and Landforms*, **29**, 907–926.

Brommer, M. B., Weltje, G. J., and Trincardi, F. 2009. Reconstruction of sediment supply from mass accumulation rates in the northern Adriatic Basin (Italy) over the past 19,000 years. *Journal of Geophysical Research*, **114**, F02008.

Brown, E. T., Stallard, R. F., Larsen, M. C., Raisbeck, G. M., and Yiou, F. 1995. Denudation rates determined from the accumulation of *in situ*-produced ^{10}Be in the Luquillo Experimental Forest, Puerto Rico. *Earth and Planetary Science Letters*, **129**, 93–202.

Brown, R. W., Summerfield, M. A., and Gleadow, A. J. W. 1994. Apatite fission track analysis: its potential for the estimation of denudation rates, and implications for models of long-term landscape development. Pages 23–53 of: Kirby, M. J. (ed), *Process Models and Theoretical Geomorphology*. New York: John Wiley and Sons Ltd.

Brunel, C., Certain, R., Sabatier, F., Robin, N., Barusseau, J.-P., Aleman, N., and Raynal, O. 2014. 20th century sediment budget trends of the Western Gulf of Lions shoreface (France): An application of an integrated method for the study of sediment coastal reservoirs. *Geomorphology*, **204**, 625–637.

Bull, W. B. 1962. *Relations of alluvial fan size and slope to drainage basin size and lithology in western Fresno County, California.* United States Geological Survey Professional Paper 450-B.

Bull, W. B. 1964. *Geomorphology of segmented alluvial fans in western Fresno County, California.* United States Geological Survey Professional Paper 353-E.

Bull, W. B. 1977. The alluvial fan environment. *Progress in Physical Geography*, **1**, 222–270.

Bull, W. B. 1984. Tectonic geomorhology. *Journal of Geological Education*, **32**, 310–324.

Bull, W. B. 1991. *Geomorphic Responses to Climate Change.* New York: Oxford University Press.

Burbank, D. W., and Anderson, R. S. 2001. *Tectonic Geomorphology.* Oxford, UK: Blackwell Publishing Ltd.

Burbank, D. W., and Vergés, J. 1994. Reconstruction of topography and related depositional systems during active thrusting. *Journal Geophysical Research*, **99**, 20,281–20,297.

Burbank, D. W., Beck, R. A., and Mulder, T. 1996. The Himalayan foreland basin. Pages 149–188 of: Yin A., and Harrison, M. (eds), *The Tectonic Evolution of Asia.* Cambridge University Press.

Burger, R. L., Fulthorpe, C. S., and Austin, J. A. 2001. Late Pleistocene channel incisions in the southern Eel River Basin, northern California: implications for tectonic vs. eustatic influences on shelf sedimentation patterns. *Marine Geology*, **177**, 317–330.

Burgess, P. M. 2016. Identifying ordered strata: Evidence, methods and meaning. *Journal Sedimentary Research*, **86**, 148–167.

Burgess, P. M., and Hovius, N. 1998. Rates of delta progradation during highstands: Consequences for timing of deposition in deep marine systems. *Journal of the Geological Society*, **155**, 217–222.

Burgess, P. M., and Prince, G. D. 2015. Non-unique stratal geometries: implications for sequence stratigraphic interpretations. *Basin Research*, **27**, 351–365.

Burkhard, M., and Sommaruga, A. 1998. Evolution of the Swiss Molasse basin: structural relations with the Alps and Jura belt. Pages 279–298 of: Mascle, A., Puigdefábregas, C., Luterbacher, H. P., and Fernández, M. (eds), *Cenozoic Foreland Basins of Western Europe.* Special Publication Geological Society London, 134.

Burt, T. P., and Allison, R. J. 2010. Sediment cascades. Pages 1–15 of: Burt, T. P., and Allison, R. J. (eds), *Sediment Cascades in the Environment: An Integrated Approach.* Chichester: John Wiley and Sons.

Busby, C., and Pérez, A. A. 2011. *Tectonics of Sedimentary Bains: Recent Advances.* London: John Wiley and Sons.

Cantuneanu, O., and Zecchin, M. 2016. Unique vs. non-unique stratal geometries: Relevance to sequence stratigraphy. *Marine and Petroleum Geology*, **78**, 184–195.

Carrétier, S., and Lucazeau, F. 2005. How does alluvial sedimentation at range fronts modify the erosional dynamics of mountain catchments? *Basin Research*, **17**, 361–341.

Carson, M. A., and Kirkby, M. J. 1972. *Hillslope Form and Process.* Cambridge, UK: Cambridge University Press.

Carter, R. M., Abbott, S. T., Fulthorpe, C. S., and Haywick, D. W. 1991. Application of global sea-level and sequence stratigraphic models in Southern Hemisphere Neogene strata from New Zealand. Pages 41–65 of: Macdonald, D. I. M. (ed), *Sedimentation, Tectonics and Eustasy: Sea-level Changes at Active Margins.* International Association Sedimentologists Special Publication 12.

Carvajal, C. R., and Steel, R. J. 2006. Thick turbidite successions from supply-dominated shelves during sea-level highstand. *Geology*, **34**, 665–668.

Carvajal, C., and Steel, R. 2009. Shelf-edge architecture and bypass of sand to deep water: Influence of sediment supply, sea level and shelf-edge processes. *Journal of Sedimentary Research*, **79**, 652–672.

Carvajal, C., and Steel, R. 2012. Source-to-sink sediment volumes within a tectono-stratigraphic model for a Laramide shelf-to-deep-water basin: Methods and results. Pages 131–151 of: Busby, C., and Azor Perez, A. (eds), *Tectonics of Sedimentary Basins: Recent Advances.* Oxford UK: Wiley-Blackwell.

Carvajal, C., Steel, R., and Petter, A. 2009. Sediment supply: the main driver of shelf-margin growth. *Earth Science Reviews*, **96**, 221–248.

Castelltort, S., and Van Den Driessche, J. 2003. How plausible are high-frequency sediment supply-driven cycles in the stratigraphic record? *Sedimentary Geology*, **157**, 3–13.

Castelltort, S., Whittaker, A. C., and Vergés, J. 2015. Tectonics, sedimentation and surface processes: from the erosional engine to basin deposition. *Earth Surface Processes and Landforms*, **40**, 2–9.

Catuneanu, O. 2006. *Principles of Sequence Stratigraphy*. Amsterdam: Elsevier.

Cavinato, G. P., Carusi, C., Dall'Asta, M., and Piacentini, T. 2002. Sedimentary and tectonic evolution of Plio-Pleistocene alluvial and lacustrine deposits of the Fucino Basin (central Italy). *Sedimentary Geology*, **148**, 29–59.

Chappell, J., and Shackleton, N. J. 1986. Oxygen isotopes and sea level. *Nature*, **324**, 137–140.

Charreau, G., Gumiaux, C., Avouac, J. P., Augier, R., Chen, J., Barrier, L., Gilder, S., Dominguez, S., Charles, N., and Wang, Q. 2009. The Neogene Xiju Formation, a diachronous prograding gravel wedge at the front of the Tien Shan: climatic and tectonic implications. *Earth and Planetary Science Letters*, **287**, 298–310.

Chester, R. 1990. *Marine Geochemistry*. Springer Science and Business Media.

Church, M., and Rood, K. 1983. *Catalogue of Alluvial River Channel Regime Data*. University of British Columbia, Department of Geography Report.

Clarke, A. 1991. Pages 119–158 of: Maines, D. (ed), *Social worlds/arenas theory as organizational theory. Social Organization and Social Progress: Essays in Honour of Anselm Strauss*. New York: Aldine de Gruyter.

Clevis, Q., de Jager, G., Nijman, W., and de Boer, P. L. 2004a. Stratigraphic signatures of translation of thrust-sheet-top basins over low-angle detachment faults. *Basin Research*, **16**, 145–163.

Clevis, Q., de Boer, P. L., and Nijman, W. 2004b. Differentiating the effect of episodic tectonism and sea level fluctuations in foreland basins filled by alluvial fans and axial deltaic systems: insights from a three-dimensional stratigraphic forward model. *Sedimentology*, **51**, 809–835.

Clift, P., and Gaedicke, P. 2002. Accelerated mass flux to the Arabian Sea during the middle to late Miocene. *Geology*, **30**, 207–210.

Clift, P. D. 2006. Controls on the erosion of Cenozoic Asia and the flux of clastic sediment to the ocean. *Earth & Planetary Science Letters*, **241**, 571–580.

Clift, P. D., and Giosan, L. 2014. Sediment fluxes and buffering in the post-glacial Indus Basin. *Basin Research*, **26**, 369–385.

Clift, P. D., Layne, G. D., and Blusztajn, J. 1994. Marine sedimentary evidence for monsoon strengthening, Tibetan uplift and drainage evolution in East Asia. Pages 255–282 of: *Continent-Ocean Interactions in the East Asian Marginal Seas*. Washington, DC: American Geophysical Union.

Clift, P. D., Shimizu, N., Layne, G. D., Blusztajn, J. 2001a. Development of the Indus Fan and its significance for the erosional history of the western Himalaya and Karakorum. *Geological Society America Bulletin*, **113**, 1039–1051.

Clift, P. D., Shimizu, N., Layne, G. D., and Blusztajn, J. 2001b. Tracing patterns of unroofing in the Early Himalaya through microprobe Pb analysis of detrital K-feldspars in the Indus Molasse, India. *Earth and Planetary Science Letters*, **188**, 475–491.

Clift, P. D., Lee, J. I., Hildebrand, P., Shimizu, N., Layne, G. D., Blusztajn, J., Blum, J. D., Garzanti, E., and Khan, A. A. 2002. Nd and Pb isotope variability in the Indus River system: Implications for sediment provenance and crustal heterogeneity in the Western Himalaya. *Earth and Planetary Science Letters*, **200**, 91–106.

Clift, P. D., Hodges, K. V., Heslop, D., Hannigan, R., Van Long, H., and Calves, G. 2008a. Correlation of Himalayan exhumation rates and Asian monsoon intensity. *Nature Geoscience*, **1**, 875–880.

Clift, P. D., Hoang, V. L., Hinton, R., Ellam, R., Hannigan, R., Tan, M. T., and Nguyen, D. A. 2008b. Evolving East Asian river systems reconstructed by trace element and Pb and Nd isotope variations in modern and ancient Red River-Song Hong sediments. *Geochemistry Geophysics Geosystems*, **9**, Q04039.

Cockburn, H. A. P., and Summerfield, M. A. 2004. Geomorphological applications of cosmogenic isotope analysis. *Progress in Physical Geography*, **28**, 1–42.

Cofaigh, C. Ó., Andrews, J. Y., Jennings, A. E., Dowdeswell, J. A., Hogan, K. A., Kilfeather, A. A., and Sheldon, C. 2013. Glacimarine lithofacies, provenance and depositional processes on a West Greenland trough-mouth fan. *Journal Quaternary Science*, **28**, 13–26.

Cogné, N., Chew, D., and Stuart, F. M. 2014. The thermal history of the western Irish offshore. *Geological Society London*, **171**, 779–792.

Colby, B. R. 1963. Fluvial sediments – a summary of source, transportation, deposition and measurement of sediment discharge. *Bulletin United States Geological Survey*, **1181-A**, 56 pages.

Coleman, J. M., Roberts, H. H., and Stone, G. W. 1998. Mississippi River delta: an overview. *Journal of Coastal Research*, **14**, 698–716.

Colombo, F. 2005. Quaternary telescopic-like alluvial fans, Andean Ranges, Argentina. Pages 69–84 of: Harvey, A. M., Mather, A. E., and Stokes, M. (eds), *Alluvial Fans: Geomorphology, Sedimentology, Dynamics*. Geological Society London Special Publication 251.

Connell, S. D., Kim, W., Paola, C., and Smith, G. A. 2012. Fluvial morphology and sediment flux steering of axial-transverse boundaries in an experimental basin. *Journal Sedimentary Research*, **82**, 310–325.

Constantine, J. A., and Dunne, T. 2008. Meander cutoff and the controls on the production of oxbow lakes. *Geology*, **36**, 23–26.

Cookman, J. L., and Flemings, P. B. 2001. STORMSED1.0: hydrodynamics and sediment transport in a 2-D, steady state, wind- and wave-driven coastal circulation model. *Computers & Geosciences*, **27**, 647–674.

Covault, J. A., and Fildani, A. 2014. Continental shelves as sediment capacitors or conveyors; source-to-sink insights from the tectonically active Oceanside shelf, southern California, USA. Pages 315–326 of: Chiocci, F. L., and Chivas, A. R. (eds), *Continental Shelves of the World: Their Evolution During the Last Glacio-Eustatic Cycle*. Geological Society of London Memoir 41.

Covault, J. A., and Graham, S. 2010. Submarine fans at all sea-level stands: Tectono-morphologic and climatic controls on terrigenous sediment delivery to the deep sea. *Geology*, **38**, 939–942.

Covault, J. A., Romans, B. W., Fildani, A., McGann, M., and Graham, S. A. 2000. Rapid climatic signal propagation from source to sink in a Southern Californian sediment routing system. *Journal of Geology*, **118**, 247–259.

Covault, J. A., Normark, W. R., Romans, B. W., and Graham, S. A. 2007. Highstand fans in the California Borderland: the overlooked deep-water depositional system. *Geology*, **35**, 783–786.

Covault, J. A., Craddock, W. H., Romans, B. W., Fildani, A., and Gosai, M. 2013. Spatial and temporal variations in landscape evolution: Historic and longer-term sediment flux through global catchments. *Journal of Geology*, **121**, 35–56.

Covault, J. A., Romans, B. W., Graham, S. A., Fildani, A., and Hilley, G. E. 2011. Terrestrial source to deep-sea sink sediment budgets at high and low sea levels: insights from tectonically active Southern California. *Geology*, **39**, 619–622.

Covey, M. 1986. The evolution of foreland basins to steady state: evidence from the western Taiwan foreland basin. Pages 77–90 of: Allen, P. A., and Homewood, P. (eds), *Foreland Basins*. Blackwell Scientific Publications, Oxford: International Association of Sedimentologists Special Publication 8.

Cowie, P. A., Gupta, S., and Dawers, N. H. 2000. Implications of fault array evolution for synrift depocentre development: Insights from a numerical fault growth model. *Basin Research*, **12**, 241–261.

Cowie, P. A., Attal, M., Tucker, G. E., Whittaker, A. C., Naylor, M., Ganas, A., and Roberts, G. P. 2006. Investigating the surface processes response to fault interaction and linkage using a numerical modelling approach. *Basin Research*, **18**, 231–266.

Crampton, S, E., and Allen, P. A. 1995. Recognition of forebulge unconformities associated with early stage foreland basin development: example from the North Alpine Foreland Basin. *Bulletin Geological Society America*, **79**, 1495–1514.

Cronin, M. F., Tozuka, T., Biastoch, A., Durgadoo, J. V., and Beal, L. M. 2013. Prevalence of strong bottom currents in the greater Agulhas system. *Geophysical Research Letters*, **40**, 1772–1776.

Cross, T. A., and Lessinger, M. A. 1998. Sediment volume partitioning: rationale for stratigraphic model evaluation and high-resolution stratigraphic correlation. Pages 171–195 of: F. Gradstein, K. Sandvik, and Milton, N. (eds), *Sequence Stratigraphy: Concepts and Applications, Special Publication*. Oslo: Norwegian Petroleum Society.

Cui, Y. T., and Parker, G. 1998. The arrested gravel front: stable gravel-sand transitions in rivers – Part 2: general numerical solution. *Journal of Hydraulic Research*, **36**, 159–182.

Culling, W. E. H. 1960. Analytical theory of erosion. *Journal of Geology*, **68**, 336–344.

Curray, J. R., Emmel, F. J., and Moore, D. G. 2002. The Bengal Fan: morphology, geometry, stratigraphy and processes. *Marine and Petroleum Geology*, **19**, 1191–1223.

Dacey, M. F., and Krumbein, W. C. 1979. Models of breakage and selection for particle size distributions. *Mathematical Geology*, **11**, 193–222.

Dade, W. B., and Friend, P. F. 1998. Grain-size, sediment transport regime and channel slope in alluvial rivers. *Journal of Geology*, **106**, 661–675.

Dade, W. B., and Verdeyen, M. E. 2007. Tectonic and climatic controls of alluvial-fan size and source-catchment relief. *Journal Geological Society London*, **164**, 353–358.

Dadson, S. J., Hovius, N., Chen, H., Hsieh, M. L., Willett, S. D., Hu, J. C., Horng, M. J., Chen, M.-C., Stark, C. P., Lague, D., and Lin, J.-C. 2003. Links between erosion, runoff variability and seismicity in the Taiwan orogen. *Nature*, **426**, 648–651.

Das, A., Krishnaswami, S., and Kumar, A. 2006. Sr and ^{87}Sr/^{86}Sr in rivers draining the Deccan Traps (India): implications to weathering, Sr fluxes and marine ^{87}Sr/^{86}Sr record around K/T. *Geochemistry, Geophysics, Geosystems*, **7**, Q06014.

Das, A., Krishnaswami, S., Sarin, M., Pande, K. 2005. Chemical weathering in the Krishna Basin and Western Ghats of the Deccan Traps, India: rates of basalt weathering and their controls. *Geochimica Cosmochimica Acta*, **69**, 2067–2084.

Davidson, S. K., and Hartley, A. J. 2010. Towards a quantitative method for estimating paleohydrology from clast size and comparison with modern rivers. *Journal Sedimentary Research*, **80**, 688–702.

Davidson, S. K., and North, C. P. 2009. Geomorphological regional curves for prediction of drainage basin area and screening modern analogues for rivers in the rock record. *Journal Sedimentary Research*, **79**, 773–792.

Dawers, N. H., and Anders, M. H. 1995. Displacement-length scaling and fault linkage. *Journal of Structural Geology*, **17**, 607–614.

Deb, D., and Sen, A. K. 2013. Rosin's law and size distribution of particles in regolith like samples - an analysis. *Planetary and Space Science*, **82**, 79–83.

DeCelles, P. G. 1988. Lithologic provenance modeling applied to the Late Cretaceous synorogenic Echo Canyon conglomerate, Utah: a case of multiple source areas. *Geology*, **16**, 1039–1043.

DeCelles, P. G., and Cavazza, W. 1999. A comparison of fluvial megafans in the Cordilleran (Upper Cretaceous) and modern Himalayan foreland basin systems. *Bulletin Geological Society of America*, **111**, 1315–1334.

DeCelles, P. G., and Giles, K. A. 1996. Foreland basin systems. *Basin Research*, **8**, 105–124.

Degens, E. T., and Ittekot, V. 1985. Particulate organic carbon: an overview. Pages 7–27 of: Degens, E. T., Kempe, S., and Herrera, R. (eds), *Transport of Carbon and Minerals in Major World Rivers, Part 3*. Mitt. Geol. Palont. Inst. Univ. Hamburg, SCOPE/UNEAP, Sonderband 58.

Dendy, F. E., and Bolton, G. C. 1976. Sediment yield-runoff-drainage basin area relationships in the United States. *Journal Soil and Water Conservation*, **32**, 264–266.

Denny, C. S. 1965. *Alluvial Fans in the Death Valley Region, California and Nevada*. United States Geological Survey Professional Paper 466.

Densmore, A. L., and Hovius, N. 2000. Topographic fingerprints of bedrock landslides. *Geology*, **28**, 371–374.

Densmore, A. L., Anderson, R. S., MacAdoo, B. G., and Ellis, M. A. 1997. Hillslope evolution by bedrock landslides. *Science*, **275**, 369–372.

Densmore, A. L., Ellis, M. A., and Anderson, R. S. 1998. Landsliding and the evolution of normal fault-bounded mountain ranges. *Journal Geophysical Research*, **103**, 15,203–15,219.

Densmore, A. L., Dawers, N. H., Gupta, S., Guidon, R., and Goldin, T. 2004. Footwall topographic development during continental extension. *Journal Geophysical Research*, **103**, F03001.

Densmore, A. L., Allen, P. A., and Simpson, G. 2007a. Development and response of a coupled catchment-fan system under changing tectonic and climatic forcing. *Journal of Geophysical Research*, **112**, 1–16.

Densmore, A. L., Gupta, S., Allen, P. A., and Dawers, N. H. 2007b. Transient landscapes at fault tips. *Journal Geophysical Research, Earth Surface*, **112**, F03S08.

DePloey, J., Kirkby, M. J., and Ahnert, F. 1991. Hillslope erosion by rainstorms - a magnitude frequency analysis. *Earth Surface Processes and Landforms*, **16**, 399–409.

Deptuck, M. E., Piper, D. J. W., Savoye, B., and Gervais, A. 2008. Dimensions and architecture of late Pleistocene submarine lobes off the northern margin of East Corsica. *Sedimentology*, **55**, 869–898.

Di Giulio, A., Ceriani, A., Ghia, E., and Zucca, F. 2003. Composition of modern stream sands derived from sedimentary source rocks in a temperate climate (Northern Apennines, Italy). *Sedimentary Geology*, **158**, 145–161.

Dickinson, W. R., and Suczek, C. 1979. Plate tectonics and sandstone compositions. *American Association Petroleum Geologists Bulletin*, **63**, 2164–2182.

Dingus, L. 1984. Effects of stratigraphic completeness on interpretations of extinction rates across the Cretaceous-Tertiary boundary. *Paleobiology*, **10**, 420–43.

Dingus, L., and Sadler, P. M. 1982. The effects of stratigraphic completeness on estimates of evolutionary rates. *Systematic Zoology*, **31**, 400–412.

Divins, D. L. 2003. NGDC Total Sediment Thickness of the World's Oceans and Marginal Seas. www.ngdc.noaa.gov/mgg/sedthick/sedthick.html.

Dorobek, S. L. 1995. Synorogenic carbonate platforms and reefs in foreland basins: Controls on stratigraphic evolution and platform/reef morphology. Pages 127–147 of: *Stratigraphic Evolution of Foreland Basins*. SEPM (Society for Sedimentary Geology) Special Publication, 52.

Dosseto, A., Bourdon, B., Gaillardet, J., Allégre, C., and Filizola, N. 2006. Time scale and conditions of weathering under tropical climate: study of the Amazon Basin with U-series. *Geochimica et Cosmochimica Acta*, **70**, 71–89.

Douglas, I. 1967. Man, vegetation, and the sediment yield of rivers. *Nature*, **215**, 925–928

Dowdeswell, J. A., Kenyon, N. H., Elverhoi, A., Laberg, J. S., Hollender, F. J., Mienert, J., and Siegert, M. J. 1996. Large-scale sedimentation on the glacier-influenced Polar North Atlantic margins: long-range side-scan sonar evidence. *Geophysical Research Letters*, **23**, 3535–3538.

Dowdeswell, J. A., Ottesen, D., Evans, J., Cofaigh, C. O., and Anderson, J. B. 2008. Submarine glacial landforms and rates of ice-stream collapse. *Geology*, **36**, 819–822.

Doyle, P., and Bennett, M. R. 1998. *Unlocking the Stratigraphic Record: Advances in Modern Stratigraphy*. Chichester, UK: John Wiley and Sons Ltd.

Drinkwater, N. J., and Pickering, K. T. 2001. Architectural elements in a high-continuity, sand-prone turbidite system, Late Precambrian Kongsfjord Formation, northern Norway: application to reservoir characterization. *Bulletin American Association Petroleum Geologists*, **85**, 1731–1757.

Drummond, C. N., and Wilkinson, B. H. 1996. Stratal thickness frequencies and the prevalence of orderedness in stratigraphic sequences. *Journal of Geology*, **104**, 1–18.

Dubille, M., and Lavé, J. 2015. Rapid grain size coarsening at sandstone/conglomerate transition: similar expression in Himalayan modern rivers and Pliocene molasse deposits. *Basin Research*, **27**, 26–42.

Duller, R. A., Whittaker, A. C., Fedele, J. J., Springett, J., Smithells, R., and Allen, P. A. 2010. From grain size to tectonics. *Journal of Geophysical Research Earth Surface*, **115**, F03022.

Duller, R. A., Whittaker, A. C., Swinehart, J. B., Armitage, J. J., Sinclair, H. D., Bair, A. R., and Allen, P. A. 2012. Abrupt landscape change post 6 Ma on the central Great Plains, USA. *Geology*, **40**, 871–874.

Dunne, T., and Leopold, L. B. 1978. *Water in Environmental Planning*. San Francisco, USA: W. H. Freeman.

Dunne, T., Mertes, L. A. K., Meade, R. H., Richey, J. E., and Forsberg, B. R. 1998. Exchanges of sediment between the flood plain and channel of the Amazon River in Brazil. *Bulletin Geological Society of America*, **110**, 450–467.

Dupont-Nivet, G., Krijgsman, W., Langereis, C. G., Abels, H. A., Dai, S., and Fang, X. M. 2007. Tibetan Plateau aridification linked to global cooling at the Eocene-Oligocene transition. *Nature*, **445**, 635–638.

Durán, R., Canals, M., Luis Sanz, J., Lastras, G., Amblas, D., and Micallef, A. 2012. Morphology and sediment dynamics of the northern Catalan continental shelf, northwestern Mediterranean Sea. *Geomorphology*, **204**, 1–20.

Dürr, H. H., Meybeck, M., Hartmann, J., Laruelle, G. G., and Roubeix, V. 2009. Global spatial distribution of natural riverine silica inputs to the coastal ocean. *Biogeosciences, Discussion*, **6**, 1345–1401.

Duval, B., Cramez, C., and Vail, P. R. 1998. Stratigraphic cycles and major marine source rocks. Pages 43–51 of: De Graciansky, P. C., Hardenbol, J., Jacquin, T., and Vail, P. R. (ed), *Mesozoic and Cenozoic Sequence Stratigraphy of European Basins*. SEPM (Society for Sedimentary Geology) Special Publication 60.

Dzhamalov, R. G., and Safronova, T. I. 2001. On estimating chemical discharge into the world ocean with groundwater. *Water Resources Research*, **29**, 680–686.

Eberl, D. D., Drits, V. A., and Srodon, J. 1998. Deducing growth mechanisms for minerals from the shapes of crystal size distributions. *American Journal of Science*, **298**, 499–533.

Egholm, D. L., Petersen, V. K., Knudsen, M. F., and Larsen, N. K. 2012. Coupling the flow of ice, water, and sediment in a glacial landscape evolution model. *Geomorphology*, **141**, 47–66.

Einsele, G., and Hinderer, M. 1997. Terrestrial sediment yield and the lifetimes of reservoirs, lakes, and larger basins. Pages 288–310 of: Schlager, W., and Kroonenberg, S. (eds), *Predictions in Geology*. Geologische Rundschau 86.

Einsele, G., Ratschbacher, L., and Wetzel, A. 1996. The Himalaya-Bengal Fan denudation-accumulation system during the last 20 Ma. *Journal of Geology*, **104**, 163–184.

Ellis, M. A., Densmore, A. L., and Anderson, R. S. 1999. Evolution of mountainous topography in the Basin and Range Province. *Basin Research*, **11**, 21–42.

Elverhoi, A., Hooke, R. LeB., and Solheim, A. 1998. Late Cenozoic erosion and sediment yield from the Svalbard-Barents Sea region: Implications for understanding erosion of glacierized basins. *Quaternary Science Reviews*, **17**, 209–241.

Erel, Y., Blum, J. D., and Roueff, E., Jiwchar, G. 2004. Lead and strontium isotopes as monitors of experimental granitoid mineral dissolution. *Geochimica Cosmochimica Acta*, **68**, 4649–4663.

Ertel, J. R., Hedges, J. I., Devol, A. H., Richey, J. A., and Ribeiro, M. de N. G., 1986. Dissolved humic substances of the Amazon River system. *Limnology and Oceanography*, **31**, 739–754.

Evan, A. T., Foltz, G. R., Zhang, D., and Vimont, D. J. 2011. Influence of African dust on ocean-atmosphere variability in the tropical Atlantic. *Nature Geoscience*, **4**, 762–765.

Farraj, A. al-, and Harvey, A. M. 2005. Morphometry and depositional style of Late Pleistocene alluvial fans: Wadi Al-Bih, northern UAE and Oman. Pages 85–94 of: Harvey, A. M., Mather, A. E., and Stoles, M. (eds), *Alluvial Fans: Geomorphology, Sedimentology, Dynamics*. Special Publication Geological Society London 251.

Fedele, J. J., and Paola, C. 2007. Similarity solutions for fluvial sediment fining by selective deposition. *Journal of Geophysical Research*, **112**, F02038.

Felletti, F., and Bersezio, R. 2010. Quantification of the degree of confinement of a turbidite-filled basin: A statistical approach based on bed thickness distribution. *Marine and Petroleum Geology*, **27**, 515–532.

Ferguson, R. 2010. Emergence of abrupt gravel to sand transitions along rivers through sorting processes. *Geology*, **31**, 159–162.

Fernandes, N. F., and Dietrich, W. F. 1997. Hillslope evolution by diffusive processes: the time scale for equilibrium adjustments. *Water Resources Research*, **33**, 1307–1318.

Filleaudeau, P. Y., Mouthereau, F., and Pik, R. 2012. Thermo-tectonic evolution of the south-central Pyrenees from rifting to orogeny: insights from detrital zircon U/Pb and (U-Th)/He thermochronometry. *Basin Research*, **24**, 401–417.

Fischer, A. G. 1986. Climatic rhythms recorded in strata. *Annual Reviews of Earth and Planetary Sciences*, **14**, 351–376.

Fischer, A. G., and Schwarzacher, W. 1984. Cretaceous bedding rhythms under orbital control? Pages 163–175 of: Berger, A. L., Imbrie, J., Hays, J., Kukla, G., and Saltzman, B. (eds), *Milankovitch and Climate: Understanding the Response to Astronomical Forcing*. Dordrecht/Boston/Lancaster, 510 pages: Reidel.

Fischer, H. B., List, E. J., Koh, C. Y., Imberger, J., and Brooks, N. H. 1979. *Mixing in Inland and Coastal Waters*. San Diego, USA: Academic Press.

Fitzgerald, P. G., Sorkhabi, R. B., Redfield, T. F., and Stump, E. 1995. Uplift and denudation of the central Alaska Range: a case-study in the use of apatite fission track thermochronology to determine absolute uplift parameters. *Journal Geophysical Research*, **100**, 21,075–21,091.

Flemming, B. W. 1981. Factors controlling shelf sediment dispersal along the southeast African continental margin. *Marine Geology*, **42**, 259–277.

Flint, S. S., and Bryant, I. D. 1993. *The Geological Modelling of Hydrocarbon Reservoirs and Outcrop Analogues*. International Association of Sedimentologists Special Publication 15.

Flood, R. D., Piper, D. J. W., and Klaus, A. and Shipboard Scientific Party 1995. *Proceedings Ocean Drilling Program Initial Report 155*. College Station, Texas, USA: Ocean Drilling Program.

Foltz, G. R., and McPhaden, M. J. 2008. Trends in Saharan dust and tropical Atlantic climate during 1980–2006. *Geophysical Research Letters*, **35**, L20802.

Ford, M. 2004. Depositional wedge-tops: interaction between low basal friction external orogenic wedges and flexural foreland basins. *Basin Research*, **16**, 361–375.

Forzoni, A., Storms, J. E. A., Whittaker, A. C., and de Jager, G. 2014. Delayed delivery from the sediment factory: modeling the impact of catchment response time to tectonics on sediment flux and fluvio-deltaic stratigraphy. *Earth Surface Processes and Landforms*, **39**, 689–704.

Fournier, F. 1960. *Climat et Erosion; la Relation entre Erosion du Sol par l'Eau et les Precipitations Atmosphériques*. Paris: Presses Universitaires de France.

France-Lanord, C., and Derry, L. A. 1997. Organic carbon burial forcing of the carbon cycle from Himalayan erosion. *Nature*, **390**, 65–67.

Francis, J. R. D. 1973. Experiments on the motion of solitary grains along the bed of a water stream. *Philosophical Transactions Royal Society London*, **332A**, 443–472.

Friedrichs, C. T., and Wright, L. D. 1986. Gravity-driven sediment transport on the continental shelf: implications for equilibrium profiles near river mouths. *Coastal Engineering*, **51**, 795–811.

Fu, B., Lin, A., Kano, K., Maruyama, T., and Guo, J. 2003. Quaternary folding of the eastern Tian Shan, northwest China. *Tectonophysics*, **369**, 79–101.

Gaillardet, J., Dupré, B., Louvat, P., and Allegrè, C. J. 1999. Global silicate weathering and CO_2 consumption rates deduced from the chemistry of large rivers. *Chemical Geology*, **159**, 3–30.

Gaillardet, J., Millot, R., and Dupré, B. 2003. Chemical denudation rates of the western Canadian orogenic belt: the Strikine terrane. *Chemical Geology*, **201**, 257–279.

Gallagher, K. 2012. Transdimensional inverse thermal history modeling for quantitative thermochronology. *Journal Geophysical Research*, **117**, B04208.

Gallagher, K., Brown, R., and Johnson, C. 1998. Fission track analysis and its application to geological problems. *Annual Review of Earth and Planetary Sciences*, **26**, 519–572.

Galloway, W. E. 2005. Gulf of Mexico Basin depositional record of Cenozoic North American drainage basin evolution. Pages 409–423 of: Blum, M., Marriott, S., and Leclair, S. (eds), *Fluvial Sedimentology VII*. International Association Sedimentologists Special Publication 35.

Galloway, W. E., Ganey-Curry, P., and Whiteaker, T. L. 2009. Regional controls from temporal and spatial distribution of continental slope and abyssal plain reservoir systems of the Gulf of Mexico Basin. *American Association Petroleum Geologists Search and Discovery Article*, **50226**.

Galloway, W. E., Whiteaker, T. L., and Ganey-Curry, P. 2011. History of Cenozoic North American drainage basin evolution, sediment yield, and accumulation in the Gulf of Mexico basin. *Geosphere*, **7**, 938–973.

Galy, A., and France-Lanord, C. 2001. Higher erosion rates in the Himalaya: geochemical constraints on riverine fluxes. *Geology*, **29**, 23–26.

Galy, V., France-Lanord, C., Beyssac, P., Faure, P., Kudrass, H., and Palhol, F. 2007. Efficient organic carbon burial in the Bengal fan sustained by the Himalayan erosional system. *Nature*, **450**, 407–410.

Galy, V., France-Lanord, C., Beyssac, O., Lartiges, B., and Rahman, M. 2011. Organic carbon cycling during Himalayan erosion: processes, fluxes and consequences for the global carbon cycle. Pages 163–179 of: Lal, R., Sivakumar, M. V. K., Faiz, M. A., Mustafizur Rahman, A. H. M., and Islam, K. R. (eds), *Climate Change and Food Security in South Asia*.

Ganti, V., Lamb, M. P., and McElroy, B. 2014. Quantitative bounds on morphodynamics and implications for reading the sedimentary record. *Nature Communications*, **5**, 3298.

Garcia-Castellanos, D., Vergés, J., Gaspar-Escribano, J., and Cloetingh, S. 2003. Interplay between tectonics, climate and fluvial transport during the Cenozoic evolution of the Ebro Basin (NE Iberia). *Journal of Geophysical Research-Solid Earth*, **108**, 8.1–8.18.

Garrels, R. M., and Mackenzie, F. T. 1971. *Evolution of Sedimentary Rocks*. New York: W. W. Norton.

Garver, J. I., Soloviev, A. V., Bullen, M. E., and Brandon, M. T. 2000. Towards a more complete record of magmatism and exhumation in continental arcs using detrital fission track thermochronometry. *Physics and Chemistry of the Earth, Part A*, **25**, 565–570.

Garzanti, E., Doglioni, C., Vezzoli, G., and Andò, S. 2007. Orogenic belts and orogenic sediment provenance. *Journal of Geology*, **115**, 315–334.

Gasparini, N. M., Tucker, G. E., and Bras, R. L. 1999. Downstream fining through selective particle sorting in an equilibrium drainage network. *Geology*, **27**, 1079–1082.

Gawthorpe, R. L., and Hurst, J. M. 1993. Transfer zones in extensional basins: their structural style and influence on drainage development and stratigraphy. *Journal Geological Society London*, **150**, 1137–1152.

Gawthorpe, R. L., and Leeder, M. R. 2000. Tectono-sedimentary evolution of active extensional basins. *Basin Research*, **12**, 195–218.

Geddes, A. 1960. The alluvial morphology of the Indo-Gangetic Plain: its mapping and geographical significance. *Institute of British Geographers, Transactions and Papers*, **28**, 253–276.

Gee, M. J. R., Masson, D. G., Watts, A. B., and Allen, P. A. 1999. The Saharan Debris Flow: an insight into the mechanics of long-runout submarine debris flows. *Sedimentology*, **46**, 317–335.

Geyer, W. R., Hill, P. S., and Kineke, G. C. 2004. The transport, transformation and dispersal of sediment by buoyant coastal flows. *Continental Shelf Research*, **24**, 927–949.

Gibbs, R. J. 1970. Mechanisms controlling world water chemistry. *Science*, **170**, 1088–1090.

Gibling, M. R. 2006. Width and thickness of fluvial channel bodies and valley fills in the geological record: A literature compilation and classification. *Journal Sedimentary Research*, **76**, 731–770.

Godard, V., Tucker, G. E., Burch Fisher, G., Burbank, D. W., and Bookhagen, B. 2013. Frequency-dependent landscape response to climatic forcing. *Geophysical Research Letters*, **40**, 859–863.

Goldstein, S. L., O'Nions, R. K., and Hamilton, P. J. 1984. A Sm-Nd isotopic study of atmospheric dusts and particulates from major river systems. *Earth and Planetary Science Letters*, **70**, 221–236.

Goldsworthy, M., Jackson, J., and Haines, J. 2002. The continuity of active fault systems in Greece. *Geophysical Journal International*, **148**, 596–618.

Goodbred, A. L., and Kuehl, S. A. 1998. Floodplain processes in the Bengal Basin and the storage of Ganges-Brahmaputra River sediment: an accretion study using ^{137}Cs and ^{210}Pb geochronology. *Sedimentary Geology*, **121**, 239–258.

Goodbred, S. and Kuehl, S., 2003. Sedimentary Geology of the Bengal Basin, Bangladesh, in relation to the Asia-Greater India collision and the evolution of the eastern Bay of Bengal. *Sedimentary Geology* (Special Issue), **155**, 175–424.

Goodbred, S. L., and Kuehl, S. A. 1999. Holocene and modern sediment budgets for the Ganges-Brahmaputra river system: Evidence for highstand dispersal to floodplain, shelf and deep-sea depocenters. *Geology*, **27**, 559–562.

Graham, S. A., Tolson, R. B., DeCelles, P. G., Ingersoll, R. V., Bargar, E., Caldwell, M., Cavazza, W., Edwards, D. P., Follo, M. F., Handschy, J. F., Lemke, L., Moxon, I., Rice, R., Smith, G. A., and White, J. 1986. Provenance modelling as a technique for analysing source terrane evolution and controls on foreland sedimentation. Pages 425–436 of: Allen, P. A., and Homewood, P. (eds), *Foreland Basins*. International Association of Sedimentologists Special Publication 8.

Graham, S. T., Famiglietti, J. S., and Maidment, D. R. 1999. Five minute, 1/2 degree, and 1 degree data sets of continental watersheds and river networks for use in regional and global hydrologic and climate system modeling studies. *Water Resources Research*, **35**, 583–587.

Gran, K. B., and Montgomery, D. R. 2005. Spatial and temporal patterns in fluvial recovery following volcanic eruptions: Channel response to basin-wide sediment loading at Mount Pinatubo, Philippines. *Geological Society America Bulletin*, **117**, 195–211.

Granger, D. E., Kirchner, J. W., and Finkel, R. 1996. Spatially averaged long-term erosion rates measured from *in situ*-produced cosmogenic nuclides in alluvial sediments. *Journal of Geology*, **104**, 249–257.

Granjeon, D., and Joseph, P. 1999. Concepts and applications of a 3-D multiple lithology, diffusive model in stratigraphic modeling. Pages 197–210 of: Harbaugh, J. W., Watney, W. L., Rankey, E. C., Slingerland, R., Goldstein, R. H., and Franseen, E. K. (eds), *Numerical Advances in Stratigraphy: Recent Advances in Stratigraphic and Sedimentologic Simulations*. Tulsa, Oklahoma: Special Publication Society for Sedimentary Geology, 62.

Grant, W. D., and Madsen, O. S. 1979. Combined wave and current interaction with a rough bottom. *Journal of Geophysical Research*, **84**, 1797–1808.

Grantham, J. H., and Velbel, M. A. 1988. The influence of climate and topography on rock-fragment abundance in modern fluvial sands of the southern Blue Ridge Mountains, North Carolina. *Journal Sedimentary Petrology*, **58**, 219–227.

Green, A. N. 2009. Sediment dynamics on the narrow, canyon-incised and current-swept shelf of the northern KwaZulu-Natal continental shelf, South Africa. *Geo-Marine Letters*, **29**, 201–219.

Griffin, J. D., Hemer, M. A., and Jones, B. G. 2008. Mobility of sediment grain size distributions on a wave dominated continental shelf, southeastern Australia. *Marine Geology*, **252**, 13–23.

Guillocheau, F., Rouby, D., Robin, C., Helm, C., Rolland, N., Le Carlier de Veslud, C., and Braun, J. 2012. Quantification and causes of the terrigeneous sediment budget at the scale of a continental margin: a new method applied to the Namibia-South Africa margin. *Basin Research*, **24**, 3–30.

Gupta, S. 1997. Himalayan drainage patterns and the origin of fluvial megafans in the Ganges foreland basins. *Geology*, **25**, 11–14.

Gupta, S., Underhill, J. R., Sharp, I. R., and Gawthorpe, R. L. 1999. Role of fault interactions in controlling synrift sediment dispersal patterns, Miocene, Abu Alaqa Group, Suez Rift, Sinai, Egypt. *Basin Research*, **11**, 167–189.

Gurnell, A., Hannah, D., and Lawler, D. 1996. Suspended sediment yield from glacier basins. *IAHS Publ.*, **236**, 97–104.

Gurnis, M. 2001. Sculpting the Earth from inside out. *Scientific American*, **284**, 40–47.

Hack, J. T. 1973. Stream profile analysis and stream gradient index. *U.S. Geological Survey Journal of Research*, **1**, 421–429.

Hajek, E. A., and Wolinsky, M. A. 2012. Simplified process modeling of river avulsion and alluvial architecture: connecting models and field data. *Sedimentary Geology*, **257–260**, 1–30.

Hallet, B., Hunter, L., and Bogen, J. 1996. Rates of erosion and sediment evacuation by glaciers: a review of field data and their implications. *Global and Planetary Change*, **12**, 213–235.

Hampson, G. J., Duller, R. A., Petter, A. L., Robinson, R. A. J., and Allen, P. A. 2014. Mass-balance constraints on stratigraphic interpretation of linked alluvial-coastal-shelfal deposits: Upper Cretaceous Castlegate Sandstone, Blackhawk Formation, Star Point Sandstone and Mancos Shale, Utah and Colorado, USA. *Journal of Sedimentary Research*, **84**, 935–960.

Hancock, G. S., Anderson, R. S., Whipple, K. X., and Wohl, E. E. 1998. Beyond power: bedrock river incision process and form. Pages 35–60 of: *Rivers over Rock, Fluvial Processes in Bedrock Channels*. Geophysical Monograph 107.

Hanks, T. C., Bucknam, R. C., LaJoie, K. R., and Wallace, R. E. 1984. Modification of wave-cut and faulting controlled landforms. *Journal Geophysical Research*, **89**, 5771–5790.

Haq, B. U. 1991. Sequence stratigraphy, sea-level change, and significance for the deep sea. Pages 3–39 of: Macdonald, D. I. M. (ed), *Sedimentation, Tectonics and Eustasy: Sea-Level Changes at Active Margins*. International Association Sedimentologists Special Publication 12.

Haq, B. U., and Al-Qahtani, A. M. 2005. Phanerozoic cycles of sea-level change on the Arabian platform. *GeoArabia*, **10**, 127–160.

Haq, B. U., and Schutter, S. R. 2008. A chronology of Paleozoic sea-level changes. *Science*, **322**, 64–68.

Haq, B. U., Hardenbol, J., and Vail, P. R. 1987. Chronology of fluctuating sea levels since the Triassic (250 Myr ago to present). *Science*, **235**, 1156–1167.

Hardenbol, J., Thierry, J., Farley, M. B., Jacquin, T., De Graciansky, P. C., and Vail, P. R. 1998. Mesozoic and Cenozoic sequence chronostratigraphic framework of European basins. Pages 3–14 of: De Graciansky, P. C., Hardenbol, J., Jacquin, T., and Vail, P. R. (eds), *Mesozoic and Cenozoic Sequence Stratigraphy of European Basins*. SEPM (Society for Sedimentary Geology) Special Publication 60.

Hardie, L. A., Bosellini, A., and Goldhammer, R. K. 1986. Repeated subaerial exposure of subtidal carbonate platforms, Triassic, northern Italy: evidence for high frequency sea level oscillations on a 10^4 year scale. *Paleoceanography*, **1**, 447–457.

Harel, M. A., Mudd, S. M., and Attal, M. 2016. Global analysis of the stream power law parameters based on worldwide ^{10}Be denudation rates. *Geomorphology*, **268**, 184–196.

Harris, C. K., and Wiberg, P. L. 2001. A two-dimensional, time-dependent model of suspended sediment transport and bed reworking for continental shelves. *Computers & Geosciences*, **27**, 675–690.

Harris, C. K., Traykovski, P., and Geyer, W. R. 2005. Flood dispersal and deposition by near-bed gravitational sediment flows and oceanographic transport: A numerical modeling study of the Eel River shelf, northern California. *Journal of Geophysical Research*, **110**, 25.1–25.16.

Harrison, C. G. A. 1988. Eustasy and epeirogeny of continents on time scales between about 1 and 100 My. *Paleoceanography*, **3**, 671–684.

Harrison, C. G. A. 2002. Power spectrum of sea level change over 15 decades of frequency. *Geochemistry Geophysics Geosystems*, **3**, 10.1029/2002GC000300.

Harrison, C. G. A., Brass, G. W., Saltzman, E., Sloan II, J., Southam, J., and Whitman, J. 1981. Sea level variations, global sedimentation rates and the hypsographic curve. *Earth and Planetary Science Letters*, **54**, 1–16.

Hartmann, W. K. 1969. Terrestrial, lunar, and interplanetary rock fragmentation. *Icarus*, **10**, 201–213.

Hartshorne, K., Hovius, N., Dade, B. W., Slingerland, R. L. 2002. Climate-driven bedrock incision in an active mountain belt. *Science*, **297**, 2036–2038.

Harvey, A. M. 2005. Differential effects of base-level, tectonic setting and climate change on Quaternary alluvial fans in the northern Great Basin, Nevada, USA. Pages 117–131 of: Harvey, A. M., Mather, A. E., and Stokes, M. (eds), *Alluvial Fans: Geomorphology, Sedimentology, Dynamics*. Geological Society London Special Publication.

Hastings, H. M., and Sugihara, G. 1993. *Fractals: A User's Guide for the Natural Sciences*. Oxford, U.K.: Oxford Science Publications.

Hay, W. W. 1998. Detrital sediment fluxes from continents to oceans. *Chemical Geology*, **145**, 287–323.

Hays, J. D., Imbrie, J., and Shackleton, N. J. 1976. Variations in the Earth's orbit: pacemaker of the ice ages. *Science*, **194**, 1121–1132.

Hedges, J. I., Clark, W. A, Quay, P. D., Richey, J. E., Devol, A. H., and Santos, U. de M. 1986. Compositions and fluxes of particulate organic material in the Amazon River. *Limnology and Oceanography*, **31**, 717–738.

Heezen, B. C., and Ewing, M. 1952. Turbidity currents and submarine slumps, and the 1929 Grand Banks earthquake. *American Journal of Science*, **250**, 849–873.

Heffern, E. L., Reiners, P. W., Naeser, C. W., and Coates, D. A. 2008. Geochronology of clinker and implications for evolution of the Powder River Basin landscape, Wyoming and Montana. *Geological Society America Reviews in Engineering Geology*, **18**, 155–175.

Heimsath, A. M., Dietrich, W. E., Nishiizumi, K., and Finkel, R. C. 1997. The soil production function and landscape equilibrium. *Nature*, **388**, 358–361.

Heins, W. A., and Kairo, S. 2007. Predicting sand character with integrated genetic analysis. Pages 345–380 of: Arribas, J., Critelli, S., and Johnsson, M. J. (eds), *Sedimentary Provenance and Petrogenesis: Perspectives from Petrography and Geochemistry*. Geological Society of America Special Paper 420.

Helland-Hansen, W., and Gjelberg, H. 2012. Towards a hierarchical classification of clinoforms. *American Association Petroleum Geologists, Annual Convention and Exhibition, Search and Discovery Article*, **90142**.

Helland-Hansen, W., and Hampson, G. J. 2009. Trajectory analysis: concepts and applications. *Basin Research*, **21**, 454–483.

Helland-Hansen, W., Sømme, T. O., Martinsen, O. J., Lunt, I., and Thurmond, J. 2016. Deciphering Earth's natural hourglasses: Perspectives on source-to-sink analysis. *Journal Sedimentary Research*, **86**, 1008–1033.

Heller, P, L., and Paola, C. 1992. The large-scale dynamics of grain size variation in alluvial basins, 2. Applications to syntectonic conglomerate. *Basin Research*, **4**, 91–102.

Henriksen, S., Hampson, G. J., Helland-Hansen, W., Johannessen, E. P., and Steel, R. J. 2011. Shelf-edge and shoreline trajectories: a dynamic approach to stratigraphic analysis. *Basin Research*, **23**, 445–453.

Hergarten, S. 2002. *Self-Organized Criticality in Earth Systems*. Berlin: Springer-Verlag.

Herman, F., Seward, D., Valla, P. G., Carter, A., Kohn, B., Willett, S. D., and Ehlers, T. A. 2013. Worldwide acceleration of mountain erosion under a cooling climate. *Nature*, **504**, 423–426.

Hetzel, R., and Hanpel, A. 2005. Slip rate variations on normal faults during glacial-interglacial changes in surface loads. *Nature*, **435**, 81–84.

Hilgen, F. 2007. Extension of the astronomically calibrated (polarity) time scale to the Miocene/Pliocene boundary. *Earth and Planetary Science Letters*, **107**, 349–368.

Hilgen, F. J., Kuiper, K. F., and Lourens, L. J. 2010. Evaluation of the astronomical time scale for the Paleocene and earliest Eocene. *Earth and Planetary Science Letters*, **300**, 139–151.

Hilton, R. G., Galy, A., Hovius, N., Chen, M. C., Horng, M. J., and Chen, H. 2008. Tropical cycle-driven erosion of the terrestrial biosphere from mountains. *Nature Geoscience*, **1**, 759–762.

Hinderer, M. 2012. From gullies to mountain belts: A review of sediment budgets at various scales. *Sedimentary Geology*, **280**, 21–59.

Hinderer, M., and Einsele, G. 2001. The world's large basins as denudation-accumulation systems and implications for their lifetimes. *Journal of Paleolimnology*, **26**, 355–372.

Hinnov, L. A., and Goldhammer, R. K. 1991. Spectral analysis of the Middle Triassic Latemar limestone. *Journal Sedimentary Petrology*, **61**, 1173–1193.

Hinnov, L. A., and Ogg, J. G. 2007. Cyclostratigraphy and the astronomical time scale. *Stratigraphy*, **4**, 239–251.

Hirst, J. P. P., and Nichols, G. J. 1986. Thrust tectonic controls on Miocene alluvial distribution patterns, southern Pyrenees. Pages 247–258 of: Allen, P. A., and Homewood, P. (eds), *Foreland Basins, Special Publication International Association Sedimentologists, 8*. Oxford, UK: Blackwell Scientific Publications.

Hoffman, P. F., and Grotzinger, J. P. 1993. Orographic precipitation, erosional unloading, and tectonic style. *Geology*, **21**, 195–198.

Hoffmann, T., Erkens, G., Cohen, K. M., Houben, P., Seidel, J., and Dikau, R. 2007. Holocene floodplain sediment storage and hillslope erosion within the Rhine catchment. *The Holocene*, **17**, 105–118.

Hogan, K. A., Dowdeswell, J. A., and Cofaigh, C. O. 2012. Glacimarine sedimentary processes and depositional environments in an embayment fed by West Greenland ice streams. *Marine Geology*, **311**, 1–16.

Holbrook, J., and Wanas, H. 2014. A fulcrum approach to assessing source-to-sink mass balance using channel paleohydrological parameters derivable from common fluvial datasets with an example from the Cretaceous of Egypt. *Journal Sedimentary Research*, **84**, 349–372.

Holeman, J. N. 1980. Erosion rates in the U.S. estimated by the Soil Conservation Service's inventory. *EOS, Transactions of the American Geophysical Union*, **61**, 954.

Homewood, P., and Allen, P. A. 1981. Wave-, tide- and current-controlled sandbodies of Miocene Molasse, western Switzerland. *Bulletin American Association Petroleum Geologists*, **65**, 2534–2545.

Homewood, P., Allen, P. A., and Williams, G. D. 1986. Dynamics of the Molasse Basin of western Switzerland. Pages 199–217 of: Allen, P. A., and Homewood, P. (eds), *Foreland Basins, Special Publication International Association Sedimentologists, 8*. Oxford, UK: Blackwell Scientific Publications.

Hooke, R. L. 1968. Steady-state relationships on arid-region alluvial fans in closed basins. *American Journal of Science*, **266**, 609–629.

Hooke, R. L., and Rohrer, W. L. 1977. Relative erodibility of source-area rock types, as determined from second order variations in alluvial fan size. *Bulletin Geological Society America*, **88**, 1177–1182.

Hooke, R. LeB. 2000. Toward a uniform theory of clastic sediment yield in fluvial systems. *Bulletin Geological Society America*, **112**, 1778–1786.

Horton, B. K., and DeCelles, P. G. 1997. The modern foreland basin system adjacent to the central Andes. *Geology*, **25**, 895–898.

Horton, R. E. 1945. Erosional development of streams and their drainage basins: hydrophysical approach to quantitative morphology. *Geological Society America Bulletin*, **56**, 275–370.

Hoth, S., Kukowski, N., and Oncken, O. 2008. Distant effects in bivergent orogenic belts – How retro-wedge erosion triggers resource formation in pro-foreland basins. *Earth and Planetary Science Letters*, **273**, 28–37.

Hovius, N. 1996. Regular spacing of drainage outlets from linear mountain belts. *Basin Research*, **8**, 29–44.

Hovius, N. 1998. Controls on sediment supply by large rivers. Pages 3–16 of: Shanley, K. W., and McCabe, P. J. (eds), *Relative Role of Eustasy, Climate and Tectonics in Continental Rocks*. Society Economic Paleontologists Mineralogists Special Publication, 59.

Hovius, N., Stark, C. P., and Allen, P. A. 2007. Sediment flux from a mountain belt derived from landslide mapping. *Geology*, **25**, 231–234.

Howard, A. D. 1987. Modelling fluvial systems: rock-, gravel- and sand-bed channels. Pages 69–94 of: Richards, K. (ed), *River Channels*. New York: Basil Blackwell.

Howard, A. D. 1997. Badland morphology and evolution: Interpretation using a simulation model. *Earth Surface Processes and Landforms*, **22**, 211–227.

Howard, A. D., and Kerby, G. 1983. Channel changes in badlands. *Geological Society America Bulletin*, **94**, 739–752.

Hsui, A. T., Rust, K. A., and Klein, G. D. 1993. A fractal analysis of Quaternary, Cenozoic-Mesozoic, and Late Pennsylvanian sea level changes. *Journal Geophysical Research*, **98**, 21,963–21,967.

Hughes Clark, J. E., Shor, A. N., Piper, D. J. W., and Mayer, L. A. 1990. Large-scale current-induced erosion and deposition in the path of the 1929 Grand Banks turbidity current. *Sedimentology*, **37**, 613–629.

Humphrey, N. F., and Heller, P. L. 1995. Natural oscillations in coupled geomorphic systems – an alternative origin for cyclic sedimentation. *Geology*, **23**, 499–502.

Hurford, A. J., and Green, P. F. 1983. The zeta age calibration of fission track dating. *Chemical Geology (Isotope Geoscience Section)*, **1**, 285–317.

Hyslip, J. P., and Vallejo, L. E. 1997. Fractal analysis of the roughness and size distribution of granular materials. *Engineering Geology*, **48**, 231–244.

Ibbeken, H. 1983. Jointed source rock and fluvial gravels controlled by Rosin's law: A grain-size study in Calabria, south Italy. *Journal of Sedimentary Petrology*, **53**, 1213–1231.

Ibbeken, H., and Schleyer, R. 1991. *Source and Sediment: A Case Study of Provenance and Mass Balance at an Active Plate Margin (Calabria, Southern Italy)*. Berlin: Springer-Verlag.

Ibbeken, H., and Schleyer, R. 2003. Relative strengths of provenance signals discriminating four source areas in Calabria, Italy. Pages 101–116 of: *Quantitative Provenance Studies in Italy*, vol. 61. Memoire Descrittive della Carta Geologica dell'Italia.

Imbrie, J., Hays, J. D., Martinson, D. G., McIntyre, A., Mix, A. C., Morley, J. J., Pisias, N. G., Prell, W. L., and Shackleton, N. J. 1984. The orbital theory of Pleistocene climate: support from a revised chronology of the marine ^{18}O record. In: Berger, A. L., Imbrie, J., Hays, J., Kukla, G., and Saltzman, B. (eds), *Milankovitch and Climate: Understanding the Response to Astronomical Forcing*. Norwell, Massachusetts: D. Reidel.

Ingersoll, R. V. 1988. Tectonics of sedimentary basins. *Geological Society of America Bulletin*, **100**, 1704–1719.

Ingersoll, R. V., and Busby, C. J. 1995. Tectonics of sedimentary basins. Pages 1–52 of: *Tectonics of Sedimentary Basins*. Oxford: Blackwell Science.

Ittekot, V. 1988. Global trends in the nature of organic matter in river suspensions. *Nature*, **322**, 436–438.

Ittekot, V., and Laane, R. W. P. M. 1991. Fate of riverine particulate organic matter. Chap. 10 of: Degens, E. T., Kempe, S., and Richey, J. E. (eds), *Biogeochemistry of Major World Rivers, SCOPE 42*. Scientific Committee on Problems of the Environment (SCOPE).

Ivy-Ochs, S., and Kober, F. 2008. Surface exposure dating with cosmogenic nuclides. *Quaternary Science Journal*, **57**, 179–209.

Jaeger, J. M., and Koppes, M. 2015. The role of the cryosphere in source-to-sink systems. *Earth Science Reviews*, **153**, 43–76.

Jago, C. F., and Barusseau, J. P. 1981. Sediment entrainment on a wave-graded shelf, Roussillon, France. *Marine Geology*, **42**, 279–299.

Jansson, P., Jacobson, D., and Hooke, R. L. 1993. Fan and playa areas in southern California and adjacent parts of Nevada. *Earth Surface Processes and Landforms*, **18**, 109–119.

Jarman, D., Agliardi, F., and Crosta, G. B. 2011. Megafans and outsize fans from catastrophic slope failures in Alpine glacial troughs: the Malser Haide and the Val Venosta cluster, Italy. Pages 253–277 of: Jaboyedoff, M. (ed), *Slope Tectonics*. Geological Society London Special Publication 351.

Jefferson, I. F., Jefferson, B. Q., Assallay, A. M., Rogers, C. D. F., and Smalley, I. J. 1997. Crushing of quartz sand to produce silt particles. *Naturwissenschaften*, **84**, 1–3.

Jerolmack, D. J., and Brzinski, T. A. III. 2010. Equivalence of abrupt grain-size transitions in alluvial rivers and eolian sand seas. *Journal of Geology*, **38**, 719–722.

Jerolmack, D. J., and Paola, C. 2010. Shredding of environmental signals by sediment transport. *Geophysical Research Letters*, **37**, L19401.

Jervey, M. T. 1988. Quantitative geological modeling of siliciclastic rock sequences and their seismic expression. Pages 47–69 of: Wilgus, C. K., Hastings, B. S., Kendall, C. G. St. C., Posamentier, H. W., Ross, C. A., and van Wagoner, J. C. (eds), *Sea-level Changes: An Integrated Approach*, vol. 42. Society of Economic Paleontologists and Mineralogists Special Publication.

Johnsson, M. J. 1993. The system controlling the composition of clastic sediments. Pages 1–19 of: Johnsson, M. J., and Basu, A. (eds), *Processes Controlling the Composition of Clastic Sediments*. Geological Society of America Special Paper 284.

Johnsson, M. J., and Meade, R. H. 1990. Chemical weathering of fluvial sediments during alluvial storage: The Macuapanim Island point bar, Solimoes River, Brazil. *Journal of Sedimentary Petrology*, **60**, 827–842.

Jolley, E. T., Turner, P., Williams, G. D., Hartley, A. J., and Flint, S. 1990. Sedimentological response of an alluvial system to Neogene thrust tectonics. *Journal Geological Society London*, **147**, 769–784.

Keil, R. G., Mayer, L. M., Quay, P. D., Richey, J. E., and Hedges, J. I. 1997. Loss of organic matter from riverine particles in deltas. *Geochimica et Cosmochimica Acta*, **61**, 1507–1511.

Kelsey, H. M., Lamberson, R., and Madej, M. A. 1987. Stochastic model for the long-term transport of stored sediment in a river channel. *Water Resources Research*, **23**, 1738–1750.

Kenyon, P. M., and Turcotte, D. L. 1985. Morphology of a delta prograding by bulk sediment transport. *Bulletin Geological Society America*, **96**, 1457–1465.

Kertznus, V., and Kneller, B. 2009. Clinoform quantification for assessing the effects of external forcing on continental margin development. *Basin Research*, **21**, 738–758.

Kesel, R, H., Dunne, K. C., McDonald, K. R., and Spicer, B. E. 1974. Lateral overbank deposition on the Mississippi River in Louisiana caused by the 1973 flooding. *Geology*, **1**, 461–464.

Kettner, A. J., Restrepo, J. D., and Syvitski, J. P. M. 2010. A spatial simulation experiment to replicate fluvial sediment fluxes within the Magdalena River Basin, Colombia. *Journal of Geology*, **118**, 363–379.

Kim, W., Paola, C., Swenson, J. B., and Voller, V. R. 2006. Shoreline response to autogenic processes of sediment storage and release in the fluvial system. *Journal of Geophysical Research*, **111**, F04013.

Kim, W., Connell, S. D., Steel, E., Smith, G. A., and Paola, C. 2011. Mass-balance control on the interaction of axial and transverse channel systems. *Geology*, **39**, 611–614.

Kirby, E., and Whipple, K. X. 2001. Quantifying differential rock uplift rates via stream profile analysis. *Geology*, **29**, 415–418.

Kirby, M. E., Lund, S. P., Anderson, M. A., and Bird, B. W. 2007. Insolation forcing of Holocene climate change in Southern California: a sediment study of Lake Elsinore. *Journal of Paleolimnology*, **38**, 395–417.

Kirchner, J. W., Finkel, R. C., Riebe, C. S., Granger, D. E., Clayton, J. L., and Megahan, W. F. 2001. Episodic mountain erosion inferred from sediment yields over 10-year and 10,000-year time scales. *Geology*, **29**, 591–594.

Kirkby, M. J., and Cox, N. J. 1995. A climatic index for soil erosion potential (CSEP) including seasonal and vegetation factors. *Catena*, **25**, 333–352.

Kittleman, L. R. 1964. Application of Rosin's distribution in size-frequency analysis of clastic rock. *Journal Sedimentary Petrology*, **34**, 483–502.

Klein, J., Giegengack, R., Middleton, R., Sharma, P., Underwood, J., and Weeks, R. A. 1986. Revealing histories of exposure using *in situ*-produced ^{26}Al and ^{10}Be in Libyan desert glass. *Radiocarbon*, **28**, 547–555.

Knighton, A. D. 1999. The gravel-sand transition in a disturbed catchment. *Geomorphology*, **27**, 325–341.

Koide, M., Soutar, A., and Goldberg, E. D. 1972. Marine geochronology with ^{210}Pb. *Earth and Planetary Science Letters*, **14**, 442–446.

Koide, M., Bruland, K. W., and Goldberg, E. D. 1973. ^{228}Th/^{232}Th and ^{210}Pb geochronologies in marine and lake sediments. *Geochimica et Cosmochimica Acta*, **37**, 1171–1187.

Komar, P. D. 1973. Computer models of delta growth due to sediment input from rivers and longshore transport. *Bulletin Geological Society of America*, **84**, 2217–2226.

Komar, P. D., and Miller, M. C. 1973. The threshold of sediment movement under oscillatory waves. *Journal Sedimentary Petrology*, **43**, 1101–1110.

Konstantinovskaya, E., and Malavieille, J. 2005. Erosion and exhumation in accretionary orogens: experimental and geological approaches. *Geochemistry, Geophysics, Geosystems*, **6**, Q02006.

Kooi, H., and Beaumont, C. 1996. Large-scale geomorphology: classical concepts reconciled and integrated with contemporary ideas via a surface processes model. *Journal Geophysical Research*, **101**, 3361–3386.

Koons, P. O. 1989. The topographic evolution of collisional mountain belts: a numerical look at the Southern Alps, New Zealand. *American Journal of Science*, **289**, 1041–1069.

Koons, P. O. 1995. Modelling the topographic evolution of collisional belts. *Annual Review of Earth and Planetary Sciences*, **23**, 375–408.

Koppes, M., and Montgomery, D. 2009. The relative efficacy of fluvial and glacial erosion over modern to orogenic time scales. *Nature Geoscience*, **2**, 644–647.

Koppes, M., Hallet, B., Rignot, A., Mouginot, J., Wellner, J. S., and Boldt, K. 2015. Observed latitudinal variations in erosion as a function of glacier dynamics. *Nature*, **526**, 100–103.

Krishnaswami, S., Singh, S. K., and Dalai, T. K. 1999. Silicate weathering in the Himalaya: role in contributing to major ions and radiogenic Sr to the Bay of Bengal. Pages 23–51 of: Somayajulu, B. L. K. (ed), *Ocean Science, Trends and Future Directions*. New Delhi: Indian National Science Academy and Akademia International.

Krishnaswami, S., Trivedi, J. R., Sarin, M. M., Ramesh, R., and Sharma, K. K. 1992. Strontium isotopes and rubidium in the Ganges-Brahmaputra river system: Weathering in the Himalaya, fluxes to the Bay of Bengal and contributions to the evolution of oceanic ^{87}Sr/^{86}Sr. *Earth and Planetary Science Letters*, **109**, 243–253.

Krishnaswamy, S., Lal, D., Martin, J. M., and Meybeck, M. 1971. Geochronology of lake sediments. *Earth and Planetary Science Letters*, **11**, 407–414.

Krumbein, W. C. 1938. Size-frequency distributions of sediments and the normal phi curve. *Journal Sedimentary Petrology*, **8**, 84–90.

Krumbein, W. C., and Tisdel, F. W. 1940. Size distributions of source rocks of sediments. *American Journal of Science*, **238**, 296–305.

Kuehl, S. A., DeMaster, D. J., and Nittrouer, C. A. 1986. Nature of sediment accumulation on the Amazon continental shelf. *Continental Shelf Research*, **6**, 209–225.

Kuehl, S. A., Levy, B. M., Moore, W. S., and Allison, M. A. 1997. Subaqueous delta of the Ganges-Brahmaputra river system. *Marine Geology*, **144**, 81–96.

Kuehl, S. A., Allison, M. A., Goodbred, S. L., and Kudrass, H. 2005. The Ganges-Brahmaputra Delta. Pages 413–434 of: *River Deltas – Concepts, Models and Examples*. Tulsa, Oklahoma: Society of Economic Paleontologists and Mineralogists, Special Publication 83.

Kuhlemann, J., Frisch, W., Szekely, B., Dunkl, I., and Kazmer, M. 2002. Post-collisional sediment budget history of the Alps; tectonic versus climatic control. *International Journal of Earth Sciences*, **9**, 818–837.

Lague, D., Hovius, N., and Davy, P. 2005. Discharge, discharge variability, and the bedrock channel profile. *Journal Geophysical Research*, **110**, F04006.

Lal, D. 1991. Cosmic ray labelling of erosion surfaces: *in situ* nuclide production rates and erosion models. *Earth and Planetary Science Letters*, **104**, 424–439.

Lambeck, K., and Chappell, J. 2001. Sea level change through the last glacial cycle. *Science*, **292**, 679–686.

Lancaster, S. T., and Bras, R. L. 2002. A simple model of river meandering and its comparison with natural channels. *Hydrological Processes*, **16**, 1–26.

Langbein, W. B., and Schumm, S. A. 1958. Yield of sediment in relation to mean annual precipitation. *American Geophysical Union Transactions*, **38**, 1076–1084.

Laske, G., and Masters, G. 1997. http://igppweb.ucsd.edu/gabi/sediment.html.

Lavé, J., and Avouac, J. P. 2001. Fluvial incision and tectonic uplift across the Himalayas of central Nepal. *Journal of Geophysical Research*, **106**, 25,561–25,591.

Lecce, S. A. 1991. Influence of lithological erodibility on alluvial fan area, western White Mountains, California and Nevada. *Earth Surface Processes and Landforms*, **16**, 11–18.

LeClair, S. F., and Bridge, J. S. 2001. Quantitative interpretation of sedimentary structures formed by river dunes. *Journal Sedimentary Research*, **71**, 713–716.

Leeder, M. R. 1999. *Sedimentology and Sedimentary Basins: From Turbulence to Tectonics*. Oxford, UK: Blackwell Publishing Ltd.

Leeder, M. R. 2011. Tectonic geomorphology: sediment systems deciphering global to local tectonics. *Sedimentology*, **58**, 2–56.

Leeder, M. R., and Mack, G. H. 2001. Lateral erosion ('toe-cutting') of alluvial fans by axial rivers: implications for basin analysis and architecture. *Journal of the Geological Society*, **158**, 885–893.

Leeder, M. R., Harris, T., and Kirkby, M. J. 1998. Sediment supply and climate change: implications for basin stratigraphy. *Basin Research*, **10**, 7–18.

Lehrmann, D. J., and Goldhammer, R. K. 1999. Secular variation in parasequence and facies stacking patterns of platform carbonates: a guide to application of stacking-patterns analysis in strata of diverse ages and settings. Pages 187–225 of: Harris, P. M., Saller, A. H., and Simo, J. A. (eds), *Advances in Carbonate Sequence Stratigraphy: Application to Reservoirs, Outcrops and Models*. Society for Sedimentary Geology Special Publication 63.

Leithold, E. L., Blair, N. E., and Wegmann, K. W. 2015. Source-to-sink sedimentary systems and global carbon burial: A river runs through it. *Earth Science Reviews*, **153**, 30–42.

Leopold, L. B., and Wolman, M. G. 1960. River meanders. *Geological Society of America Bulletin*, **71**, 769–794.

Lerman, A. 1988. Weathering rates and major transport processes: An introduction. Pages 1–10 of: *Physical and Chemical Weathering in Geochemical Cycles*.

Li, G., West, A. J., Densmore, A. L., Jin, Z., Parker, R. N., and Hilton, R. G. 2014. Seismic mountain building: Landslides associated with the 2008 Wenchuan earthquake in the context of a generalized model for earthquake volume balance. *Geochemistry, Geophysics, Geosystems*, **15**, doi:10.1002/2013GC005067.

Lin, W., and Bhattacharya, J. P. 2017. Estimation of source-to-sink mass balance by a fulcrum approach using channel paleohydrologic parameters of the Cretaceous Dunvegan Formation, Canada. *Journal Sedimentary Research* **87**, 97–116.

Lithgow-Bertelloni, C., and Gurnis, M. 1997. Cenozoic subsidence and uplift of continents from time-varying dynamic topography. *Geology*, **25**, 735–738.

Lobo, F. J., and Ridente, D. 2014. Stratigraphic architecture and spatio-temporal variability of high-frequency (Milankovitch) depositional cycles on modern continental margins: An overview. *Marine Geology*, **352**, 215–247.

Lopez-Blanco, M., Marzo, M., Burbank, D. W., Vergés, J., Roca, E., Anadon, P., and Pina, J. 2010. Tectonic and climatic controls on the development of foreland fan deltas: Montserrat and Sant Llorenc del Munt systems (middle Eocene, Ebro Basin, NE Spain). *Sedimentary Geology*, **138**, 17–39.

Ludwig, W., Probst, J. L., and Kempe, S. 1996. Predicting the oceanic input of organic carbon by continental erosion. *Global Biogeochemical Cycles*, **10**, 23–41.

Lutjeharms, J. R. E. 2006. The ocean environment off south-eastern Africa: a review. *South African Journal of Science, Coelacanth Research*, **102**, 419–426.

Ma, Y. 2009. *Continental Shelf Sediment Transport and Depositional Processes on an Energetic, Active Margin: the Waiapu River Shelf, New Zealand*. PhD thesis, College of William and Mary in Virginia.

Mackin, J. H. 1948. Concept of the graded river. *Bulletin Geological Society America*, **59**, 463–512.

Maizels, J. 1986. Modeling of palaeohydrologic change during deglaciation. *Géographie Physique et Quaternaire*, **40**, 263–277.

Malamud, B. D., and Turcotte, D. L. 2006. The applicability of power-law frequency statistics to floods. *Journal of Hydrology*, **322**, 168–180.

Maldonado, A. 1972. El Delta del Ebro. Estudio sedimentologico y estratigrafico. *Bol. Estratigrafia, Univ. Barcelona*, **1**, 1–486.

Maldonado, A., Swift, D. J. P., Young, R. A., Han, G., Nittrouer, C. A., DeMaster, D. J., Rey, J., Palomo, C., Acosta, J., Ballester, A., and Castellvi, J. 1983. Sedimentation on the Valencia Continental Shelf: preliminary results. *Continental Shelf Research*, **2**, 195–211.

Malinverno, A. 1997. On the power-law size distribution of turbidite beds. *Basin Research*, **9**, 263–274.

Malmon, D. V., Dunne, T., and Reneau, S. L. 2003. Stochastic theory of particle trajectories through alluvial valley floors. *Journal of Geology*, **111**, 525–542.

Malmon, D. V., Reneau, S. L., Dunne, T., Katzman, D., and Drakis, P. G. 2005. Influence of sediment storage on downstream delivery of contaminated sediment. *Water Resources Research*, **41**, W05008.

Mancktelow, N. S., and Grasemann, B. 1997. Time-dependent effects of heat advection and topography on cooling histories during erosion. *Tectonophysics*, **270**, 167–195.

Mandelbrot, B. B. 1983. *The Fractal Geometry of Nature*. New York: Henry Holt and Co.

Mange, M. A., and Maurer, H. F. W. 1992. *Heavy Minerals in Colour*. London: Chapman and Hall.

Mange, M. A., and Wright, D. T. 2007. *Heavy Minerals in Use*. Amsterdam: Elsevier Science, Developments in Sedimentology 58.

Marr, J. G., Swenson, J. B., Paola, C., and Voller, V. R. 2000. A two-diffusion model of fluvial stratigraphy in closed depositional basins. *Basin Research*, **12**, 381–398.

Martin, J., Paola, C., Abreu, V., Neal, J., and Sheets, B. 2009. Sequence stratigraphy of experimental strata under known conditions of differential subsidence and variable base-level. *Bulletin American Association Petroleum Geologists*, **93**, 503–533.

Martin, J.-M., and Meybeck, M. 1979. Elemental mass-balance of material carried by major world rivers. *Marine Chemistry*, **7**, 173–206.

Marzo, M., Nijman, W., and Puigdefàbregas, C. 1988. Architecture of the Castissent fluvial sheet sandstones, Eocene, south Pyrenees, Spain. *Sedimentology*, **35**, 719–738.

Mayer, L. M. 1994. Surface area control of organic carbon accumulation in continental shelf sediments. *Geochimica et Cosmochimica Acta*, **58**, 1271–1284.

McCave, I. N., and Tucholke, B. E. 1986. Deep current-controlled sedimentation in the western North Atlantic. Pages 451–468 of: Vogt, P. R., and Tucholke, B. E. (eds), *The Geology of North America, The Western North Atlantic Region, Decade of North American Geology*. Boulder, Colorado: Geological Society of America.

McEwen, M. C., Fessenden, F. W., and Rogers, J. J. W. 1959. Texture and composition of some weathered granites and slightly transported arkosic sands. *Journal of Sedimentary Petrology*, **29**, 477–492.

McLeod, A. E., Dawers, N. H., and Underhill, J. R. 2000. The propagation and linkage of normal faults; insights from the Strathspey-Brent-Statfjord fault array, northern North Sea. *Basin Research*, **12**, 263–284.

McMillan, M. E., Angevine, C. L., and Heller, P. L. 2002. Postdepositional tilt of the Miocene-Pliocene Ogallala Group on the western Great Plains: Evidence of late Cenozoic uplift of the Rocky Mountains. *Geology*, **30**, 63–66.

Meade, R. H. 1972. Fate of river sediments on Atlantic coast of U.S. *EOS, Transactions, American Geophysical Union*, **53**, 369.

Meade, R. H. 1982. Sources, sinks and storage of river sediment in the Atlantic drainage of the United States. *Journal of Geology*, **90**, 235–252.

Meade, R. H., Dunne, T., Richey, J. E., Santos, U. de M., and Salati, E. 1985. Storage and remobilization of suspended sediment in the lower Amazon River of Brazil. *Science*, **228**, 488–490.

Mertes, L. A. K., and Warrick, J. A. 2001. Measuring flood output from 110 coastal watersheds in California with field measurements and Sea-WiFS. *Geology*, **29**, 659–662.

Métivier, F., and Gaudemer, Y. 1999. Stability of output fluxes of large rivers in South and East Asia during the last 2 million years: implications for floodplain processes. *Basin Research*, **11**, 293–304.

Métivier, F., Gaudemer, Y., Tapponier, P., and Klein, M. 1999. Mass accumulation rates in Asia during the Cenozoic. *Geophysical Journal International*, **137**, 280–318.

Métivier, F., Meunier, P., Crave, A., Chaduteau, C., Ye, B., and Liu, G. 2004. Transport dynamics and morphology of a high mountain stream during the peak flow season: the Urumqi River (Chinese Tian Shan). *River Flow*, **1**, 761–777.

Meunier, P., Métivier, F., Lajeunesse, E., Meriaux, A. S., and Faure, J. 2006. Flow pattern and sediment transport in a braided river: the 'torrent de St. Pierre' (French Alps). *Journal of Hydrology*, **330**, 496–505.

Meybeck, M. 1976. Total mineral dissolved transport by world major rivers. *Hydrological Science Bulletin*, **21**, 65–284.

Meybeck, M. 1982. Carbon, nitrogen, and phosphorus transport by world rivers. *American Journal of Science*, **282**, 401–450.

Meybeck, M. 1986. Composition chimique naturelle des eaux courantes françaises. *Sci. Géol. Bulletin*, **39**, 3–77.

Meybeck, M. 1987. Global chemical weathering of surficial rocks estimated from river dissolved loads. *American Journal of Science*, **287**, 401–428.

Meybeck, M., and Ragu, A. 1996. *River Discharge to the Oceans: An Assessment of Suspended Solids, Major Ions and Nutrients*. Nairobi, Kenya: Division of the Environment, Information, Assessment/Water Branch, United Nations Environment Programme.

Meybeck, M., and Ragu, A. 1997. Presenting the GEMS-GLORI, a compendium of world river discharge to the ocean. Pages 3–14 of: *Freshwater Contamination, Proceedings of the Rabat Symposium S4, April-May 1997*. International Association of Hydrological Science IAHS, 243.

Miall, A. D. 1991. Hierarchies of architectural units in terrigenous clastic rocks, and their relationship to sedimentation rate. Pages 6–12 of: Miall, A. D., and Tyler, N. (eds), *The Three-Dimensional Facies Architecture of Terrigenous Clastic Sediments and Its Implications for Hydrocarbon Discovery and Recovery*. Tulsa, Oklahoma: Society of Economic Paleontologists and Mineralogists, Concepts in Sedimentology and Paleontology, 3.

Miall, A. D. 1994. Sequence stratigraphy and chronostratigraphy: problems of definition and precision in correlation, and their implications for global eustasy. *Geoscience Canada*, **21**, 1–26.

Miall, A. D. 2010. *The Geology of Stratigraphic Sequences, Second Edition*. Berlin, Heidelberg: Springer–Verlag.

Miall, A. D. 2014. Updating uniformitarianism: stratigraphy as just a set of 'frozen accidents'. Pages 11–36 of: Smith, D. G., Bailey, R. J., Burgess, P. M., and Fraser, A. J. (eds), *Strata and Time: Probing the Gaps in Our Understanding*. Geological Society London Special Publication 404.

Miall, A. D. 2016. *Stratigraphy: A Modern Synthesis*. Switzerland: Springer International Publishing AG.

Michael, N. A., Whittaker, A. C., and Allen, P. A. 2013. The functioning of sediment routing systems using a mass balance approach: Example from the Eocene of the southern Pyrenees. *Journal of Geology*, **121**, 581–606.

Michael, N. A., Whittaker, A. C., Carter, A., and Allen, P. A. 2014a. Volumetric budget and grain-size fractionation of a geological sediment routing system: Eocene Escanilla Formation, South-Central Pyrenees. *Bulletin Geological Society of America*, **126**, 585–599.

Michael, N. A., Carter, A., Whittaker, A. C., and Allen, P. A. 2014b. Erosion rates in the source region of an ancient sediment routing system: comparison of depositional volumes with thermochronometric estimates. *Journal Geological Society London*, **171**, 401–412.

Middlekoop, H., and Asselman, N. E. M. 1998. Spatial variability of floodplain sedimentation at the event scale in the Rhine-Meuse delta, the Netherlands. *Earth Surface Processes and Landforms*, **23**, 561–573.

Middleton, G. V. 1976. Hydraulic interpretation of sand size distributions. *Journal of Geology*, **84**, 405–426.

Miller, W. R. 2002. *Influence of Rock Composition on the Geochemistry of Stream and Spring Waters from Mountainous Watershed in the Gunnison, Uncomphagre, and Grand Mesa National Forests, Colorado*. Denver, Colorado: U.S. Geological Survey Professional Paper 1667.

Milliman, J. D. 1995. Sediment discharge to the ocean of small mountainous rivers: the New Guinea example. *Geo-Marine Letters*, **15**, 127–133.

Milliman, J. D., and Farnsworth, E. L. 2011. *River Discharge to the Coastal Ocean: A Global Synthesis*. Cambridge, UK: Cambridge University Press.

Milliman, J. D., and Meade, R. H. 1983. Worldwide delivery of river sediment to the oceans. *Journal of Geology*, **91**, 1–21.

Milliman, J. D., and Syvitski, J. P. M. 1992. Geomorphic/tectonic control of sediment discharge to the ocean: the importance of small mountainous rivers. *Journal of Geology*, **100**, 525–544.

Milliman, J. D., Summerhayes, C. P., and Barretto, H. T. 1975. Quaternary sedimentation on the Amazon continental margin: a model. *Geological Society of America Bulletin*, **86**, 610–614.

Milliman, J. D., Huang-Ting, S., Zuo-Sheng, Y., and Meade, R. H. 1985. Transport and deposition of river sediment in the Changjiang estuary and adjacent continental shelf. *Continental Shelf Research*, **4**, 37–45.

Mitchell, S. G., and Reiners, P. W. 2003. Influence of wildfires on apatite and zircon (U-Th)/He ages. *Geology*, **31**, 1025–1028.

Molnar, P., and England, P. 1990. Late Cenozoic uplift of mountain ranges and global climate change: chicken or egg? *Nature*, **346**, 29–34.

Molnar, P., Anderson, R. S., Kier, G., and Rose, J. 2006. Relationships among probability distributions of stream discharges in floods, climate, bedload transport, and river incision. *Journal Geophysical Research*, **111**, F02001.

Montgomery, D. R. 1984. Valley incision and uplift of mountain peaks. *Journal Geophysical Research*, **99**, 13,913–13,921.

Montgomery, D. R., and Brandon, M. T. 2002. Topographic controls on erosion rates in tectonically active mountain ranges. *Earth and Planetary Science Letters*, **201**, 481–489.

Montgomery, D. R., and Dietrich, W. E. 1988. Where do channels begin? *Nature*, **336**, 232–234.

Montgomery, D. R., and Dietrich, W. E. 1992. Channel initiation and the problem of landscape scale. *Science*, **255**, 826–830.

Moore, W. S. 1996. Large groundwater inputs to coastal waters revealed by 226-Ra enrichment. *Nature*, **380**, 612–614.

Mulder, T., and Syvitski, J. P. M. 1996. Climatic and morphologic relationships of rivers: implications of sea level fluctuations on river loads. *Journal of Geology*, **104**, 509–523.

Mulder, T., Savoye, B., Piper, D. J. W., and Syvitski, J. P. M. 1998. The Var submarine sedimentary system: Understanding Holocene sediment delivery processes and their importance to the geological record. Pages 145–166 of: Stoker, M. S., Evans, D., and Cramp, A. (eds), *Geological Processes on Continental Margins: Sedimentation, Mass-Wasting and Stability*. Geological Society of London Special Publication 129.

Müller, G. 1966. The new Rhine delta in Lake Constance. Pages 107–124 of: Shirley, M. L. (ed), *Deltas in Their Geologic Framework*. Houston, Texas Houston Geological Society.

Muller, R. A., and MacDonald, G. J. 1997. Glacial cycles and astronomical forcing. *Science*, **277**, 215–218.

Mundil, R., Brack, P., Meier, M., Rieber, H., and Oberli, F. 1996. High resolution U-Pb dating of Middle Triassic volcaniclastics: time-scale calibration and verification of tuning parameters for carbonate sedimentation. *Earth and Planetary Science Letters*, **141**, 137–151.

Murray, A. B., and Paola, C. 1994. A cellular model of braided rivers. *Nature*, **371**, 54–57.

Muto, T., and Steel, R. J. 2000. The accommodation concept in sequence stratigraphy: some dimensional problems and possible redefinition. *Sedimentary Geology*, **130**, 1–10.

Muto, T., and Steel, R. J. 1997. Principles of regression and transgression: the nature of the interplay between accommodation and sediment supply. *Journal Sedimentary Research*, **67**, 994–1000.

Muto, T., and Swenson, J. B. 2006. Autogenic attainment of large-scale alluvial grade with steady sea level fall: an analog tank/flume experiment. *Geology*, **34**, 161–164.

Muto, T., Steel, R. J., and Swenson, J. B. 2007. Autostratigraphy: A framework norm for genetic stratigraphy. *Journal Sedimentary Research*, **77**, 2–12.

Naeser, C. W. 1967. The use of apatite and sphene for fission track age determination. Bulletin Geological Society of America Bulletin, **78**, 15–23.

Najman, Y. 2005. The detrital record of orogenesis: A review of approaches and techniques used in the Himalayan sedimentary basins. *Earth Science Reviews*, **74**, 1–72.

Neal, J., and Abreu, V. 2009. Sequence stratigraphy hierarchy and the accommodation succession method. *Geology*, **37**, 779–782.

Nesbit, W. H., Fedo, C. M., and Young, G. M. 1997. Quartz and feldspar stability, steady and non steady-state weathering, and petrogenesis of siliciclastic sands and muds. *Journal of Geology*, **105**, 173–191.

Newell, N. D. 1962. Paleontological gaps and geochronology. *Journal of Paleontology*, **36**, 592–610.

Nicholson, U., Poynter, S., Clift, P. D., and Macdonald, D. I. M. 2014. Tying catchment to basin in a giant sediment routing system: a source-to-sink study of the Neogene-Recent Amur River and its delta in the North Sakhalin Basin. In Scott, R. A., Smyth, H. R., Morton, A. C., and Richardson, N. (eds) *Sediment Provenance Studies in Hydrocarbon Exploration and Production*. Geological Society London Special Publication 386, 163–193.

Nittrouer, C. A., Kuehl, S. A., DeMaster, D. J., and Kowsmann, R. O. 1986. The deltaic nature of Amazon shelf sedimentation. *Bulletin Geological Society America*, **97**, 444–458.

Nittrouer, C. A., Kuehl, S. A., Figueiredo, A. G., Allison, M. A., Sommerfield, C. K., Rine, J. M., Faria, E. C., and Silveira, O. M. 1996. The geological record preserved by Amazon shelf sedimentation. *Continental Shelf Research*, **16**, 817–841.

Nittrouer, C. A., Austin, J. A., Field, M. E., Kravitz, J. H., Syvitski, J. P. M., and Wiberg, P. L. 2007. *Continental Margin Sedimentation: From Sediment Transport to Sequence Stratigraphy*. International Association of Sedimentologists, Blackwell Publishing.

Noller, J. S., Sowers, J. M., and Lettis, W. R. 2000. *Quaternary Geochronology: Methods and Applications*. Washington, DC: American Geophysical Union.

Normark, W. R., Piper, D. J. W., and Sliter, R. 2006. Sea-level and tectonic control on middle to late Pleistocene turbidite systems in Santa Monica Basin, offshore California. *Sedimentology*, **53**, 867–897.

Nyberg, B., and Howell, J. A. 2015. Is the present the key to the past? A global characterization of modern sedimentary basins. *Geology*, **43**, 643–646.

Nygård, A., Sejrup, H. P., Haflidason, H., Lekens, W. A. H., Clark, C. D., and Bigg, G. R. 2007. Extreme sediment and ice discharge from marine-based ice streams: New evidence from the North Sea. *Geology*, **35**, 395–398.

Oberlander, T. M. 1985. Origin of drainage transverse to structures in orogens. Pages 155–182 of: *Tectonic Geomorphology. The Binghampton Symposia in Geomorphology, International Series, 15*. London: Allen and Unwin.

O'Grady, D. B., Syvitski, J. P. M., Pratson, L. F., and Sarg, J. F. 2000. Categorizing the morphological variability of siliciclastic passive continental margins. *Geology*, **28**, 207–210.

Olsen, P. E. 1984. Periodicity of lake-level cycles in the Late Triassic Lockatong Formation of the Newark Basin (Newark Supergroup, New Jersey and Pennsylvania). Pages 129–146 of: Berger, A. L., Imbrie, J., Hays, J., Kukla, G., and Saltzman, B. (eds), *Milankovitch and Climate: Understanding the Response to Astronomical Forcing*. Dordrecht/Boston/Lancaster: Reidel.

Ori, G. G., and Friend, P. F. 1984. Sedimentary basins, formed and carried piggyback on active thrust sheets. *Geology*, **12**, 475–478.

Ori, G. G., Roveri, M., and Valloni, F. 1986. Plio-Pleistocene sedimentation in the Apenninic Adriatic foredeep (central Adriatic Sea, Italy). Pages 183–198 of: *Foreland Basins*. Oxford, UK: Special Publication International Association Sedimentologists, 8.

Orive, E., Elliott, M., and de Jong, V. N. (editors). 2002. *Nutrients and Eutrophication in Estuaries and Coastal Waters*. Developments in Hydrobiology 164. Springer, Science and Business Media, B. V.

Ottesen, D., and Dowdeswell, J. A. 2009. An inter-ice-stream glacial margin: submarine landforms and a geomorphic model based on marine geophysical data from Svalbard. *Bulletin Geological Society America*, **121**, 1647–1665.

Paike, H., Norris, R. D., Herrie, J. O., Wilson, P., Coxall, H. K., Lear, C. H., Shackleton, N., Tripati, A., and Wade, B. 2006. The heartbeat of the Oligocene climate system. *Science*, **314**, 1894–1898.

Palanques, A., Guillén, J., Puig, P., and Durrieu de Madron, X. 2008. Storm-driven shelf-to-canyon suspended sediment transport at the southwestern Gulf of Lions. *Continental Shelf Research*, **28**, 1947–1956.

Palomares, M., and Arribas, J. 1993. Modern stream sands from compound crystalline sources: composition and sand generation index. Pages 313–322 of: *Processes Controlling the Composition of Clastic Sediments*. Geological Society America Special Paper 284.

Paola, C. 2000. Quantitative models of sedimentary basin filling. *Sedimentology*, **47**, Supplement 1, 121–178.

Paola, C., and Martin, J. M. 2012. Mass-balance effects in depositional systems. *Journal of Sedimentary Research*, **82**, 435–450.

Paola, C., and Mohrig, D. 1996. Palaeohydraulics revisited: Paleoslope estimation in coarse-grained braided rivers. *Basin Research*, **8**, 243–254.

Paola, C., and Seal, R. 1995. Grain-size patchiness as a cause of selective deposition and downstream fining. *Water Resources Research*, **31**, 1395–1407.

Paola, C., and Voller, V. R. 2005. A generalized Exner equation for sediment mass balance. *Journal Geophysical Research-Earth Surface*, **110**, F04014.

Paola, C., Heller, P. L., and Angevine, C. L. 1992. The large-scale dynamics of grain-size variation in alluvial basins, 1: Theory. *Basin Research*, **4**, 73–90.

Paola, C., Foufoula, E., Dietrich, W. E., Hondzo, M., Mohrig, D., Parker, G., Rodriguez-Iturbe, I., Voller, V., and Wilcock, P. 2006. Toward a unified science of the Earth's surface: Opportunities for synthesis among hydrology, geomorphology, geochemistry and ecology. *Water Resources Research*, **42**, W03S10.

Paola, C., Straub, K., Mohrig, D., and Reinhardt, L. 2009. The 'unreasonable effectiveness' of stratigraphic and geomorphic experiments. *Earth Science Reviews*, **97**, 1–43.

Parker, G. 1978a. Self-formed straight rivers with equilibrium banks and mobile bed. Part 1. The sand-silt river. *Journal of Fluid Mechanics*, **89**, 109–125.

Parker, G. 1978b. Self-formed straight rivers with equilibrium banks and mobile bed. Part 2. The gravel river. *Journal of Fluid Mechanics*, **89**, 127–146.

Parker, G., and Cui, Y. T. 1998. The arrested gravel front: stable gravel-sand transitions in rivers – Part 1: simplified analytical solution. *Journal of Hydraulic research*, **36**, 75–100.

Parker, G., Paola, C., Whipple, K. X., and Mohrig, D. C. 1998. Alluvial fans formed by channellized fluvial and sheet flow, 1: Theory. *Journal of Hydraulic Engineering*, **124**, 985–995.

Parsons, A. J., Michael, N., Whittaker, A. C., Duller, R. A., and Allen, P. A. 2012. Grain size trends reveal the late orogenic tectonic and erosional history of the south-central Pyrenees, Spain. *Journal of the Geological Society London*, **109**, 111–114.

Passega, R., Rizzini, A., and Borghetti, G. 1967. Transport of sediment by waves, Adriatic coastal shelf, Italy. *Bulletin American Association Petroleum Geologists*, **51**, 1304–1319.

Patruno, S., Hampson, G. J., and Jackson, C. A.-L. 2015. Quantitative characterisation of deltaic and subaqueous clinoforms. *Earth Science Reviews*, **142**, 79–119.

Patterson, M. O., McKay, T., Naish, T., Escutia, C., Jimenez-Espejo, F. J., Raymo, M. E., Meyers, S. R. Tauxe, L., and Brinkhuis, H. Integrated Ocean Drilling Expedition 318 Scientists. 2014. Orbital forcing of the East Atlantic ice sheet during the Pliocene and early Pleistocene. *Nature Geoscience*, **7**, 841–847.

Paull, C. K., Mitts, P., Ussler, W., Keaten, R., and Greene, H. G. 2005. Trail of sand in upper Monterey Canyon, offshore California. *Bulletin Geological Society of America*, **117**, 1134–1145.

Pazzaglia, F. J., and Brandon, M. T. 1996. Macrogeomorphic evolution of the post-Triassic Appalachian mountains determined by deconvolution of the offshore basin sedimentary record. *Basin Research*, **8**, 255–278.

Pearce, A. J., and Watson, A. J. 1986. Effects of earthquake-induced landslides on sediment budget and transport over a 50-yr period. *Geology*, **14**, 52–55.

Pelletier, J. 2008. *Quantitative Modeling of Earth Surface Processes*. Cambridge, UK: Cambridge University Press.

Pelletier, J. D. 2004. The influence of piedmont deposition on the time scale of mountain belt denudation. *Geophysical Research Letters*, **31**, L15502.

Pepin, E., Carrétier, S., and Herail, G. 2010. Erosion dynamics in a coupled catchment-fan system with constant external forcing. *Geomorphology*, **122**, 78–90.

Petter, A. L., Steel, R. J., Mohrig, D., Kim, W., and Carvajal, C. 2013. Estimation of the paleoflux of terrestrial-derived solids across ancient basin margins using the stratigraphic record. *Bulletin Geological Society America*, **125**, 578–593.

Pettijohn, F. J., Potter, P. E., and Siever, R. 1987. *Sand and Sandstone, 2nd edition*. New York: Springer-Verlag.

Peucker-Ehrenbrink, B. 2009. Land2Sea database of river discharge, basin sizes, annual water discharges, and suspended sediment fluxes. *Geochemistry, Geophysics, Geosystems*, **10**, Q06014.

Pillans, B., Chappell, J., and Naish, T. R. 1998. A review of the Milankovitch climatic beat: template for Plio-Pleistocene sea-level changes and sequence stratigraphy. *Sedimentary Geology*, **122**, 5–21.

Pinet, P., and Souriau, M. 1988. Continental erosion and large-scale relief. *Tectonics*, **7**, 563–82.

Pirmez, C., Pratson, L. F., and Steckler, M. S. 1998. Clinoform development by advection-diffusion of suspended sediment: Modeling and comparison to natural systems. *Journal Geophysical Research*, **103**, 24,141–24,157.

Pivnik, D. A. 1990. Thrust-generated fan-delta deposition: Little Muddy Creek conglomerate, SW Wyoming. *Journal of Sedimentary Petrology*, **60**, 489–503.

Pizzuto, J. E. 1987. Sediment diffusion during overbank flows. *Sedimentology*, **34**, 301–317.

Plint, A. G., and Wadsworth, J. A. 2003. Sedimentology and paleogeomorphology of four large valley systems incising delta plains, western Canada Foreland Basin: implications for mid-Cretaceous sea-level changes. *Sedimentology*, **50**, 1147–1186.

Plotnick, R. E. 1986. A fractal model for the distribution of stratigraphic hiatuses. *Journal of Geology*, **94**, 885–890.

Porebski, S. J., and Steel, R. J. 2003. Shelf-margin deltas: their stratigraphic significance and relation to deepwater sands. *Earth Science Reviews*, **62**, 283–326.

Portenga, E. W., and Bierman, P. R. 2011. Understanding Earth's eroding surface with [10]Be. *GSA Today*, **21**, 4–10.

Posamentier, H. W., and Vail, P. R. 1988. Eustatic controls on clastic deposition, II, sequence and systems tract models. Pages 125–154 of: Wilgus, C. K., Hastings, B. S., Kendall, C. G. St. C., Posamentier, H. W., Ross, C. A. and von Wagoner, J. C., C. K. Wilgus (ed), *Sea Level Changes: An Integrated Approach*. Special Publication Society Economic Palaeontologists and Mineralogists 42.

Posamentier, H. W., Erskine, R. D., and Mitchum, R. M. Jr. 1991. Models for submarine fan deposition within a sequence stratigraphic framework. Pages 127–136 of: Weimer, P., and Link, M. H. (eds), *Seismic Facies and Sedimentary Processes of Submarine Fans and Turbidite Systems* New York: Springer-Verlag.

Potter, P. E., and Pettijohn, F. J. 1977. *Paleocurrents and Basin Analysis, 2nd edition*. Berlin Heidelberg: Springer-Verlag.

Prather, B. E. 2003. Controls on reservoir distribution, architecture and stratigraphic trapping in slope settings. *Marine and Petroleum Geology*, **20**, 529–545.

Preto, N., Hinnov, L. A., Hardie, L. A., and De Zanche, V. 2001. Middle Triassic orbital signature recorded in the shallow marine Latemar carbonate buildup (Dolomites, Italy). *Geology*, **29**, 1123–1126.

Pritchard, D., Roberts, G. G., White, N. J., and Richardson, C. N. 2009. Uplift histories from river profiles. *Geophysical Research Letters*, **36**, L24301.

Prizomwala, S. P., Bhatt, N., and Basavaiah, N. 2014. Understanding the sediment routing system along the Gulf of Katchchh coast, western India: Significance of small ephemeral rivers. *Journal Earth System Science*, **123**, 121–133.

Puig, P., Ogston, A. S., Mullenbach, B. I., Nittrouer, C. A., Parsons, J. D., and Sternberg, R. W. 2004. Storm-induced sediment gravity flows at the head of the Eel submarine canyon, northern California margin. *Journal Geophysical Research*, **109**, http://dx.doi.org/10.1029/2003JC001918.

Puig, P., Palanques, A., and Martín, J. 2014. Contemporary sediment-transport processes in submarine canyons. *Annual Review of Marine Science*, **6**, 53–77.

Pujalte, V., Baceta, J. I., and Schmitz, B. 2015. A massive input of coarse-grained siliciclastics in the Pyrenean Basin during the PETM: the missing ingredient in a coeval change in hydrological regime. *Climate of the Past*, **11**, 1653–1672.

Ramos, E., Busquets, P., and Vergés, J. 2002. Interplay between longitudinal fluvial and transverse alluvial fan systems and growing thrusts in a piggyback basin (SE Pyrenees). Pages 105–131 of: *Geology of Growth Strata*. Sedimentary Geology, 146.

Raymo, M. E., and Ruddiman, W. F. 1992. Tectonic forcing of Late Cenozoic climate. *Nature*, **359**, 117–122.

Reading, H. G., and Richards, M. 1994. Turbidite systems in deep-water basin margins, classified by grain size and feeder system. *Bulletin American Association Petroleum Geologists*, **78**, 792–822.

Rebesco, M., Hernádez-Molina, F. J., Van Rooij, D., and Wahlin, A. 2014. Contourites and associated sediments controlled by deep-water circulation processes: State-of-the-art and future considerations. *Marine Geology*, **352**, 111–154.

Reid, S. K., and Dorobek, S. L. 1993. Sequence stratigraphy and evolution of a progradational foreland carbonate ramp, Lower Mississippian Mission Canyon Formation and stratigraphic equivalents. Pages 327–352 of: Loucks, R. G., and Sarg, J. F. (eds), *Carbonate Sequence Stratigraphy, Recent Developments and Applications*. American Association Petroleum Geologists Memoir 57.

Reineck, H. E. 1960. Uber Zeitlücken in rezenten Flachsee-Sedimenten. *Geologisches Rundschau*, **48**, 149–161.

Reiners, P. W. 2002. (U-Th)/He chronometry experiences a renaissance. *EOS, Transactions, American Geophysical Union*, **83**, 26–27.

Reiners, P. W., Ehlers, T. A., Mitchell, S. G., and Montgomery, D. R. 2003. Coupled spatial variations in precipitation and long-term erosion rates across the Washington Cascades. *Nature*, **426**, 645–647.

Reiners, P. W., Ehlers, T. A., and Zeitler, P. K. 2005. Past, present and future of thermochronology. Pages 1–18 of: Reiners, P. W., and Ehlers, T. A. (eds), *Low-Temperature Thermochronology: Techniques, Interpretations, and Applications, Reviews in Mineralogy and Geochemistry* 58.

Repka, J. L., Anderson, R. S., and Finkel, R. C. 1997. Cosmogenic dating of fluvial terraces, Fremont River, Utah. *Earth and Planetary Science Letters*, **152**, 59–73.

Restrepo, J. D., and Kjerfve, B. 2000. Magdalena River: interannual variability and revised water discharge and sediment load estimates. *Journal of Hydrology*, **235**, 137–149.

Restrepo, J. D., Kjerfve, B., Hermelin, M., and Restrepo, J. C. 2006a. Factors controlling sediment yield in a major South American drainage basin: the Magdalena River, Colombia. *Journal of Hydrology*, **316**, 213–232.

Restrepo, J. D., Zapata, P., Diaz, J. M., Garzon-Ferreira, J., and Garcia, C. B. 2006b. Fluvial fluxes into the Caribbean Sea and their impact on coastal ecosystems: The Magdalena River, Colombia. *Global and Planetary Change*, **50**, 33–49.

Ricci Lucchi, F. 1986. The Oligocene to Recent foreland basins of the northern Apennines. Pages 105–140 of: *Foreland Basins, Special Publication International Association Sedimentologists*, 8. Oxford, UK: Blackwell Scientific Publications.

Richey, J. E., Brook, J. T., Naiman, T. J., Wissmar, R. C., and Stallard, R. F. 1980. Organic carbon: oxidation and transport in the Amazon River. *Science*, **207**, 1348–1351.

Richter, F. M., Rowley, D. B., and DePaolo, D. J. 1992. Sr isotope evolution of seawater: the role of tectonics. *Earth and Planetary Science Letters*, **109**, 11–23.

Ricken, W. 1991. Time span assessment – an overview. Pages 773–794 of: Einsele, G., Ricken, W., and Seilacher, A. (eds), *Cycles and Events in Stratigraphy*. Berlin: Springer-Verlag.

Ridente, D. 2016. Releasing the sequence stratigraphy paradigm: Overview and perspectives. In: Burgess, P. M., Allen, P. A., and Steel, R. J. (eds), *The Future of Sequence Stratigraphy*. Journal Geological Society London Special Publication, 173.

Rittner, M., Vermeesch, P., Carter, A., Bird, A., Stevens, T., Garzanti, E., Andò, S., Vezzoli, G., Dutt, R., Xu, Z., and Lu, H. 2016. The provenance of Taklimakan desert sand. *Earth and Planetary Science Letters*, **437**, 127–137.

Roberts, G. G., and White, N. J. 2010. Estimating uplift rate histories from river profiles using African examples. *Journal Geophysical Research*, **115**, B02406.

Roberts, G. G., White, N. J., Martin-Brandis, G. L., Crosby, A. G. 2012. An uplift history of the Colorado Plateau and its surroundings from inverse modelling of longitudinal river profiles. *Tectonics*, **31**, TC4022.

Robinson, R. A. J., and Slingerland, R. L. 1998. Origin of fluvial grain size trends in a foreland basin: The Pocono Formation on the central Appalachian Basin. *Journal of Sedimentary Research*, **68**, 473–486.

Roering, J. J., Kirchner, J. W., and Dietrich, W. E. 1999. Evidence for nonlinear, diffusive sediment transport and implications for landscape morphology. *Water Resources Research*, **35**, 853–870.

Rohais, S., Bonnet, S., and Eschard, R. 2012. Sedimentary record of tectonic and climatic erosional perturbations in an experimental coupled catchment-fan system. *Basin Research*, **24**, 198–212.

Romans, B. W., and Graham, S. A. 2013. A deep-time perspective of land-ocean linkages in the sedimentary record. *Annual Review of Marine Science*, **5**, 69–94.

Romans, B. W., Normark, W. R., McGann, M. M., Covault, J. A., and Graham, S. A. 2009. Coarse-grained sediment delivery and distribution in the Holocene Santa Monica Basin, California: Implications for evaluating source-to-sink flux at millennial time scales. *Bulletin Geological Society America*, **121**, 1394–1408.

Romans, B. W., Castelltort, S., Covault, J. A., Fildani, A., and Walsh, J. P. 2015. Environmental signal propagation in sedimentary systems across time scales. *Earth Science Reviews*, **153**, 7–29.

Ronov, A. B., and Yaroshevskiy, A. A. 1972. Earth's crust geochemistry. Pages 243–254 of: Fairbridge, R. (ed), *Encyclopaedia of Geochemistry and Environmental Sciences*, New York: Van Nostrand.

Ronov, A. B., and Yaroshevskiy, A. A. 1976. A new model for the chemical structure of the Earth's crust. *Geochemistry International*, **13**, 89–121.

Rosenbloom, N. A., and Anderson, R. A. 1994. Hillslope and channel evolution in a marine terraced landscape, Santa Cruz, California. *Journal Geophysical Research*, **99**, 14,013–14,029.

Rosendahl, B. R., Reynolds, D. J., Lorber, P. M., Burgess, C. F., McGill, J., Scott, D., Lambiase, J. J., and Derksen, S. J. 1986. Structural expressions of rifting: lessons from Lake Tanganyika, Africa. Pages 29–43 of: Frostick, L. E., Renaut, R. W., Reid, I., and Tiercelin, J. J. (eds), *Sedimentation in the African Rifts*. Special Publication Geological Society of London, 25.

Rothman, D. H., and Grotzinger, J. P. 1995. Scaling properties of gravity-driven sediments. *Nonlinear Processes in Geophysics*, **2**, 178–185.

Rothman, D. H., Grotzinger, J. P., and Flemings, P. 1994. Scaling in turbidite deposition. *Journal Sedimentary Petrology*, **A24**, 59–67.

Rouby, D., Bonnet, S., Guillocheau, F., Gallagher, K., Robin, C., Biancotto, F., Dauteuil, O., and Braun, J. 2009. Sediment supply to the Orange sedimentary system over the last 150 My: An evaluation from sedimentation/denudation balance. *Marine and Petroleum Geology*, **26**, 782–794.

Rubin, D. M., and McCullough, D. S. 1980. Single and superimposed bedforms: a synthesis of San Francisco Bay and flume observations. *Sedimentary Geology*, **26**, 207–231.

Ruddiman, W. F. 2006. What is the timing of orbital-scale monsoon changes? *Quaternary Science Reviews*, **25**, 657–658.

Ruhl, K. W., and Hodges, K. V. 2005. The use of detrital mineral cooling ages to evaluate steady state assumptions in active orogens: an example from the central Nepalese Himaya. *Tectonics*, **24**, TC4015.

Sabatier, F., Maillet, G., Provensal, M., Fleury, T.-J., Suanez, S., and Vella, C. 2006. Sediment budget of the Rhone delta shoreface since the middle of the 19th century. *Marine Geology*, **234**, 143–157.

Sadler, P. M. 1981. Sedimentation rates and the completeness of stratigraphic sections. *Journal of Geology*, **89**, 569–584.

Sadler, P. M., and Jerolmack, D. J. 2015. Scaling laws for aggradation, denudation and progradation rates: the case for time-scale invariance at sediment sources and sinks. Pages 69–88 in: Smith, D. G., Bailey, R. J., Burgess, P. M., and Fraser, A. J. (eds), *Strata and Time: Probing the Gaps in Our Understanding*. Special Publication Geological Society of London 404.

Sadler, P. M., and Strauss, D. J. 1990. Estimation of completeness of stratigraphical sections using empirical data and theoretical models. *Journal of the Geological Society London*, **147**, 471–485.

Sambrook Smith, G. H., and Ferguson, R. I. 1996. The gravel-sand transition: flume study of channel response to reduced slope. *Geomorphology*, **16**, 147–159.

Sarmiento, J. L., and Sundquist, E. T. 1992. Revised budget for the oceanic uptake of anthropogenic carbon dioxide. *Nature*, **356**, 589–593.

Schaller, M., von Blanckenburg, F., Hovius, N., and Kubik, P. W. 2001. Large-scale erosion rates from *in situ*-produced cosmogenic nuclides in European river sediments. *Earth and Planetary Science Letters*, **188**, 3–4.

Schaller, M., von Blanckenburg, F., Hovius, N., Veldkamp, A., and van Meindert, W. 2004. Paleoerosion rates from cosmogenic Be-10 in a 1.3 Ma terrace sequence: response of the River Meuse to changes in climate and rock uplift. *Journal of Geology*, **112**, 127–144.

Schindel, D. E. 1980. Microstratigraphic sampling and the limits of paleontological resolution. *Paleobiology*, **6**, 408–426.

Schlager, W. 2004. Fractal nature of stratigraphic sequences. *Geology*, **32**, 185–188.

Schlager, W. 2010. Ordered hierarchy versus scale invariance in sequence stratigraphy. *International Journal of Earth Science (Geologisches Rundschau)*, **99**, Supplement 1, S139–S151.

Schlische, R. W. 1991. Half-graben basin filling models: New constraints on continental extensional basin development. *Basin Research*, **3**, 123–141.

Schlunegger, F., and Hinderer, M. 2003. Pleistocene-Holocene climate change, re-establishment of fluvial drainage network and increase in relief in the Swiss Alps. *Terra Nova*, **15**, 88–95.

Schlunegger, F., and Norton, K. P. 2015. Climate vs. tectonics: the competing roles of Late Oligocene warming and Alpine orogenesis in constructing alluvial fan megasequences in the North Alpine Foreland Basin. *Basin Research*, **27**, 230–245.

Schlünz, B., Schneider, R. R., Müller, P. J., Showers, W. J., and Wefer, G. 1999. Terrestrial organic carbon accumulation on the Amazon deep sea fan during the last glacial sea level low stand. *Chemical Geology*, **159**, 263–281.

Schmitz, B., and Pujalte, V. 2007. Abrupt increase in seasonal extreme precipitation at the Paleocene-Eocene boundary. *Geology*, **35**, 215–218.

Schröder, K. W., and Theune, C. 1984. Festoffabtrag und Stauraumsverlandung in Mitteleuropa. *Wasserwirtschaft*, **74**, 374–379.

Schroeder, M. 1991. *Fractals, Chaos, Power Laws: Minutes from an Infinite Paradise*. New York: Freeman and Co.

Schubel, J. R., and Carter, H. H. 1976. Suspended sediment budget for Chesapeake Bay. Pages 48–62 of: Wiley, M. (ed), *Estuarine Processes, vol. 2, Circulation, Sediments and Transfer of Material in the Estuary*. New York: Academic Press.

Schumer, R., and Jerolmack, D. J. 2009. Real and apparent changes in sediment deposition rates through time. *Journal of Geophysical Research*, **114**, F00A06.

Schumer, R., Jerolmack, D. J., and McElroy, B. 2011. The stratigraphic filter and bias in measurement of geological rates. *Journal of Geophysical Research*, **38**, L11405.

Schumm, S. A. 1968. Speculations regarding paleohydrological controls on terrestrial sedimentation. *Bulletin Geological Society America*, **79**, 1573–1588.

Schumm, S. A. 1977. *The Fluvial System*. New York: Wiley.

Schwarzacher, W. 1987. Astronomically controlled cycles in the Lower Tertiary of Gubbio (Italy). *Earth and Planetary Science Letters*, **84**, 22–26.

Schwarzacher, W. 2000. Repetitions and cycles in stratigraphy. *Earth Science Reviews*, **50**, 51–75.

Seal, R., Paola, C., Parker, G., Southard, J., and Wilcock, P. 1997. Experiments on downstream fining of gravel: I. Narrow-channel runs. *Journal Hydraulic Engineering*, **123**, 874–884.

Seidl, M. A., Dietrich, W. R., and Kirchner, J. W. 1994. Longitudinal profile development into bedrock: An analysis of Hawaiian channels. *Journal of Geology*, **102**, 457–474.

Shanley, K. W., and McCabe, P. 1994. Perspectives on the sequence stratigraphy of continental strata. *Bulletin American Association Petroleum Geologists*, **78**, 544–568.

Sharland, P. R., Archer, R., Casey, D. M., Hall, S. H., Heward, A. P., Horbury, A. D., and Simmons, M. D. 2001. *Arabian Plate Sequence Stratigraphy*. Bahrain: GeoArabia Special Publication 2.

Sharland, P. R., Casey, D. M., Davies, R. B., Simmons, M. D., and Sutcliffe, O. E. 2004. Arabian plate sequence stratigraphy. *GeoArabia*, **9**, 199–214.

Sharp, I. R., Gawthorpe, R. L., Armstrong, B., and Underhill, J. R. 2000. Propagation history and passive rotation of mesoscale normal faults: implications for syn-rift stratigraphic development. *Basin Research*, **12**, 285–306.

Sheets, B. A., Hickson, T. A., and Paola, C. 2002. Assembling the stratigraphic record: Depositional patterns and time-scales in an experimental alluvial basin. *Basin Research*, **14**, 287–301.

Shepard, F. P., and Dill, R. F. 1966. *Submarine Canyons and other Sea Valleys*. Rand McNally, USA.

Shepard, F. P., Marshall, N. F., and McLoughlin, P. A. 1974. 'Internal waves' advancing along submarine canyons. *Science*, **183**, 195–198.

Showers, W. J., and Angle, D. G. 1986. Stable isotopic characterisation of organic carbon accumulation on the Amazon continental shelf. *Continental Shelf Research*, **6**, 227–244.

Showers, W. J., and Bevis, M. 1988. Amazon cone isotope stratigraphy: evidence for the source of the tropical freshwater spike. *Palaeogeography, Palaeoclimatology, Palaeoecology*, **64**, 189–199.

Sibley, D. F., and Wilband, J. T. 1977. Chemical balance of the Earth's crust. *Geochimica et Cosmochimica Acta*, **41**, 545–554.

Simoes, M., Braun, J., and Bonnet, S. 2010. Continental-scale erosion and transport laws: A new approach to quantitatively investigate macroscale landscapes and associated sediment fluxes over the geological past. *Geochemistry, Geophysics, Geosystems*, **11**, Q09001.

Simon, Q., Hillaire-Marcel, C., St-Onge, G., and Andrews, J. T. 2014. North-eastern Laurentide, western Greenland and southern Innuitian ice stream dynamics during the last glacial cycle. *Journal Quaternary Science*, **29**, 14–26.

Simpson, G. 2004a. Role of river incision in enhancing deformation. *Geology*, **32**, 341–344.

Simpson, G. 2004b. Dynamic interactions between erosion, deposition, and three-dimensional deformation in compressional fold belt settings. *Journal Geophysical Research*, **109**, F03007.

Simpson, G., and Castelltort, S. 2012. Model shows that rivers transmit high-frequency climate cycles to the sedimentary record. *Geology*, **40**, 1131–1134.

Simpson, G. D. H. 2006a. A dynamic model to investigate coupling between erosion, deposition, and three-dimensional (thin plate) deformation. *Journal Geophysical Research*, **109**, F02006.

Simpson, G. D. H. 2006b. How and to what extent does the emergence of orogens above sea level influence their tectonic development? *Terra Nova*, **18**, 447–451.

Simpson, G. D. H. 2006c. Modelling interactions between fold-thrust belt deformation, foreland flexure and surface mass transport. *Basin Research*, **18**, 1–19.

Simpson, G. D. H. 2010. Influence of the mechanical behaviour of brittle-ductile fold-thrust belts on the development of foreland basins. *Basin Research*, **22**, 139–156.

Simpson, G. D. H., and Schlunegger, F. 2003. Topographic evolution and morphology of surfaces evolving in response to coupled fluvial and hillslope sediment transport. *Journal of Geophysical Research-Solid Earth*, **108**, doi:10.1029/2002JB002162.

Sinclair, H. D. 1997. Tectono-stratigraphic model for underfilled peripheral foreland basins: An Alpine perspective. *Bulletin Geological Society America*, **109**, 324–346.

Sinclair, H. D. 2012. Thrust wedge/foreland basin systems. Pages 522–537 of: Busby, C., and Azor, A. (eds), *Tectonics of Sedimentary Basins: Recent Advances*. Wiley-Blackwell.

Sinclair, H. D., and Allen, P. A. 1992. Vertical versus horizontal motions in the Alpine orogenic wedge: stratigraphic response in the foreland basin. *Basin Research*, **4**, 215–232.

Sinclair, H. D., and Cowie, P. A. 2003. Basin floor topography and the scaling of turbidites. *Journal of Geology*, **111**, 277–299.

Sinclair, H. D., and Tomasso, M. 2002. Depositional evolution of confined turbidite basins. *Journal Sedimentary Research*, **72**, 451–456.

Sinclair, H. D., Coakley, B., Allen, P. A., and Watts, A. B. 1991. Simulation of foreland basin stratigraphy using a diffusion model of mountain belt uplift and erosion: An example from the central Alps, Switzerland. *Tectonics*, **10**, 599–620.

Sinclair, H. D., Gibson, M., Naylor, M., and Morris, R. G. 2005. Asymmetric growth of the Pyrenees revealed through measurement and modelling of orogenic fluxes. *American Journal of Science*, **305**, 369–406.

Sklar, L., and Dietrich, W. R. 1998. River longitudinal profiles and bedrock incision models: Stream power and the influence of sediment supply. Pages 237–260 of: Tinkler, K. J., and Wohl, E. E. (eds), *Rivers over Rock: Fluvial Processes in Bedrock Channels*. American Geophysical Union Geophysical Monograph 107.

Slaymaker, O. 2003. The sediment budget as conceptual framework and management tool. *Hydrobiologia*, **494**, 71–82.

Small, E. E., Anderson, R. S., Finkel, R. S., and Repka, J. 1997. Erosion rates of summit flats using cosmogenic radionuclides. *Earth and Planetary Science Letters*, **150**, 423–425.

Smalley, I. J., Kumar, R., O'Hara Dhand, K., Jefferson, I. F., and Evans, R. D. 2005. The formation of silt material for terrestrial sediments: particularly loess and dust. *Sedimentary Geology*, **179**, 321–328.

Smith, G., and Ferguson, R. 1995. The gravel sand transition along river channels. *Journal Sedimentary Research*, **65**, 423–430.

Smith, T. R., and Bretherton, F. P. 1972. Stability and the conservation of mass in drainage basin evolution. *Water Resources Research*, **8**, 1506–1529.

Snyder, N. P., Whipple, K. X., Tucker, G. E., and Merrits, D. J. 2000. Landscape to tectonic forcing: digital elevation model analysis of stream profiles in the Mendocino triple junction region, northern California. *Geological Society America Bulletin*, **112**, 1250–1263.

Sømme, T., Jackson, C., Lunt, I., and Martinsen, O. J. 2010. Source-to-sink in rift basins - Predicting reservoir distribution in ancient subsurface systems. *Search and Discovery Article, American Asssociation Petroleum Geologists*, 10258.

Sømme, T. O., Helland-Hansen, W., Martinsen, O. J., and Thurmond, J. B. 2009. Relationships between morphological and sedimentological parameters in source-to-sink systems: a basis for predicting semi-quantitative characteristics in subsurface systems. *Basin Research*, **21**, 361–388.

Sømme, T. O., Piper, D. J. W., Deptuck, M. E., and Helland Hansen, W. 2011. Linking onshore-offshore sediment dispersal in the Golo source-to-sink system (Corsica, France) during the Late Quaternary. *Journal of Sedimentary Research*, **81**, 118–137.

Sommerfield, C. K., and Nittrouer, C. A. 1999. Modern accumulation rates and a sediment budget for the Eel shelf: A flood-dominated depositional environment. *Marine Geology*, **154**, 227–241.

Sommerfield, C. K., Nittrouer, C. A., and Alexander, C. R. 1999. [7]Be as a tracer of flood sedimentation on the northern California continental margin. *Continental Shelf Research*, **19**, 335–361.

Specht, T. D., and Rosendahl, B. R. 1989. Architecture of the Lake Malawi Rift, East Africa. *Journal African Earth Sciences*, **8**, 355–382.

Stallard, R. F., and Edmond, J. M. 1981. Geochemistry of the Amazon 1: Precipitation chemistry and the marine contribution to the dissolved load at the time of peak discharge. *Journal Geophysical Research, Oceans*, **86**, 9844–9858.

Stallard, R. F., and Edmond, J. M. 1983. Geochemistry of the Amazon 2: The influence of geology and weathering environment on the dissolved load. *Journal of Geophysical Research, Oceans*, **88**, 9671–9688.

Stallard, R. F., and Edmond, J. M. 1987. Geochemistry of the Amazon 3: Weathering chemistry and limits to dissolved inputs. *Journal Geophysical Research, Oceans*, **92**, 8293–8302.

Steel, R. J., Carvajal, C., Petter, A., and Uroza, C. 2009. Shelf and shelf-margin growth in scenarios of rising and falling sea level. Pages 47–71 of: Hampson, G. J., Steel, R. J., Burgess, P. M., and Dalrymple, R. W. (eds), *Recent Advances in Models of Siliciclastic Shallow-Marine Stratigraphy*. Society for Sedimentary Geology Special Publication 90.

Stefan, J. 1891. Uber die Theore der Eisbildung, insbesondere über die Eisbildung im Polarmeere. *Ann. Physik Chem.*, **42**, 269–286.

Steidtmann, J. R., and Schmitt, J. G. 1988. Provenance and dispersal of tectogenic sediments in thin-skinned, thrusted terrains. Pages 353–366 of: Kleinspehn, K. L., and Paola, C. (eds), *New Perspectives in Basin Analysis*. New York: Springer-Verlag.

Stock, J. D., and Montgomery, D. R. 1999. Geologic constraints on bedrock river incision using the stream power rule. *Journal Geophysical Research*, **104**, 4983–4993.

Stock, J. M., Ehlers, T. A., and Farley, K. A. 2006. Where does sediment come from? Quantifying catchment erosion with detrital apatite (U-Th)/He thermochronometry. *Geology*, **34**, 725–728.

Stöckli, D. F., Farley, K. A., and Dumitru, T. A. 2000. Calibration of the apatite (U-Th)/He thermochronometer on an exhumed fault block, White Mountains, California. *Geology*, **28**, 983–986.

Stracke, A., Bizimis, M., and Salters, V. J. M. 2003. Recycling oceanic crust: Quantitative constraints. *Geochemistry, Geophysics, Geosystems*, **4**, 8003.

Strahler, A. N. 1952. Hypsometric (area-altitude) analysis of erosional topology. *Geological Society America Bulletin*, **63**, 1117–1142.

Strahler, A. N. 1957. Quantitative analysis of watershed geomorphology. *Transactions American Geophysical Union*, **38**, 913–920.

Strakhov, N. M. 1967. *Principles of Lithogenesis, Vol. 1*. Edinburgh: Oliver and Boyd.

Straub, K. M., Paola, C., Mohrig, D., Wolinsky, M. A., and George, T. 2009. Compensational stacking of channelized sedimentary deposits. *Journal Sedimentary Research*, **79**, 673–688.

Strong, N., Sheets, B. A., Hickson, T. A., and Paola, C. 2005. A mass balance framework for quantifying downstream changes in fluvial architecture. Pages 243–253 of: Blum, M. D., Marriott, S. B., and Leclair, S. F. (eds), *Fluvial Sedimentology VII*, vol. 35. International Association Sedimentologists Special Publication.

Stumm, W., and Morgan, J. J. 1996. *Aquatic Chemistry: Chemical Equilibria and Rates in Natural waters, 3rd edition*. New York: Wiley-Interscience.

Stüwe, K., White, L., and Brown, R. 1994. The influence of eroding topography on steady state isotherms; applications to fission track analysis. *Earth and Planetary Science Letters*, **124**, 63–74.

Subramanian, V., and Ittekkot, V. 1991. Carbon transport by the Himalayan rivers. Chap. 7 of: Degens, E. T., Kempe, S., and Richey, J. E. (eds), *Biogeochemistry of Major World Rivers, SCOPE 42*. Scientific Committee on Problems of the Environment (SCOPE).

Summerfield, M. A. 1991. *Global Geomorphology*. London: Longman.

Summerfield, M. A., and Hulton, N. J. 1994. Natural controls of fluvial denudation rates in major world drainage basins. *Journal of Geophysical Research*, **99**, 13,871–13,883.

Sun, T., Meakin, P., and Jossang, T. 1996. A simulation model for meandering rivers. *Water Resources Research*, **37**, 2937–2954.

Sun, T., Meakin, P., and Jossang, T. 2001. A computer model for meandering rivers with multiple bed load sediment sizes. 2. Computer simulations. *Water Resources Research*, **37**, 2243–2258.

Swenson, J. B., Voller, V. R., Paola, C., Parker, G., and Marr, J. G. 2000. Fluvio-deltaic sedimentation: A generalized Stefan problem. *European Journal Applied Mathematics*, **11**, 433–452.

Swenson, J. B., Paola, C., Pratson, L., Voller, V. R., and Murray, A. B. 2005. Fluvial and marine controls on combined subaerial and subaqueous delta progradation: morphodynamic modeling of compound clinoform development. *Journal of Geophysical Research*, **110**, F02013.

Sylvester, Z. 2007. Turbidite bed thickness distributions: methods and pitfalls of analysis and modelling. *Sedimentology*, **54**, 847–870.

Syvitski, J. P. M. 1989. On the deposition of sediment within glacier-influenced fjords: oceanographic controls. *Marine Geology*, **85**, 301–329.

Syvitski, J. P. M., and Alcott, J. M. 1993. GRAIN2: predictions of particle size seaward of river mouths. *Computers and Geosciences*, **19**, 399–446.

Syvitski, J. P. M., and Hutton, E. W. H. 2001. 2D SEDFLUX 1.0C: an advanced process-response numerical model for the fill of sedimentary basins. *Computers & Geosciences*, **27**, 731–753.

Syvitski, J. P. M., and Kettner, A. J. 2008. Scaling sediment flux across landscapes. Pages 1–8 of: *Sediment Dynamics in Changing Environments*. Proceedings of a Symposium held in Christchurch, New Zealand, December 2008, IAHS Publ. 325.

Syvitski, J. P. M., and Milliman, J. D. 2007. Geology, geography, and humans battle for dominance over the delivery of fluvial sediment to the coastal ocean. *Journal of Geology*, **115**, 1–19.

Syvitski, J. P. M., and Morehead, M. D. 1999. Estimating river-sediment discharge to the ocean: application to the Eel margin, northern California. *Marine Geology*, **154**, 13–28.

Syvitski, J. P. M., and Saito, Y. 2007. Morphodynamics of deltas under the influence of humans. *Global and Planetary Change*, **57**, 261–282.

Syvitski, J. P. M., Pratson, L., and Morehead, M. 1997. EARTHWORKS: a large spatial scale numerical model to study the flux of sediment to ocean basins and reworking of deposits over various time scales. *AGU 1997 Fall Meeting EOS supplement*, **78**, F258.

Syvitski, J. P. M., Peckham, S. D., Hilberman, R., and Mulder, T. 2003. Predicting the terrestrial flux of sediment to the global ocean: a planetary perspective. *Sedimentary Geology*, **162**, 5–24.

Syvitski, J. P. M., Vörösmarty, C. J., Kettner, A. J., and Green, P. 2005. Impact of humans on the flux of terrestrial sediment to the global ocean. *Science*, **308**, 376–380.

Syvitski, J. P. M., Weaver, P. E., Berne, S., Nitrouer, C. A., Trincardi, F., and Canals, M. (eds). 2004. Strata Formation on European Margins, *Oceanography*, **17**, 4.

Talling, P. J. 2001. On the frequency distribution of turbidite thickness. *Sedimentology*, **48**, 1297–1331.

Talling, P. J., Lawton, T. F., Burbank, D. W., and Hobbs, R. S. 1995. Evolution of latest Cretaceous-Eocene nonmarine deposystems in the Axhandle piggyback basin of central Utah. *Bulletin Geological Society America*, **107**, 297–315.

Talling, P. J., Stewart, M. D., Stark, C. P., Gupta, S., and Vincent, S. J. 1997. Regular spacing of drainage outlets from linear fault blocks. *Basin Research*, **9**, 275–302.

Talling, P. J., Wynn, R. B., Masson, D. G., et al. 2007. Onset of submarine debris flow deposition far from original giant landslide. *Nature*, **450**, 541–544.

Talling, P. J., Clare, M., Urlaub, M., Pope, E., Hunt, J. E., and Watt, S. F. L. 2014. Large submarine landslides on continental slopes: Geohazards, methane release, and climate change. *Oceanography*, **27**, 32–45.

Taylor, A. S., and Lasaga, A.C. 1999. The role of basalt weathering in the Sr isotope budget of the oceans. *Chemical Geology*, **161**, 199–214.

Tesi, T., Goni, M., Langon, L., and Miserocchi, S. 2010. Reexposure and advection of ^{14}C-depleted organic carbon from old deposits at the upper continental slope. *Global Biogeochemical Cycles*, **24**, GB4002.

Tinker, J., de Wit, M., and Brown, R. 2008. Linking source and sink: evaluating the balance between onshore erosion and offshore sediment accumulation since Gondwana breakup, South Africa. *Tectonophysics*, **455**, 94–103.

Tipper, E., Bickle, M., and Galy, A. West, J. A., Pomiés, C., and Chapman, H. J.. 2006. The short term climatic sensitivity of carbonate and silicate weathering fluxes: insight from seasonal variations in river chemistry. *Geochimica Cosmochimica Acta*, **70**, 2737–2754.

Toro-Escobar, C. M., Parker, G., and Paola, C. 1996. Transfer function for the deposition of poorly sorted gravel in response to stream bed aggradation. *Journal of Hydraulic Research*, **34**, 35–53.

Tripathy, G. R., and Singh, S. K. 2010. Chemical erosion rates of river basins of the Ganga system in the Himalaya: reanalysis based on inversion of dissolved major ions, Sr, and ^{87}Sr/^{86}Sr. *Geochemistry, Geophysics, Geosystems*, **11**, Q03013.

Tripathy, G. R., Singh, S. K., and Krishnasami, S. 2011. Sr and Nd isotopes as tracers of chemical and physical erosion. Pages 521–552 of: Baskaran, M. (ed), *Handbook of Environmental Isotope Geochemistry, Advances in Isotope Geochemistry*. Berlin-Heidelberg: Springer-Verlag.

Tucker, G. E., and Bras, R. L. 2000. A stochastic approach to modelling the role of rainfall variability in drainage basin evolution. *Water Resources Research*, **36**, 1953–1964.

Tucker, G. E., and Slingerland, R. 1997. Drainage basin responses to climate change. *Water Resources Research*, **33**, 2031–2047.

Tucker, G. E., and Slingerland, R. L. 1996. Predicting sediment flux from fold and thrust belts. *Basin Research*, **8**, 329–349.

Turcotte, D. L. 1997. *Fractals and Chaos in Geology and Geophysics, Second Edition*. Cambridge UK: Cambridge University Press.

Turcotte, D. L., and Greene, L. 1993. A scale-invariant approach to flood-frequency analysis. *Stochastic Hydrology and Hydraulics*, **7**, 33–40.

Turcotte, D. L., and Schubert, G. 2002. *Geodynamics, Second Edition*. Cambridge University Press.

Turowski, J. M., Rickenmann, D., and Dadson, S. J. 2010. The partitioning of the total sediment load of a river into suspended load and bedload: a review of empirical data. *Sedimentology*, **57**, 1126–1146.

Tyler, S. W., and Wheatcraft, S. W. 1992. Fractal scaling of soil particle-size distributions: Analysis and limitations. *Soil Science Society of America Journal*, **56**, 362–369.

Vail, P. R., Mitchum, R. M., and Thompson, S. 1977. Seismic stratigraphy and global changes of sea level, Part 4, Global cycles of relative changes of sea level. Pages 83–97 of: Payton, C. E. (ed), *Seismic Stratigraphy: Applications to Hydrocarbon Exploration*. American Association of Petroleum Geologists Memoirs 26.

Vail, P. R., Audemard, F., Bowman, S. A., Eisner, P. N., and Perez-Cruz, H. 1991. The stratigraphic signature of tectonics, eustasy, and sedimentation. Pages 617–659 of: Einsele, G., Ricken, W., and Seilacher, A. (eds), *Cycles and Events in Stratigraphy*. Berlin: Springer.

Van den Berg van Saparoea, A.-P., and Postma, G. 2008. Control of climate change on the yield of river systems. Pages 15–33 of: *Recent Advances in Models of Siliciclastic Shallow-Marine Stratigraphy*. Tulsa, Oklahoma: Society Economic Paleontologists and Mineralogists Special Publication 90.

Van der Zwan, C. J. 2002. The impact of Milankovitch-scale climate forcing on sediment supply. *Sedimentary Geology*, **147**, 271–294.

Van der Zwan, C. J., and Brugman, W. A. 1999. Biosignals from the EA Field, Nigeria. Pages 291–301 of: Jones, R. W., and Simmons, M. D. (eds), *Biostratigraphy in Production and Development Geology*. Special Publication Geological Society London 152.

Van Houten, F. B. 1964. Cyclic lacustrine sedimentation, Upper Triassic Lockatong Formation, central New Jersey and adjacent Pennsylvania. *Kansas State Geological Survey Bulletin*, **169**, 497–531.

Van Rijn, L. C. 1984. Sediment transport, II: Suspended load transport. *Journal Hydraulic Engineering*, **110**, 1431–1456.

Van Wagoner, J. C., Mitchum, R. M., Campion, K. M., and Rahmanian, V. D. 1990. *Siliciclastic Sequence Stratigraphy in Well Logs, Cores, and Outcrops: Concepts for High-Resolution Correlation of Time and Facies*. American Association Petroleum Geologists Methods in Exploration 7.

Veizer, J. 1989. Strontium isotopes in sea water through time. *Annual Reviews Earth and Planetary Sciences*, **17**, 141–168.

Ver, L. M. B., Mackenzie, F. T., and Lerman, A. 1999. Carbon cycle in the coastal zone: effects of global perturbations and change in the last three centuries. *Chemical Geology*, **159**, 283–304.

Vergés, J. 2007. Drainage responses to oblique and lateral ramps: a review. Pages 29–47 of: Nichols, G., Williams, E., and Paola, C. (eds), *Sedimentary Processes, Environments and Basins: A Tribute to Peter Friend* Special Publication International Association Sedimentologists 38. Oxford: Blackwell Publishing.

Vermeesch, P. 2013. Multi-sample comparison of detrital age distributions. *Chemical Geology*, **341**, 140–146.

Vernon, A. J., van der Beek, P. A., Sinclair, H. D., Persano, C., Foeken, J., and Stuart, F. 2009. Variable late Neogene exhumation of the Central Alps: Low-temperature thermochronology from the Aar Massif, Switzerland, and the Lepontine Dome, Italy. *Tectonics*, **28**, TC5004.

Vezzoli, G. 2004. Erosion in thhe Western Alps (Dora Baltea Basin): 2, Quantifying sediment yield. *Sedimentary Geology*, **171**, 247–259.

Vezzoli, G., Garzanti, E., and Monguzzi, S. 2004. Erosion in the Western Alps (Dora Baltea Basin): 1. Quantifying sediment provenance. *Sedimentary Geology*, **171**, 227–246.

Vidondo, B., Prairie, Y. T., Blanco, J. M., and Duarte, C. M. 1997. Some aspects of the analysis of size spectra in aquatic ecology. *Limnology and Oceanography*, **42**, 184–192.

Vincent, S. J. 2001. The Sis palaeovalley: a record of proximal fluvial sedimentation and drainage basin development in response to Pyrenean mountain building. *Sedimentology*, **48**, 1235–1276.

von Blanckenburg, F. 2005. The control mechanisms of erosion and weathering at basin scale from cosmogenic nuclides in river sediment. *Earth and Planetary Science Letters*, **237**, 462–479.

Vörösmarty, C. J., Fekete, B. M., Meybeck, M., and Lammers, R. B. 2000. Global system of rivers: Its role in organizing continental land mass and defining land-to-ocean linkages. *Global Biogeochemical Cycles*, **14**, 599–621.

Walcott, R. C., and Summerfield, M. A. 2009. Universality and variability in basin outlet spacing: implications for the two-dimensional form of drainage basins. *Basin Research*, **21**, 147–155.

Walford, H. L., White, N. J., and Sydow, J. C. 2005. Solid sediment load history of the Zambezi delta. *Earth and Planetary Science Letters*, **238**, 49–63.

Wallace, R. E. 1978. Geometry and rates of changes of fault-generated range fronts, north-central Nevada. *United States Geological Survey Journal of Research*, **6**, 637–650.

Walsh, J. P., and Nittrouer, C. A. 2003. Contrasting styles of off-shelf sediment accumulation in New Guinea. *Marine Geology*, **196**, 105–125.

Walsh, J. P., and Nittrouer, C. A. 2007. Understanding fine-grained river-sediment dispersal on continental margins. *Marine Geology*, **263**, 34–45.

Walsh, J. P., Wiberg, P., and Aalto, R. (eds). 2015. Source-to-sink systems: Sediment and solute transfer on the earth surface, *Earth Science Reviews*, **153**, 1–334.

Wang, H. J., Yang, S. Z., Satio, Y., Liu, J. P., and Sun, X. 2006. Interannual and seasonal variation of the Huanghe (Yellow River) water discharge over the past 50 years: Connections to impacts from ENSO events and dams. *Global and Planetary Change*, **50**, 212–225.

Wang, J., Zhangdong, J., Hilton, R. G., Zhang, F., Li, G., Densmore, A. L., Gröcke, D. R., Xu, X., and West, A. J. 2016. Earthquake-triggered increase in biospheric carbon export from a mountain belt. *Geology*, G37533.1.

Wang, Y., Straub, K. M., and Hajek, E. A. 2011. Scale-dependent compensational stacking: an estimate of autogenic time scales in channelized sedimentary deposits. *Geology*, **39**, 811–814.

Warrick, J. A. 2014. Eel River margin source-to-sink sediment budgets: Revisited. *Marine Geology*, **351**, 25–37.

Warrick, J. A., and Fong, D. A. 2004. Dispersal scaling from the world's rivers. *Geophysical Research Letters*, **31**, L04301.

Warrick, J. A., and Milliman, J. D. 2003. Hyperpycnal sediment discharge from semi-arid southern California rivers: Implications for coastal sediment budgets. *Geology*, **31**, 781–784.

Warrick, J. A., and Rubin, D. M. 2007. Suspended-sediment rating curve response to urbanization and wildfire, Santa Ana River, California. *Journal Geophysical Research*, **112**, 1–15.

Waters, J. V., Jones, S. J., and Armstrong, H. A. 2010. Climatic controls on late Pleistocene alluvial fans, Cyprus. *Geomorphology*, **115**, 228–251.

Weaver, P. P. E., Rothwell, R. G., Ebbing, J., Gunn, D., and Hunter, P. M. 1992. Correlation, frequency of emplacement and source directions of megaturbidites on the Madeira Abyssal Plain. *Marine Geology*, **109**, 1–20.

Weedon, G. P. 1989. The detection and illustration of regular sedimentary cycles using Walsh power spectra and filtering, with examples from the Lias of Switzerland. *Journal Geological Society London*, **146**, 133–144.

Weedon, G. P., and Jenkyns, H. C. 1990. Regular and irregular climatic cycles and the Belemnite Marls (Pliensbachian, Lower Jurassic, Wessex Basin). *Journal of the Geological Society*, **147**, 915–918.

Weiguo, L., Bhattacharya, J. P., and Yingmin, W. 2011. Delta symmetry: Concepts, characteristics, and depositional models. *Petroleum Science*, **8**, 278–289.

Weissmann, G. S., Hartley, A. J., Nichols, G. J., Scuderi, L. A., Olson, M., Buehler, H., and Banteah, R. 2010. Fluvial form in modern continental sedimentary basins: Distributive fluvial systems. *Geology*, **38**, 39–42.

Weissmann, G. S., Hartley, A. J., Scuderi, L. A., J., Nichols. G., Owen, A., Wright, S., Felicia, A. L., Holland, F., and Anaya, F. M. L. 2015. Fluvial geomorphic elements in modern sedimentary basins and their potential preservation in the rock record: A review. *Geomorphology*, **250**, 187–219.

Wells, T., Willgoose, G. R., and Hancock, G. R. 2008. Modeling weathering pathways and processes of the fragmentation of salt weathered quartz-chlorite schist. *Journal of Geophysical Research*, **113**, F01014.

Weltje, G. J. 1994. *Provenance and Dispersal of Sand-sized Sediment: Reconstruction of Dispersal Patterns and Sources of Sand-sized Sediments by Means of Inverse Modelling Techniques.* The Netherlands: PhD thesis, Utrecht University, Geologie Ultraiectina 121.

Weltje, G. J. 2012. Quantitative models of sediment generation and provenance: State of the art and future developments. *Sedimentary Geology*, **280**, 4–20.

Weltje, G. J., and Brommer, M. B. 2011. Sediment budget modelling of multi-sourced basin-fills: application to recent deposits of the western Adriatic mud wedge (Italy). *Basin Research*, **23**, 291–308.

Weltje, G. J., and de Boer, P. L. 1993. Astronomically induced paleoclimatic oscillations reflected in Pliocene turbidite deposits on Corfu (Greece): Implications for the interpretation of higher order cyclicity in ancient turbidite systems. *Geology*, **21**, 307–310.

Weltje, G. J., and Prins, M. 2003. Muddled or mixed? Inferring palaeoclimate from size distributions of deep-sea clastics. *Sedimentary Geology*, **162**, 39–62.

Weltje, G. J., and von Eynatten, H. 2004. Quantitative provenance analysis of sediments: review and outlook. *Sedimentary Geology*, **171**, 1–11.

Weltje, G. J., Meijer, X. D., and de Boer, P. L. 1998. Stratigraphic inversion of siliciclastic basin fills: a note on the distinction between supply signals resulting from tectonic and climatic forcing. *Basin Research*, **10**, 129–153.

Werner, B. T. 1999. Complexity in natural landform patterns. *Science, Viewpoint*, **284**, 102–104.

Wetzel, A. 1993. The transfer of river load to deep-sea fans: A quantitative approach. *Bulletin American Association Petroleum Geologists*, **77**, 1679–1692.

Wetzel, R. G. 1975. Organic carbon cycle and detritus. Pages 583–621 of: Wetzel, R. G. (ed), *Limnology*. Philadelphia, USA: W. B. Saunders Co.

Wheatcroft, R. A., and Borgeld, J. C. 2000. Oceanic flood deposits on the northern California coast: Large-scale distribution and small-scale physical properties. *Continental Shelf Research*, **20**, 2163–2190.

Wheatcroft, R. A., and Sommerfield, C. K. 2005. River sediment flux and shelf sediment accumulation rates on the Pacific Northwest margin. *Continental Shelf Research*, **25**, 311–332.

Wheeler, H. E. 1958. Time-stratigraphy. *American Association of Petroleum Geologists Bulletin*, **42**, 1047–1063.

Wheeler, H. E. 1959. Stratigraphic units in time and space. *American Journal of Science*, **257**, 692–706.

Whipple, K. X. 2001. Fluvial landscape response time: how plausible is steady state denudation? *American Journal of Science*, **301**, 313–325.

Whipple, K. X. 2009. The influence of climate on the tectonic evolution of mountain belts. *Nature Geoscience*, **2**, 97–104.

Whipple, K. X., and Trayler, C. R. 1996. Tectonic control on fan size: the importance of spatially variable subsidence rates. *Basin Research*, **8**, 351–366.

Whipple, K. X., and Tucker, G. E. 1999. Dynamics of the stream power river incision model: Implications for height limits of mountain ranges, landscape response timescales, and research needs. *Journal Geophysical Research*, **104**, 17,661–17,674.

Whitchurch, A. L., Carter, A., Sinclair, H. D., Duller, R. A., Whittaker, A. C., and Allen, P. A. 2011. Sediment routing system evolution within a diachronously uplifting orogen: insights from detrital zircon thermochronological analyses from the south-central Pyrenees. *American Journal of Science*, **311**, 442–482.

White, A. F., and Brantley, S. L. 1995. Chemical weathering rates of silicate minerals: an overview. Pages 1–22 of: *Chemical Weathering Rates of Silicate Minerals*. Reviews of Mineralogy 31.

Whittaker, A. C. 2012. How do landscapes record tectonics and climate? *Lithosphere*, **4**, 160–164.

Whittaker, A. C., and Boulton, S. J. 2012. Tectonic and climatic controls on knickpoint retreat rates and landscape response times. *Journal Geophysical Research*, **117**, 1–19.

Whittaker, A. C., and Walker, A. S. 2015. Geomorphic constraints on fault throw rates and linkage times: Examples from the northern Gulf of Evia, Greece. *Journal Geophysical Research Earth Surface*, **120**, 137–158.

Whittaker, A. C., Attal, M., and Allen, P. A. 2010. Characterising the origin, nature and fate of sediment exported from catchments perturbed by active tectonics. *Basin Research*, **22**, 809–828.

Whittaker, A. C., Attal, M., and Tucker, G. E. 2007. Contrasting transient and steady-state rivers crossing active normal faults: New field observations from the Central Apennines, Italy. *Basin Research*, **19**, 529–556.

Whittaker, A. C., Duller, R. A., Springett, J., Smithells, R., Whitchurch, A. L., and Allen, P. A. 2011. Decoding downstream trends in stratigraphic grain size as a function of tectonic subsidence and sediment supply. *Geological Society of America Bulletin*, **123**, 1363–1382.

Wilgus, C. K., Hastings, B. S., Kendall, C. G. St. C., Posamentier, H. W., Ross, C. A., and Van Wagoner, J. C. (eds). 1988. *Sea-level Changes: An Integrated Approach*. Tulsa, Oklahoma: Society of Economic Paleontologists and Mineralogists Special Publication 42.

Wilkinson, B. H., and McElroy, B. J. 2007. The impact of humans on continental erosion and sedimentation. *Geological Society of America Bulletin*, **19**, 140–156.

Wilkinson, B. H., Drummond, C. N., Rothman, E. D., and Diedrich, N. W. 1997. Stratal order in peritidal carbonate sequences. *Journal Sedimentary Research*, **67**, 1068–1082.

Willenbring, J. K., Codilean, A. T., and McElroy, B. 2013. Earth is (mostly) flat: Apportionment of the flux of continental sediment over millennial time scales. *Geology*, **41**, 343–346.

Willett, S. D. 1999. Orogeny and orography: the effects of erosion on the structure of mountain belts. *Journal Geophysical Research*, **104**, 28957–28981.

Willett, S. D., and Brandon, M. T. 2002. On steady states in mountain belts. *Geology*, **30**, 175–178.

Williams, D. F. 1988. Evidence for and against sea-level changes from the stable isotopic record of the Cenozoic. Pages 31–36 of: Wilgus, C. K., Hastings, B. S., Ross, C. A., Posamentier, H., Van Wagoner, J. C., and Kendall, C. G. St. C. (eds), *Sea-level Changes: An Integrated Approach.* Society for Sedimentary Geology, Special Publication 42.

Wilson, L. 1973. Variations in mean annual sediment yield as a function of mean annual precipitation. *American Journal Science*, **273**, 335–349.

Wittmann, H., and von Blanckenburg, F. 2009. Cosmogenic nuclide budgeting of floodplain sediment transfer. *Geomorphology*, **109**, 246–256.

Wittmann, H., von Blanckenburg, F., Guyot, J. L., Maurice, L., and Kubik, P. W. 2009. From source to sink: Preserving the cosmogenic ^{10}Be-derived denudation rate signal of the Bolivian Andes in sediment of the Beni and Mamore foreland basins. *Earth and Planetary Science Letters*, **288**, 463–474.

Wobus, C. W., Tucker, G. E., and Anderson, R. S. 2010. Does climate change create distinctive patterns of landscape incision? *Journal Geophysical Research*, **115**, F04008.

Wolf, R., Farley, K., and Silver, L. 1996. Assessment of (U-Th)/He thermochronometry: the low temperature history of the San Jacinto Mountains, California. *Geology*, **25**, 65–68.

Wolf, R., Farley, K. A., and Kass, D. M. 1998. Modelling the temperature sensitivity of the apatite U-Th/He thermochronometer. *Chemical Geology*, **148**, 105–114.

Woodcock, N. H. 2004. Life span and fate of basins. *Geology*, **32**, 685–688.

Wright, L. D. 1977. Sediment transport and deposition at river mouths: a synthesis. *Bulletin Geological Society of America*, **88**, 857–868.

Wright, L. D., and Coleman, J. M. 1972. River delta morphology: Wave climate and the role of the subaqueous profile. *Science*, **176**, 282–284.

Wright, L. D., and Coleman, J. M. 1973. Variations in morphology of major river deltas as functions of ocean wave and river discharge regimes. *Bulletin American Association of Petroleum Geologists*, **57**, 370–398.

Wu, Q., Borkovee, M., and Sticher, H. 1993. On particle size distributions in soils. *Soil Science Society of America*, **57**, 883–890.

Wynn, R. B., Weaver, P. P. E., Masson, D. G., and Stow, D. A. V. 2002. Turbidite depositional architecture across three interconnected deep-water basins on the northwest African margin. *Sedimentology*, **49**, 669–695.

Xiao, X. Y., Shen, J., Wang, S. M., Xiao, H. F., and Tong, G. B. 2010. The variation of the southwest monsoon from the high resolution pollen record in Heqing Basin, Yunnan Province, China for the last 2.78 Ma. *Paleogeography, Paleoclimatology, Paleoecology*, **287**, 45–57.

Xie, X., and Heller, P. L. 2009. Plate tectonics and basin subsidence history. *Bulletin of Geological Society America*, **121**, 55–64.

Xu, J. P., Noble, M. A., and Rosenfeld, L. K. 2004. *In-situ* measurements of velocity structure within turbidity currents. *Geophysical Research Letters*, **31**, L09311.

Xu, K. H., Milliman, J. D., Yang, Z., and Xu, H. 2007. Climatic and anthropogenic impacts on water and sediment discharges from the Yangtze River (Changjiang), 1950–2005. Pages 609–626 of: Gupta, A. (ed), *Large Rivers: Geomorphology and Management.* Chichester, UK: John Wiley.

Yoo, D. G., and Park, S. C. 2000. High-resolution seismic study as a tool for sequence stratigraphic evidence of high-frequency sea-level changes: latest Pleistocene-Holocene example from the Korea Strait. *Journal Sedimentary Research*, **70**, 296–309.

Yu, J., Sui, F., Liu, H., and Wang, Y. 2008. Recognition of Milankovitch cycles in the stratigraphic record: application of the CWT and the FFT to well-log data. *Journal China University Mining and Technology*, **18**, 594–598.

Zachos, J., Opdyke, B., Quinn, T., Jones, C. E., and Halliday, A. N. 1999. Early Cenozoic glaciation, Antarctic weathering, and seawater ^{87}Sr/^{86}Sr: is there a link? *Chemical Geology*, **161**, 165–180.

Zeitler, P. K., Herczig, A. L., McDougall, I., and Honda, M. 1987. U-Th-He dating of apatite: a potential thermochronometer. *Geochimica et Cosmochimica Acta*, **51**, 2865–2868.

Zhang, X., Drake, N., and Wainwright, J. 2002. Scaling land surface parameters for global-scale soil erosion estimation. *Water Resources Research*, **38**, 1180.

Zhang, Y., Swift, D. J. P., Fan, S., Niederoda, A. W., and Reed, C. W. 1999. Two-dimensional numerical modeling of storm deposition on the northern California shelf. *Marine Geology*, **154**, 155–167.

Zuffa, G. G. 1985. *Provenance of Arenites*. Dordrecht: D. Reidel Publishing Co.

Zuffa, G. G. 1987. Unravelling hinterland and offshore paleogeography from deep-water arenites. Pages 39–61 of: Leggett, J. K., and Zuffa, G. G. (eds), *Marine Clastic Sedimentology*. London: Graham and Trotman.

Zuhlke, R., Bechstadt, T., and Mundil, R. 2003. Sub-Milankovitch and Milankovitch forcing on a model Mesozoic carbonate platform - the Latemar (Middle Triassic, Italy). *Terra Nova*, **15**, 69–80.